Praise for *Essential Math for AI*

Technology and AI markets are like a river, where some parts are moving faster than others. Successfully applying AI requires the skill of assessing the direction of the flow and complementing it with a strong foundation, which this book enables, in an engaging, delightful, and inclusive way. Hala has made math fun for a spectrum of participants in the AI-enabled future!

*—Adri Purkayastha, Group Head, AI Operational Risk and
Digital Risk Analytics, BNP Paribas*

Texts on artificial intelligence are usually either technical manuscripts written by experts for other experts, or cursory, math-free introductions catered to general audiences. This book takes a refreshing third path by introducing the mathematical foundations for readers in business, data, and similar fields without advanced mathematics degrees. The author weaves elegant equations and pithy observations throughout, all the while asking the reader to consider the very serious implications artificial intelligence has on society. I recommend *Essential Math for AI* to anyone looking for a rigorous treatment of AI fundamentals viewed through a practical lens.

—George Mount, Data Analyst and Educator

Hala has done a great job in explaining crucial mathematical concepts. This is a must-read for every serious machine learning practitioner. You'd love the field more once you go through the book.

—Umang Sharma, Senior Data Scientist and Author

To understand artificial intelligence, one needs to understand the relationship between math and AI. Dr. Nelson made this easy by giving us the foundation on which the symbiotic relationship between the two disciplines is built.

*—Huan Nguyen, Rear Admiral (Ret.),
Cyber Engineering, NAVSEA*

Essential Math for AI

*Next-Level Mathematics for Efficient and
Successful AI Systems*

Hala Nelson

Beijing · Boston · Farnham · Sebastopol · Tokyo

Essential Math for AI

by Hala Nelson

Published by O'Reilly Media, Inc., 1005 Gravenstein Highway North, Sebastopol, CA 95472.

O'Reilly books may be purchased for educational, business, or sales promotional use. Online editions are also available for most titles (*http://oreilly.com*). For more information, contact our corporate/institutional sales department: 800-998-9938 or *corporate@oreilly.com*.

Acquisitions Editor: Aaron Black	**Indexer:** nSight, Inc.
Development Editor: Angela Rufino	**Interior Designer:** David Futato
Production Editor: Kristen Brown	**Cover Designer:** Karen Montgomery
Copyeditor: Sonia Saruba	**Illustrator:** Kate Dullea
Proofreader: JM Olejarz	

January 2023:   First Edition

Revision History for the First Edition

2023-01-04:   First Release
2023-02-03:   Second Release

See *http://oreilly.com/catalog/errata.csp?isbn=9781098107635* for release details.

978-1-098-10763-5

[LSI]

Table of Contents

Preface

Why I Wrote This Book

AI is built on mathematical models. We need to know how.

I wrote this book in purely colloquial language, leaving most of the technical details out. It is a math book about AI with very few mathematical formulas and equations, no theorems, no proofs, and no coding. My goal is *to not keep this important knowledge in the hands of the very few elite, and to attract more people to technical fields.* I believe that many people get turned off by math before they ever get a chance to know that they might love it and be naturally good at it. This also happens in college or in graduate school, where many students switch their majors from math, or start a Ph.D. and never finish it. The reason is not that they do not have the ability, but that they saw no motivation or an end goal for learning torturous methods and techniques that did not seem to transfer to anything useful in their lives. It is like going to a strenuous mental gym every day only for the sake of going there. No one even wants to go to a real gym every day (this is a biased statement, but you get the point). In math, formalizing objects into functions, spaces, measure spaces, and entire mathematical fields comes *after* motivation, not *before*. Unfortunately, it gets taught in reverse, with formality first and then, if we are lucky, some motivation.

The most beautiful thing about math is that it has the expressive ability to connect seemingly disparate things together. A field as big and as consequential as AI not only builds on math, as that is a given; it also needs the binding ability that only math can provide in order to tell its big story concisely. In this book I will extract the math required for AI in a way that does not deviate at all from the real-life AI application in mind. It is infeasible to go through existing tools in detail and not fall into an encyclopedic and overwhelming treatment. What I do instead is try to teach you *how to think* about these tools and *view them from above*, as a means to an end that we can tweak and adjust when we need to. I hope that you will get out of this book a way of seeing how things relate to each other and why we develop or use certain methods among others. In a way, this book provides a platform that launches you to whatever area you find interesting or want to specialize in.

Another goal of this book is to democratize mathematics, and to build more confidence to ask about how things work. Common answers such as "It's complicated mathematics," "It's complicated technology," or "It's complex models," are no longer satisfying, especially since the technologies that build on mathematical models currently affect every aspect of our lives. We do not need to be experts in every field in mathematics (no one is) in order to understand how things are built and why they operate the way they do. There is one thing about mathematical models that everyone needs to know: they *always* give an answer. They always spit out a number. A model that is vetted, validated, and backed with sound theory gives an answer. Also, a model that is *complete trash* gives an answer. Both compute mathematical functions. Saying that our decisions are based on mathematical models and algorithms does not make them sacred. What are the models built on? What are their assumptions? Limitations? Data they were trained on? Tested on? What variables did they take into account? And what did they leave out? Do they have a feedback loop for improvement, ground truths to compare to and improve on? Is there any theory backing them up? We need to be *transparent* with this information when the models are ours, and ask for it when the models are deciding our livelihoods for us.

The unorthodox organization of the topics in this book is intentional. I wanted to avoid getting stuck in math details before getting to the applicable stuff. My stand on this is that we do not ever need to dive into background material unless we happen to be personally practicing something, and that background material becomes an unfulfilled gap in our knowledge that is stopping us from making progress. Only then it is worth investing serious time to learn the intricate details of things. It is much more important to see how it all ties together and where everything fits. In other words, this book provides a map for how everything between math and AI interacts nicely together.

I also want to make a note to newcomers about the era of large data sets. Before working with large data, real or simulated, structured or unstructured, we might have taken computers and the internet for granted. If we came up with a model or needed to run analytics on small and curated data sets, we might have assumed that our machine's hardware would handle the computations, or that the internet would just give more curated data when we needed it, or more information about similar models. The reality and limitations to access data, errors in the data, errors in the outputs of queries, hardware limitations, storage, data flow between devices, and vectorizing unstructured data such as natural language or images and movies hits us really hard. That is when we start getting into parallel computing, cloud computing, data management, databases, data structures, data architectures, and data engineering in order to understand the compute infrastructure that allows us to run our models. What kind of infrastructure do we have? How is it structured? How did it evolve? Where is it headed? What is the architecture like, including the involved solid materials? How do these materials work? And what is all the fuss about quantum computing? We should not view the software as separate from the hardware, or our models separate from the infrastructure that allows us to simulate them. This book focuses only on the math, the AI models, and some data. There are neither exercises nor coding. In other words, we focus on the *soft*, the *intellectual*, and the *I do not need to touch anything* side of things. But we need to keep learning until we are able to comprehend the technology that powers many aspects of our lives as the one interconnected body that it actually is: hardware, software, sensors and measuring devices, data warehouses, connecting cables, wireless hubs, satellites, communication centers, physical and software security measures, and mathematical models.

Who Is This Book For?

I wrote this book for:

- A person who knows math but wants to get into AI, machine learning, and data science.
- A person who practices AI, data science, and machine learning but wants to brush up on their mathematical thinking and get up-to-date with the mathematical ideas behind the state-of-the-art models.
- Undergraduate or early graduate students in math, data science, computer science, operations research, science, engineering, or other domains who have an interest in AI.
- People in management positions who want to integrate AI and data analytics into their operations but want a deeper understanding of how the models that they might end up basing their decisions on actually work.
- Data analysts who are primarily doing business intelligence, and are now, like the rest of the world, driven into *AI-powered business intelligence*. They want to know what that actually means before adopting it into business decisions.
- People who care about the ethical challenges that AI might pose to the world and want to understand the inner workings of the models so that they can argue for or against certain issues such as autonomous weapons, targeted advertising, data management, etc.
- Educators who want to put together courses on math and AI.
- Any person who is curious about AI.

Who Is This Book Not For?

This book is not for a person who likes to sit down and do many exercises to master a particular mathematical technique or method, a person who likes to write and prove theorems, or a person who wants to learn coding and development. This is not a math textbook. There are many excellent textbooks that teach calculus, linear algebra, and probability (but few books relate this math to AI). That said, this book has many in-text pointers to the relevant books and scientific publications for readers who want to dive into technicalities, rigorous statements, and proofs. This is also not a coding book. The emphasis is on concepts, intuition, and general understanding, rather than on implementing and developing the technology.

How Will the Math Be Presented in This Book?

Writing a book is ultimately a decision-making process: how to organize the material of the subject matter in a way that is most insightful into the field, and how to choose what and what not to elaborate on. I will detail some math in a few places, and I will omit details in others. This is on purpose, as my goal is to not get distracted from telling the story of:

Which math do we use, why do we need it, and where exactly do we use it in AI?

I always define the AI context, with many applications. Then I talk about the related mathematics, sometimes with details and other times only with the general way of thinking. Whenever I skip details, I point out the relevant questions that we should be asking and how to go about finding answers. I showcase the math, the AI, and the models as one connected entity. I dive deeper into math only if it must be part of the foundation. Even then, I favor intuition over formality. The price I pay here is that on very few occasions, I might use some technical terms before defining them, secretly hoping that you might have encountered these terms before. In this sense, I adopt AI's *transformer* philosophy (see Google Brain's 2017 article: "Attention Is All You Need" (*https://oreil.ly/bJ0is*)) for natural language understanding: the model learns word meanings from their context. So when you encounter a technical term that I have not defined before, focus on the term's surrounding environment. Over the course of the section within which it appears, you will have a very good intuition about its meaning. The other option, of course, is to google it. Overall, I avoid jargon and I use zero acronyms.

Since this book lies at the intersection of math, data science, AI, machine learning, and philosophy, I wrote it expecting a diverse audience with drastically different skill sets and backgrounds. For this reason, depending on the topic, the same material might feel trivial to some but complicated to others. I hope I do not insult anyone's mind in the process. That said, this is a risk that I am willing to take, so that all readers will find useful things to learn from this book. For example, mathematicians will learn the AI application, and data scientists and AI practitioners will learn deeper math.

The sections go in and out of technical difficulty, so if a section gets too confusing, make a mental note of its existence and skip to the next section. You can come back to what you skipped later.

Most of the chapters are independent, so readers can jump straight to their topics of interest. When chapters are related to other chapters, I point that out. Since I try to make each chapter as self-contained as possible, I may repeat a few explanations across different chapters. I push the probability chapter to all the way near the end (Chapter 11), but I use and talk about probability distributions all the time (especially the joint probability distribution of the features of a data set). The idea is to get used to the language of probability and how it relates to AI models before learning its grammar, so when we get to learning the grammar, we have a good idea of the context that it fits in.

I believe that there are two types of learners: those who learn the specifics and the details, then slowly start formulating a bigger picture and a map for how things fit together; and those who first need to understand the big picture and how things relate to each other, then dive into the details only when needed. Both are equally important, and the difference is only in someone's type of brain and natural inclination. I tend to fit more into the second category, and this book is a reflection of that: how does it all look from above, and how do math and AI interact with each other? The result might feel like a whirlwind of topics, but you'll come out on the other side with a great knowledge base for both math and AI, plus a healthy dose of confidence.

When my dad taught me to drive, he sat in the passenger's seat and asked me to drive. Ten minutes in, the road became a cliffside road. He asked me to stop, got out of the car, then said: "Now drive, just try not to fall off the cliff, don't be afraid, I am watching" (like that was going to help). I did not fall off the cliff, and in fact I love cliffside roads the most. Now tie this to training self-driving cars by reinforcement learning, with the distinction that the cost of falling off the cliff would've been minus infinity for me. I could not afford that; I am a real person in a real car, not a simulation.

This is how you'll do math and AI in this book. There are no introductions, conclusions, definitions, theorems, exercises, or anything of the like. There is immersion.

You are already in it. You know it. Now drive.

Infographic

I accompany this book with an infographic, visually tying all the topics together. You can also find this on the book's GitHub page (*https://github.com/halanelson/Essential-Math-For-AI*).

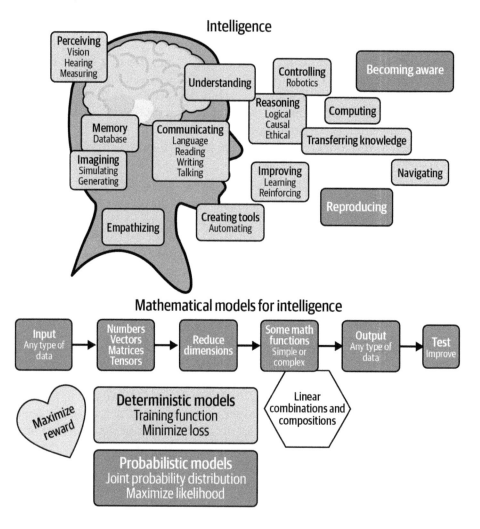

What Math Background Is Expected from You to Be Able to Read This Book?

This book is self-contained in the sense that we motivate everything that we need to use. I do hope that you have been exposed to calculus and some linear algebra, including vector and matrix operations, such as addition, multiplication, and some matrix decompositions. I also hope that you know what a function is and how it maps an input to an output. Most of what we do mathematically in AI involves constructing a function, evaluating a function, optimizing a function, or composing a bunch of functions. You need to know about derivatives (these measure how fast things change) and the chain rule for derivatives. You do not necessarily need to know how to compute them for each function, as computers, Python, Desmos, and/or Wolfram|Alpha mathematics do a lot for us nowadays, but you need to know their meaning. Some exposure to probabilistic and statistical thinking are helpful as well. If you do not know any of the above, that is totally fine. You might have to sit down and do some examples (from some other books) on your own to familiarize yourself with certain concepts. The trick here is to know *when* to look up the things that you do not know...*only when you need them*, meaning only when you encounter a term that you do not understand, and you have a good idea of the context within which it appeared. If you are truly starting from scratch, you are not too far behind. This book tries to avoid technicalities at all costs.

Overview of the Chapters

We have a total of 14 chapters.

If you are a person who cares for math and the AI technology as they relate to ethics, policy, societal impact, and the various implications, opportunities, and challenges, then read Chapters 1 and 14 first. If you do not care for those, then we make the case that you should. In this book, we treat math as the binding agent of seemingly disparate topics, rather than the usual presentation of math as an oasis of complicated formulas, theorems, and Greek letters.

Chapter 13 might feel separate from the book if you've never encountered differential equations (ODEs and PDEs), but you will appreciate it if you are into mathematical modeling, the physical and natural sciences, simulation, or mathematical analysis, and you would like to know how AI can benefit your field, and in turn how differential equations can benefit AI. Countless scientific feats build on differential equations, so we cannot leave them out when we are at a dawn of a computational technology that has the potential to address many of the field's long-standing problems. This chapter is not essential for AI per se, but it is essential for our general understanding of mathematics as a whole, and for building theoretical foundations for AI and neural operators.

The rest of the chapters are essential for AI, machine learning, and data science. There is no optimal location for Chapter 6 on the singular value decomposition (the essential math for principal component analysis and latent semantic analysis, and a great method for dimension reduction). Let your natural curiosity dictate when you read this chapter: before or after whichever chapter you feel would be the most fitting. It all depends on your background and which industry or academic discipline you happen to come from.

Let's briefly overview Chapters 1 through 14:

Chapter 1, "Why Learn the Mathematics of AI?"
Artificial intelligence is here. It has already penetrated many areas of our lives, is involved in making important decisions, and soon will be employed in every sector of our society and operations. The technology is advancing very fast and its investments are skyrocketing. What is artificial intelligence? What is it able to do? What are its limitations? Where is it headed? And most importantly, how does it work, and why should we really care about knowing how it works? In this introductory chapter we briefly survey important AI applications, the problems usually encountered by companies trying to integrate AI into their systems, incidents that happen when systems are not well implemented, and the math typically used in AI solutions.

Chapter 2, "Data, Data, Data"
This chapter highlights the fact that data is central to AI. It explains the differences between concepts that are usually a source of confusion: structured and unstructured data, linear and nonlinear models, real and simulated data, deterministic functions and random variables, discrete and continuous distributions, prior probabilities, posterior probabilities, and likelihood functions. It also provides a map for the probability and statistics needed for AI without diving into any details, and introduces the most popular probability distributions.

Chapter 3, "Fitting Functions to Data"
At the core of many popular machine learning models, including the highly successful neural networks that brought artificial intelligence back into the popular spotlight since 2012, lies a very simple mathematical problem: fit a given set of data points into an appropriate function, then make sure this function performs well on new data. This chapter highlights this widely useful fact with a real data set and other simple examples. We discuss regression, logistic regression, support vector machines, and other popular machine learning techniques, with one unifying theme: training function, loss function, and optimization.

Chapter 4, "Optimization for Neural Networks"
Neural networks are modeled after the brain cortex, which involves millions of neurons arranged in a layered structure. The brain learns by reinforcing neuron connections when faced with a concept it has seen before, and weakening connections if it learns new information that undoes or contradicts previously learned concepts. Machines only understand numbers. Mathematically, stronger connections correspond to larger numbers (weights), and weaker connections correspond to smaller numbers. This chapter explains the optimization and backpropagation steps used when training neural networks, similar to how

learning happens in our brain (not that humans fully understand this). It also walks through various regularization techniques, explaining their advantages, disadvantages, and use cases. Furthermore, we explain the intuition behind approximation theory and the universal approximation theorem for neural networks.

Chapter 5, "Convolutional Neural Networks and Computer Vision"
Convolutional neural networks are widely popular for computer vision and natural language processing. In this chapter we start with the convolution and cross-correlation operations, then survey their uses in systems design and filtering signals and images. Then we integrate convolution with neural networks to extract higher-order features from images.

Chapter 6, "Singular Value Decomposition: Image Processing, Natural Language Processing, and Social Media"
Diagonal matrices behave like scalar numbers and hence are highly desirable. Singular value decomposition is a crucially important method from linear algebra that transforms a dense matrix into a diagonal matrix. In the process, it reveals the action of a matrix on space itself: rotating and/or reflecting, stretching and/or squeezing. We can apply this simple process to *any* matrix of numbers. This wide applicability, along with the ability to dramatically reduce the dimensions while retaining essential information, make singular value decomposition popular in the fields of data science, AI, and machine learning. It is the essential mathematics behind principal component analysis and latent semantic analysis. This chapter walks through singular value decomposition along with its most relevant and up-to-date applications.

Chapter 7, "Natural Language and Finance AI: Vectorization and Time Series"
We present the mathematics in this chapter in the context of natural language models, such as identifying topics, machine translation, and attention models. The main barrier to overcome is moving from words and sentences that carry meaning to low-dimensional vectors of numbers that a machine can process. We discuss state-of-the-art models such as Google Brain's transformer (starting in 2017), among others, while we keep our attention only the relevant math. Time series data and models (such as recurrent neural networks) appear naturally here. We briefly introduce finance AI, as it overlaps with natural language both in terms of modeling and how the two fields feed into each other.

Chapter 8, "Probabilistic Generative Models"
Machine-generated images, including those of humans, are becoming increasingly realistic. It is very hard nowadays to tell whether an image of a model in the fashion industry is that of a real person or a computer-generated image. We have generative adversarial networks (GANs) and other generative models to thank for this progress, where it is harder to draw a line between the virtual and the

real. Generative adversarial networks are designed to repeat a simple mathematical process using two neural networks until the machine itself cannot tell the difference between a real image and a computer-generated one, hence the "very close to reality" success. Game theory and zero-sum games occur naturally here, as the two neural networks "compete" against each other. This chapter surveys generative models, which mimic imagination in the human mind. These models have a wide range of applications, from augmenting data sets to completing masked human faces to high energy physics, such as simulating data sets similar to those produced at the CERN Large Hadron Collider.

Chapter 9, "Graph Models"

Graphs and networks are everywhere: cities and roadmaps, airports and connecting flights, the World Wide Web, the cloud (in computing), molecular networks, our nervous system, social networks, terrorist organization networks, even various machine learning models and artificial neural networks. Data that has a natural graph structure can be better understood by a mechanism that exploits and preserves that structure, building functions that operate directly on graphs, as opposed to embedding graph data into existing machine learning models that attempt to artificially reshape the data before analyzing it. This is the same reason convolutional neural networks are successful with image data, recurrent neural networks are successful with sequential data, and so on. The mathematics behind graph neural networks is a marriage among graph theory, computing, and neural networks. This chapter surveys this mathematics in the context of many applications.

Chapter 10, "Operations Research"

Another suitable name for operations research would be *optimization for logistics*. This chapter introduces the reader to problems at the intersection of AI and operations research, such as supply chain, traveling salesman, scheduling and staffing, queuing, and other problems whose defining features are high dimensionality, complexity, and the need to balance competing interests and limited resources. The math required to address these problems draws from optimization, game theory, duality, graph theory, dynamic programming, and algorithms.

Chapter 11, "Probability"

Probability theory provides a systematic way to quantify randomness and uncertainty. It generalizes logic to situations that are of paramount importance in artificial intelligence: when information and knowledge are uncertain. This chapter highlights the essential probability used in AI applications: Bayesian networks and causal modeling, paradoxes, large random matrices, stochastic processes, Markov chains, and reinforcement learning. It ends with rigorous probability theory, which demystifies measure theory and introduces interested readers to the universal approximation theorem for neural networks.

Chapter 12, "Mathematical Logic"

This important topic is positioned near the end to not interrupt the book's natural flow. Designing agents that are able to gather knowledge, reason logically about the environment within which they exist, and make inferences and good decisions based on this logical reasoning is at the heart of artificial intelligence. This chapter briefly surveys propositional logic, first-order logic, probabilistic logic, fuzzy logic, and temporal logic, within an intelligent knowledge-based agent.

Chapter 13, "Artificial Intelligence and Partial Differential Equations"

Differential equations model countless phenomena in the real world, from air turbulence to galaxies to the stock market to the behavior of materials and population growth. Realistic models are usually very hard to solve and require a tremendous amount of computational power when relying on traditional numerical techniques. AI has recently stepped in to accelerate solving differential equations. The first part of this chapter acts as a crash course on differential equations, highlighting the most important topics and arming the reader with a bird's-eye view of the subject. The second part explores new AI-based methods that simplify the whole process of solving differential equations. These have the potential to unlock long-standing problems in the natural sciences, finance, and other fields.

Chapter 14, "Artificial Intelligence, Ethics, Mathematics, Law, and Policy"

I believe this chapter should be the first chapter in any book on artificial intelligence; however, this topic is so wide and deep that it needs multiple books to cover it completely. This chapter only scratches the surface and summarizes various ethical issues associated with artificial intelligence, including: equity, fairness, bias, inclusivity, transparency, policy, regulation, privacy, weaponization, and security. It presents each problem along with possible solutions (mathematical or with policy and regulation).

My Favorite Books on AI

There are many excellent and incredibly insightful books on AI and on topics intimately related to the field. The following is not even close to being an exhaustive list. Some are technical books heavy on mathematics, and others are either introductory or completely nontechnical. Some are code-oriented (Python 3) and others are not. I have learned a lot from all of them:

- Brunton, Steven L. and J. Nathan Kutz, *Data-Driven Science and Engineering: Machine Learning, Dynamical Systems and Control* (Cambridge University Press, 2022)
- Crawford, Kate, *Atlas of AI* (Yale University Press, 2021)

- Ford, Martin, *Architects of Intelligence* (Packt Publishing, 2018)
- Géron, Aurélien, *Hands-On Machine Learning with Scikit-Learn, Keras and TensorFlow* (O'Reilly, 2022)
- Goodfellow, Ian, Yoshua Bengio, and Aaron Courville, *Deep Learning* (MIT Press, 2016)
- Grus, Joel, *Data Science from Scratch* (O'Reilly, 2019)
- Hawkins, Jeff, *A Thousand Brains* (Basic Books, 2021)
- Izenman, Alan J., *Modern Multivariate Statistical Techniques* (Springer, 2013)
- Jones, Herbert, *Data Science: The Ultimate Guide to Data Analytics, Data Mining, Data Warehousing, Data Visualization, Regression Analysis, Database Querying, Big Data for Business and Machine Learning for Beginners* (Bravex Publications, 2020)
- Kleppmann, Martin, *Designing Data-Intensive Applications* (O'Reilly, 2017)
- Lakshmanan, Valliappa, Sara Robinson, and Michael Munn, *Machine Learning Design Patterns* (O'Reilly, 2020)
- Lane, Hobson, Hannes Hapke, and Cole Howard, *Natural Language Processing in Action* (Manning, 2019)
- Lee, Kai-Fu, *AI Superpowers* (Houghton Mifflin Harcourt, 2018)
- Macey, Tobias, ed., *97 Things Every Data Engineer Should Know* (O'Reilly, 2021)
- Marr, Bernard and Matt Ward, *Artificial Intelligence in Practice* (Wiley, 2019)
- Moroney, Laurence, *AI and Machine Learning for Coders* (O'Reilly, 2021)
- Mount, George, *Advancing into Analytics: From Excel to Python and R* (O'Reilly, 2021)
- Norvig, Peter and Stuart Russell, *Artificial Intelligence: A Modern Approach* (Pearson, 2021)
- Pearl, Judea, *The Book of Why* (Basic Books, 2020)
- Planche, Benjamin and Eliot Andres, *Hands-On Computer Vision with Tensor-Flow2* (Packt Publishing, 2019)
- Potters, Marc, and Jean-Philippe Bouchaud, *A First Course in Random Matrix Theory for Physicists, Engineers, and Data Scientists* (Cambridge University Press, 2020)
- Rosenthal, Jeffrey S., *A First Look at Rigorous Probability Theory* (World Scientific Publishing, 2016)
- Roshak, Michael, *Artificial Intelligence for IoT Cookbook* (Packt Publishing, 2021)
- Strang, Gilbert, *Linear Algebra and Learning from Data* (Wellesley Cambridge Press, 2019)

- Stone, James V., *Artificial Intelligence Engines* (Sebtel Press, 2020)
- Stone, James V., *Bayes' Rule, A Tutorial Introduction to Bayesian Analysis* (Sebtel Press, 2013)
- Stone, James V., *Information Theory: A Tutorial Introduction* (Sebtel Press, 2015)
- Vajjala, Sowmya et al., *Practical Natural Language Processing* (O'Reilly, 2020)
- Van der Hofstad, Remco, *Random Graphs and Complex Networks* (Cambridge, 2017)
- Vershynin, Roman, *High-Dimensional Probability: An Introduction with Applications in Data Science* (Cambridge University Press, 2018)

Conventions Used in This Book

The following typographical conventions are used in this book:

Italic
Indicates new terms, URLs, email addresses, filenames, and file extensions.

`Constant width`
Used for program listings, as well as within paragraphs to refer to program elements such as variable or function names, databases, data types, environment variables, statements, and keywords.

`Constant width bold`
Shows commands or other text that should be typed literally by the user.

`Constant width italic`
Shows text that should be replaced with user-supplied values or by values determined by context.

 This element signifies a tip or suggestion.

 This element signifies a general note.

This element indicates a warning or caution.

Using Code Examples

The very few code examples that we have in this book are available for download at *https://github.com/halanelson/Essential-Math-For-AI*.

If you have a technical question or a problem using the code examples, please send an email to *bookquestions@oreilly.com*.

This book is here to help you get your job done. In general, if example code is offered with this book, you may use it in your programs and documentation. You do not need to contact us for permission unless you're reproducing a significant portion of the code. For example, writing a program that uses several chunks of code from this book does not require permission. Selling or distributing examples from O'Reilly books does require permission. Answering a question by citing this book and quoting example code does not require permission. Incorporating a significant amount of example code from this book into your product's documentation does require permission.

We appreciate, but generally do not require, attribution. An attribution usually includes the title, author, publisher, and ISBN. For example: "*Essential Math for AI* by Hala Nelson (O'Reilly). Copyright 2023 Hala Nelson, 978-1-098-10763-5."

If you feel your use of code examples falls outside fair use or the permission given above, feel free to contact us at *permissions@oreilly.com*.

O'Reilly Online Learning

 For more than 40 years, *O'Reilly Media* has provided technology and business training, knowledge, and insight to help companies succeed.

Our unique network of experts and innovators share their knowledge and expertise through books, articles, and our online learning platform. O'Reilly's online learning platform gives you on-demand access to live training courses, in-depth learning paths, interactive coding environments, and a vast collection of text and video from O'Reilly and 200+ other publishers. For more information, visit *https://oreilly.com*.

How to Contact Us

Please address comments and questions concerning this book to the publisher:

O'Reilly Media, Inc.
1005 Gravenstein Highway North
Sebastopol, CA 95472
800-998-9938 (in the United States or Canada)
707-829-0515 (international or local)
707-829-0104 (fax)

We have a web page for this book, where we list errata, examples, and any additional information. You can access this page at *https://oreil.ly/essentialMathAI*.

Email *bookquestions@oreilly.com* to comment or ask technical questions about this book.

For news and information about our books and courses, visit *https://oreilly.com*.

Find us on LinkedIn: *https://linkedin.com/company/oreilly-media*

Follow us on Twitter: *https://twitter.com/oreillymedia*

Watch us on YouTube: *https://www.youtube.com/oreillymedia*

Acknowledgments

My dad, Yousef Zein, who taught me math, and made sure to always remind me: *Don't think that the best thing we gave you in this life is land, or money. These come and go. Humans create money, buy assets, and create more money. What we did give you is a brain, a really good brain. That's your real asset, so go out and use it.* I love your brain, this book is for you, dad.

My mom, Samira Hamdan, who taught me both English and philosophy, and who gave up everything to make sure we were happy and successful. I wrote this book in English, not my native language, thanks to you, mom.

My daughter, Sary, who kept me alive during the most vulnerable times, and who is the joy of my life.

My husband, Keith, who gives me the love, passion, and stability that allow me to be myself, and to do so many things, some of them unwise, like writing a five-hundred-or-so-page book on math and AI. I love you.

My sister, Rasha, who is my soulmate. This says it all.

My brother, Haitham, who went against all our cultural norms and traditions to support me.

The memory of my uncle Omar Zein, who also taught me philosophy, and who made me fall in love with the mysteries of the human mind.

My friends Sharon and Jamie, who let me write massive portions of this book at their house, and were great editors any time I asked.

My lifetime friend Oren, who on top of being one of the best friends anyone can wish for, agreed to read and review this book.

My friend Huan Nguyen (*https://oreil.ly/XdWpu*), whose story should be its own book, and who also took the time to read and review this book. Thank you, admiral.

My friend and colleague John Webb, who read every chapter word by word, and provided his invaluable pure math perspective.

My wonderful friends Deb, Pankaj, Jamie, Tamar, Sajida, Jamila, Jen, Mattias, and Karen, who are part of my family. I love life with you.

My mentors Robert Kohn (New York University) and John Schotland (Yale University), to whom I owe reaching many milestones in my career. I learned a great deal from you.

The memory of Peter, whose impact was monumental, and who will forever inspire me.

The reviewers of this book, who took time and care despite their busy schedules to make it much better. Thank you for your great expertise and for generously giving me your unique perspectives from all your different domains.

All the waiters and waitresses in many cities in the world, who tolerated me sitting at my laptop at their restaurants for hours and hours and hours, writing this book. I got so much energy and happiness from you.

My incredible, *patient*, cheerful, and always supportive editors, Angela Rufino and Kristen Brown.

CHAPTER 1

Why Learn the Mathematics of AI?

It is not until someone said, "It is intelligent," that I stopped searching, and paid attention.
—H.

Artificial intelligence, known as AI, is here. It has penetrated multiple aspects of our lives and is increasingly involved in making very important decisions. Soon it will be employed in every sector of our society, powering most of our daily operations. The technology is advancing very fast and its investments are skyrocketing. At the same time, it feels like we are in the middle of an AI frenzy. Every day we hear about a new AI accomplishment. AI beats the best human player at a Go game. AI outperforms human vision in classification tasks. AI makes deep fakes. AI generates high energy physics data. AI solves difficult partial differential equations that model the natural phenomena of the world. Self-driving cars are on the roads. Delivery drones are hovering in some parts of the world.

We also hear about AI's seemingly unlimited potential. AI will revolutionize health-care and education. AI will eliminate global hunger. AI will fight climate change. AI will save endangered species. AI will battle disease. AI will optimize the supply chain. AI will unravel the origins of life. AI will map the observable universe. Our cities and homes will be smart. Eventually, we cross into science fiction territory. Humans will upload their brains into computers. Humans will be *enhanced* by AI. Finally, the voices of fear and skepticism emerge: AI will take over and destroy humanity.

Amid this frenzy, where the lines between reality, speculation, exaggeration, aspiration, and pure fiction are blurred, we must first define AI, at least within the context of this book. We will then discuss some of its limitations, where it is headed, and set the stage for the mathematics that is used in today's AI. My hope is that when you understand the mathematics, you will be able to look at the subject from a relatively deep perspective, and the blurring lines between fiction, reality, and everything in between will become more clear. You will also learn the main ideas behind state-of-the-art math in AI, arming you with the confidence needed to use, improve, or even create entirely new AI systems.

What Is AI?

I have yet to come across a unified definition of AI. If we ask two AI experts, we hear two different answers. Even if we ask the same expert on two different days, they might come up with two different definitions. The reason for this inconsistency and seeming inability to define AI is that until now it has not been clear what the definition of the *I* is. What is intelligence? What makes us human and unique? What makes us conscious of our own existence? How do neurons in our brain aggregate tiny electric impulses and translate them into images, sounds, feelings, and thoughts? These are vast topics that have fascinated philosophers, anthropologists, and neuroscientists for centuries. I will not attempt to go there in this book. I will, however, address artificial intelligence in terms of an AI agent and list the following defining principles for the purposes of this book. In 2022, an AI agent can be one or more of the following:

- An AI agent can be pure software or have a physical robotic body.
- An AI agent can be geared toward a specific task, or be a flexible agent exploring and manipulating its environment, building knowledge with or without a specific aim.
- An AI agent *learns* with experience, that is, it gets better at performing a task with more practice at that task.
- An AI agent perceives its environment, then builds, updates, and/or evolves a model for this environment.
- An AI agent perceives, models, analyzes, and makes decisions that lead to accomplishing its goal. This goal can be predefined and fixed, or variable and changing with more input.
- An AI agent understands cause and effect, and can tell the difference between patterns and causes.

Whenever a mathematical model for AI is inspired by the way our brain works, I will point out the analogy, hence keeping AI and human intelligence in comparison, without having to define either. Even though today's AI is nowhere close to human intelligence, except for specific tasks such as image classification, AlphaGo, etc., so many human brains have recently converged to develop AI that the field is bound to grow and have breakthroughs in the coming years.

It is also important to note that some people use the terms artificial intelligence, machine learning, and data science interchangeably. These three domains overlap but they are not the same. The fourth very important but slightly less hyped area is that of robotics, where physical parts and motor skills must be integrated into the learning and reasoning processes, merging mechanical engineering, electrical engineering, and bioengineering with information and computer engineering. One fast way to think about the interconnectivity of these fields is: *data fuels machine learning algorithms that in turn power many popular AI and/or robotics systems.* The mathematics in this book is useful, in different proportions, for all four domains.

Why Is AI So Popular Now?

In the past decade, AI has sprung into worldwide attention due to the successful combination of following factors:

Generation and digitization of massive amounts of data
This may include text data, images, videos, health records, e-commerce, network, and sensor data. Social media and the Internet of Things have played a very significant role here with their continuous streaming of great volumes of data.

Advances in computational power
This occurs through parallel and distributed computing as well as innovations in hardware, allowing for efficient and relatively cheap processing of large volumes of complex structured and unstructured data.

Recent success of neural networks in making sense of big data
AI has surpassed human performance in certain tasks such as image recognition and the Go game. When AlexNet (*https://oreil.ly/GubT1*) won the ImageNet Large Scale Visual Recognition Challenge (*https://oreil.ly/1LchH*) in 2012, it spurred a myriad of activity in convolutional neural networks (supported by graphical processing units), and in 2015, PReLU-Net (ResNet (*https://oreil.ly/0TYPA*)) was the first to outperform humans in image classification.

When we examine these factors, we realize that today's AI is not the same as science fiction AI. Today's AI is centered around big data (all kinds of data), machine learning algorithms, and is heavily geared toward performing one task extremely well, as opposed to developing and adapting varied intelligence types and goals as a response to the surrounding environment.

What Is AI Able to Do?

There are many more areas and industries where AI can be successfully applied than there are AI experts who are well suited to respond to this ever-growing need. Humans have always strived for automating processes, and AI carries a great promise to do exactly that, at a massive scale. Large and small companies have volumes of raw data that they would like to analyze and turn into insights for profits, optimal strategies, and allocation of resources. The health industry suffers a severe shortage of doctors, and AI has innumerable applications and unlimited potential there. Worldwide financial systems, stock markets, and banking industries have always depended heavily on our ability to make good predictions, and have suffered tremendously when those predictions failed. Scientific research has progressed significantly with our increasing ability to compute, and today we are at a new dawn where advances in AI enable computations at scales thought impossible a few decades ago.

Efficient systems and operations are needed everywhere, from the power grid, transportation, and the supply chain to forest and wildlife preservation, battling world hunger, disease, and climate change. Automation is even sought after in AI itself, where an AI system spontaneously decides on the optimal pipelines, algorithms, and parameters, readily producing the desired outcomes for given tasks, thus eliminating the need for human supervision altogether.

An AI Agent's Specific Tasks

In this book, as I work through the math, I will focus on popular application areas of AI, in the context of an AI agent's specified tasks. Nevertheless, the beneficial mathematical ideas and techniques are readily transferable across different application domains. The reason for this seeming easiness and wide applicability is that we happen to be at the age of AI *implementation*, in the sense that the main ideas for addressing certain tasks have already been developed, and with only a little tweaking, they can be *implemented* across various industries and domains. Our AI topics and/or tasks include:

Simulated and real data
> Our AI agent processes data, provides insights, and makes decisions based on that data (using mathematics and algorithms).

The brain neocortex
> Neural networks in AI are modeled after the neocortex, or the *new brain*. This is the part of our brain responsible for high functions such as perception, memory, abstract thought, language, voluntary physical action, decision making, imagination, and consciousness. The neocortex has many layers, six of which are mostly distinguishable. It is flexible and has a tremendous learning ability. The *old brain* and the *reptilian brain* lie below the neocortex, and are responsible for emotions and more basic and primitive survival functions such as breathing, regulating the heartbeat, fear, aggression, sexual urges, and others. The old brain keeps records of actions and experiences that lead to favorable or unfavorable feelings, creating our emotional memory that influences our behavior and future actions. Our AI agent, in a very basic way, emulates the neocortex and sometimes the old brain.

Computer vision

Our AI agent senses and recognizes its environment through cameras, sensors, etc. It peeks into everything, from our daily pictures and videos, to our MRI scans, and all the way into images of distant galaxies.

Natural language processing

Our AI agent communicates with its environment and automates tedious and time-consuming tasks such as text summarization, language translation, sentiment analysis, document classification and ranking, captioning images, and chatting with users.

Financial systems

Our AI agent detects fraud in our daily transactions, assesses loan risks, and provides 24-hour feedback and insights about our financial habits.

Networks and graphs

Our AI agent processes network and graph data, such as animal social networks, infrastructure networks, professional collaboration networks, economic networks, transportation networks, biological networks, and many others.

Social media

Our AI agent has social media to thank for providing the large amount of data necessary for its learning. In return, our AI agent attempts to characterize social media users, identifying their patterns, behaviors, and active networks.

The supply chain

Our AI agent is an optimizing expert. It helps us predict optimal resource needs and allocation strategies at each level of the production chain. It also finds ways to end world hunger.

Scheduling and staffing

Our AI agent facilitates our daily operations.

Weather forecasting

Our AI agent solves partial differential equations used in weather forecasting and prediction.

Climate change

Our AI agent attempts to fight climate change.

Education

Our AI agent delivers personalized learning experiences.

Ethics

Our AI agent strives to be fair, equitable, inclusive, transparent, unbiased, and protective of data security and privacy.

What Are AI's Limitations?

Along with the impressive accomplishments of AI and its great promise to enhance or revolutionize entire industries, there are some real limitations that the field needs to overcome. Some of the most pressing limitations are:

Intelligence

Current AI is not even remotely close to being *intelligent* in the sense that we humans consider ourselves uniquely intelligent. Even though AI has outperformed humans in innumerable tasks, it cannot naturally switch and adapt to new tasks. For example, an AI system trained to recognize humans in images cannot recognize cats without retraining, or generate text without changing its architecture and algorithms. In the context of the three types of AI, we have thus far only partially accomplished *artificial narrow intelligence*, which has a narrow range of abilities. We have accomplished neither *artificial general intelligence*, on par with human abilities, nor *artificial super intelligence*, which is more capable than humans'. Moreover, machines today are incapable of experiencing any of the beautiful human emotions, such as love, closeness, happiness, pride, dignity, caring, sadness, loss, and many others. Mimicking emotions is different than experiencing and genuinely providing them. In this sense, machines are nowhere close to replacing humans.

Large volumes of labeled data

Most popular AI applications need large volumes of labeled data, for example, MRI images can be labeled cancer or not-cancer, YouTube videos can be labeled safe for children or unsafe, or house prices can be available with the house district, number of bedrooms, median family income, and other features—in this case the house price is the label. The limitation is that the data required to train a system is usually not readily available, and not cheap to obtain, label, maintain, or warehouse. A substantial amount of data is confidential, unorganized, unstructured, biased, incomplete, and unlabeled. Obtaining the data, curating it, preprocessing it, and labeling it become major obstacles requiring large time and resource investments.

Multiple methods and hyperparameters

For a certain AI task, there are sometimes many methods, or algorithms, to accomplish it. Each task, data set, and/or algorithm has parameters, called hyperparameters, that can be tuned during implementation, and it is not always clear what the best values for these hyperparameters are. The variety of methods and hyperparameters available to tackle a specific AI task mean that different methods can produce extremely different results, and it is up to humans to assess which methods' decisions to rely on. In some applications, such as which dress styles to recommend for a certain customer, these discrepancies may be inconsequential. In other areas, AI-based decisions can be life-changing: a patient is told they do not have a certain disease, while in fact they do; an inmate is mislabeled as highly likely to reoffend and gets their parole denied as a consequence; or a loan gets rejected for a qualified person. Research is ongoing on how to address these issues, and I will expand on them as we progress through the book.

Resource limitations

Human abilities and potential are limited to our brainpower, the capacity of our biological bodies, and the resources available on Earth and in the universe that we are able to manipulate. These are again limited by the power and capacity of our brains. AI systems are similarly limited by the computing power and hardware capability of the systems supporting the AI software. Recent studies have suggested that computation-intensive deep learning is approaching its computational limits, and new ideas are needed to improve algorithm and hardware efficiency, or discover entirely new methods. Progress in AI has heavily depended on large increases in computing power. This power, however, is not unlimited, is extremely costly for large systems processing massive data sets, and has a substantial carbon footprint that cannot be ignored, for example, the power required to run and cool down data warehouses, individual devices, keep the cloud connected, etc. Moreover, data and algorithmic software do not exist in the vacuum. Devices such as computers, phones, tablets, batteries, and the warehouses and systems needed to store, transfer, and process data and algorithms are made of real physical materials harvested from Earth. It took Earth millions of years to make some of these materials, and the type of infinite supply required to forever sustain these technologies is just not there.

Security costs

Security, privacy, and adversarial attacks remain a primary concern for AI, especially with the advent of interconnected systems. A lot of research and resources are being allocated to address these important issues. Since most of the current AI is software and most of the data is digital, the arms race in this area is never-ending. This means that AI systems need to be constantly monitored and updated, requiring more expensive-to-hire AI and cybersecurity specialists, probably at a cost that defeats the initial purpose of automation at scale.

Broader impacts

The AI research and implementation industries have thus far viewed themselves as slightly separate from the economical, social, and security consequences of their advancing technologies. Usually these ethical, social, and security implications of the AI work are acknowledged as important and needing to be attended to, but beyond the scope of the work itself. As AI becomes widely deployed and its impacts on the fabric and nature of society, markets, and potential threats are felt more strongly, the field as a whole has to become more intentional in the way it attends to these issues of paramount importance. In this sense, the AI development community has been limited in the resources it allocates to addressing the broader impacts of the implementation and deployment of its new technologies.

What Happens When AI Systems Fail?

A very important part of learning about AI is learning about its incidents and failures. This helps us foresee and avoid similar outcomes when designing our own AI, *before* deploying out into the real world. If the AI fails after being deployed, the consequences can be extremely undesirable, dangerous, or even lethal.

One online repository for AI failures, called the AI Incident Database (*https://incident database.ai*), contains more than a thousand such incidents. Examples from this website include:

- A self-driving car kills a pedestrian.
- Self-driving cars lose contact with their company's server for a full 20 minutes and all stall at once in the streets of San Francisco (June 28 and May 18 of 2022).
- A trading algorithm causes a market *flash crash* where billions of dollars automatically transfer between parties.
- A facial recognition system causes an innocent person to be arrested.
- Microsoft's infamous chatbot Tay is shut down only 16 hours after its release, since it quickly learned and tweeted offensive, racist, and highly inflammatory remarks.

Such bad outcomes can be mitigated but require a deep understanding of how these systems work, at all levels of production, as well as of the environment and users they are deployed for. Understanding the mathematics behind AI is one crucial step in this discerning process.

Where Is AI Headed?

To be able to answer, or speculate on, where AI is headed, it is best to recall the field's original goal since its inception: mimic human intelligence. This field was conceived in the fifties. Examining its journey over the past seventy years *might* tell us something about its future direction. Moreover, studying the history of the field and its trends enables us to have a bird's-eye view of AI, putting everything in context and providing a better perspective. This also makes learning the mathematics involved in AI a less overwhelming experience. The following is a very brief and nontechnical overview of AI's evolution and its eventual thrust into the limelight thanks to the recent impressive progress of deep learning.

In the beginning, AI research attempted to mimic intelligence using rules and logic. The idea was that all we needed to do is feed machines facts and logical rules of reasoning about these facts (we will see examples of this logical structure in Chapter 12). There was no emphasis on the learning process. The challenge here was that, in order to capture human knowledge, there are too many rules and constraints to be tractable for a coder, and the approach seemed unfeasible.

In the late 1990s and the early 2000s, various machine learning methods became popular. Instead of programming the rules, and making conclusions and decisions based on these preprogrammed rules, machine learning infers the rules from the data. The more data a machine learning system is able to handle and process, the better its performance. Data and the ability to process and *learn from* large amounts of data economically and efficiently became the main goals. Popular machine learning algorithms in that time period were support vector machines, Bayesian networks, evolutionary algorithms, decision trees, random forests, regression, logistic regression, and others. These algorithms are still popular now.

After 2010, and particularly in 2012, a tidal wave of neural networks and deep learning took over after the success of AlexNet's convolutional neural network in image recognition.

Most recently, in the last five years, reinforcement learning gained popularity after DeepMind's AlphaGo beat the world champion in the very complicated ancient Chinese game of Go.

Note that this glimpse of history is very rough: regression has been around since Legendre and Gauss in the very early 1800s, and the first artificial neurons and neural networks were formulated in the late 1940s and early 1950s with the works of neurophysiologist Warren McCulloch, mathematician Walter Pitts, and psychologists Donald Hebb and Frank Rosenblatt. The Turing Test, originally called the Imitation Game, was introduced in 1950 by Alan Turing, a computer scientist, cryptanalyst, mathematician, and theoretical biologist, in his paper "Computing Machinery and Intelligence" (*https://oreil.ly/bJp5a*). Turing proposed that a machine possesses artificial intelligence if its responses are indistinguishable from those of a human. Thus, a machine is considered intelligent if it is able to imitate human responses. The Turing Test, however, for a person outside the field of computer science, sounds limiting in its definition of intelligence, and I wonder if the Turing Test might have inadvertently limited the goals or the direction of AI research.

Even though machines are able to mimic human intelligence in some specific tasks, the original goal of replicating human intelligence has not been accomplished yet, so it might be safe to assume that is where the field is headed, even though it could involve rediscovering old ideas or inventing entirely new ones. The current level of investment in the area, combined with the explosion in research and public interest, are bound to produce new breakthroughs. Nonetheless, breakthroughs brought about by recent AI advancements are already revolutionizing entire industries eager to implement these technologies. These contemporary AI advancements involve plenty of important mathematics that we will be exploring throughout this book.

Who Are the Current Main Contributors to the AI Field?

The main AI race has been between the United States, Europe, and China. Some of the world leaders in the technology industry have been Google and its parent company Alphabet, Amazon, Facebook, Microsoft, Nvidia, and IBM in the United States, DeepMind in the UK and the United States (owned by Alphabet), and Baidu and Tencent in China. There are major contributors from the academic world as well, but these are too many to enumerate. If you are new to the field, it is good to know the names of the big players, their histories and contributions, and the kinds of goals they are currently pursuing. It is also valuable to learn about the controversies, if any, surrounding their work. This general knowledge comes in handy as you navigate through and gain more experience in AI.

What Math Is Typically Involved in AI?

When I say the word "math," what topics and subjects come to your mind?

Whether you are a math expert or a beginner, whatever math topic that you thought of to answer the question is most likely involved in AI. Here is a commonly used list of the most useful math subjects for AI implementation: calculus, linear algebra, optimization, probability, and statistics; however, you do not need to be an expert in all of these fields to succeed in AI. What you do need is a deep understanding of certain useful topics drawn from these math subjects. Depending on your specific application area, you might need special topics from: random matrix theory, graph theory, game theory, differential equations, and operations research.

In this book we will walk through these topics without presenting a textbook on each one. AI application and implementation are the unifying themes for these varied and intimately interacting mathematical subjects. Using this approach, I might offend some math experts by simplifying a lot of technical definitions or omitting whole theorems and delicate details, and I might as well offend AI or specialized industry experts, again omitting details involved in certain applications and implementations. The goal, however, is to keep the book simple and readable, while at the same time covering most of the math topics that are important for AI applications. Interested readers who want to dive deeper into the math or the AI field can then read more involved books on the particular area they want to focus on. My hope is that this book is a concise summary and a thorough overview, hence a reader can afterward branch out confidently to whatever AI math field or AI application area interests them.

Summary and Looking Ahead

Human intelligence reveals itself in perception, vision, communication through natural language, reasoning, decision making, collaboration, empathy, modeling and manipulating the surrounding environment, transfer of skills and knowledge across populations and generations, and generalization of innate and learned skills into new and uncharted domains. Artificial intelligence aspires to replicate all aspects of human intelligence. In its current state, AI addresses only one or few aspects of intelligence at a time. Even with this limitation, AI has been able to accomplish impressive feats, such as modeling protein folding and predicting protein structures, which are the building blocks of life. The implications of this one AI application (among many) for understanding the nature of life and battling all kinds of diseases are boundless.

When you enter the AI field, it is important to remain mindful of which aspect of intelligence you are developing or using. Is it perception? Vision? Natural language? Navigation? Control? Reasoning? Which mathematics to focus on and why then follow naturally, since you already know where in the AI field you are situated. It will then be easy to attend to the mathematical methods and tools used by the community developing that particular aspect of AI. The recipe in this book is similar: first the AI type and application, then the math.

In this chapter, we addressed general questions. What is AI? What is AI able to do? What are AI's limitations? Where is AI headed? How does AI work? We also briefly surveyed important AI applications, the problems usually encountered by companies trying to integrate AI into their systems, incidents that happen when systems are not well implemented, and the math subjects typically needed for AI implementations.

In the next chapter, we dive into data and affirm its intimate relationship to AI. When we talk data, we also talk data distributions, and that plunges us straight into probability theory and statistics.

CHAPTER 2

Data, Data, Data

Maybe if I know where it all came from, and why, I would know where it's all headed, and why.

—H.

Data is the fuel that powers most AI systems. In this chapter, we will understand how data, and devising methods for extracting useful and actionable information from data, is at the heart of *perception AI*.

Perception AI is based on statistical learning from data, where an AI agent, or a machine, perceives data from its environment, then detects patterns within this data, allowing it to draw conclusions and/or make decisions.

Perception AI is different from the three other types of AI:

Understanding AI
> Where an AI system understands that the image it classified as a chair serves the function of sitting, the image it classified as cancer means that the person is sick and needs further medical attention, or the textbook it read about linear algebra can be used to extract useful information from data.

Control AI
> This has to do with controlling the physical parts of the AI agent in order to navigate spaces, open doors, serve coffee, etc. Robotics have made significant progress in this area. We need to augment robots with "brains" that include perception AI and understanding AI, and connect those to the control AI. Ideally, like humans, control AI then learns from its physical interactions with its environment by passing that information to its perception and understanding systems, which in turn pass control commands to the agent's control systems.

Awareness AI

Where an AI agent has an inner experience similar to the human experience. Since we do not know yet how to mathematically define awareness, we do not visit this concept at all in this book.

Ideally, true human-like intelligence combines all four aspects: perception, understanding, control, and awareness. The main focus of this chapter and the next few chapters is perception AI. Recall that AI and data have become intertwined to the extent that it is now common, though erroneous, to use the terms data science and AI synonymously.

Data for AI

At the core of many popular machine learning models, including the highly successful neural networks that brought artificial intelligence back into the popular spotlight with AlexNet in 2012, lies a very simple mathematical problem:

Fit a given set of data points into an appropriate function (mapping an input to an output) that picks up on the important signals in the data and ignores the noise, then make sure this function performs well on new data.

Complexity and challenges, however, arise from various sources:

Hypothesis and features

Neither the true function that generated the data nor all the features it actually depends on are known. We simply observe the data, then try to estimate a hypothetical function that generated it. Our function tries to learn which features of the data are important for our predictions, classifications, decisions, or general purposes. It also learns how these features interact in order to produce the observed results. One of the great potentials of AI in this context is its ability to pick up on subtle interactions between features of data that humans do not usually pick up on, since we are very good at observing strong features but may miss more subtle ones. For example, we as humans can tell that a person's monthly income affects their ability to pay back a loan, but we might not observe that their daily commute, or morning routine, may have a nontrivial effect on that as well. Some feature interactions are much simpler than others, such as linear interactions. Others are more complex and are nonlinear. From a mathematical point of view, whether our feature interactions are simple (linear) or complex (nonlinear), we still have the same goal: find the hypothetical function that fits your data and is able to make good predictions on new data. One extra complication arises here: there are many hypothetical functions that can *fit* the same data set—how do we know which ones to choose?

Performance

Even after computing a hypothetical function that fits our data, how do we know whether it will perform well on new and unseen data? How do we know which performance measure to choose, and how to monitor this performance after deploying into the real world? Real-world data and scenarios do not come to us all labeled with ground truths, so we cannot easily measure whether our AI system is doing well and making correct or appropriate predictions and decisions. We do not know what to measure the AI system's results against. If real-world data and scenarios were labeled with ground truths, then we would all be out of business since we would know what to do in every situation, there would be peace on Earth, and we would live happily ever after (not really, I wish it was that simple).

Volume

Almost everything in the AI field is very high-dimensional! The number of data instances, observed features, and unknown weights to be computed could be in the millions, and the required computation steps in the billions. Efficient storage, transport, exploration, preprocessing, structuring, and computation on such volumes of data become center goals. In addition, exploring the landscapes of the involved high-dimensional mathematical functions is a nontrivial endeavor.

Structure

Most of the data created by the modern world is unstructured. It is not organized in easy-to-query tables that contain labeled fields such as names, phone numbers, genders, ages, zip codes, house prices, income levels, etc. Unstructured data is everywhere: posts on social media, user activity, word documents, PDF files, images, audio and video files, collaboration software data, traffic or seismic or weather data, GPS, military movement, emails, instant messenger, mobile chat data, and many others. Some of these examples, such as email data, can be considered semistructured, since emails come with headings that include the email's metadata: From, To, Date, Time, Subject, Content-Type, Spam Status, etc. Moreover, large volumes of important data are not available in digital format and are fragmented over multiple and noncommunicating databases. Examples here include historical military data, museum archives, and hospital records. Presently, there is great momentum toward digitalizing our world and our cities to leverage more AI applications. Overall, it is easier to draw insights from structured and labeled data than from unstructured data. Mining unstructured data requires innovative techniques that are currently driving forces in the fields of data science, machine learning, and artificial intelligence.

Real Data Versus Simulated Data

When we work with data, it is very important to know the difference between real data and simulated data. Both types of data are extremely valuable for human discovery and progress.

Real data

This data is collected through real-world observations, using measuring devices, sensors, surveys, structured forms like medical questionnaires, telescopes, imaging devices, websites, stock markets, controlled experiments, etc. This data is often imperfect and noisy due to inaccuracies and failures in measuring methods and instruments. Mathematically, *we do not know the exact function or probability distribution that generated the real data*, but we can hypothesize about it using models, theories, and simulations. We can then test our models, and finally use them to make predictions.

Simulated data

This is data generated using a *known* function or randomly sampled from a *known* probability distribution. Here, we have our known mathematical function(s), or *model*, and we plug numerical values into the model to generate our data points. Examples are plentiful: numerical solutions of partial differential equations modeling all kinds of natural phenomena on all kinds of scales, such as turbulent flows, protein folding, heat diffusion, chemical reactions, planetary motion, fractured materials, traffic, and even and even enhancing Disney movie animations, such as simulating natural water movement in *Moana* or Elsa's hair movement in *Frozen*.

In this chapter, we present two examples about human height and weight data to demonstrate the difference between real and simulated data. In the first example, we visit an online public database, then download and explore two real data sets containing measurements of the heights and weights of real individuals. In the second example, we simulate our own data set of heights and weights *based on a function that we hypothesize*: we assume that the weight of an individual depends linearly on their height. This means that when we plot the weight data against the height data, we expect to see a straight, or flat, visual pattern.

Mathematical Models: Linear Versus Nonlinear

Linear dependencies model flatness in the world, like one-dimensional straight lines, two-dimensional flat surfaces (called *planes*), and higher-dimensional hyperplanes. The graph of a linear function, which models a linear dependency, is forever flat and does not bend. Every time you see a flat object, like a table, a rod, a ceiling, or a bunch of data points huddled together around a straight line or a flat surface, know that their representative function is linear. Anything that isn't flat is nonlinear, so functions whose graphs bend are nonlinear, and data points that congregate around bending curves or surfaces are generated by nonlinear functions.

The formula for a linear function, representing a linear dependency of the function output on the *features*, or *variables*, is very easy to write down. The features appear in the formula as just themselves, with no powers or roots, and are not embedded in any other functions, such as denominators of fractions, sine, cosine, exponential, logarithmic, or other calculus functions. They can only be multiplied by *scalars* (real or complex numbers, not vectors or matrices), and added to or subtracted from each other. For example, a function that depends linearly on three features x_1, x_2, and x_3 can be written as:

$$f(x_1, x_2, x_3) = \omega_0 + \omega_1 x_1 + \omega_2 x_2 + \omega_3 x_3$$

where the parameters $\omega_0, \omega_1, \omega_2$, and ω_3 are scalar numbers. The parameters or *weights* ω_1, ω_2, and ω_3 linearly combine the features, and produce the outcome of $f(x_1, x_2, x_3)$ after adding the bias term ω_0. In other words, the outcome is produced as a result of linear interactions between the features x_1, x_2, and x_3, plus bias.

The formula for a nonlinear function, representing a nonlinear dependency of the function output on the features, is very easy to spot as well. One or more features appear in the function formula with a power other than one, or multiplied or divided by other features, or embedded in some other calculus functions, such as sines, cosines, exponentials, logarithms, etc. The following are three examples of functions depending nonlinearly on three features x_1, x_2, and x_3:

$$f(x_1, x_2, x_3) = \omega_0 + \omega_1 \sqrt{x_1} + \omega_2 \frac{x_2}{x_3}$$

$$f(x_1, x_2, x_3) = \omega_0 + \omega_1 x_1^2 + \omega_2 x_2^2 + \omega_3 x_3^2$$

$$f(x_1, x_2, x_3) = \omega_1 e^{x_1} + \omega_2 e^{x_2} + \omega_3 \cos(x_3)$$

As you can tell, we can come up with all kinds of nonlinear functions, and the possibilities related to what we can do and how much of the world we can model using nonlinear interactions are limitless. In fact, neural networks are successful because of their ability to pick up on the relevant *nonlinear* interactions between the data features.

We will use the previous notation and terminology throughout the book, so you will become very familiar with terms like linear combination, weights, features, and linear and nonlinear interactions between features.

An Example of Real Data

You can find the Python code to investigate the data and produce the figures in the following two examples at the book's GitHub page (*https://github.com/halanel son/Essential-Math-For-AI*).

Structured Data

The two data sets for height, weight, and gender that we will work with here are examples of *structured* data sets. They come organized in rows and columns. Columns contain the features, such as weight, height, gender, health index, etc. Rows contain the feature scores for each data instance, in this case, each person. On the other hand, data sets that are a bunch of audio files, Facebook posts, images, or videos are all examples of unstructured data sets.

I downloaded two data sets from the Kaggle website (*https://www.kaggle.com*) for data scientists. Both data sets contain height, weight, and gender information for a certain number of individuals. My goal is to learn how the weight of a person depends on their height. Mathematically, I want to write a formula for the weight as a function of one feature, the height:

$$weight = f(height)$$

so that if I am given the height of a *new* person, I would be able to *predict* their weight. Of course, there are other features than height that a person's weight depends on, such as their gender, eating habits, workout habits, genetic predisposition, etc. However, for the data sets that I downloaded, we only have height, weight, and gender data available. Unless we want to look for more detailed data sets, or go out and collect new data, we have to work with what we have. Moreover, the goal of this example is only to illustrate the difference between real data and simulated data. We can work with more involved data sets with larger numbers of features when we have more involved goals.

For the first data set (*https://oreil.ly/pxgwe*), I plot the weight column against the height column in Figure 2-1, and obtain something that seems to have no pattern at all!

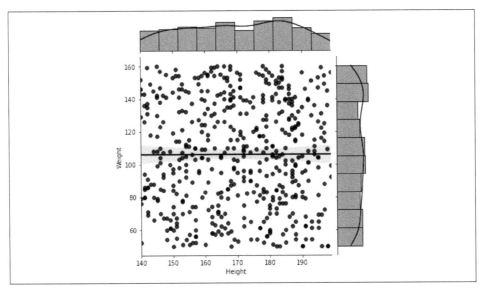

Figure 2-1. When plotting the weight against the height for the first data set (https:// oreil.ly/aaSEl), we cannot detect a pattern. The graphs on the top and on the right of the scatterplot show the respective histograms and empirical distributions of the height and weight data.

For the second data set (*https://oreil.ly/8bE36*), I do the same, and I can visually observe an obvious linear dependency in Figure 2-2. The data points seem to congregate around a straight line!

So what is going on? Why does my first real data set reflect no dependency between the height and weight of a person whatsoever, but my second one reflects a linear dependency? We need to look deeper into the data.

This is one of the many challenges of working with real data. We do not know what function generated the data, and why it looks the way it looks. We investigate, gain insights, detect patterns, if any, and we propose a hypothesis function. Then we test our hypothesis, and if it performs well based on our measures of performance, which have to be thoughtfully crafted, we deploy it into the real world. We make predictions using our deployed model, until new data tells us that our hypothesis is no longer valid, in which case we investigate the updated data and formulate a new hypothesis. This process and feedback loop keeps going for as long as our models are in business.

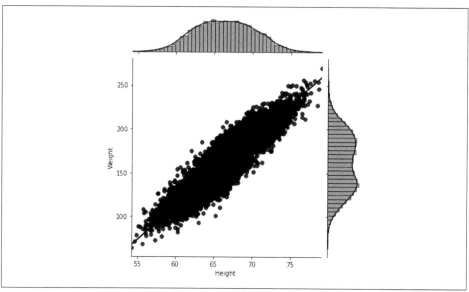

Figure 2-2. When plotting the weight against the height for the second data set (https://oreil.ly/rZNBS), we observe a linear pattern. Note that the empirical distribution of the weight data is plotted on the righthand side of the figure, and the empirical distribution of the height data is plotted at the top of the figure. Both seem to have two peaks (bimodal), suggesting the existence of a mixture distribution. In fact, both the height and the weight data sets can be modeled using a mixture of two normal distributions, called Gaussian mixtures, representing mixing the underlying distributions for the male and female data. So if we plot the data for either the female or male subpopulations alone, as in Figure 2-6, we observe that the height and weight data are normally distributed (bell-shaped).

Before moving on to simulated data, let's explain why the first data set seemed to have no insight at all about the relationship between the height and the weight of an individual. Upon further inspection, we notice that the data set has an overrepresentation of individuals with Index scores 4 and 5, referring to obesity and extreme obesity. So, I decided to split the data by Index score, and plot the weight against the height for all individuals with similar Index scores. This time around, a linear dependency between the height and the weight is evident in Figure 2-3, and the mystery is resolved. This might feel like we are cheating our way to linearity, by conditioning on individuals' Index scores. All is fair game in the name of data exploration.

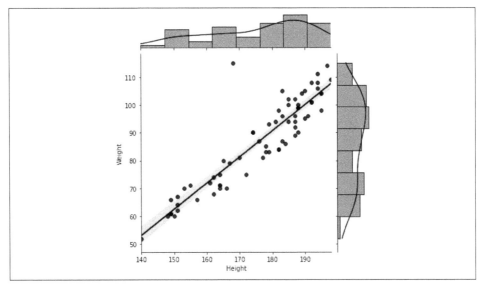

Figure 2-3. When plotting the weight against the height for individuals with similar Index scores in the first data set (https://oreil.ly/0cPnT), we observe a linear pattern. This figure shows the weight against the height for individuals with Index score 3.

Now we can safely go ahead and hypothesize that the weight depends linearly on the height:

$$weight = \omega_0 + \omega_1 \times height$$

Of course, we are left with the task of finding appropriate values for the parameters ω_0 and ω_1. Chapter 3 teaches us how to do exactly that. In fact, the bulk of activity in machine learning, including deep learning, is about *learning* these ω's *from the data*. In our very simple example, we only have two ω's to learn, since we only had one feature, the height, and we *assumed* linear dependency after observing a linear pattern in the real data. In the next few chapters, we will encounter some deep learning networks with millions of ω's to learn, yet we will see that the mathematical structure of the problem is in fact the same exact structure that we will learn in Chapter 3.

An Example of Simulated Data

In this example, I simulate my own height-weight data set. Simulating our own data circumvents the trouble of searching for data from the web, the real world, or even building a lab to obtain controlled measurements. This is incredibly valuable when the required data is not available or very expensive to obtain. It also helps test different scenarios by only changing numbers in a function, as opposed to, say, creating new materials or building labs and running new experiments. Simulating data is so convenient because all we need is a mathematical function, a probability distribution if we want to involve randomness and/or noise, and a computer.

Let's again assume linear dependency between the height and the weight, so the function that we will use is:

$$weight = \omega_0 + \omega_1 \times height$$

For us to be able to simulate numerical (*height,weight*) pairs, or data points, we must assume numerical values for the parameters ω_0 and ω_1. Without having insights from real data about the correct choices for these ω's, we are left with making educated guesses from the context of the problem and experimenting with different values. Note that for the height-weight case in this example, we happen to have real data that we can use to learn appropriate values for the ω's, and one of the goals of Chapter 3 is to learn how to do that. However, in many other scenarios, we do not have real data, so the only way to go is experimenting with various numerical values for these ω's.

In the following simulations, we set $\omega_0 = -314.5$ and $\omega_1 = 7.07$, so the function becomes:

$$weight = -314.5 + 7.07 \times height$$

Now we can generate as many numerical (*height,weight*) pairs as we want. For example, plugging $height = 60$ in the formula for the weight function, we get $weight = -314.5 + 7.07 \times 60 = 109.7$. So our linear model *predicts* that a person whose height is 60 inches weighs 109.7 pounds, and the data point that we can plot on the height-weight graph has coordinates (60,109.7). In Figure 2-4, we generate 5,000 of these data points: we choose 5,000 values for the height between 54 and 79 inches and plug them into the weight function. We notice that the graph in Figure 2-4 is a perfect straight line, with no noise or variation in the simulated data, since we did not incorporate those into our linear model.

That's a hallmark of simulated data: it does what the function(s) it was generated from does. If we understand the function (called *model*) that we used to build our simulation, and if our computation does not accumulate too many numerical errors and/or very large numbers that go rogue, then we understand the data that our model generates, and we can use this data in any way we see fit. There isn't much space for surprises. In our example, our proposed function is linear, so its equation is that of a straight line, and as you see in Figure 2-4, the generated data lies perfectly on this straight line.

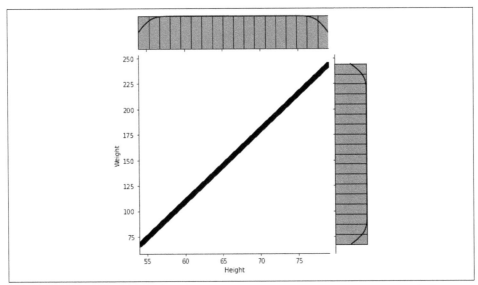

Figure 2-4. Simulated data: we generated 5,000 (height, weight) points using the linear function weight = − 314.5 + 7.07 × height

What if we want to simulate more realistic data for height and weight? Then we can sample the height values from a more realistic distribution for the heights of a human population: the bell-shaped normal distribution! Again, we *know* the probability distribution that we are sampling from, which is different from the case for real data. After we sample the height values, we plug those into the linear model for the weight, then we add some noise, since we want our simulated data to be realistic. Since noise has a random nature, we must also pick the probability distribution it will be sampled from. We again choose the bell-shaped normal distribution, but we could have chosen the uniform distribution to model uniform random fluctuations. Our more realistic height-weight model becomes:

$$weight = -314.5 + 7.07 \times height + noise$$

We obtain the results seen in Figure 2-5.

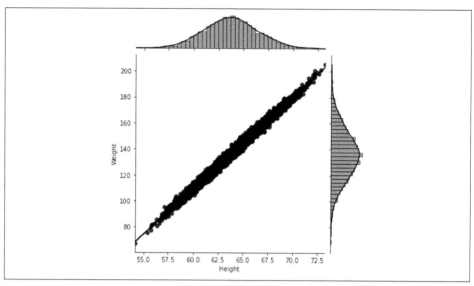

Figure 2-5. Simulated data: we generated 5,000 (height, weight) points using the linear function $weight = -314.5 + 7.07 \times height$. The height points are normally distributed and we added normally distributed noise as well. Note that the distributions of the weight and height data at the righthand side and at the top of the figure, respectively, are normally distributed. This is not surprising since we designed them that way in our simulation.

Now compare Figure 2-5 containing our simulated height-weight data to Figure 2-6 containing real height-weight data of 5,000 females from the second data set (*https:// oreil.ly/rZNBS*) that we used. Not too bad, given that it only took five minutes of code writing to generate this data, as opposed to collecting real data! Had we spent more time tweaking the values of our ω's, and the parameters for the normally distributed noise that we added (mean and standard deviation), we would've obtained an even better-looking simulated data set. However, we will leave this simulation here since very soon our whole focus will be on *learning* the appropriate parameter values for our hypothesized models.

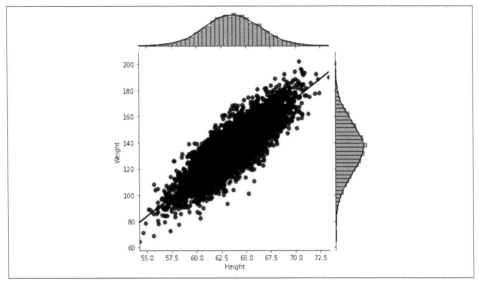

Figure 2-6. Real data: weight data plotted against the height data of the 5,000 females in the second data set (https://oreil.ly/rZNBS). Note that the distributions of the female weight and height data at the righthand side and at the top of the figure, respectively, are normally distributed. For more details, see the book's GitHub page (https:// github.com/halanelson/Essential-Math-For-AI) for more details.

Mathematical Models: Simulations and AI

We can always adjust our mathematical models to make them more realistic. We are the designers, so we get to decide what goes into these models. It is often the case that the more a model mimics nature, the more mathematical objects get incorporated within it. Therefore, while building a mathematical model, the usual trade-off is between getting closer to reality, and the model's simplicity and accessibility for mathematical analysis and computation. Different designers come up with different mathematical models, and some capture certain phenomena better than others. These models keep improving and evolving as the quest to capture natural behaviors continues. Thankfully, our computational capabilities have dramatically improved in the past decades, enabling us to create and test more involved and realistic mathematical models.

Nature is at the same time very finely detailed and enormously vast. Interactions in nature range from the subatomic quantum realm all the way to the intergalactic scale. We, as humans, are forever trying to understand nature and capture its intricate components with their numerous interconnections and interplays. Our reasons for this are varied. They include pure curiosity about the origins of life and the universe, creating new technologies, enhancing communication systems, designing drugs and discovering cures for diseases, building weapons and defense systems, and traveling to distant planets and perhaps inhabiting them in the future. Mathematical models provide an excellent and almost miraculous way to describe nature with all its details using only numbers, functions, equations, and invoking quantified randomness through probability when faced with uncertainty. Computer simulations of these mathematical models enable us to investigate and visualize various simple and complex behaviors of the modeled systems or phenomena. In turn, insights from computer simulations aid in model enhancement and design, in addition to supplying deeper mathematical insights. This incredibly positive feedback cycle makes mathematical modeling and simulations an indispensable tool that is enhanced greatly with our increased computational power.

It is a mystery of the universe that its various phenomena can be accurately modeled using the abstract language of mathematics, and it is a marvel of the human mind that it can discover and comprehend mathematics, and build powerful technological devices that are useful for all kinds of applications. Equally impressive is that these devices, at their core, are doing nothing but computing or transmitting mathematics, more specifically, a bunch of zeros and ones.

The fact that humans are able to generalize their understanding of simple numbers all the way to building and applying mathematical models for natural phenomena at all kinds of scales is a spectacular example of *generalization* of learned knowledge, and is a hallmark of human intelligence. In the AI field, a common goal for both general AI (human-like and super-AI) and narrow AI (specific task oriented) is *generalization*: the ability of an AI agent to *generalize* learned abilities to new and unknown situations. In Chapter 3, we will understand this principle for narrow and task-oriented AI: *an AI agent learns from data, then produces good predictions for new and unseen data.*

AI interacts in three ways with mathematical models and simulations:

Mathematical models and simulations create data for AI systems to train on.
Self-driving cars are considered by some to be a benchmark for AI. It will be inconvenient to let intelligent car prototypes drive off cliffs, hit pedestrians, or crash into new work zones before the car's AI system learns that these are unfavorable events that must be avoided. Training on simulated data is especially valuable here, as simulations can create all kinds of hazardous virtual situations for a car to train on before releasing it out on the roads. Similarly, simulated data is tremendously helpful for training AI systems for rovers on Mars, drug discovery, materials design, weather forecasting, aviation, military training, and so on.

AI enhances existing mathematical models and simulations.
AI has great potential to assist in areas that have traditionally been difficult and limiting for mathematical models and simulations, such as learning appropriate

values for the parameters involved in the models, appropriate probability distributions, mesh shapes and sizes when discretizing equations (fine meshes capture fine details and delicate behaviors at various spatial and time scales), and scaling computational methods to longer times or to larger domains with complicated shapes. Fields like navigation, aviation, finance, materials science, fluid dynamics, operations research, molecular and nuclear sciences, atmospheric and ocean sciences, astrophysics, physical and cyber security, and many others rely heavily on mathematical modeling and simulations. Integrating AI capabilities into these domains is starting to take place with very positive outcomes. We will come across examples of AI enhancing simulations in later chapters of this book.

AI itself is a mathematical model and simulation.

One of the big aspirations of AI is to *computationally* replicate human intelligence. Successful machine learning systems, including neural networks with all their architectures and variations, are mathematical models aimed at simulating tasks that humans associate with intelligence, such as vision, pattern recognition and generalization, communication through natural language, and logical reasoning. Understanding, emotional experience, empathy, and collaboration are also associated with intelligence and have contributed tremendously to the success and domination of humankind, so we must also find ways to replicate them if we want to achieve general AI while gaining a deeper understanding of the nature of intelligence and the workings of the human brain. Efforts in these areas are already on the way. What we want to keep in mind is that in all these areas, what machines are doing is *computing*. Machines *compute* meanings of documents for natural language processing, combine and *compute* digital image pixels for computer vision, convert audio signals to vectors of numbers and *compute* new audio for human-machine interaction, and so on. It is then easy to see how software AI is one big mathematical model and simulation. This will become more evident as we progress in this book.

Where Do We Get Our Data From?

When I first decided to enter the AI field, I wanted to apply my mathematical knowledge to help solve real-world problems that I felt passionate about. I grew up in a war-torn country, and saw many problems erupt, disrupt, then eventually dissipate or get resolved, either by direct fixes or by the human network adjusting around them and settling into completely new (unstable) equilibria. Common problems in war were sudden and massive disruptions to different supply chains, sudden destruction of large parts of the power grid, sudden paralysis of entire road networks by targeted bombing of certain bridges, sudden emergence of terrorist networks, black markets, trafficking, inflation, and poverty. The number of problems that math can help solve in these scenarios, including war tactics and strategy, is limitless. From the safety of the United States, my Ph.D. in mathematics, and my tenure at a university, I started approaching companies, government agencies, and the military, looking for real projects with real data to work on. I offered to help find solutions to their problems for free. What I did not know, and learned the hard way, was that getting real data was the biggest hurdle. There are many regulations, privacy issues, institutional review boards, and other obstacles standing in the way. Even after jumping through all these hoops, the companies, institutions, and organizations tend to hold on to their data, even when they know they are not making the best use of it, and one almost has to beg in order to get real data. It turned out that the experience I had was not unique. The same had happened to many others in the field.

This story is not meant to discourage you from getting the real data that you need to train your AI systems. The point is to not get surprised and disheartened if you encounter hesitation and resistance from the owners of the data that you need. Keep asking, and someone will be willing to take that one leap of faith.

Sometimes the data you need is available publicly on the web. For the simple models in this chapter, I am using data sets from the Kaggle website (*https://www.kaggle.com*). There are other great public data repositories, which I will not list here, but a simple Google search with keywords like "best data repositories" will return excellent results. Some repositories are geared toward computer vision, others toward natural language processing, audio generation and transcription, scientific research, and so on.

Crawling the web to acquire data is common, but you have to abide by the rules of the websites you are crawling. You also have to learn how to crawl (some people say that the difference between data scientists and statisticians is that data scientists know how to hack!). Some websites require you to obtain written permission before you crawl. For example, if you are interested in social media user behavior, or in collaboration networks, you can crawl social media and professional networks—Facebook, Instagram, YouTube, Flickr, LinkedIn, etc.—for statistics on user accounts, such as the number of friends or connections, likes, comments, and activity on these sites. You will end up with very large data sets with hundreds of thousands of records, which you can then do your computations on.

To gain an intuitive understanding of how data gets integrated into AI, and the type of data that goes into various systems, while at the same time avoiding feeling overwhelmed by all the information and the data that is out there, it is beneficial to develop a habit of exploring the data sets that successful AI systems were trained on, if they are available. You do not have to download them and work on them. Browsing the data set, its metadata, what features and labels (if any) it comes with, etc., is enough to get you comfortable with data. For example, DeepMind's WaveNet (*https://oreil.ly/TI5N8*) (2016), which we will learn about in Chapter 7, is a neural network that generates raw machine audio, with realistic-sounding human voices or enjoyable pieces of music. It accomplishes tasks like text-to-audio conversion with natural-sounding human voice connotations, even with a specific person's voice if the network gets conditioned on this person's voice. We will understand the mathematical meaning of conditioning when we study WaveNet in Chapter 7. For now, think of it as a restriction imposed artificially on a problem so it restricts its results to a certain set of outcomes. So what data was WaveNet trained on? For multispeaker audio generation that is not conditioned on text, WaveNet was trained on a data set of audio files consisting of 44 hours of audio from 109 different speakers: the English Multispeaker Corpus from CSTR Voice Cloning Toolkit (*https://oreil.ly/lvLPX*) (2012). For converting text to speech, WaveNet was trained on the North American English data set (*https://oreil.ly/rY4qS*), which contains 24 hours of speech data, and on the Mandarin Chinese data set (*https://oreil.ly/FGcrO*), which has 34.8 hours of speech data. For generating music, WaveNet was trained on the YouTube Piano data set (*https://oreil.ly/Jwwxm*), which has 60 hours of solo piano music obtained from YouTube videos, and the MagnaTagATune data set (*https://oreil.ly/7fnPa*) (2009), which consists of about 200 hours of music audio, where each 29-second clip is labeled with 188 tags describing the genre, instrumentation, tempo, volume, mood, and various other labels for the music. Labeled data is extremely valuable for AI systems, because it provides a ground truth to measure the output of your hypothesis function against. We will learn this in the next few sections.

How about the famous image classification (for computer vision) AlexNet (2012)? What data was its convolutional neural network trained and tested on? AlexNet was trained on ImageNet (*https://oreil.ly/Yq1pJ*), a data set containing millions of images (scraped from the internet) and labeled (crowdsourced human labelers) with thousands of classes.

Note that all of these examples were all examples of unstructured data.

If the data that a certain system was trained on is not publicly available, it is good to look up the published paper on the system or its documentation and read about how the required data was obtained. That alone will teach you a lot.

Before moving on to doing mathematics, keep in mind the following takeaways:

- AI systems need digital data.
- Sometimes, the data you need is not easy to acquire.
- There is a movement to digitalize our whole world.

The Vocabulary of Data Distributions, Probability, and Statistics

When you enter a new field, the first thing you want to learn is the vocabulary of that field. It is similar to learning a new language. You can learn it in a classroom, and suffer, or you can travel to a country that speaks the language, and listen to frequently used terms. You don't have to know what "bonjour" means in French. But while you are in France, you notice that people say it to each other all the time, so you start saying it as well. Sometimes you will not use it in the right context, like when you have to say "bonsoir" instead of "bonjour." But slowly, as you find yourself staying longer in France, you will be using the right vocabulary in the right context.

One more advantage of learning the vocabulary as fast as you can, without necessarily mastering any of the details, is that different fields refer to the same concepts with different terms, since there is a massive vocabulary collision out there. This ends up being a big source of confusion, therefore the embodiment of *language barriers*. When you learn the common vocabulary of the field, you will realize that you might already know the concepts, except that now you have new names for them.

The vocabulary terms from probability and statistics that you want to know for the purposes of AI applications are not too many. I will define each term once we get to use it, but note that the goal of probability theory is to make deterministic statements about random or stochastic quantities and events, since humans hate uncertainty and like their world to be controllable and predictable. Watch for the following language from the fields of probability and statistics whenever you are reading about AI, machine learning, or data science. Again, you do not have to know any of the definitions yet; you just need to hear the terms discussed in the following sections and be familiar with the way they progress after each other.

Random Variables

It all starts with *random variables*. Math people talk about functions nonstop. Functions have certain or deterministic outcomes. When you evaluate a function, you know exactly what value it will return. Evaluate the function x^2 at 3, and you are certain you will get $3^2 = 9$. Random variables, on the other hand, do not have deterministic outcomes. Their outcomes are uncertain, unpredictable, or stochastic. When you call a random variable, you do not know, before you actually see the outcome, what value it will return. Since you cannot aim for certainty anymore, what you can instead aim for is quantifying how likely it is to get an outcome. For example, when you roll a die, you can confidently say that your chance to get the outcome 4 is 1/6, assuming that the die you rolled is not tampered with. You never know ahead of time what outcome you will get before you actually roll the die. If you did, then casinos would run out of business, and the finance sector would eliminate its entire predictive analytics and risk management departments. Just like deterministic functions, a random variable can return outcomes from a discrete set (discrete random variable) or from the continuum (continuous random variable). The key distinction between a random variable and a function is in the randomness versus the certainty of the outcomes.

Probability Distributions

After random variables we define *probability density functions* for continuous random variables and *probability mass functions* for discrete random variables. We call both *distributions* in order to add to our confusion. Usually, whether a distribution represents a discrete or a continuous random variable is understood from the context. Using this terminology, we sometimes say that one random variable, whether continuous or discrete, is *sampled* from a probability distribution, and multiple random variables are sampled from a *joint probability distribution*. In practice, it is rare that we know the full joint probability distribution of all the random variables involved in our data. When we do, or if we are able to *learn it from the data*, it is a powerful thing.

Marginal Probabilities

Marginal probability distributions sit literally on the margins of a joint probability distribution (if we represent the joint probability distribution with a table containing the probabilities of all the combined states of the involved variables; see, for example, the first table on this Wikipedia page (*https://oreil.ly/11WiO*)). In this setting, you are lucky enough to have access to the full joint probability distribution of multiple random variables, and you are interested in finding out the probability distribution of only one or few of them. You can find these *marginal probability distributions* easily using the *sum rule* for probabilities, for example:

$$p(x) = \Sigma_{y \,\in\, \text{all states of } y}\, p(x,y).$$

The Uniform and the Normal Distributions

The *uniform distribution* and the *normal distribution* are the most popular continuous distributions, so we start with them. The normal distribution and the fundamental *central limit theorem* from probability theory are intimately related. There are many other useful distributions representing the different random variables involved in our data, but we do not need them right away, so we postpone them until we need to use them.

Conditional Probabilities and Bayes' Theorem

The moment we start dealing with multiple random variables (such as our gender, height, weight, and health index data), which is almost always the case, we introduce *conditional probabilities, Bayes' Rule or Theorem*, and the *product or chain rule* for conditional probabilities, along with the concepts of *independent and conditionally independent random variables* (knowing the value of one does not change the probability of the other).

Conditional Probabilities and Joint Distributions

Both conditional probabilities and joint distributions involve multiple random variables, so it makes sense that they have something to do with each other. Slice the graph of a joint probability distribution (when we fix the value of one of the variables) and we get a conditional probability distribution (see Figure 2-7 later on).

Bayes' Rule Versus Joint Probability Distribution

It is very important to keep the following in mind: if we happen to have access to the full joint probability distribution of all the multiple random variables that we care for in our setting, then we would not need Bayes' Rule. In other words, Bayes' Rule helps us calculate the desired conditional probabilities *when we do not have access* to the full joint probability distribution of the involved random variables.

Prior Distribution, Posterior Distribution, and Likelihood Function

From logical and mathematical standpoints, we can define conditional probabilities, then move on smoothly with our calculations and our lives. Practitioners, however, give different names to different conditional probabilities, depending on whether they are conditioning on data that has been observed or on weights (also called parameters) that they still need to estimate. The vocabulary words here are: *prior distribution* (general probability distribution for the weights of our model prior to observing any data), *posterior distribution* (probability distribution for the weights given the observed data), and the *likelihood function* (function encoding the probability of observing a data point given a particular weight distribution). All of these can be related through the Bayes' Rule, as well as through the joint distribution.

We Say Likelihood Function not Likelihood Distribution

We refer to the likelihood as a function and not as a distribution because probability distributions *must* add up to one (or integrate to one if we are dealing with continuous random variables), but the likelihood function does not necessarily add up to one (or integrate to one in the case of continuous random variables).

Mixtures of Distributions

We can mix probability distributions and produce *mixtures of distributions. Gaussian mixtures* are pretty famous. The earlier height data that contains measurements for both males and females is a good example of a Gaussian mixture.

Sums and Products of Random Variables

We can add or multiply random variables sampled from simple distributions to produce new random variables with more complex distributions, representing more complex random events. The natural question that's usually investigated here is: *what is the distribution of the sum random variable or the product random variable?*

Using Graphs to Represent Joint Probability Distributions

Finally, we use directed and undirected graph representations (diagrams) to efficiently decompose joint probability distributions. This makes our computational life much cheaper and tractable.

Expectation, Mean, Variance, and Uncertainty

Four quantities are central to probability, statistics, and data science: *expectation* and *mean*, quantifying an average value, and the *variance* and *standard deviation*, quantifying the spread around that average value, hence encoding uncertainty. Our goal is to have control over the variance in order to reduce the uncertainty. The larger the variance, the more error you can commit when using your average value to make predictions. Therefore, when you explore the field, you often notice that mathematical statements, inequalities, and theorems mostly involve some control over the expectation and variance of any quantities that involve randomness.

When we have one random variable with a corresponding probability distribution, we calculate the *expectation* (expected average outcome), *variance* (expected squared distance from the expected average), and *standard deviation* (expected distance from the average). For data that has been already sampled or observed, for example, our height and weight data above, we calculate the *sample mean* (average value), variance (average squared distance from the mean), and standard deviation (average distance from the mean, so this measures the spread around the mean). So if the data we care for has not been sampled or observed yet, we speculate on it using the language of *expectations*, but if we already have an observed or measured sample, we calculate its *statistics*. Naturally we are interested in how far off our speculations are from our computed statistics for the observed data, and what happens in the limiting (but idealistic) case where we can in fact measure data for the entire population. The *law of large numbers* answers that for us and tells us that in this limiting case (when the sample size goes to infinity), our expectation matches the sample mean.

Covariance and Correlation

When we have two or more random variables, we calculate the *covariance, correlation*, and *covariance matrix*. This is when the field of linear algebra with its language of vectors, matrices, and matrix decompositions (such as eigenvalues and singular value decompositions) gets married to the field of probability and statistics. The variance of each random variable sits on the diagonal of the covariance matrix, and the covariances of each possible pair sit off the diagonal. The covariance matrix is symmetric. When you diagonalize it, using standard linear algebra techniques, you *uncorrelate* the involved random variables.

Meanwhile, we pause and make sure we know the difference between independence and zero covariance. Covariance and correlation are all about capturing a linear relationship between two random variables. Correlation works on *normalized* random variables, so that we can still detect linear relationships even if random variables or data measurements have vastly different scales. When you normalize a quantity, its scale doesn't matter anymore. It wouldn't matter whether it is measured on a scale of millions or on a 0.001 scale. Covariance works on unnormalized random variables. Life is not all linear. Independence is stronger than zero covariance.

Markov Process

Markov processes are very important for AI's reinforcement learning paradigm. They are characterized by all possible states of a system, a set of all possible actions that can be performed by an agent (move left, move right, etc.), a matrix containing the transition probabilities between all states, the probability distribution for what states an agent will transition to after taking a certain action, and a reward function, which we want to maximize. Two popular examples from AI include board games and a smart thermostat such as Nest. We will go over these in Chapter 11.

Normalizing, Scaling, and/or Standardizing a Random Variable or Data Set

This is one of the many cases where there is a vocabulary collision. Normalizing, scaling, and standardizing are used synonymously in various contexts. The goal is always the same. Subtract a number (shift) from the data or from all possible outcomes of a random variable, then divide by a constant number (scale). If you subtract the mean of your data sample (or the expectation of your random variable) and divide by their standard deviation, then you get new *standardized* or *normalized* data values (or a new standardized or normalized random variable) that have a mean equal to zero (or expectation zero) and standard deviation equal to one. If instead you subtract the minimum and divide by the range (max value minus min value), then you get new data values or a new random variable with outcomes all between zero and one. Sometimes we talk about normalizing vectors of numbers. In this case, what we mean is we divide every number in our vector by the length of the vector itself, so that we obtain a new vector of length one. So whether we say we are normalizing, scaling, or standardizing a collection of numbers, the goal is to try to control the values of these numbers, center them around zero, and/or restrict their spread to be less than or equal to one while at the same time preserving their inherent variability.

Common Examples

Mathematicians like to express probability concepts in terms of flipping coins, rolling dice, drawing balls from urns, drawing cards from decks, trains arriving at stations, customers calling a hotline, customers clicking on an ad or a website link, diseases and their symptoms, criminal trials and evidence, and the time until something happens, such as a machine failing. Do not be surprised that these examples are everywhere, as they generalize nicely to many other real-life situations.

In addition to this map of probability theory, we will borrow very few terms and functions from statistical mechanics (for example, the *partition function*) and information theory (for example, signal versus noise, entropy, and the *cross-entropy function*). We will explain these when we encounter them in later chapters.

Continuous Distributions Versus Discrete Distributions (Density Versus Mass)

When we deal with continuous distributions, it is important to use terms like observing or sampling a data point *near* or *around* a certain value instead of observing or sampling an *exact* value. In fact, the probability of observing an exact value in this case is zero.

When our numbers are in the continuum, there is no discrete separation between one value and the next value. Real numbers have an infinite precision. For example, if I measure the height of a male and I get 6 feet, I wouldn't know whether my measurement is exactly 6 feet or 6.00000000785 feet or 5.9999111134255 feet. It's better to set my observation in an interval around 6 feet, for example 5.95 < *height* < 6.05, then quantify the probability of observing a height between 5.95 and 6.05 feet.

We do not have such a worry for discrete random variables, as we can easily separate the possible values from each other. For example, when we roll a die, our possible values are 1, 2, 3, 4, 5, or 6. So we can confidently assert that the probability of rolling an exact 5 is 1/6. Moreover, a discrete random variable can have nonnumerical outcomes; for example, when we flip a coin, our possible values are heads or tails. A continuous random variable can only have numerical outcomes.

Because of that reasoning, when we have a continuous random variable, we define its probability *density* function, not its probability *mass* function, as in the case of discrete random variables. A density specifies how much of a substance is present within a certain length or area or volume of space (depending on the dimension we're in). To find the mass of a substance in a specified region, we multiply the density by the length, area, or volume of the considered region. If we are given the density per an infinitesimally small region, then we must integrate over the whole region to find the mass within that region, because an integral is akin to a sum over infinitely many infinitesimally small regions.

We will elaborate on these ideas and mathematically formalize them in Chapter 11. For now, we stress the following:

- If we only have one continuous random variable, such as the height of males in a certain population, then we use a one-dimensional probability density function to represent its probability distribution: $f(x_1)$. To find the probability of the height being between 5.95 < *height* < 6.05, we integrate the probability density function $f(x_1)$ over the interval (5.95, 6.05), and we write:

$$P(5.95 < height < 6.05) = \int_{5.95}^{6.05} f(x_1)dx_1$$

- If we have two continuous random variables, such as the height and weight of males in a certain population, or the true height and the measured height of a person (which usually includes random noise), then we use a two-dimensional probability density function to represent their *joint probability distribution*: $f(x_1,x_2)$. So in order to find the joint probability that the height is between 5.95 < *height* < 6.05 *and* the weight is between 160 < *weight* < 175, we *double integrate* the joint probability density function $f(x_1,x_2)$, assuming that we know the formula for $f(x_1,x_2)$, over the intervals (5.95, 6.05) and (160,175), and we write:

$$P(5.95 < height < 6.05, 160 < weight < 175) = \int_{160}^{175} \int_{5.95}^{6.05} f(x_1,x_2)dx_1 dx_2$$

- If we have more than two continuous random variables, then we use a higher-dimensional probability density function to represent their joint distribution. For example, if we have the height, weight, and blood pressure of males in a certain population, then we use a three-dimensional joint probability distribution function: $f(x_1,x_2,x_3)$. Similar to the reasoning explained previously, to find the joint probability that the first random variable is between $a < x_1 < b$, the second random variable is between $c < x_2 < d$, and the third random variable is between $e < x_3 < f$, we *triple integrate* the joint probability density function over the intervals (a,b), (c,d), and (e,f), and we write:

$$P(a < x_1 < b, c < x_2 < d, e < x_3 < f) = \int_e^f \int_c^d \int_a^b f(x_1,x_2,x_3)dx_1 dx_2 dx_3$$

Not all our worries (that things will add up mathematically) are eliminated even after defining the probability density function for a continuous random variable. Again, the culprit is the infinite precision of real numbers. If we allow *all sets* to have probability, we encounter paradoxes in the sense that we can construct disjoint sets (such as fractal-shaped sets or sets formulated by transforming the set of rational numbers) whose probabilities add up to more than one! One must admit that these sets are pathological and must be carefully constructed by a person who has plenty of time on their hands; however, they exist and they produce paradoxes. *Measure theory* in mathematics steps in and provides a mathematical framework where we can work with probability density functions without encountering paradoxes. It defines sets of measure zero (these occupy no volume in the space that we are working in), then gives us plenty of theorems that allow us to do our computations *almost everywhere*, that is, except on sets of measure zero. This turns out to be more than enough for our applications.

The Power of the Joint Probability Density Function

Having access to the joint probability distribution of many random variables is a powerful but rare thing. The reason is that the joint probability distribution encodes within it the probability distribution of each separate random variable (marginal distributions), as well as all the possible co-occurrences (and the conditional probabilities) that we ever encounter between these random variables. This is akin to seeing a whole town from above, rather than being inside the town and observing only one intersection between two or more alleys.

If the random variables are independent, then the joint distribution is simply the product of each of their individual distributions. However, when the random variables are not independent, such as the height and the weight of a person, or the observed height of a person (which includes measurement noise) and the true height of a person (which doesn't include noise), accessing the joint distribution is much more difficult and expensive storage-wise. The joint distribution in the case of dependent random variables is not separable, so we cannot only store each of its parts alone. We need to store every value for every co-occurrence between the two or more variables. This exponential increase in storage requirements (and computations or search spaces) as you increase the number of dependent random variables is one embodiment of the infamous *curse of dimensionality*.

When we *slice* a joint probability density function, say $f(x_1,x_2)$, meaning when we fix one of the random variables (or more in higher dimensions) to be an exact value, we retrieve a distribution proportional to the *posterior probability distribution* (probability distribution of the model parameters given the observations), which we are usually interested in. For example, slice through $f(x_1,x_2)$ at $x_1 = a$, and we get $f(a, x_2)$, which happens to be proportional to the probability distribution $f(x_2 | x_1 = a)$ (see Figure 2-7).

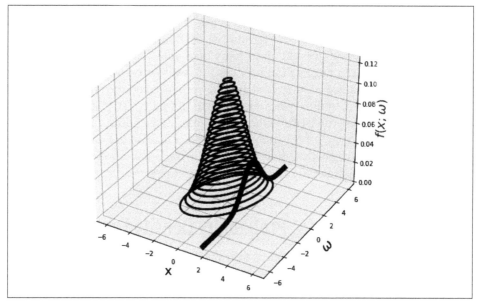

Figure 2-7. Slicing through the joint probability distribution

Again, this is in the luxurious case that we know the joint probability distribution, otherwise, we use Bayes' Rule to obtain the same posterior probability distribution (using the prior distribution and the likelihood function).

In some AI applications, the AI system learns the joint probability distribution by separating it into a product of conditional probabilities using the product rule for probabilities. Once it learns the joint distribution, it then samples from it to generate new and interesting data. DeepMind's WaveNet does that in its process of generating raw audio.

The next sections introduce the most useful probability distributions for AI applications. Two ubiquitous continuous distributions are the *uniform distribution* and the *normal distribution* (also known as the *Gaussian distribution*), so we start there. Refer to the Jupyter Notebook (*https://github.com/halanelson/Essential-Math-For-AI*) for reproducing figures and more details.

Distribution of Data: The Uniform Distribution

To intuitively understand the uniform distribution, let's give an example of a *nonuniform* distribution, which we have already seen earlier in this chapter. In our real height-weight data sets, we cannot use the *uniform distribution* to model the height data. We also cannot use it to model the weight data. The reason is that human heights and weights are not evenly distributed. In the general population, it is not *equally likely* to encounter a person with height around 7 feet as it is to encounter a person with height around 5 feet 6 inches.

The *uniform distribution* only models data that is evenly distributed. If we have an interval (x_{min}, x_{max}) containing all the values in the continuum between x_{min} and x_{max} of our data, and our data is uniformly distributed over our interval, then the probability of observing a data point near any particular value in our interval is the same for all values in this interval. That is, if our interval is $(0,1)$, it is equally likely to pick a point near 0.2 as it is to pick a point near 0.75.

The probability density function for the uniform distribution is therefore constant. For one random variable x over an interval (x_{min}, x_{max}), the formula for the probability density function for the continuous uniform distribution is given by:

$$f(x; x_{min}, x_{max}) = \frac{1}{x_{max} - x_{min}} \text{ for } x_{min} < x < x_{max}$$

and zero otherwise.

Let's plot the probability density function for the uniform distribution over an interval (x_{min}, x_{max}). The graph in Figure 2-8 is a straight segment because uniformly distributed data, whether real or simulated, appears evenly distributed across the entire interval under consideration. No data values within the interval are more favored to appear than others.

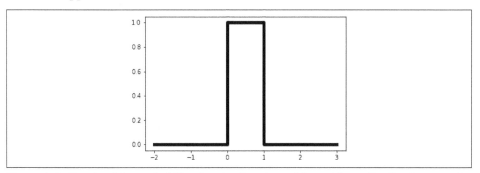

Figure 2-8. Graph of the probability density function of the uniform distribution over the interval [0,1]

The uniform distribution is extremely useful in computer simulations for generating random numbers *from any other probability distribution*. If you peek into the random number generators that Python uses, you would see the uniform distribution used somewhere in the underlying algorithms.

Distribution of Data: The Bell-Shaped Normal (Gaussian) Distribution

A continuous probability distribution better suited to model human height data (when restricted to one gender) is the *bell-shaped normal distribution*, also called the *Gaussian distribution*. Samples from the normal distribution tend to congregate around an average value where the distribution peaks, called the *mean* μ, then dwindle symmetrically as we get farther away from the mean. How far from the mean the distribution spreads out as it dwindles down is controlled by the second parameter of the normal distribution, called the *standard deviation* σ. About 68% of the data falls within one standard deviation of the mean, 95% of the data falls within two standard deviations of the mean, and about 99.7% of the data falls within three standard deviations of the mean (Figure 2-9).

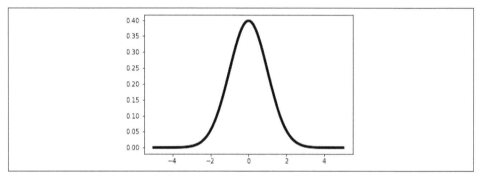

Figure 2-9. Graph of the probability density function of the bell-shaped normal distribution with parameters $\mu = 0$ and $\sigma = 1$

Values near the mean are more likely to be picked (or to occur, or to be observed) when we are sampling data from the normal distribution, and values that are very small ($\rightarrow -\infty$) or very large ($\rightarrow \infty$) are less likely to be picked. This peaking near the mean value and decaying on the outer skirts of the distribution give this distribution its famous bell shape. Note that there are other bell-shaped continuous distributions out there, but the normal distribution is the most prevalent. It has a neat mathematical justification for this well-deserved fame, based on an important theorem in probability theory called the *central limit theorem* (CLT).

The central limit theorem states that the average of many independent random variables that all have the same distribution (not necessarily the normal distribution) is normally distributed. This explains why the normal distribution appears everywhere in society and nature. It models baby birth weights, student grade distributions, countries' income distributions, distribution of blood pressure measurements, etc. There are special statistical tests that help us determine whether a real data set can be modeled using the normal distribution. We will expand on these ideas later in Chapter 11.

If you happen to find yourself in a situation where you are uncertain and have no prior knowledge about which distribution to use for your application, the normal distribution is usually a reasonable choice. In fact, among all choices of distributions with the same variance, the normal distribution is the choice with maximum *uncertainty*, so it does in fact encode the least amount of prior knowledge into your model.

The formula for the probability density function of the normal distribution for one random variable x (univariate) with mean μ and standard deviation σ is:

$$g(x;\mu,\sigma) = \frac{1}{\sqrt{2\pi\sigma^2}}e^{-\frac{(x-\mu)^2}{2\sigma^2}}$$

and its graph for $\mu = 0$ and $\sigma = 1$ is plotted in Figure 2-9.

The formula for the probability density function for the normal distribution of two random variables x and y (bivariate) is:

$$g(x,y;\mu_1,\sigma_1,\mu_2,\sigma_2,\rho) = \frac{1}{\sqrt{(2\pi)^2 \det\begin{pmatrix} \sigma_1^2 & \rho\sigma_1\sigma_2 \\ \rho\sigma_1\sigma_2 & \sigma_2^2 \end{pmatrix}}}e^{-\left\{\frac{1}{2}(x-\mu_1\ y-\mu_2)\begin{pmatrix} \sigma_1^2 & \rho\sigma_1\sigma_2 \\ \rho\sigma_1\sigma_2 & \sigma_2^2 \end{pmatrix}^{-1}\begin{pmatrix} x-\mu_1 \\ y-\mu_2 \end{pmatrix}\right\}}$$

and its graph is plotted in Figure 2-10.

We can write the above bivariate formula in more compact notation using the language of linear algebra:

$$g(x,y;\mu,\Sigma) = \frac{1}{\sqrt{(2\pi)^2 \det(\Sigma)}}e^{-\left\{\frac{1}{2}(u-\mu)^T\Sigma^{-1}(u-\mu)\right\}}$$

In Figure 2-11, we sample 6,000 points from the bivariate normal distribution. Points near the center are more likely to be picked, and points away from the center are less likely to be picked. The lines roughly trace the contour lines of the normal distribution, had we only observed the sample points without knowing which distribution they were sampled from.

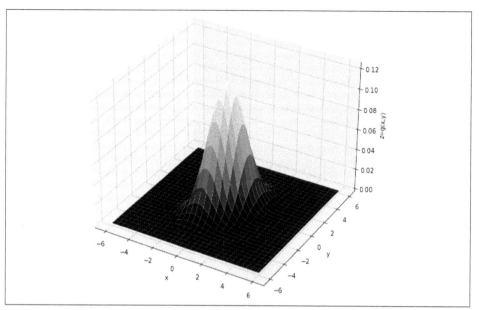

Figure 2-10. Graph of the probability density function of the bell-shaped bivariate normal distribution

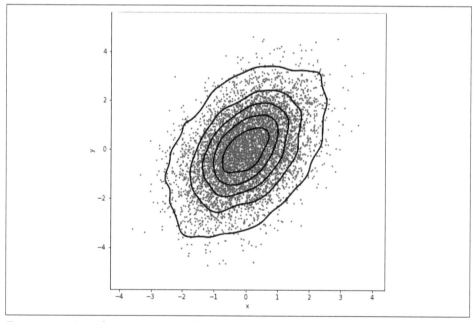

Figure 2-11. Sampling 6,000 points from the bivariate normal distribution

Let's pause and compare the formula of the probability density function for the bivariate normal distribution to the formula of the probability density function for the univariate normal distribution:

- When there is only one random variable, we only have one mean μ and one standard deviation σ.

- When there are two random variables, we have two means $\begin{pmatrix} \mu_1 \\ \mu_2 \end{pmatrix}$ and two standard deviations $\begin{pmatrix} \sigma_1 \\ \sigma_2 \end{pmatrix}$. The product σ^2 will be replaced by the covariance matrix $\Sigma = \begin{pmatrix} \sigma_1^2 & \rho\sigma_1\sigma_2 \\ \rho\sigma_1\sigma_2 & \sigma_2^2 \end{pmatrix}$ and its determinant. ρ is the correlation between the two random variables, which is the covariance of the two normalized versions of the random variables.

The same exact formula for the probability density function of the bivariate normal distribution generalizes to any dimension, where we have many random variables instead of only two random variables. For example, if we have 100 random variables, representing 100 features in a data set, the mean vector in the formula will have 100 entries in it, and the covariance matrix will have the size 100×100, with the variance of each random variable on the diagonal and the covariance of each of the 4,950 pairs of random variables off the diagonal.

Distribution of Data: Other Important and Commonly Used Distributions

Almost everything you did not understand in this chapter will be revisited many times throughout the book, and Chapter 11 focuses exclusively on probability. The concepts will get reinforced as they appear again and again in various interesting contexts. Our goal for this chapter is to get exposed to the vocabulary of probability and statistics, and have a guiding map for the important ideas that frequently appear in AI applications. We also want to acquire a good probabilistic intuition for the following chapters without having to take a deep dive and delay our progress for no necessary reason.

There are many probability distributions out there. Each models a different type of real-world scenario. The uniform and normal distributions are very common, but we have other important distributions that frequently appear in the AI field. Recall that our goal is to model the world around us in order to make good designs, predictions, and/or decisions. Probability distributions help us make predictions when our models involve randomness or when we are uncertain about our outcomes.

When we study distributions, one frustrating part is that most of them have weird names that provide zero intuition about what kind of phenomena a given distribution would be useful for. This makes us either expend extra mental energy to memorize these names, or keep a distribution cheat sheet in our pocket. I prefer to keep a cheat sheet. Another frustrating part is that most textbook examples involve flipping a coin, rolling a die, or drawing colored balls from urns. This leaves us with no real-life examples or motivation to understand the subject, as I never met anyone walking around flipping coins and counting heads or tails, except for Two-Face (aka Harvey Dent) in *The Dark Knight* (a really good 2008 movie, where the Joker [played by Heath Ledger] says some of my favorite and profound statements about randomness and chance, like this one: "The world is cruel. And the only morality in a cruel world…is chance. Unbiased. Unprejudiced. Fair."). I will try to amend this as much as I can in this book, pointing to as many real-world examples as my page limit allows.

Some of the following distributions are mathematically related to each other, or follow naturally from others. We will explore these relationships in Chapter 10. For now, let's name a popular distribution, state whether it is discrete (predicts a count of something that we care for) or continuous (predicts a quantity that exists in the continuum, such as the time needed to elapse before something happens; careful, this is not the number of hours, since the number of hours is discrete, but it is the length of the time period), state the parameters that control it, and state its defining properties that are useful for our AI applications:

Binomial distribution

   This is discrete. It represents the probability of obtaining a certain number of successes when repeating one experiment, independently, multiple times. Its controlling parameters are n, the number of experiments we perform, and p, the *predefined* probability of success. Real-world examples include predicting the number of patients that will develop side effects for a vaccine or a new medication in a clinical trial, the number of ad clicks that will result in a purchase, and the number of customers that will default on their monthly credit card payments. When we model examples from the real world using a probability distribution that requires independent trials, it means that we *are assuming* independence even if the real-world trials are not really independent. It is good etiquette to point our models' assumptions.

Poisson distribution

   This is discrete. It predicts the number of rare events that will occur in a given period of time. These events are independent or weakly dependent, meaning that the occurrence of the event once does not affect the probability of its next occurrence in the same time period. They also occur at a *known and constant* average rate λ. Thus, we know the average rate, and we want to predict *how many* of these events will happen during a certain time period. The Poisson distribution's controlling parameter is the predefined rare event rate λ. Real-world examples include predicting the number of babies born in a given hour, the number of people in a population who age past 98, the number of alpha particles discharged from a radioactive system during a certain time period, the number of duplicate bills sent out by the IRS, the number of a not-too-popular product sold on a particular day, the number of typos that one page of this book contains, the number of defective items produced by a certain machine on a certain day, the number of people entering a store at a certain hour, the number of car crashes an insurance company needs to cover within a certain time period, and the number of earthquakes happening within a particular time period.

Geometric distribution

This is discrete. It predicts the number of trials needed before we obtain a success when performing independent trials, each with a *known* probability p for success. The controlling parameter here is obviously the probability p for success. Real-world examples include estimating the number of weeks that a company can function without experiencing a network failure, the number of hours a machine can function before producing a defective item, or the number of people we need to interview before meeting someone who opposes a certain political bill that we want to pass. Again, for these real-world examples, we might be assuming independence if modeling using the geometric distribution, while in reality the trials might not be independent.

Exponential distribution

This is continuous. If we happen to know that a certain event occurs at a constant rate λ, then exponential distribution predicts the waiting time until this event occurs. It is *memoryless*, in the sense that the remaining lifetime of an item that belongs to this exponential distribution is also exponential. The controlling parameter is the constant rate λ. Real-world examples include the amount of time we have to wait until an earthquake occurs, the time until someone defaults on a loan, the time until a machine part fails, or the time before a terrorist attack strikes. This is very useful for the reliability field, where the reliability of a certain machine part is calculated, hence statements such as a 10-year guarantee, etc.

Weibull distribution

This is continuous. It is widely used in engineering in the field of predicting product lifetimes (10-year warranty statements are appropriate here as well). Here, a product consists of many parts, and if any of its parts fail, then the product stops working. For example, a car will not work if the battery fails, or if a fuse in the gearbox burns out. A Weibull distribution provides a good approximation for the lifetime of a car before it stops working, after accounting for its many parts and their *weakest link* (assuming we are not maintaining the car and resetting the clock). It is controlled by three parameters: shape, scale, and location. The exponential distribution is a special case of this distribution, because the exponential distribution has a constant rate of event occurrence, but the Weibull distribution can model rates of occurrence that increase or decrease with time.

Log-normal distribution
This is continuous. If we take the logarithms of each value provided in this distribution, we get normally distributed data. Meaning that in the beginning, your data might not appear normally distributed, but if you try transforming it using the log function, you will see normally distributed data. This is a good distribution to use when encountering skewed data with low mean value, large variance, and *assuming only positive values.* Just like the normal distribution appears when you average many independent samples of a random variable (using the central limit theorem), the log-normal distribution appears when you take the *product* of many positive sample values. Mathematically, this is due to an awesome property of log functions: the log of a product is a sum of the logs. This distribution is controlled by three parameters: shape, scale, and location. Real-world examples include the volume of gas in a petroleum reserve, and the ratio of the price of a security at the end of one day to its price at the end of the day before.

Chi-squared distribution
This is continuous. It is a distribution for the sum of squares of normally distributed independent random variables. You might wonder why would we care about squaring normally distributed random variables, then adding them up. The answer is that this is how we usually compute the variance of a random variable or of a data sample, and one of our main goals is controlling the variance in order to lower our uncertainties. There are two types of significance tests associated with this distribution: the goodness of fit test, which measures how far off our expectation is from our observation, and the independence and homogeneity of data features test.

Pareto distribution
This is continuous. It is useful for many real-world applications, such as the time to complete a job assigned to a supercomputer (think machine learning computations), the household income level in a certain population, the number of friends in a social network, and the file size of internet traffic. This distribution is controlled by only one parameter α, and it is *heavy tailed* (its tail is heavier than the exponential distribution).

Let's throw in few other distributions before moving on, without fussing about any of the details. These are all more or less related to the aforementioned distributions:

Student's t-distribution
Continuous, similar to the normal distribution, but used when the sample size is small and the population variance is unknown.

Beta distribution
    Continuous, produces random values in a given interval.

Cauchy distribution
    Continuous, pathological because neither its mean nor its variance is defined, can be obtained using the tangents of randomly chosen angles.

Gamma distribution
    Continuous, has to do with the waiting time until n independent events occur, instead of only one event, as in the exponential distribution.

Negative binomial distribution
    Discrete, has to do with the number of independent trials needed to obtain a certain number of successes.

Hypergeometric distribution
    Discrete, similar to the binomial but the trials are not independent.

Negative hypergeometric distribution
    Discrete, captures the number of dependent trials needed before we obtain a certain number of successes.

The Various Uses of the Word "Distribution"

You might have already noticed that the word *distribution* refers to many different (but related) concepts, depending on the context. This inconsistent use of the same word could be a source of confusion and an immediate turnoff for some people who are trying to enter the field.

Let's list the different concepts that the word distribution refers to, so that we easily recognize its intended meaning in a given context:

- If you have real data, such as the height-weight data in this chapter, and plot the histogram of one feature of your data set, such as the height, then you get the *empirical distribution* of the height data. You usually do not know the underlying probability density function of the height of the entire population, also called distribution, since the real data you have is only a sample of that population. So you try to estimate it, or model it, using the probability distributions given by probability theory. For the height and weight features, when separated by gender, a Gaussian distribution is appropriate.

- If you have a discrete random variable, the word distribution could refer to either its probability mass function or its cumulative distribution function (which specifies the probability that the random variable is less than or equal to a certain value, $f(x) = prob(X \leq x)$).

- If you have a continuous random variable, the word distribution could refer to either its probability density function or its cumulative distribution function, whose integral gives the probability that the random variable is less than or equal to a certain value.

- If you have multiple random variables (discrete, continuous, or a mix of both), then the word distribution refers to their joint probability distribution.

A common goal is to establish an appropriate correspondence between an idealized mathematical function, such as a random variable with an appropriate distribution, and real observed data or phenomena, with an observed empirical distribution. When working with real data, each feature of the data set can be modeled using a random variable. So in a way, a mathematical random variable with its corresponding distribution is an idealized version of our measured or observed feature.

Finally, distributions appear everywhere in AI applications. We will encounter them plenty of times in the next chapters, for example, the distribution of the weights at each layer of a neural network, and the distribution of the noise and errors committed by various machine learning models.

A/B Testing

Before leaving this chapter, we make a tiny detour into the world of A/B testing, also called *split testing*, or *randomized single-blind* or *double-blind trials*. We make this detour because this is a topic important for data scientists: countless companies rely on data from A/B tests to increase engagement, revenue, and customer satisfaction. Microsoft, Amazon, LinkedIn, Google, and others each conduct thousands of A/B tests annually.

The idea of an A/B test is simple: split the population into two groups. Roll out a version of something you want to test (a new web page design, a different font size, a new medicine, a new political ad) to one group, the test group, and keep the other group as a control group. Compare the data between the two groups.

The test is *single blind* if the subjects do not know which group they belong to (some do not even know that they are in a test at all), but the experimenters know. The test is *double blind* if neither the experimenters nor the subjects know which group they are interacting with.

Summary and Looking Ahead

In this chapter, we emphasized the fact that data is central to AI. We also clarified the differences between concepts that are usually a source of confusion: structured and unstructured data, linear and nonlinear models, real and simulated data, deterministic functions and random variables, discrete and continuous distributions, and posterior probabilities and likelihood functions. We also provided a map for the probability and statistics needed for AI without diving into any of the details, and we introduced the most popular probability distributions.

If you find yourself lost in some new probability concept, you might want to consult the map provided in this chapter and see how that concept fits within the big picture of probability theory, and most importantly, how it relates to AI. Without knowing how a particular mathematical concept relates to AI, you are left with having some tool that you know how to turn on, but you have no idea what it is used for.

We have not yet mentioned *random matrices* and *high-dimensional probability*. In these fields, probability theory, with its constant tracking of distributions, expectations, and variances of any relevant random quantities, merges with linear algebra, with its hyperfocus on eigenvalues and various matrix decompositions. These fields are very important for the extremely high-dimensional data that is involved in AI applications. We discuss them in Chapter 11 on probability.

In the next chapter, we learn how to fit our data into a function, then use this function to make predictions and/or decisions. Mathematically, we find the weights (the ω's) that characterize the strengths of various interactions between the features of our data. When we characterize the involved types of interactions (the formula of the fitting function, called the *learning* or *training* function) and the strengths of these interactions (the values of the ω's), we can make our predictions. In AI, this one concept of *characterizing the fitting function with its suitable weight values* can be used successfully for computer vision, natural language processing, predictive analytics (like house prices, time until maintenance, etc.), and many other applications.

CHAPTER 3

Fitting Functions to Data

Today it fits. Tomorrow?

   —H.

In this chapter, we introduce the core mathematical ideas lying at the heart of many AI applications, including the mathematical engines of neural networks. Our goal is to internalize the following structure of the machine learning part of an AI problem:

Identify the problem
   The problem depends on the specific use case: classify images, classify documents, predict house prices, detect fraud or anomalies, recommend the next product, predict the likelihood of a criminal reoffending, predict the internal structure of a building given external images, convert speech to text, generate audio, generate images, generate video, etc.

Acquire the appropriate data
   This is about training our models to do the right thing. We say that our models *learn* from the data. Make sure this data is clean, complete, and if necessary, depending on the specific model we are implementing, transformed (normalized, standardized, some features aggregated, etc.). This step is usually way more time-consuming than implementing and training the machine learning models.

Create a hypothesis function
   We use the terms hypothesis function, *learning function*, *prediction function*, *training function*, and *model* interchangeably. Our main assumption is that this input/output mathematical function explains the observed data, and it can be used later to make predictions on new data. We give our model features, like a person's daily habits, and it returns a prediction, like this person's likelihood to pay back a loan. In this chapter, we will give our model the length measurements of a fish, and it will return its weight.

Find the numerical values of weights

We will encounter many models (including neural networks) where our training function has unknown parameters called *weights*. The goal is to find the numerical values of these weights using the data. After we find these weight values, we can use the trained function to make predictions by plugging the features of a new data point into the formula of the trained function.

Create an error function

To find the values of the unknown weights, we create another function called the *error function, cost function, objective function,* or *loss function* (everything in the AI field has three or more names). This function has to measure some sort of distance between the ground truth and our predictions. Naturally, we want our predictions to be as close to the ground truth as possible, so we search for weight values that minimize our loss function. Mathematically, we solve a minimization problem. The field of *mathematical optimization* is essential to AI.

Decide on mathematical formulas

Throughout this process, we are the engineers, so we get to decide on the mathematical formulas for training functions, loss functions, optimization methods, and computer implementations. Different engineers decide on different processes, with different performance results, and that is OK. The judge, in the end, is the performance of the deployed model, and contrary to popular belief, mathematical models are flexible and can be tweaked and altered when needed. It is crucial to monitor performance after deployment.

Find a way to search for minimizers

Since our goal is to find the weight values that minimize the error between our predictions and ground truths, we need to find an efficient mathematical way to search for these *minimizers*: those special weight values that produce the least error. The *gradient descent* method plays a key role here. This powerful yet simple method involves calculating *one derivative* of our error function. This is one reason we spent half of our calculus classes calculating derivatives (and the gradient: this is one derivative in higher dimensions). There are other methods that require computing two derivatives. We will encounter them and comment on the benefits and the downsides of using higher-order methods.

Use the backpropagation algorithm

When data sets are enormous and our model happens to be a layered neural network, we need an efficient way to calculate this one derivative. The *backpropagation algorithm* steps in at this point. We will walk through gradient descent and backpropagation in Chapter 4.

Regularize a function

If our learning function fits the given data too well, then it will not perform well on new data. A function with too good of a fit with the data picks up on the noise in the data as well as the signal (for example, the function on the left in Figure 3-1). We do not want to pick up on noise. This is where *regularization* helps. There are multiple mathematical ways to regularize a function, which means to make it smoother and less oscillatory and erratic. In general, a function that follows the noise in the data oscillates too much. We want more regular functions. We visit regularization techniques in Chapter 4.

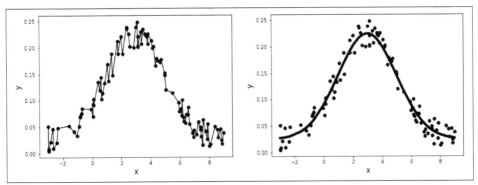

Figure 3-1. Left: the fitting function fits the data perfectly; however, it is not a good prediction function since it fits the noise in the data instead of the main signal. Right: a more regular function fitting the same data set. Using this function will give better predictions than the function in the left subplot, even though the function in the left subplot matches the data points better.

In the following sections, we explore this structure of an AI problem with real, but simple, data sets. We will see in subsequent chapters how the same concepts generalize to much more involved tasks.

Traditional and Very Useful Machine Learning Models

All the data used in this chapter is *labeled* with ground truths, and the goal of our models is to *predict* the labels of new (unseen) and unlabeled data. This is *supervised* learning.

In the next few sections, we fit training functions into our labeled data using the following popular machine learning models. While you may hear so much about the latest and greatest developments in AI, in a typical business setting you are probably better off starting with these more traditional models:

Linear regression
> Predict a numerical value.

Logistic regression
> Classify into two classes (binary classification).

Softmax regression
> Classify into multiple classes.

Support vector machines
> Classify into two classes, or regression (predict a numerical value).

Decision trees
> Classify into any number of classes, or regression (predict a numerical value).

Random forests
> Classify into any number of classes, or regression (predict a numerical value).

Ensembles of models
> Bundle up the results of many models by averaging the prediction values, voting for the most popular class, or some other bundling mechanism.

k-means clustering
> Classify into any number of classes, or regression

We try multiple models on the same data sets to compare performance. In the real world, it is rare that any model ever gets deployed without having been compared with many other models. This is the nature of the computation-heavy AI industry, and why we need parallel computing, which enables us to train multiple models at once (except for models that build and improve on the results of other models, like in the case of *stacking*; for those we cannot use parallel computing).

Before we dive into any machine learning models, it is extremely important to note that it has been reported again and again that only about 5% of a data scientist's time, and/or an AI researcher's time, is spent on training machine learning models. The majority of the time is consumed by acquiring data, cleaning data, organizing data, creating appropriate pipelines for data, etc., *before* feeding the data into machine learning models. So machine learning is only one step in the production process, and it is an easy step once the data is ready to train the model. We will discover how these machine learning models work: most of the mathematics we need resides in these models. AI researchers are always trying to enhance machine learning models and automatically fit them into production pipelines. It is therefore important for us to eventually learn about the whole pipeline, from raw data (including its storage, hardware, query protocols, etc.) to deployment to monitoring. Learning machine learning is only one piece of a bigger and more interesting story.

We must start with *regression* since the ideas of regression are so fundamental for most of the AI models and applications that will follow. Only for *linear regression* do we find our minimizing weights using an *analytical* method, giving an explicit formula for the desired weights directly in terms of the training data set and its target labels. The simplicity of the linear regression model allows for this explicit analytical solution. Most other models do not have such explicit solutions, and we have to find their minimizers using numerical methods, among which the gradient descent is extremely popular.

In regression and many other upcoming models, including the neural networks of the next few chapters, watch for the following progression in the modeling process:

1. The training function
2. The loss function
3. Optimization

Numerical Solutions Versus Analytical Solutions

It is important to be aware of the differences between numerical solutions and analytical solutions of mathematical problems. A mathematical problem can be anything, such as:

- Find the minimizer of some function.
- Find the best way to go from destination A to destination B, with a constrained budget.
- Find the best way to design and query a data warehouse.
- Find the solution of a mathematical equation (where a lefthand side with math stuff equals a righthand side with math stuff). These equations could be algebraic equations, ordinary differential equations, partial differential equations, integro-differential equations, systems of equations, or any sort of mathematical equations. Their solutions could be static or evolving in time. They could model anything from the physical, biological, socioeconomic, or natural worlds.

Here is the vocabulary:

Numerical
    Has to do with numbers

Analytical
    Has to do with analysis

As a rule of thumb, numerical solutions are much easier to obtain and much more accessible than analytical solutions, provided that we have enough computational power to simulate and compute these solutions. All we usually need to do is discretize some continuous spaces and/or functions (change the continuum into a bunch of points), albeit sometimes in very clever ways, and evaluate functions on these discrete quantities. The only problem with numerical solutions is that they are only approximate solutions. Unless they are backed by estimates on how far off they are from the true analytical solutions and how fast they converge to these true solutions, which in turn requires mathematical background and analysis, numerical solutions are not exact. They do, however, provide incredibly useful insights about the true solutions. In many cases, numerical solutions are the only ones available, and many scientific and engineering fields would not have advanced at all had they not relied on numerical solutions of complex problems. If those fields waited for analytical solutions and proofs to happen, or, in other words, for mathematical theory to catch up, they would have had very slow progress.

Analytical solutions, on the other hand, are exact, robust, and have a whole mathematical theory backing them up. They come accompanied with theorems and proofs. When analytical solutions are available, they are very powerful. They are, however, not easily accessible, sometimes impossible to obtain, and they do require deep knowledge and domain expertise in fields such as calculus, mathematical analysis, algebra, theory of differential equations, etc. Analytical methods, however, are extremely valuable for describing important properties of solutions (even when explicit solutions are not available), guiding numerical techniques, and providing ground truths to compare approximate numerical methods against (in the lucky cases when these analytical solutions are available).

Some researchers are purely analytical and theoretical, others are purely numerical and computational, and the best place to exist is somewhere near the intersection, where we have a decent understanding of the analytical and the numerical aspects of our mathematical problems.

Regression: Predict a Numerical Value

A quick search on the Kaggle website (*https://www.kaggle.com*) for data sets for regression returns many excellent data sets and related notebooks. I randomly chose a simple Fish Market (*https://oreil.ly/yaV96*) data set that we will use to explain our upcoming mathematics. Our goal is to build a model that predicts the weight of a fish given its five different length measurements, or features, labeled in the data set as Length1, Length2, Length3, Height, and Width (see Figure 3-2). For the sake of simplicity, we choose not to incorporate the categorical feature, Species, into this model, even though we could (and that would give us better predictions, since a fish's type is a good predictor of its weight). If we choose to include the Species feature, then we would have to convert its values into numerical values, using *one-hot encoding*, which means exactly what it sounds like: assign a code for each fish made up of ones and zeros based on its category (type). Our Species feature has seven categories: Perch, Bream, Roach, Pike, Smelt, Parkki, and Whitefish. So if our fish is a Pike, we would code its species as (0,0,0,1,0,0,0), and if it is a Bream we would code its species as (0,1,0,0,0,0,0). Of course this adds seven more dimensions to our feature space and seven more weights to train.

	Species	Weight	Length1	Length2	Length3	Height	Width
0	Bream	242.0	23.2	25.4	30.0	11.5200	4.0200
1	Bream	290.0	24.0	26.3	31.2	12.4800	4.3056
2	Bream	340.0	23.9	26.5	31.1	12.3778	4.6961
3	Bream	363.0	26.3	29.0	33.5	12.7300	4.4555
4	Bream	430.0	26.5	29.0	34.0	12.4440	5.1340

Figure 3-2. The first five rows of the Fish Market data set downloaded from Kaggle (https://oreil.ly/URygY). The Weight column is the target feature, and our goal is to build a model that predicts the weight of a new fish given its length measurements.

Let's save ink space and relabel our five features as x_1, x_2, x_3, x_4, and x_5, then write the fish weight as a function of these five features, $y = f(x_1,x_2,x_3,x_4,x_5)$. This way, once we settle on an acceptable formula for this function, all we have to do is input the feature values for a certain fish and our function will output the predicted weight of that fish.

This section builds a foundation for everything to come, so it is important to first see how it is organized:

Training function
- Parametric models versus nonparametric models.

Loss function
- The predicted value versus the true value.
- The absolute value distance versus the squared distance.
- Functions with singularities (pointy points).
- For linear regression, the loss function is the mean squared error.
- Vectors in this book are always column vectors.
- The training, validation, and test subsets.
- When the training data has highly correlated features.

Optimization
- Convex landscapes versus nonconvex landscapes.
- How do we locate minimizers of functions?
- Calculus in a nutshell.
- A one-dimensional optimization example.
- Derivatives of linear algebra expressions that we use all the time.
- Minimizing the mean squared error loss function.
- Caution: multiplying large matrices by each other is very expensive—multiply matrices by vectors instead.
- Caution: we never want to fit the training data too well.

Training Function

A quick exploration of the data, as in plotting the weight against the various length features, allows us to assume a linear model (even though a nonlinear one could be better in this case). That is, we assume that the weight depends linearly on the length features (see Figure 3-3).

This means that the weight of a fish, y, can be computed using a *linear combination* of its five different length measurements, plus a bias term ω_0, giving the following *training function*:

$$y = \omega_0 + \omega_1 x_1 + \omega_2 x_2 + \omega_3 x_3 + \omega_4 x_4 + \omega_5 x_5$$

After our major decision in the modeling process to use a linear training function $f(x_1, x_2, x_3, x_4, x_5)$, all we have to do is to find the appropriate values of the parameters ω_0, ω_1, ω_2, ω_3, ω_4, and ω_5. We will *learn* the best values for our ω's from the data. The process of using the data to find the appropriate ω's is called *training* the model. A *trained* model is then a model where the ω values have been decided on.

In general, training functions, whether linear or nonlinear, including those representing neural networks, have unknown parameters, ω's, that we need to learn from the given data. For linear models, each parameter gives each feature a certain weight in the prediction process. So if the value of ω_2 is larger than the value of ω_5, then the second feature plays a more important role than the fifth feature in our prediction, assuming that the second and fifth features have comparable scales. This is one of the reasons it is good to scale or normalize the data before training the model. If, on the other hand, the value ω_3 associated with the third feature dies, meaning becomes zero or negligible, then the third feature can be omitted from the data set, as it plays no role in our predictions. Therefore, learning our ω's from the data allows us to mathematically compute the contribution of each feature to our predictions (or the importance of feature combinations if some features were combined during the data preparation stage, before training). In other words, the models learn how the data features interact and how strong these interactions are. The moral is that through a trained learning function, we can quantify how features come together to produce both observed and yet-to-be observed results.

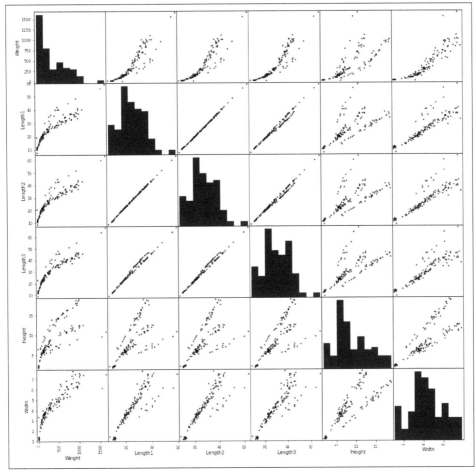

Figure 3-3. Scatterplots of the Fish Market numerical features. For more details, see the book's GitHub page (https://github.com/halanelson/Essential-Math-For-AI), or some of the public notebooks on Kaggle (https://oreil.ly/jFgxF) associated with this data set.

Parametric Models Versus Nonparametric Models

A model that has parameters (we are calling them weights) pre-built into its formula, such as the ω's in our current linear regression model:

$$y = \omega_0 + \omega_1 x_1 + \omega_2 x_2 + \omega_3 x_3 + \omega_4 x_4 + \omega_5 x_5$$

(and later the ω's of neural networks) is called a *parametric model*. This means that we fix the formula of the training function ahead of the actual training, and all the training does is solve for the parameters that are involved in the formula. Fixing the formula ahead of time is analogous to specifying the *family* that a training function belongs to, and finding the parameter values specifies the exact member of that family that best explains the data.

Nonparametric models, such as decision trees and random forests that we will discuss later in this chapter, do not specify the formula for the training function with its parameters ahead of time. So when we train a nonparametric model, we do not know how many parameters the trained model will end up having. The model *adapts* to the data and determines the required amount of parameters depending on the data. Careful here, the bells of overfitting are ringing! Recall that we don't want our models to adapt to the data too much, because they might not generalize well to unseen data. These models are usually accompanied with techniques that help them avoid overfitting.

Both parametric and nonparametric models have other parameters called *hyperparameters* that also need to be tuned during the training process. These, however, are not built into the formula of the training function (and don't end up in the formula of a nonparametric model either). We will encounter plenty of hyperparameters throughout the book.

Loss Function

We have convinced ourselves that the next logical step is finding suitable values for the ω's that appear in the training function (of our linear parametric model), using the data that we have. To do that, we need to *optimize an appropriate loss function*.

The predicted value versus the true value

Suppose we assign some random numerical values for each of our unknown ω_0, ω_1, ω_2, ω_3, ω_4, and ω_5—say, for example, $\omega_0 = -3$, $\omega_1 = 4$, $\omega_2 = 0.2$, $\omega_3 = 0.03$, $\omega_4 = 0.4$, and $\omega_5 = 0.5$. Then the formula for the linear training function $y = \omega_0 + \omega_1 x_1 + \omega_2 x_2 + \omega_3 x_3 + \omega_4 x_4 + \omega_5 x_5$ becomes:

$$y = -3 + 4x_1 + 0.2x_2 + 0.03x_3 + 0.4x_4 + 0.5x_5$$

and is ready to make predictions. Plug in numerical values for the length features of the ith fish, then obtain a predicted value for the weight of this fish. For example, the first fish in our data set is a bream and has length measurements $x_1^1 = 23.2$, $x_2^1 = 25.4$, $x_3^1 = 30$, $x_4^1 = 11.52$, and $x_5^1 = 4.02$. Plugging these into the training function, we get the prediction for the weight of this fish:

$$
\begin{aligned}
y_{predict}^1 &= \omega_0 + \omega_1 x_1^1 + \omega_2 x_2^1 + \omega_3 x_3^1 + \omega_4 x_4^1 + \omega_5 x_5^1 \\
&= -3 + 4(23.2) + 0.2(25.4) + 0.03(30) + 0.4(11.52) + 0.5(4.02) \\
&= 102.398 \text{ grams}
\end{aligned}
$$

In general, for the ith fish, we have:

$$
y_{predict}^i = \omega_0 + \omega_1 x_1^i + \omega_2 x_2^i + \omega_3 x_3^i + \omega_4 x_4^i + \omega_5 x_5^i
$$

The fish under consideration, however, has a certain *true* weight, y_{true}^i, which is its label if it belongs in the labeled data set. For the first fish in our data set, the true weight is $y_{true}^1 = 242$ grams. Our linear model with randomly chosen ω values predicted 102.398 grams. This is of course pretty far off, since we did not calibrate the ω values at all. In any case, we can measure the error between the weight predicted by our model and the true weight, then find ways to do better in terms of our choices for the ω's.

The absolute value distance versus the squared distance

One of the nice things about mathematics is that it has multiple ways to measure how far things are from each other, using different distance metrics. For example, we can naively measure the distance between two quantities as being 1 if they are different, and 0 if they are the same, encoding the words: different-1, similar-0. Of course, using such a naive metric, we lose a ton of information, since the distance between quantities such as 2 and 10 will be equal to the distance between 2 and 1 million, namely 1.

There are some distance metrics that are popular in machine learning. We first introduce the two most commonly used:

- The absolute value distance: $|y_{predict} - y_{true}|$, stemming from the calculus function $|x|$.

- The squared distance: $|y_{predict} - y_{true}|^2$, stemming from the calculus function $|x|^2$ (which is the same as x^2 for scalar quantities). Of course, this will square the units as well.

Inspecting the graphs of the functions $|x|$ and x^2 in Figure 3-4, we notice a great difference in function smoothness at the point (0,0). The function $|x|$ has a corner at that point, rendering it undifferentiable at x = 0. This *singularity* of $|x|$ at x = 0 turns many practitioners (and mathematicians!) away from incorporating this function, or functions with similar singularities, into their models. However, let's engrave the following into our brains.

Mathematical models are flexible. When we encounter a hurdle we dig deeper, understand what's going on, then we work around the hurdle.

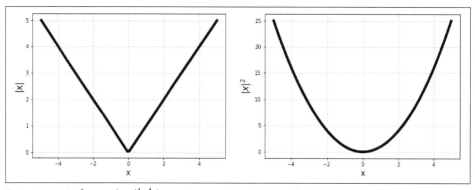

Figure 3-4. Left: graph of $|x|$ has a corner at x = 0, rendering its derivative undefined at that point. Right: graph of $|x|^2$ is smooth at x = 0, so its derivative has no problems there.

Other than the difference in the *regularity* of the functions $|x|$ and $|x|^2$ (meaning whether they have derivatives at all points or not), there is one more point that we need to pay attention to before deciding to incorporate either function into our error formula: *if a number is large, then its square is even larger.* This simple observation means that if we decide to measure the error using squared distances between true values and predicted values, then our method will be *more sensitive to the outliers* in the data. One messed-up outlier might skew our whole prediction function toward it, and hence away from the more prevalent patterns in the data. Ideally, we would have taken care of outliers and decided whether we should keep them or not during the data preparation step, before feeding the data into any machine learning model.

One last difference between $|x|$ (and similar piecewise linear functions) and x^2 (and similar nonlinear but differentiable functions) is that the derivative of $|x|$ is very easy:

1 if $x > 0$, -1 if $x < 0$ (and undefined if $x = 0$)

In a model that involves billions of computational steps, this property where there is *no need to evaluate anything* when using the derivative of $|x|$ proves to be extremely valuable. Derivatives of functions that are neither linear nor piecewise linear usually involve evaluations (because they also have x's in their formulas and not only constants, like in the piecewise linear case), which can be expensive in big data settings.

Functions with singularities

In general, graphs of *differentiable* functions do not have cusps, kinks, corners, or anything pointy. If they do have such *singularities*, then the function has no derivative at these points. The reason is that at a pointy point, you can draw two different tangent lines to the graph of the function, depending on whether you decide to draw the tangent line to the left or to the right of the point (see Figure 3-5). Recall that the derivative of a function at a point is the slope of the tangent line to the graph of the function at that point. If there are two *different* slopes, then we cannot define the derivative at the point.

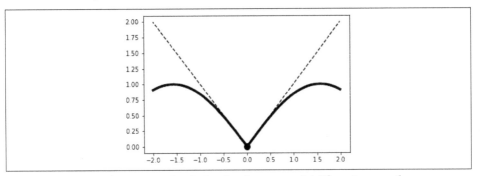

Figure 3-5. At singular points, the derivative does not exist. There is more than one possible slope of the tangent at such points.

This discontinuity in the slope of the tangent creates a problem for methods that rely on evaluating the derivative of the function, such as the gradient descent method. The problem here is twofold:

Undefined derivative
    What derivative value should you use? If you happen to land at a quirky pointy point, then the method doesn't know what to do, since there is no defined derivative there. Some people assign a value for the derivative at that point (called the *subgradient* or the *subdifferential*) and move on. In reality, what are the odds that we will be unlucky enough to land exactly at that one horrible point? Unless the landscape of the function looks like the rough terrain of the Alps (actually, many do), the numerical method might manage to avoid them.

Instability

The other problem is instability. Since the value of the derivative jumps so abruptly as you traverse the landscape of the function across this point, a method using this derivative will abruptly change value as well, creating instabilities if you are trying to converge somewhere. Imagine you are hiking down the Swiss Alps in Figure 3-6 (the landscape of the loss function) and your destination is that pretty little town that you can see down in the valley (the place with the lowest error value). Then suddenly you get carried away by some alien (the alien is the mathematical search method relying on this abruptly changing derivative) to the *other* side of the mountain, where you cannot see your destination anymore. In fact, now all you can see in the valley are some ugly shrubs and an extremely narrow canyon that can trap you if your method carries you there. Your convergence to your original destination is now unstable, if not totally lost.

Figure 3-6. Swiss Alps: optimization is similar to hiking the landscape of a function

Nevertheless, functions with such singularities are used all the time in machine learning. We will encounter them in the formulas of some neural network training functions (rectified linear unit function—who names these?), in some loss functions (absolute value distance), and in some regularizing terms (lasso regression—who names these too?).

For linear regression, the loss function is the mean squared error

Back to the main goal for this section: constructing an error function, also called the *loss function*, which encodes how much error our model commits when making its predictions, and must be made small.

For linear regression, we use the *mean squared error function*. This function averages over the squared distance errors between the prediction and the true value for m data points (we will mention which data points to include here shortly):

$$\text{Mean Squared Error} = \frac{1}{m}\left(\left|y_{predict}^1 - y_{true}^1\right|^2 + \left|y_{predict}^2 - y_{true}^2\right|^2 + \cdots \right.$$
$$\left. + \left|y_{predict}^m - y_{true}^m\right|^2\right)$$

Let's write the above expression more compactly using the sum notation:

$$\text{Mean Squared Error} = \frac{1}{m}\Sigma_{i=1}^m \left|y_{predict}^i - y_{true}^i\right|^2$$

Now we get into the great habit of using the even more compact linear algebra notation of vectors and matrices. This habit proves extremely handy in the field, as we don't want to drown while trying to keep track of indices. Indices can sneak up into our rosy dreams of understanding everything and quickly transform them into very scary nightmares. Another important reason to use the compact linear algebra notation is that both the software and the hardware built for machine learning models are optimized for matrix and *tensor* (think of an object made of layered matrices, like a three-dimensional box instead of a flat square) computations. Moreover, the beautiful field of numerical linear algebra has worked through many potential problems and paved the way for us to enjoy the fast methods to perform all kinds of matrix computations.

Using linear algebra notation, we can write the mean squared error as:

$$\text{Mean Squared Error} = \frac{1}{m}\left(\vec{y}_{predict} - \vec{y}_{true}\right)^t\left(\vec{y}_{predict} - \vec{y}_{true}\right)$$
$$= \frac{1}{m}\left\|\vec{y}_{predict} - \vec{y}_{true}\right\|_{l^2}^2$$

The last equality introduces the l^2 *norm* of a vector, which by definition is just the $\sqrt{\text{sum of squares of its components}}$.

Take-home idea: *the loss function that we constructed encodes the difference between the predictions and the ground truths for the data points involved the training process, measured in some norm—a mathematical entity that acts as a distance.* There are many other norms that we could have used, but the l^2 *norm* is pretty popular.

Notation: Vectors in this book are always column vectors

To be consistent in notation throughout the book, *all* vectors are column vectors. So if a vector $\vec{v}$ has four components, the symbol $\vec{v}$ stands for $\begin{pmatrix} v_1 \\ v_2 \\ v_3 \\ v_4 \end{pmatrix}$.

The transpose of a vector $\vec{v}$ is, then, always a row vector. The transpose of the above vector with four components is $\vec{v}^{\,t} = \begin{pmatrix} v_1 & v_2 & v_3 & v_4 \end{pmatrix}$.

We might also use the dot product notation (also called the scalar product because we multiply two vectors but our answer is a scalar number). The dot product of two vectors $\vec{a}.\vec{b}$ is the same thing as $\vec{a}^{\,t}\vec{b}$. In essence, $\vec{a}^{\,t}\vec{b}$ thinks of a column vector as a matrix of shape, *length of the vector by 1*, and its transpose as a matrix of shape, *1 by length of the vector*.

Suppose now that $\vec{a}$ and $\vec{b}$ have four components, then:

$$\vec{a}^{\,t}\vec{b} = \begin{pmatrix} a_1 & a_2 & a_3 & a_4 \end{pmatrix} \begin{pmatrix} b_1 \\ b_2 \\ b_3 \\ b_4 \end{pmatrix} = a_1 b_1 + a_2 b_2 + a_3 b_3 + a_4 b_4 = \Sigma_{i=1}^{4} a_i b_i$$

Moreover,

$$\| \vec{a} \|_{l^2}^2 = \vec{a}^{\,t}\vec{a} = a_1^2 + a_2^2 + a_3^2 + a_4^2$$

Similarly,

$$\| \vec{b} \|_{l^2}^2 = \vec{b}^{\,t}\vec{b} = b_1^2 + b_2^2 + b_3^2 + b_4^2$$

This way, we use matrix notation throughout, and we only put an arrow above a letter to indicate that we are dealing with a column vector.

The training, validation, and test subsets

Which data points do we include in our loss function? Do we include the whole data set, some small batches of it, or even only one point? Are we measuring this mean squared error for the data points in the *training subset*, the *validation subset*, or the *test subset*? And what are these subsets anyway?

In practice, we split a data set into three subsets:

Training subset
This is the subset of the data that we use to fit our training function. This means that the data points in this subset are the ones that get incorporated into our loss function (by plugging their feature values and label into the $y_{predict}$ and the y_{true} of the loss function).

Validation subset
The data points in this subset are used in multiple ways:

- The common description is that we use this subset to *tune the hyperparameters of the machine learning model*. The hyperparameters are any parameters in the machine learning model that *are not* the ω's of the training function that we are trying to solve for. In machine learning, there are many of these, and their values affect the results and the performance of the model. Examples of hyperparameters include (you don't have to know what these are yet): the learning rate that appears in the gradient descent method; the hyperparameter that determines the width of the margin in support vector machine methods; the percentage of original data split into training, validation, and test subsets; the batch size when doing randomized batch gradient descent; weight decay hyperparameters such as those used in ridge, lasso, and elastic net regression; hyperparameters that come with momentum methods such as gradient descent with momentum and ADAM (these have terms that accelerate the convergence of the method toward the minimum, and these terms are multiplied by hyperparameters that need to be tuned before testing and deployment); the number of *epochs* during the optimization process (the number of passes over the entire training subset that the optimizer has seen); and the architecture of a neural network (such as the number of layers, the width of each layer, etc.).

- The validation subset also helps us know when to stop optimizing *before* overfitting our training subset.

- It also serves as a test set to compare the performance of different machine learning models on the same data set, for example, comparing the performance of a linear regression model to a random forest to a neural network.

Test subset
After deciding on the best model to use (or averaging or aggregating the results of multiple models) and training the model, we use this untouched subset of the data as a last-stage test for our model before deployment into the real world. Since the model has not seen any of the data points in this subset before (which

means it has not included any of them in the optimization process), it can be considered as the closest analog to a real-world situation. This allows us to judge the performance of our model before we start applying it to completely new real-world data.

Recap

Let's recap a little before moving forward:

- Our current machine learning model is called *linear regression.*
- Our training function is linear with the formula:

$$y = w_0 + w_1 x_1 + w_2 x_2 + w_3 x_3 + w_4 x_4 + w_5 x_5$$

  The x's are the features, and the w's are the unknown weights or parameters.
- If we plug in the feature values of a particular data point–for example, the tenth data point–into the formula of the training function, we get our model's prediction for this point:

$$y_{predict}^{10} = w_0 + w_1 x_1^{10} + w_2 x_2^{10} + w_3 x_3^{10} + w_4 x_4^{10} + w_5 x_5^{10}$$

  The superscript 10 indicates that these are values corresponding to the tenth data point.
- Our loss function is the mean squared error function with the formula:

$$\text{Mean Squared Error} = \frac{1}{m}\left(\vec{y}_{predict} - \vec{y}_{true}\right)^t \left(\vec{y}_{predict} - \vec{y}_{true}\right)$$
$$= \frac{1}{m} \left\| \vec{y}_{predict} - \vec{y}_{true} \right\|_{l^2}^2$$

- We want to find the values of the w's that minimize this loss function. So the next step must be solving a minimization (optimization) problem.

To make our optimization life much easier, we will once again employ the convenient notation of linear algebra (vectors and matrices). This allows us to include the entire training subset of the data as a matrix in the formula of the loss function, and do our computations immediately on the training subset, as opposed to computing on each data point separately. This little notation maneuver saves us from a lot of mistakes, pain, and tedious calculations with many components that are difficult to keep track of on very large data sets.

First, write the prediction of our model corresponding to each data point of the training subset:

$$y^1_{predict} = 1\omega_0 + \omega_1 x^1_1 + \omega_2 x^1_2 + \omega_3 x^1_3 + \omega_4 x^1_4 + \omega_5 x^1_5$$
$$y^2_{predict} = 1\omega_0 + \omega_1 x^2_1 + \omega_2 x^2_2 + \omega_3 x^2_3 + \omega_4 x^2_4 + \omega_5 x^2_5$$
$$\vdots$$
$$y^m_{predict} = 1\omega_0 + \omega_1 x^m_1 + \omega_2 x^m_2 + \omega_3 x^m_3 + \omega_4 x^m_4 + \omega_5 x^m_5$$

We can easily arrange this system as:

$$\begin{pmatrix} y^1_{predict} \\ y^2_{predict} \\ \vdots \\ y^m_{predict} \end{pmatrix} = \begin{pmatrix} 1 \\ 1 \\ \vdots \\ 1 \end{pmatrix} \omega_0 + \begin{pmatrix} x^1_1 \\ x^2_1 \\ \vdots \\ x^m_1 \end{pmatrix} \omega_1 + \begin{pmatrix} x^1_1 \\ x^2_2 \\ \vdots \\ x^m_2 \end{pmatrix} \omega_2 + \begin{pmatrix} x^1_3 \\ x^2_3 \\ \vdots \\ x^m_3 \end{pmatrix} \omega_3 + \begin{pmatrix} x^1_4 \\ x^2_4 \\ \vdots \\ x^m_4 \end{pmatrix} \omega_4 + \begin{pmatrix} x^1_5 \\ x^2_5 \\ \vdots \\ x^m_5 \end{pmatrix} \omega_5$$

Or even better:

$$\begin{pmatrix} y^1_{predict} \\ y^2_{predict} \\ \vdots \\ y^m_{predict} \end{pmatrix} = \begin{pmatrix} 1 & x^1_1 & x^1_2 & x^1_3 & x^1_4 & x^1_5 \\ 1 & x^2_1 & x^2_2 & x^2_3 & x^2_4 & x^2_5 \\ \vdots & & & & & \\ 1 & x^m_1 & x^m_2 & x^m_3 & x^m_4 & x^m_5 \end{pmatrix} \begin{pmatrix} \omega_0 \\ \omega_1 \\ \omega_2 \\ \omega_3 \\ \omega_4 \\ \omega_5 \end{pmatrix}$$

The vector on the lefthand side of that equation is $\vec{y}_{predict}$, the matrix on the righthand side is the training subset X augmented with the vector of ones, and the last vector on the righthand side has all the unknown weights packed neatly into it. Call this vector $\vec{\omega}$, then write $\vec{y}_{predict}$ compactly in terms of the training subset and $\vec{\omega}$ as:

$$\vec{y}_{predict} = X\vec{\omega}$$

Now the formula of the mean squared error loss function, which we wrote before as:

$$\text{Mean Squared Error} = \frac{1}{m}\left(\vec{y}_{predict} - \vec{y}_{true}\right)^t\left(\vec{y}_{predict} - \vec{y}_{true}\right)$$
$$= \frac{1}{m} \| \vec{y}_{predict} - \vec{y}_{true} \|^2_{l^2}$$

becomes:

$$\text{Mean Squared Error} = \frac{1}{m}\left(X\vec{\omega} - \vec{y}_{true}\right)^t\left(X\vec{\omega} - \vec{y}_{true}\right) = \frac{1}{m} \parallel X\vec{\omega} - \vec{y}_{true} \parallel_{l^2}^2$$

We are now ready to find the $\vec{\omega}$ that minimizes the neatly written loss function. For that, we have to visit the rich and beautiful mathematical field of optimization.

When the training data has highly correlated features

Inspecting the training matrix (augmented with the vector of ones):

$$X = \begin{pmatrix} 1 & x_1^1 & x_2^1 & x_3^1 & x_4^1 & x_5^1 \\ 1 & x_1^2 & x_2^2 & x_3^2 & x_4^2 & x_5^2 \\ \vdots & & & & & \\ 1 & x_1^m & x_2^m & x_3^m & x_4^m & x_5^m \end{pmatrix}$$

that appears in the vector $\vec{y}_{predict} = X\vec{\omega}$, the formula of the mean squared error loss function, and later the formula that determines the unknown $\vec{\omega}$ (also called *the normal equation*):

$$\vec{\omega} = \left(X^tX\right)^{-1}X^t\vec{y}_{true}$$

we can see how our model might have a problem if two or more features (*x* columns) of the data are highly correlated. This means that there is a strong linear relationship between the features, so one of these features can be determined (or nearly determined) using a linear combination of the others. Thus, the corresponding feature columns are *not linearly independent* (or close to not being linearly independent). For matrices, this is a problem, since it indicates that the matrix either cannot be inverted or is *ill conditioned*. Ill-conditioned matrices produce large instabilities in computations, since slight variations in the training data (which must be assumed) produce large variations in the model parameters and hence render its predictions unreliable.

We desire well-conditioned matrices in our computations, so we must get rid of the sources of ill conditioning. When we have highly correlated features, one possible avenue is to include only one of them in our model, as the others do not add much information. Another solution is to apply dimension reduction techniques such as principal component analysis, which we will encounter in Chapter 11. The

Fish Market data set has highly correlated features, and the accompanying Jupyter Notebook addresses those.

That said, it is important to note that some machine learning models, such as decision trees and random forests (discussed soon) are not affected by correlated featured, while others, such as the current linear regression model, and the next logistic regression and support vector machine models, are negatively affected by them. As for neural network models, even though they can *learn* the correlations involved in the data features during training, they perform better when these redundancies are taken care of ahead of time, in addition to saving computation cost and time.

Optimization

Optimization means finding the optimal, best, maximal, minimal, or extreme solution.

We wrote a linear training function:

$$y = \omega_0 + \omega_1 x_1 + \omega_2 x_2 + \omega_3 x_3 + \omega_4 x_4 + \omega_5 x_5$$

and we left the values of its six parameters ω_0, ω_1, ω_2, ω_3, ω_4, and ω_5 unknown. The goal is to find the values that make our training function *best fit the training data subset*, where the word *best* is quantified using the loss function. This function provides a measure of how far the prediction made by the model's training function is from the ground truth. We want this loss function to be small, so we solve a minimization problem.

We are not going to sit there and try out every possible ω value until we find the combination that gives the least loss. Even if we did, we wouldn't know when to stop, since we wouldn't know whether there are other *better* values. We must have prior knowledge about the landscape of the loss function and take advantage of its mathematical properties. The analogy is hiking down the Swiss Alps with a blindfold versus hiking with no blindfold and a detailed map (Figure 3-7 shows the rough terrain of the Swiss Alps). Instead of searching the landscape of the loss function for minimizers with a blindfold, we tap into the field of *optimization*. Optimization is a beautiful branch of mathematics that provides various methods to efficiently search for and locate optimizers of functions and their corresponding optimal values.

The optimization problem in this chapter and in the next few looks like:

$$\min_{\vec{\omega}} \text{loss function}$$

Figure 3-7. Swiss Alps: optimization is similar to hiking the landscape of a function. The destination is the bottom of the lowest valley (minimization) or the top of the highest peak (maximization). We need two things: the coordinates of the minimizing or maximizing points, and the height of the landscape at those points.

For the current linear regression model, this is:

$$\min_{\vec{\omega}} \frac{1}{m}\left(X\vec{\omega} - \vec{y}_{true}\right)^{t}\left(X\vec{\omega} - \vec{y}_{true}\right) = \min_{\vec{\omega}} \frac{1}{m} \| X\vec{\omega} - \vec{y}_{true} \|_{l^2}^{2}$$

When we do math, we should never lose track of what it is that we know and what it is that we are looking for. Otherwise we would run the risk of getting trapped in a circular logic. In the formula just mentioned, we know:

m

The number of instances in the training subset

X

The training subset augmented with a vector of ones

$\vec{y}_{true}$

The vector of labels corresponding to the training subset

And we are looking for:

- The minimizing $\vec{\omega}$

- The minimum value of the loss function at the minimizing $\vec{\omega}$

Convex landscapes versus nonconvex landscapes

The *easiest* functions to deal with and the easiest equations to solve are linear. Unfortunately, most of the functions (and equations) that we deal with are nonlinear. At the same time, this is not too unfortunate since linear life is flat, boring, unstimulating, and uneventful. When the function we have at hand is full-blown nonlinear, we sometimes *linearize* it near certain points that we care for. The idea here is that even though the full picture of the function might be nonlinear, we may be able to approximate it with a linear function in the locality that we focus on. In other words, in a very small neighborhood, the nonlinear function might look and behave linearly, albeit the said neighborhood might be infinitesimally small. For an analogy, think about how Earth looks (and behaves in terms of calculating distances, etc.) flatly from our own locality, and how we can only see its nonlinear shape from high up. When we want to linearize a function near a point, we approximate it by its tangent space near that point (this is its tangent line if it is a function of one variable, tangent plane if it is a function of two variables, and tangent hyperplane if it is a function of three or more variables). For this, we need to calculate one derivative of the function with respect to all of its variables, since this gives us the slope (which measures the inclination) of the approximating flat space.

The sad news is that linearizing near one point may not be enough, and we may want to use linear approximations at multiple locations. Thankfully, that is doable, since we all we have to do computationally is to evaluate one derivative at several points. This leads us to the *next easiest* functions to deal with (after linear functions): piecewise linear functions, which are linear but only in piecewise structures, or linear except at isolated points or locations. The field of *linear programming* deals with such functions, where the functions to optimize are linear, and the boundaries of the domains where the optimization happens are piecewise linear (they are intersections of half spaces).

When our goal is optimization, the best functions to deal with are either linear (where the field of linear programming helps us) or convex (where we do not worry about getting stuck at local minima, and where we have good inequalities that help us with the analysis).

One important type of function to keep in mind, which appears in machine learning, is a function that is the maximum of two or more convex functions. These functions are always convex. Recall that linear functions are flat, so they are at the same time convex and concave. This is useful since some functions are defined as the maxima of linear functions: those are not guaranteed to be linear (they are piecewise linear), but are guaranteed to be convex. That is, even though we lose linearity when we take the maximum of linear functions, we are compensated with convexity.

The Rectified Linear Unit function (ReLU) that is used as a nonlinear activation function in neural networks is an example of a function defined as the maximum of two linear functions: $ReLU(x) = max(0,x)$. Another example is the hinge loss function used for support vector machines: $H(x) = max(0, 1 - tx)$ where t is either 1 or −1.

Note that the minimum of a family of convex functions is not guaranteed to be convex; it can have a double well. However, their maximum is definitely convex.

There is one more relationship between linearity and convexity. If we have a convex function (nonlinear since linear would be trivial), then the maximum of all the linear functions that stay below our function is exactly equal to it. In other words, convexity replaces linearity, in the sense that when linearity is not available, but convexity is, we can replace our convex function with the maximum of all the linear functions whose graph lies below our function's graph (see Figure 3-8). Recall that the graph of a convex function lies above the graph of its tangent at any point, and that the tangents are linear. This gives us a direct path to exploiting the simplicity of linear functions when we have convex functions. We have equality when we consider the maximum of *all* the tangents, and only approximation when we consider the maximum of the tangents at a few points.

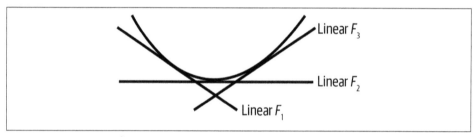

Figure 3-8. A convex function is equal to the maximum of all of its tangents

Figures 3-9 and 3-10 show the general landscapes of nonlinear convex and nonconvex functions, respectively. Overall, the landscape of a convex function is good for minimization problems. We have no fear of getting stuck at local minima since any local minimum is also a global minimum for a convex function. The landscape of a nonconvex function has peaks, valleys, and saddle points. A minimization problem on such a landscape runs the risk of getting stuck at the local minima and never finding the global minima.

Finally, make sure you know the distinction among a convex function, a convex set, and a convex optimization problem, which optimizes a convex function over a convex set.

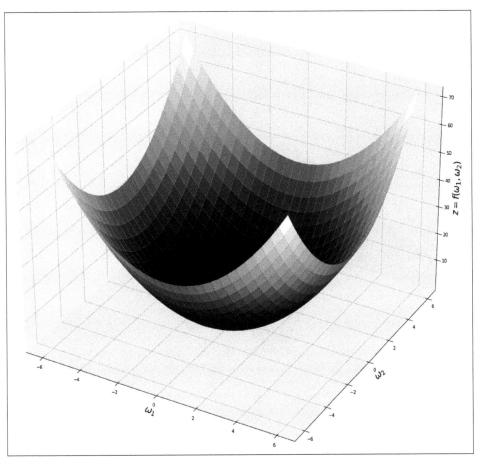

Figure 3-9. The landscape of a convex function is good for minimization problems. We have no fear of getting stuck at the local minima since any local minimum is also a global minimum for a convex function.

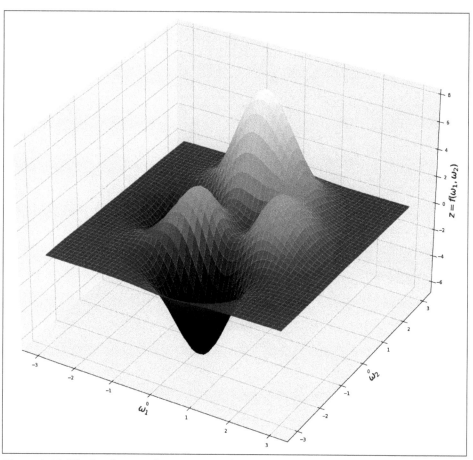

Figure 3-10. The landscape of a nonconvex function has peaks, valleys, and saddle points. A minimization problem on such a landscape runs the risk of getting stuck at a local minima and never finding the global minima.

How do we locate minimizers of functions?

In general, there are two approaches to locating minimizers (and/or maximizers) of functions. The trade-off is usually between:

1. Calculating only one derivative and converging to the minimum slowly (though there are acceleration methods to speed up the convergence). These are called *gradient* methods. The gradient is one derivative of a function of several variables. For example, our loss function is a function of several ω's (or of one vector $\vec{\omega}$).

2. Calculating two derivatives (computationally much more expensive, which is a big turnoff, especially when we have thousands of parameters) and converging to

the minimum faster. Computation costs can be saved a little by approximating the second derivative instead of computing it exactly. Second derivative methods are called *Newton's* methods. The *Hessian* (the matrix of second derivatives) or an approximation of the Hessian appears in these methods.

We never need to go beyond calculating two derivatives.

But why are the first and second derivatives of a function so important for locating its optimizers? The concise answer is that the first derivative contains information on how fast a function increases or decreases at a point (so if you follow its direction, you might ascend to the maximum or descend to the minimum), and the second derivative contains information on the shape of the *bowl* of the function—if it curves up or curves down.

One key idea from calculus remains fundamental: minimizers (and/or maximizers) happen at *critical points* (defined as the points where one derivative of our function is either equal to zero or does not exist) or at boundary points. So to locate these optimizers, we must search through *both* the boundary points (if our search space has a boundary) *and* the interior critical points.

How do we locate the critical points in the interior of our search space?

Approach 1
  We follow these steps:

  - Find one derivative of our function (not too bad, we all did it in calculus).
  - Set it equal to zero (we can all write the symbols equal and zero).
  - Solve for the w's that make our derivative zero (this is the bad step!).

For functions whose derivatives are linear, such as our mean squared error loss function, it is sort of easy to solve for these w's. The field of linear algebra was especially built to help solve linear systems of equations. The field of numerical linear algebra was built to help solve realistic and large systems of linear equations where ill conditioning is prevalent. We have many tools at our disposal (and software packages) when our systems are linear.

On the other hand, when our equations are nonlinear, finding solutions is an entirely different story. It becomes a hit-or-miss game, with mostly misses! Here's a short example that illustrates the difference between solving a linear and a nonlinear equation:

Solving a linear equation
  Find w such that $0.002w - 5 = 0$.

  Solution: Moving the 5 over to the other side, then dividing by 0.002, we get $w = 5/0.002 = 2500$. Done.

Solving a nonlinear equation

Find ω such that $0.002 \sin(\omega) - 5\omega^2 + e^\omega = 0$.

Solution: Yes, I am out of here. We need a numerical method! (See Figure 3-11 for a graphical approximation of the solution of this nonlinear equation.)

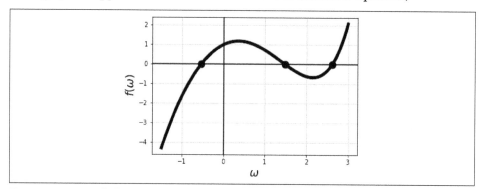

Figure 3-11. It is difficult to solve nonlinear equations. Here, we plot $f(\omega) = 0.002 \sin(\omega) - 5\omega^2 + e^\omega$ *and approximate its three roots (points where* $f(\omega) = 0$) *on the graph.*

There are many numerical techniques devoted solely to finding solutions of nonlinear equations (and entire fields devoted to numerically solving nonlinear ordinary and partial differential equations). These methods find approximate solutions, then provide bounds on how far off the numerical solutions are from the exact analytical solutions. They usually construct a sequence that converges to the analytical solution under certain conditions. Some methods converge faster than others, and are better suited for certain problems than others.

Approach 2

Another option is to follow the gradient direction to descend toward the minimum or ascend toward the maximum.

To understand these gradient-type methods, think of hiking down a mountain (or skiing down the mountain if the method is accelerated or has momentum). We start at a random point in our search space, and that sets us at an initial height level on the landscape of the function. Now the method moves us to a new point in the search space, and hopefully, at this new location, we end up at a new height level that is *lower* than the height level we came from. Hence, we would have *descended*. We repeat this and ideally, if the terrain of the function cooperates, this sequence of points will converge toward the minimizer of the function that we are looking for.

Of course, for functions with landscapes that have many peaks and valleys, where we start–or in other words, how to initialize–matters, since we could descend down an entirely different valley than the one where we want to end up. We might end up at a local minimum instead of a global minimum.

Functions that are convex and bounded below are shaped like a salad bowl, so with those we do not worry about getting stuck at local minima and away from global minima. There could be another source of worry with convex functions: when the shape of the bowl of the function is too narrow, our method might become painfully slow to converge. We will go over this in detail in Chapter 4.

Both Approach 1 and Approach 2 are useful and popular. Sometimes, we have no option but to use one or the other, depending on how fast each method converges for our particular setting, how *regular* the function we are trying to optimize is (how many well-behaved derivatives it has), etc. Other times, it is just a matter of taste. For linear regression's mean squared error loss function, both types of methods work, so we will use Approach 1, only because we will use gradient descent methods for *all* the other loss functions in this book.

We must mention that the hiking-down-the-mountain analogy for descent methods is excellent but a tad bit misleading. When we humans hike down a mountain, we physically belong in the same three-dimensional space that our mountain landscape exists in, meaning we are at a certain elevation and we are able to descend to a location at a lower elevation, even with a blindfold, and even when it is too foggy and we can only descend one tiny step at a time. We sense the elevation, then move downhill. Numerical descent methods, on the other hand, do not search for the minimum in the same space dimension as the one the landscape of the function is embedded in. Instead, they search on the ground level, one dimension *below* the landscape (see Figure 3-12). This makes descending toward the minimum much more difficult, since at ground level we can move from any point to any other without knowing what height level exists above us, until we evaluate the function itself at the point and find the height. So our method might accidentally move us from one ground point with a certain elevation above us to another ground point with a *higher* elevation, hence farther away from the minimum. This is why it is important to locate, on ground level, a direction that quickly *decreases* the function height, and how far we can move in that direction on ground level (step size) while *still decreasing the function height above us*. The step size is also called the *learning rate hyperparameter*, which we will encounter every time we use a descent method.

Back to our main goal: we want to find the best $\vec{w}$ for our training function, so we must minimize our mean squared error loss function using Approach 1: take one derivative of the loss function and set it equal to zero, then solve for the vector $\vec{w}$. For this, we need to master doing calculus on linear algebra expressions. Let's revisit our calculus class first.

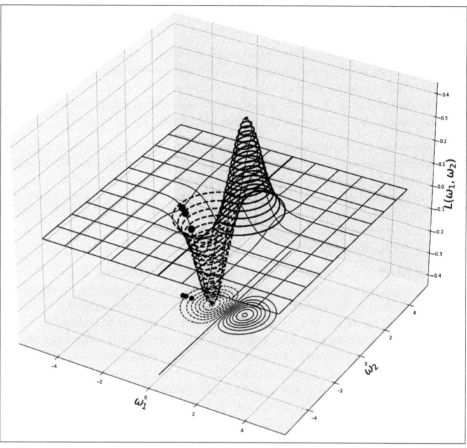

Figure 3-12. Searching for the minimum happens on ground level and not directly on the landscape of the function

Calculus in a nutshell

In a first course on calculus, we learn about functions of single variables ($f(\omega)$), their graphs, and how to evaluate them at certain points. Then we learn about the most important operation in mathematical analysis: the limit. From the limit concept, we define continuity and discontinuity of functions, derivative at a point $f'(\omega)$ (limit of slopes of secants through a point), and integral over a domain (limit of sums of mini regions determined by the function over the domain). We end the class with the fundamental theorem of calculus, relating integration and differentiation as inverse operations. One of the key properties of the derivative is that it determines how fast a function increases or decreases at a certain point; hence, it plays a crucial role in locating a minimum and/or maximum of a function in the interior of its domain (boundary points are separate).

In a multivariable calculus course, which is usually a third course in calculus, many ideas transfer from single-variable calculus, including the importance of the derivative, now called the *gradient* because we have several variables, in locating any interior minima and/or maxima. The gradient $\nabla\left(f(\vec{w})\right)$ of $f(\vec{w})$ is the derivative of the function with respect to the vector $\vec{w}$ of variables.

In deep learning, the unknown weights are organized in matrices, not in vectors, so we need to take the derivative of a function $f(W)$ with respect to a matrix W of variables.

For our purposes in AI, the function whose derivative we need to calculate is the loss function, which has the training function built into it. By the *chain rule for derivatives*, we would also need to calculate the derivative of the training function with respect to the w's.

Let's demonstrate using one simple example from single-variable calculus, then immediately transition to taking derivatives of linear algebra expressions.

A one-dimensional optimization example

Problem: Find the minimizer(s) and the minimum value (if any) of the function $f(w) = 3 + (0.5w - 2)^2$ on the interval $[-1,6]$.

One impossibly long way to go about this is to try out the *infinitely many* values of w between -1 and 6, and choose the w's that give the lowest f value. Another way is to use our calculus knowledge that optimizers (minimizers and/or maximizers) happen either at critical points (where the derivative is either nonexistent or zero) or at boundary points. For reference, see Figure 3-13.

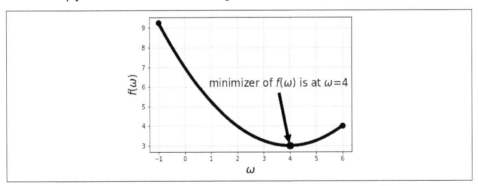

Figure 3-13. The minimum value of the function $f(w) = 3 + (0.5w - 2)^2$ on the interval $[-1,6]$ is 3 and happens at the critical point $w = 4$. At this critical point, the derivative of the function is zero, meaning that if we draw a tangent line, it will be horizontal.

Our boundary points are –1 and 6, so we evaluate our function at these points first: $f(-1) = 3 + (0.5(-1) - 2)^2 = 9.25$ and $f(6) = 3 + (0.5(6) - 2)^2 = 4$. Obviously, –1 is not a minimizer since $f(6) < f(-1)$, so this boundary point gets out of the competition and now only boundary point 6 is competing with interior critical point(s). In order to find our critical points, we inspect the derivative of the function in the interior of the interval [–1,6]: $f'(\omega) = 0 + 2(0.5\omega - 2) * 0.5 = 0.25(0.5\omega - 2)$. Setting this derivative equal to zero, we have $0.25(0.5\omega - 2) = 0$, implying that $\omega = 4$. Thus, we only found one critical point $\omega = 4$ in the interior of the interval [–1,6]. At this special point, the value of the function is $f(4) = 3 + (0.5(4) - 2)^2 = 3$. Since the value of f is the lowest here, we have obviously found the winner of our minimization competition, namely, $\omega = 4$ with the minimum f value equal to 3.

Derivatives of linear algebra expressions that we use all the time

It is efficient to calculate derivatives directly on expressions involving vectors and matrices, without having to resolve them into their components. The following two are popular:

1. When a and ω are scalars and a is constant, the derivative of $f(\omega) = a\omega$ is $f'(\omega) = a$. When $\vec{a}$ and $\vec{\omega}$ are vectors (of the same length) and the entries of $\vec{a}$ are constant, then the gradient of $f(\vec{\omega}) = \vec{a}^t\vec{\omega}$ is $\nabla f(\vec{\omega}) = \vec{a}$. Similarly, the gradient of $f(\vec{\omega}) = \vec{\omega}^t\vec{a}$ is $\nabla f(\vec{\omega}) = \vec{a}$.

2. When s is scalar and constant and ω is scalar, then the derivative of the quadratic function $f(\omega) = s\omega^2$ is $f'(\omega) = 2s\omega$. The analogous high-dimensional case is when S is a symmetric matrix with constant entries, then the function $f(\vec{\omega}) = \vec{\omega}^t S\vec{\omega}$ is quadratic and its gradient is $\nabla f(\vec{\omega}) = 2S\vec{\omega}$.

Minimizing the mean squared error loss function

We are finally ready to minimize the mean squared error loss function:

$$L(\vec{\omega}) = \frac{1}{m}\left(X\vec{\omega} - \vec{y}_{true}\right)^t\left(X\vec{\omega} - \vec{y}_{true}\right)$$

Let's open that expression up before taking its gradient and setting it equal to zero:

$$L\left(\vec{\omega}\right) = \frac{1}{m}\left(\left(X\vec{\omega}\right)^t - \vec{y}^{\,t}_{true}\right)\left(X\vec{\omega} - \vec{y}_{true}\right)$$

$$= \frac{1}{m}\left(\vec{\omega}^{\,t} X^t - \vec{y}^{\,t}_{true}\right)\left(X\vec{\omega} - \vec{y}_{true}\right)$$

$$= \frac{1}{m}\left(\vec{\omega}^{\,t} X^t X\vec{\omega} - \vec{\omega}^{\,t} X^t \vec{y}_{true} - \vec{y}^{\,t}_{true} X\vec{\omega} + \vec{y}^{\,t}_{true}\vec{y}_{true}\right)$$

$$= \frac{1}{m}\left(\vec{\omega}^{\,t} S\vec{\omega} - \vec{\omega}^{\,t}\vec{a} - \vec{a}^{\,t}\vec{\omega} + \vec{y}^{\,t}_{true}\vec{y}_{true}\right),$$

where in the last step we set $X^t X = S$ and $X^t \vec{y}_{true} = \vec{a}$. Next, take the gradient of the last expression with respect to $\vec{\omega}$ and set it equal to zero. When calculating the gradient, we use what we just learned about differentiating linear algebra expressions:

$$\nabla L\left(\vec{\omega}\right) = \frac{1}{m}\left(2S\vec{\omega} - \vec{a} - \vec{a} + 0\right)$$

$$= \vec{0}$$

Now it is easy to solve for $\vec{\omega}$:

$$\frac{1}{m}\left(2S\vec{\omega} - 2\vec{a}\right) = \vec{0}$$

so:

$$2S\vec{\omega} = 2\vec{a}$$

which gives:

$$\vec{\omega} = S^{-1}\vec{a}$$

Now recall that we set $S = X^t X$ and $\vec{a} = X^t y_{true}$, so let's rewrite our minimizing $\vec{\omega}$ in terms of the training set X (augmented with ones) and the corresponding labels vector $\vec{y}_{true}$:

$$\vec{\omega} = \left(X^t X\right)^{-1} X^t \vec{y}_{true}$$

For the Fish Market data set, this ends up being (see the accompanying Jupyter notebook):

$$\vec{\omega} = \begin{pmatrix} \omega_0 \\ \omega_1 \\ \omega_2 \\ \omega_3 \\ \omega_4 \\ \omega_5 \end{pmatrix} = \begin{pmatrix} -475.19929130109716 \\ 82.84970118 \\ -28.85952426 \\ -28.50769512 \\ 29.82981435 \\ 30.97250278 \end{pmatrix}$$

Multiplying Large Matrices by Each Other Is Very Expensive; Multiply Matrices by Vectors Instead

Try to avoid multiplying matrices by each other at all costs; instead, multiply your matrices with *vectors*. For example, in the normal equation $\vec{\omega} = \left(X^t X\right)^{-1} X^t \vec{y}_{true}$, compute $X^t \vec{y}_{true}$ first, and avoid computing $\left(X^t X\right)^{-1}$ altogether. The way around this is to solve instead the linear system $X\vec{\omega} = \vec{y}_{true}$ using the pseudoinverse of X (check the accompanying Jupyter notebook). We will discuss the the pseudoinverse in Chapter 11, but for now, it allows us to invert (which is equivalent to divide by) matrices that do not have an inverse.

We just located the weights vector $\vec{\omega}$ that gives the best fit between our training data and the linear regression training function:

$$f\left(\vec{\omega}; \vec{x}\right) = \omega_0 + \omega_1 x_1 + \omega_2 x_2 + \omega_3 x_3 + \omega_4 x_4 + \omega_5 x_5$$

We used an analytical method (compute the gradient of the loss function and set it equal to zero) to derive the solution given by the normal equation. This is one of the very rare instances where we are able to derive an analytical solution. All other methods for finding the minimizing $\vec{\omega}$ will be numerical.

We Never Want to Fit the Training Data Too Well

The $\vec{\omega} = \left(X^t X\right)^{-1} X^t \vec{y}_{true}$ that we calculated gives the ω values that make the training function *best fit* the training data, but too good of a fit means that the training function might also be picking up on the noise and not only on the signal in the data. So the solution just mentioned, or even the minimization problem itself, needs to be modified in order to *not get too good of a fit*. *Regularization* or *early stopping* are helpful here. We will spend some time on those in Chapter 4.

This was the long way to regression. We had to pass through calculus and linear algebra on the way, because we are just starting. Presenting the upcoming machine learning models—logistic regression, support vector machines, decision trees, and random forests—will be faster, since all we do is apply the exact same ideas to different functions.

Logistic Regression: Classify into Two Classes

Logistic regression is mainly used for classification tasks. We first explain how we can use this model for binary classification tasks (classify into two classes, such as cancer/not cancer, safe for children/not safe, likely to pay back a loan/unlikely, etc.). Then we will generalize the model into classifying into multiple classes (for example, classify handwritten images of digits into 0, 1, 2, 3, 4, 5, 6, 7, 8, or 9). Again, we have the same mathematical setup:

1. Training function
2. Loss function
3. Optimization

Training Function

Similar to linear regression, the training function for logistic regression computes a linear combination of the features and adds a constant bias term, but instead of outputting the result as is, it passes it through the *logistic function*, whose graph is plotted in Figure 3-14, and whose formula is:

$$\sigma(s) = \frac{1}{1 + e^{-s}}$$

This is a function that only takes values between 0 and 1, so its output can be interpreted as a probability of a data point belonging to a certain class. If the output is less than 0.5, then classify the data point as belonging to the first class, and if the output is greater than 0.5, then classify the data point in the other class. The number 0.5 is the *threshold* where the decision to classify the data point is made.

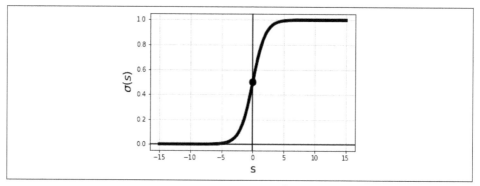

Figure 3-14. Graph of the logistic function $\sigma(s) = \dfrac{1}{1 + e^{-s}}$. Note that this function can be evaluated at any s and always outputs a number between 0 and 1, hence its output can be interpreted as a probability.

Therefore, the training function here ends up being a linear combination of features, plus bias, composed first with the logistic function, then finally composed with a thresholding function:

$$y = Thresh(\sigma(\omega_0 + \omega_1 x_1 + \cdots + \omega_n x_n))$$

Similar to the linear regression case, the ω's are the unknowns for which we need to optimize our loss function. Just like linear regression, the number of these unknowns is equal to the number of data features, plus one for the bias term. For tasks like classifying images, each pixel is a feature, so we could have thousands of those.

Loss Function

Let's design a good *loss function* for classification. We are the engineers and we want to penalize wrongly classified training data points. In our labeled data set, if an instance belongs in a class, then its $y_{true} = 1$, and if it doesn't, then its $y_{true} = 0$.

We want our training function to output $y_{predict} = 1$ for training instances that belong in the positive class (whose y_{true} is also 1). Successful ω values should give a high value of t (result of the linear combination step) to go into the logistic function, hence assigning high probability for positive instances and passing the 0.5 threshold to obtain $y_{predict} = 1$. Therefore, if the linear combination plus bias step gives a low t value while $y_{true} = 1$, penalize it.

Similarly, successful weight values should give a low t value to go into the logistic function for training instances that do not belong in the class (their true $y_{true} = 0$). Therefore, if the linear combination plus bias step gives a high t value while $y_{true} = 0$, penalize it.

So how do we find a loss function that penalizes a wrongly classified training data point? Both false positives and false negatives should be penalized. Recall that the outputs of this classification model are either 1 or 0:

- Think of a calculus function that rewards 1 and penalizes 0: $- \log(s)$ (see Figure 3-15).
- Think of a calculus function that penalizes 1 and rewards 0: $- \log(1 - s)$ (see Figure 3-15).

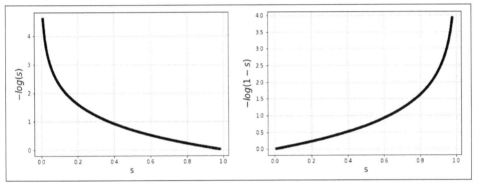

Figure 3-15. Left: graph of the function $f(s) = - \log(s)$. This function assigns high values for numbers near 0 and low values for numbers near 1. Right: graph of the function $f(s) = - \log(1 - s)$. This function assigns high values for numbers near 1 and low values for numbers near 0.

Now focus on the output of the logistic function $\sigma(s)$ for the current choice of w's:

- If $\sigma(s)$ is less than 0.5 (model prediction is $y_{predict} = 0$) but the true $y_{true} = 1$ (a false negative), make the model pay by penalizing $- \log(\sigma(s))$. If instead $\sigma(s) > 0.5$, the model prediction is $y_{predict} = 1$ (a true positive), $- \log(\sigma(s))$ is small, so no high penalty is paid.
- Similarly, if $\sigma(s)$ is more than 0.5, but the the true $y_{true} = 0$ (a false positive), make the model pay by penalizing $- \log(1 - \sigma(s))$. Again, for a true negative no high penalty is paid either.

Therefore, we can write the cost for misclassifying one training instance $\left(x_1^i, x_2^i, \cdots, x_n^i; y_{true}\right)$ as:

$$cost = \begin{cases} -\log\left(\sigma(s)\right) \text{ if } y_{true} = 1 \\ -\log\left(1 - \sigma(s)\right) \text{ if } y_{true} = 0 \end{cases}$$
$$= -y_{true} \log\left(\sigma(s)\right) - \left(1 - y_{true}\right) \log\left(1 - \sigma(s)\right)$$

Finally, the loss function is the average cost over m training instances, giving us the formula for the popular *cross-entropy loss function*:

$$L\left(\overrightarrow{\omega}\right) = -\frac{1}{m} \sum_{i=1}^{m} y_{true}^i \log\left(\sigma\left(\omega_0 + \omega_1 x_1^i + \cdots + \omega_n x_n^i\right)\right) +$$
$$\left(1 - y_{true}^i\right) \log\left(1 - \sigma\left(\omega_0 + \omega_1 x_1^i + \cdots + \omega_n x_n^i\right)\right)$$

Optimization

Unlike the linear regression case, if we decide to minimize the loss function by setting $\nabla L(\omega) = 0$, there is no closed form solution formula for the ω's. The good news is that this function is convex, so the gradient descent in Chapter 4 (or stochastic or mini-batch gradient descents) is guaranteed to find a minimum (if the *learning rate* is not too large and if we wait long enough).

Softmax Regression: Classify into Multiple Classes

We can easily generalize the logistic regression idea to classify into multiple classes. A famous example for such a nonbinary classification task is classifying images of the 10 handwritten numerals 0, 1, 2, 3, 4, 5, 6, 7, 8, and 9 using the MNIST data set (*https://oreil.ly/HQL5F*).[1] This data set contains 70,000 images of handwritten numerals (see Figure 3-16 for samples of these images), split into a 60,000-image training subset and a 10,000-image test subset. Each image is labeled with the class it belongs to, which is one of the 10 numerals.

This data set also contains results from many classifying models, including linear classifiers, *k-nearest neighbors*, *decision trees*, *support vector machines* with various *kernels*, and neural networks with various architectures, along with references for the corresponding papers and their years of publication. It is interesting to see the progress in performance as the years go by and the methods evolve.

---

1 At the time of writing, this site may request credentials, but you can access it using these instructions from Hacker News (*https://oreil.ly/Csj-0*).

Figure 3-16. Sample images from the MNIST data set (image source (https://oreil.ly/0H32W))

Do Not Confuse Classifying into Multiple Classes with Multioutput Models

Softmax regression predicts one class at a time, so we cannot use it to classify, for example, five people in the same image. Instead, we can use it to check whether a given Facebook image is a picture of me, my sister, my brother, my husband, or my daughter. An image passed into the softmax regression model can have only one of the five of us, or the model's classification would be less obvious. This means that our classes have to be mutually exclusive. So when Facebook automatically tags five people in the same image, they are using a multioutput model, not a softmax regression model.

Suppose we have the features of a data point and we want to use this information to classify the data point into one of k possible classes. The following training function, loss function, and optimization process should be clear by now.

Features of Image Data

For grayscale images, each pixel intensity is a feature, so images usually have thousands of features. Grayscale images are usually represented as two-dimensional matrices of numbers, with pixel intensities as the matrix entries. Color images come in three channels, red, green, and blue, where each channel is again represented as a two-dimensional matrix of numbers, and the channels are stacked on top of each other, forming three layers of two-dimensional matrices. This structure is called a *tensor*. Check out the notebook on processing images at this book's GitHub page (*https://github.com/halanelson/Essential-Math-For-AI*) that illustrates how we can work with grayscale and color images in Python.

Training Function

The first step is always the same: linearly combine the features and add a constant bias term. In logistic regression, when we only had two classes, we passed the result into the logistic function of the formula:

$$\sigma(s) = \frac{1}{1 + e^{-s}} = \frac{1}{1 + \frac{1}{e^s}} = \frac{e^s}{1 + e^s} = \frac{e^s}{e^0 + e^s}$$

which we interpreted as the probability of the data point belonging in the class of interest or not. Note that we rewrote the formula for the logistic function as $\sigma(s) = \frac{e^s}{e^0 + e^s}$ to highlight the fact that it captures two probabilities, one for each class. In other words, $\sigma(s)$ gives the probability that a data point is in the class of interest, and $1 - \sigma(s) = \frac{e^0}{e^0 + e^s}$ gives the probability that the data point is not in the class.

When we have multiple classes instead of only two, then for the same data point, we repeat same process multiple times: one time for each class. Each class has its own bias and set of weights that linearly combine the features, thus given a data point with feature values $x_1, x_2, \ldots$, and x_n, we compute k different linear combinations plus biases:

$$s^1 = \omega_0^1 + \omega_1^1 x_1 + \omega_2^1 x_2 + \cdots + \omega_n^1 x_n$$
$$s^2 = \omega_0^2 + \omega_1^2 x_1 + \omega_2^2 x_2 + \cdots + \omega_n^2 x_n$$
$$\vdots$$
$$s^k = \omega_0^k + \omega_1^k x_1 + \omega_2^k x_2 + \cdots + \omega_n^k x_n$$

Get into Good Habits

You want to get into the good habit of keeping track of how many unknown ω's end up in the formula for your training function. Recall that these are the ω's that we find via minimizing a loss function. The other good habit is having an efficient and consistent way to organize them throughout your model (in a vector, matrix, etc.). In the softmax case, when we have k classes, and n features for each data point, we end up with $k \times n$ ω's for the linear combinations, then k biases, for a total of $k \times n + k$ unknown ω's. For example, if we use a softmax regression model to classify images in the MNIST data set (*https://oreil.ly/wJfei*) of handwritten numerals, each image has 28×28 pixels, meaning 784 features, and we want to classify them into 10 classes, so we end up having to optimize for 7850 ω's. For both the linear and logistic regression models, we only had $n + 1$ unknown ω's that we needed to optimize for.

Next we pass each of these k results into the *softmax function*, which generalizes the logistic function from two to multiple classes, and we also interpret it as a probability. The formula for the softmax function looks like:

$$\sigma(s^j) = \frac{e^{s^j}}{e^{s^1} + e^{s^2} + \cdots + e^{s^k}}$$

This way, the same data point will get k probability scores, one score corresponding to each class. Finally, we classify the data point as belonging to the class where it obtained the largest probability score.

Aggregating all of the above, we obtain the final formula of the training function that we can now use for classification (that is, after we find the optimal ω values by minimizing an appropriate loss function):

$$y = j \text{ such that } \sigma\left(\omega_0^j + \omega_1^j x_1 + \cdots + \omega_n^j x_n\right) \text{ is maximal}$$

Note that for this training function, all we have to do is input the data features (the x values), and it returns one class number: j.

The Logistic and Softmax Functions and Statistical Mechanics

If you are familiar with statistical mechanics, you might have noticed that the logistic and softmax functions calculate probabilities in the same way the *partition function* from the field of statistical mechanics calculates the probability of finding a system in a certain state.

Loss Function

We derived the cross-entropy loss function in the case of logistic regression:

$$L\left(\overrightarrow{\omega}\right) = -\frac{1}{m}\sum_{i=1}^{m} y_{true}^i \log\left(\sigma\left(\omega_0 + \omega_1 x_1^i + \cdots + \omega_n x_n^i\right)\right) +$$

$$\left(1 - y_{true}^i\right) \log\left(1 - \sigma\left(\omega_0 + \omega_1 x_1^i + \cdots + \omega_n x_n^i\right)\right)$$

using:

$$cost = \begin{cases} -\log\left(\sigma(s)\right) \text{ if } y_{true} = 1 \\ -\log\left(1 - \sigma(s)\right) \text{ if } y_{true} = 0 \end{cases}$$
$$= -y_{true} \log\left(\sigma(s)\right) - \left(1 - y_{true}\right) \log\left(1 - \sigma(s)\right)$$

Now we generalize the same logic to multiple classes. Let's use the notation that $y_{true,i} = 1$ if a certain data point belongs in the ith class, and is zero otherwise. Then we have the cost associated with misclassifying a certain data point as:

$$cost = \begin{cases} -\log\left(\sigma\left(s^1\right)\right) \text{ if } y_{true,1} = 1 \\ -\log\left(\sigma\left(s^2\right)\right) \text{ if } y_{true,2} = 1 \\ -\log\left(\sigma\left(s^3\right)\right) \text{ if } y_{true,3} = 1 \\ \vdots \\ -\log\left(\sigma\left(s^k\right)\right) \text{ if } y_{true,k} = 1 \end{cases}$$
$$= -y_{true,1} \log\left(\sigma\left(s^1\right)\right) - \cdots - y_{true,k} \log\left(\sigma\left(s^k\right)\right)$$

Averaging over all the m data points in the training set, we obtain the *generalized cross-entropy loss function*, generalizing it from the case of only two classes to the case of multiple classes:

$$L\left(\overrightarrow{\omega}\right) = -\frac{1}{m}\Sigma_{i=1}^{m} y_{true,1}^i \log\left(\sigma\left(\omega_0^1 + \omega_1^1 x_1^i + \cdots + \omega_n^1 x_n^i\right)\right) + y_{true,2}^i$$
$$\log\left(\sigma\left(\omega_0^2 + \omega_1^2 x_1^i + \cdots + \omega_n^2 x_n^i\right)\right) + \cdots + y_{true,k}^i \log\left(\sigma\left(\omega_0^k + \omega_1^k x_1^i + \cdots + \omega_n^k x_n^i\right)\right)$$

Optimization

Now that we have a formula for the loss function, we can search for its minimizing ω's. As with most of the loss functions that we will encounter, there is no explicit formula for the minimizers of this loss function in terms of the training set and their target labels, so we settle for finding the minimizers using numerical methods, in particular: gradient descent, stochastic gradient descent, or mini-batch gradient descent (see Chapter 4). Again, the generalized cross-entropy loss function has its convexity working to our advantage in the minimization process, so we are guaranteed to find our sought-after ω's.

Cross Entropy and Information Theory

The cross-entropy concept is borrowed from information theory. We will elaborate on this when discussing decision trees later in this chapter. For now, keep the following quantity in mind, where p is the probability of an event occurring:

$$\log\left(\frac{1}{p}\right) = -\log(p)$$

That quantity is large when p is small, therefore, it quantifies bigger *surprise* for *less probable* events.

Incorporating These Models into the Last Layer of a Neural Network

The linear regression model makes its predictions by appropriately linearly combining data features, then adding bias. The logistic regression and the softmax regression models make their classifications by appropriately linearly combining data features, adding bias, then passing the result into a probability scoring function. In these simple models, the features of the data are only linearly combined, hence, these models are weak in terms of picking up on potentially important nonlinear interactions among the data features. Neural network models incorporate nonlinear *activation functions* into their training functions, do this over multiple layers, and hence are better equipped to detect nonlinear and more complex relationships. The last layer of a neural network is its output layer. The layer right before the last layer spits out some higher-order features and inputs them into the last layer. If we want our network to classify data into multiple classes, then we can make our last layer a softmax layer; if we want it to classify into two classes, then our last layer can be a logistic regression layer; and if we want the network to predict numerical values, then we can make its last layer a regression layer. We will see examples of these in Chapter 5.

Other Popular Machine Learning Techniques and Ensembles of Techniques

After regression and logistic regression, it is important to branch out into the machine learning community and learn the ideas behind some of the most popular techniques for classification and regression tasks. *Support vector machines, decision trees*, and *random forests* are very powerful and popular, and are able to perform both classification and regression tasks. The natural question is then, when do we use a specific machine learning method, including linear and logistic regression, and later neural networks? How do we know which method to use and base our conclusions and predictions on? These are the types of questions where the mathematical analysis of the machine learning models helps.

Since the mathematical analysis of each method, including the types of data sets it is usually best suited for, is only now gaining serious attention after the recent increase in resource allocation for research in AI, machine learning, and data science, the current practice is to try out each method on the same data set and use the one with the best results. That is, assuming we have the required computational and time resources to try out different machine learning techniques. Even better, if you do have the time and resources to train various machine learning models (parallel computing is perfect here), then it is good to utilize *ensemble methods*. These combine the results of different machine learning models, either by averaging or by voting, ironically, yet mathematically sound, giving better results than the best individual performers, *and even when the best performers are weak performers!*

One example of an ensemble is a random forest: it is an ensemble of decision trees.

When basing our predictions on ensembles, industry terms like *bagging* (or *bootstrap aggregating*), *pasting*, *boosting* (such as *ADA boost* and *gradient boosting*), *stacking*, and *random patches* appear. Bagging and pasting train *the same* machine learning model on different random subsets of the training set. Bagging samples instances from the training set with replacement, and pasting samples instances from the training set without replacement. *Random patches* sample from the feature space as well, training a machine learning model on a random subset of the features at a time. This is very helpful when the data set has many, many features, such as images (where each pixel is a feature). *Stacking* learns the prediction mechanism of the ensemble instead of simple voting or averaging.

Support Vector Machines

Support vector machine is an extremely popular machine learning method that's able to perform classification and regression tasks with both linear (flat) and nonlinear (curved) decision boundaries.

For classification, this method seeks to separate the labeled data using a widest possible margin, resulting in an optimal *highway* of separation as opposed to a thin line of separation. Let's explain how support vector machines classify labeled data instances in the context of this chapter's structure of training function, loss function, and optimization.

Training function

Once again we linearly combine the features of a data point with unknown weights ω's and add bias ω_0. We then pass the answer through the *sign* function. If the linear combination of features plus bias is a positive number, return 1 (or classify in the first class), and if it is negative, return –1 (or classify in the other). So the formula for the training function becomes:

$$f\left(\vec{\omega}; \vec{x}\right) = sign\left(\vec{\omega}^{t}\vec{x} + \omega_0\right)$$

Loss function

We must design a loss function that penalizes misclassified points. For logistic regression, we used the cross-entropy loss function. For support vector machines, our loss function is based on a function called the *hinge loss function*:

$$\max\left(0, 1 - y_{true}\left(\vec{\omega}^{t}\vec{x} + \omega_0\right)\right)$$

Let's see how the hinge loss function penalizes errors in classification. First, recall that y_{true} is either 1 or –1, depending on whether the data point belongs in the positive or the negative class.

- If for a certain data point y_{true} is 1 but $\vec{\omega}^{t}\vec{x} + \omega_0 < 0$, the training function will misclassify it and give us $y_{predict} = -1$, and the hinge loss function's value will be $1 - (1)\left(\vec{\omega}^{t}\vec{x} + \omega_0\right) > 1$, which is a high penalty when your goal is to minimize.

- If, on the other hand, y_{true} is 1 and $\vec{\omega}^{t}\vec{x} + \omega_0 > 0$, the training function will correctly classify it and give us $y_{predict} = 1$. The hinge loss function, however, is designed in such a way that it would still penalize us if $\vec{\omega}^{t}\vec{x} + \omega_0 < 1$, and its value will be $1 - (1)\left(\vec{\omega}^{t}\vec{x} + \omega_0\right)$, which is now less than 1 but still bigger than 0.

- Only when y_{true} is 1 and $\vec{\omega}^{t}\vec{x} + \omega_0 > 1$ (the training function will still correctly classify this point and give $y_{predict} = 1$) will the hinge loss function value be 0, since it will be the maximum between 0 and a negative quantity.

- The same logic applies when y_{true} is −1. The hinge loss function will penalize a lot for a wrong prediction, and a little for a right prediction when it doesn't have far enough *margin* from the *zero divider* (a margin bigger than 1). The hinge loss function will return 0 only when the prediction is right and the point is at a distance larger than 1 from the 0 divider.

- Note that the 0 divider has the equation $\vec{w}^t\vec{x} + w_0 = 0$, and the margin edges have equations $\vec{w}^t\vec{x} + w_0 = -1$ and $\vec{w}^t\vec{x} + w_0 = 1$. The distance between the margin edges is easy to calculate as $\frac{2}{\|w\|_2}$. So if we want to increase this margin width, we have to decrease $\|w\|_2$; thus, this term must enter the loss function, along with the hinge loss function, which penalizes both misclassified points and points within the margin boundaries.

Now if we average the hinge loss over all the m data points in the training set, and add $\|w\|_2^2$, we obtain the formula for the loss function that is commonly used for support vector machines:

$$L(\vec{w}) = \frac{1}{m}\Sigma_{i=1}^m \max\left(0,1 - y_{true}^i\left(\vec{w}^t\vec{x}^i + w_0\right)\right) + \lambda\|w\|_2^2$$

Optimization

Our goal now is to search for the $\vec{w}$ that minimizes the loss function. Let's observe this loss function for a minute:

- It has two terms: $\frac{1}{m}\Sigma_{i=1}^m \max\left(0,1 - y_{true}^i\left(\vec{w}^t\vec{x}^i + w_0\right)\right)$ and $\lambda\|\vec{w}\|_2^2$. Whenever we have more than one term in an optimization problem, it is most likely that they are competing terms, in the sense that the same w values that make the first term small and thus happy might make the second term big and thus sad. So it is a push-and-pull game between the two terms as we search for the $\vec{w}$ that optimizes their sum.

- The λ that appears with the $\lambda\|\vec{w}\|_2^2$ term is an example of a model hyperparameter that we can tune during the validation stage of the training process. Note that controlling the value of λ helps us control the width of the margin this way: if we choose a large λ value, the optimizer will get busy choosing $\vec{w}$ with very low $\|w\|_2^2$, to compensate for that large λ, and the first term of the loss function will get less attention. But recall that a smaller $\|w\|_2$ means a larger margin!

- The $\lambda\|\vec{w}\|_2^2$ term can also be thought of as a *regularization term*, which we will discuss in Chapter 4.

- This loss function is convex and bounded below by 0, so its minimization problem is not too bad: we don't have to worry about getting stuck at local minima. The first term has a singularity, but as mentioned before, we can define its subgradient at the singular point, then apply a descent method.

Some optimization problems can be reformulated, and instead of solving the original *primal* problem, we end up solving its *dual* problem! Usually, one is easier to solve than the other. We can think of the dual problem as another optimization problem living in a parallel universe of the primal problem. The universes meet at the optimizer. Hence solving one problem automatically gives the solution of the other. We study duality when we study optimization. Of particular interest and huge application are linear and quadratic optimization, also known as linear and quadratic programming. The minimization problem that we currently have:

$$\min_{\vec{\omega}} \frac{1}{m} \Sigma_{i=1}^{m} \max\left(0, 1 - y_{true}^{i}\left(\vec{\omega}^{t}\vec{x}^{i} + \omega_0\right)\right) + \lambda \parallel \omega \parallel_{2}^{2}$$

is an example of quadratic programming, and it has a dual problem formulation that turns out to be easier to optimize than the primal (especially when the number of features is high):

$$\max_{\vec{\alpha}} \Sigma_{j=1}^{m} \alpha_j - \frac{1}{2} \Sigma_{j=1}^{m} \Sigma_{k=1}^{m} \alpha_j \alpha_k y_{true}^{j} y_{true}^{k} \left(\left(\vec{x}^{j}\right)^{t}\vec{x}^{k}\right)$$

subject to the constraints $\alpha_j \geq 0$ and $\Sigma_{j=1}^{m} \alpha_j y_{true}^{j} = 0$. Writing that formula is usually straightforward when we learn about primal and dual problems, so we skip the derivation in favor of not interrupting our flow.

Quadratic programming is a very well-developed field, and there are many software packages that can solve this problem. Once we find the maximizing $\vec{\alpha}$, we can find the vector $\vec{\omega}$ that minimizes the primal problem using $\vec{\omega} = \Sigma_{j=1}^{m} \alpha_j y_{true}^{i} \vec{x}^{j}$. Once we have our $\vec{\omega}$, we can classify new data points using our now trained function:

$$f\left(\vec{x}_{new}\right) = sign\left(\vec{\omega}^{t}\vec{x}_{new} + \omega_0\right)$$

$$= sign\left(\Sigma_{j} \alpha_j y^{i}\left(\vec{x}^{j}\right)^{t}\vec{x}_{new} + \omega_0\right)$$

If you want to avoid quadratic programming, there is another very fast method called *coordinate descent* that solves the dual problem and works very well with large data sets with a high number of features.

The kernel trick

We can now transition the same ideas to nonlinear classification. Let's first observe this important note about the dual problem: the data points appear only in pairs, more specifically, only in a scalar product, namely, $\left(\overrightarrow{x}^{j}\right)^{t}\overrightarrow{x}^{k}$. Similarly, they only appear as a scalar product in the trained function. This simple observation allows for magic:

- If we find a function $K\left(\overrightarrow{x}^{j},\overrightarrow{x}^{k}\right)$ that acts on pairs of data points, and it happens to output the scalar product of *a transformation* of the data points into some higher dimensional space, namely, $K\left(\overrightarrow{x}^{j},\overrightarrow{x}^{k}\right)=\overrightarrow{\phi\left(\overrightarrow{x}^{j}\right)}^{t}\phi\left(\overrightarrow{x}^{k}\right)$ (we don't even have to know what the actual transformation $\overrightarrow{\phi}$ into higher dimensions is), then we can solve the same exact dual problem in higher dimensions by replacing the scalar product in the formula of the dual problem with $K\left(\overrightarrow{x}^{j},\overrightarrow{x}^{k}\right)$.

- The intuition here is that data that is nonlinearly separable in lower dimensions is almost always linearly separable in higher dimensions. So transform all the data points to higher dimensions, then separate. The kernel trick solves the linear classification problem in higher dimensions *without* transforming each point. The kernel itself evaluates the dot product of transformed data without transforming the data. Pretty cool stuff.

Examples of kernel functions include (note that the transformation $\overrightarrow{\phi}$ is nowhere to be seen):

- $K\left(\overrightarrow{x}^{j},\overrightarrow{x}^{k}\right)=\left(\left(\overrightarrow{x}^{j}\right)^{t}\overrightarrow{x}^{k}\right)^{2}$

- The polynomial kernel: $K\left(\overrightarrow{x}^{j},\overrightarrow{x}^{k}\right)=\left(1+\left(\overrightarrow{x}^{j}\right)^{t}\overrightarrow{x}^{k}\right)^{d}$

- The Gaussian kernel: $K\left(\overrightarrow{x}^{j},\overrightarrow{x}^{k}\right)=e^{-\gamma|x_{j}-x_{k}|^{2}}$

Decision Trees

Staying with our driving theme for this chapter that everything is a function, a decision tree, in essence, is a function that takes Boolean variables as an input (these are variables that can only assume *true* [or 1] or *false* [or 0] values) such as: is the feature > 5, is the feature = sunny, is the feature = man, etc. It outputs a *decision* such as: approve the loan, classify as covid19, return 25, etc. Instead of adding or multiplying boolean variables, we use the logical *or*, *and*, and *not* operators.

But what if our features are not given in the original data set as Boolean variables? Then we must transform them to Boolean variables before feeding them into the model to make predictions. For example, the decision tree in Figure 3-17 was trained on the Fish Market data set. It is a regression tree. The tree takes raw data, but the function representing the tree actually operates on new variables, which are the original data features transformed into Boolean variables:

1. $a1 = (\text{Width} \leq 5.117)$

2. $a2 = (\text{Length3} \leq 59.55)$

3. $a3 = (\text{Length3} \leq 41.1)$

4. $a4 = (\text{Length3} \leq 34.9)$

5. $a5 = (\text{Length3} \leq 27.95)$

6. $a6 = (\text{Length3} \leq 21.25)$

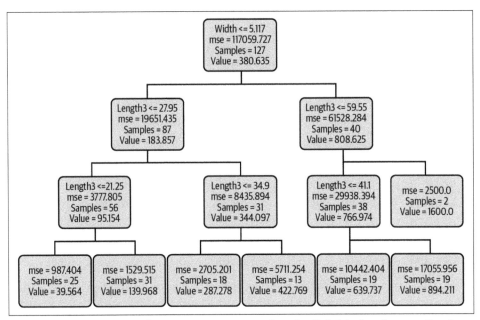

Figure 3-17. A regression decision tree constructed on the Fish Market data set. See the accompanying Jupyter notebook for details.

Now the function representing the decision tree in Figure 3-17 is:

$$f(a1,a2,a3,a4,a5,a6) = (a1 \text{ and } a5 \text{ and } a6) \times 39.584 + (a1 \text{ and } a5 \text{ and not } a6)$$
$$\times 139.968 + (a1 \text{ and not } a5 \text{ and } a4) \times 287.278 + (a1 \text{ and not } a5 \text{ and not } a4)$$
$$\times 422.769 + (\text{not } a1 \text{ and } a2 \text{ and } a3) \times 639.737 + (\text{not } a1 \text{ and } a2 \text{ and not } a3)$$
$$\times 824.211 + (\text{not } a1 \text{ and not } a2) \times 1600$$

Note that unlike the training functions that we've encountered in this chapter so far, this function has no parameters w's that we need to solve for. This is called a *nonparametric model*, and it doesn't fix the shape of the function ahead of time. This gives it the flexibility to *grow* with the data, or in other words, adapt to the data. Of course, with this high adaptability to the data comes the high risk of overfitting the data. Thankfully there are ways around this, some of which we list here without any elaboration: pruning the tree after growing it, restricting the number of layers, setting a minimum number of data instances per node, or using an ensemble of trees instead of one tree, called a *random forest*, discussed later.

One very important observation: the decision tree *decided* to split over only two features of the original data set, namely the Width and Length3 features. Decision trees are designed to keep the more important features (those providing the most information that contribute to our prediction) closer to the root. Therefore, decision trees can help in feature selection, where we select the most important features to contribute to our final model's predictions.

It is no wonder that the Width and Length3 features ended up being the most important for predicting the weight of the fish. The correlation matrix in Figure 3-18 and the scatterplots in Figure 3-3 show extremely strong correlation between all the length features. This means that the information they provide is redundant, and including all of them in our prediction models will increase computation costs and lower performance.

	Weight	Length1	Length2	Length3	Height	Width
Weight	1.000000	0.908678	0.911888	0.917883	0.747700	0.896036
Length1	0.908678	1.000000	0.999493	0.991731	0.637844	0.870414
Length2	0.911888	0.999493	1.000000	0.993869	0.653291	0.877268
Length3	0.917883	0.991731	0.993869	1.000000	0.716450	0.882716
Height	0.747700	0.637844	0.653291	0.716450	1.000000	0.802115
Width	0.896036	0.870414	0.877268	0.882716	0.802115	1.000000

Figure 3-18. Correlation matrix for the Fish Market data set. There is an extremely strong correlation between all the length features.

Feature Selection

We just introduced the very important topic of feature selection. Real-world data sets come with many features, and some of them may provide redundant information, while others are not important at all for predicting our target label. Including irrelevant and redundant features in a machine learning model increases computational cost and lowers its performance. We just saw that decision trees are one way to help select the important features. Another way is a regularization technique called lasso regression, which we will introduce in Chapter 4. There are statistical tests that test for feature dependencies on each other. The *F-test* tests for linear dependencies (this gives higher scores for correlated features, but correlations alone are deceptive), and *mutual information* tests for nonlinear dependencies. These provide a measure of how much a feature contributes to determining the target label, and hence aid in feature selection by keeping the most promising features. We can also test for feature dependencies on each other, as well as their correlations and scatterplots. *Variance* thresholding removes features with little to no variance, on the premise that if a feature does not vary much within itself, it has little predictive power.

How do we train a decision tree on a data set? What function do we optimize? There are two functions that are usually optimized when *growing* decision trees: the entropy and the Gini impurity. Using one or the other doesn't make much difference in the resultant trees. We develop these next.

Entropy and Gini impurity

Here, we decide to split a node of the tree on the feature that is evaluated as the *most important*. Entropy and Gini impurity are two popular ways to measure importance of a feature. They are not mathematically equivalent, but they both work and provide reasonable decision trees. Gini impurity is usually less expensive to compute, so it is the default in software packages, but you have the option to change from the default setting to entropy. Using Gini impurity tends to produce less-balanced trees when there are classes with much higher frequency than others. These classes end up isolated in their own branches. However, in many cases, using either entropy or Gini impurity does not provide much difference in the resulting decision trees.

With the *entropy* approach, we look for the feature split that provides the *maximal information gain* (we'll give its formula shortly). Information gain is borrowed from information theory, and it has to do with the concept of entropy. Entropy, in turn, is borrowed from thermodynamics and statistical physics, and it quantifies the amount of disorder in a certain system.

With the *Gini impurity* approach, we look for the feature split that provides children nodes with lowest average Gini impurity (we'll also give its formula shortly).

To maximize information gain (or minimize Gini impurity), the algorithm that grows a decision tree has to go over each feature of the training data subset and calculate the information gain (or Gini impurity), accomplished whether the tree uses that particular feature as a node to split on, then choose the feature that provides the highest information gain (or children nodes with lowest average Gini impurity). Moreover, if the feature has real numerical values, the algorithm has to decide on *what question to ask at the node*, meaning what feature value to split on; for example, is $x_5 < 0.1$? The algorithm has to do this sequentially at each layer of the tree, calculating information gain (or Gini impurities) over the features of the data instances in each node, and sometimes on each split value possibility. This is easier to understand with examples. But first, we write the formulas for the entropy, information gain, and Gini impurity.

Entropy and information gain

The easiest way to understand the entropy formula is to rely on the intuition that if an event is highly probable, then there is little surprise associated with it happening. So when $p(event)$ is large, its surprise is low. We can mathematically encode this with a function that decreases when its probability increases. The calculus function $\log \frac{1}{x}$ works and has the additional property that the surprises of independent events add up. Therefore, we can define:

$$Surprise(event) = \log \frac{1}{p(event)} = - \log (p(event))$$

Now the entropy of a random variable (which in our case is a particular feature in our training data set) is defined as the *expected surprise* associated with the random variable, so we must add up the surprises of each possible outcome of the random variable (surprise of each value of the feature in question) multiplied by their respective probabilities, obtaining:

$$Entropy(X) = - p(outcome_1) \log (p(outcome_1)) - p(outcome_2) \log (p(outcome_2)) - \cdots - p(outcome_n) \log (p(outcome_n))$$

The entropy for one feature of our training data that assumes a bunch of values is:

$$Entropy(Feature) = - p(value_1) \log (p(value_1)) - p(value_2) \log (p(value_2)) - \cdots - p(value_n) \log (p(value_n))$$

Since our goal is to select to split on a feature that provides large information gain about the outcome (the label or the target feature), let's first calculate the entropy of the outcome feature.

Binary output. Assume for simplicity that this is a binary classification problem, so the outcome feature only has two values: positive (in the class) and negative (not in the class).

If we let p be the number of positive instances in the target feature and n be the number of negative ones, then $p + n = m$ will be the number of instances in the training data subset. Now the probability to select a positive instance from that target column will be $\frac{p}{m} = \frac{p}{p+n}$, and the probability to select a negative instance is similarly $\frac{n}{m} = \frac{n}{p+n}$.

Thus, the entropy of the outcome feature (without leveraging any information from the other features) is:

$$Entropy(\text{Outcome Feature})$$
$$= -p(positive) \log\left(p(positive)\right) - p(negative) \log\left(p(negative)\right)$$
$$= -\frac{p}{p+n} \log\left(\frac{p}{p+n}\right) - \frac{n}{p+n} \log\left(\frac{n}{p+n}\right)$$

Next, we leverage information from one other feature and calculate the difference in entropy of the outcome feature, which we expect to decrease as we gain more information (more information generally results in less surprise).

Suppose we choose Feature A to split a node of our decision tree on. Suppose Feature A assumes four values, and has k_1 instances with $value_1$, of these p_1 are labeled positive as their target outcome, and n_1 are labeled negative as their target outcome, so $p_1 + n_1 = k_1$. Similarly, Feature A has k_2 instances with $value_2$, of these p_2 are labeled positive as their target outcome, and n_2 are labeled negative as their target outcome, so $p_2 + n_2 = k_2$. The same applies for $value_3$ and $value_4$ of Feature A. Note that $k_1 + k_2 + k_3 + k_4 = m$, the total number of instances in the training subset of the data set.

We can think of each value $value_k$ of Feature A as a random variable in its own respect, with p_k positive outcomes and n_k negative outcomes, so we can calculate its entropy (expected surprise):

$$\text{Entropy}(value_1) = -\frac{p_1}{p_1 + n_1} \log\left(\frac{p_1}{p_1 + n_1}\right) - \frac{n_1}{p_1 + n_1} \log\left(\frac{n_1}{p_1 + n_1}\right)$$

$$\text{Entropy}(value_2) = -\frac{p_2}{p_2 + n_2} \log\left(\frac{p_2}{p_2 + n_2}\right) - \frac{n_2}{p_2 + n_2} \log\left(\frac{n_2}{p_2 + n_2}\right)$$

$$\text{Entropy}(value_3) = -\frac{p_3}{p_3 + n_3} \log\left(\frac{p_3}{p_3 + n_3}\right) - \frac{n_3}{p_3 + n_3} \log\left(\frac{n_3}{p_3 + n_3}\right)$$

$$\text{Entropy}(value_4) = -\frac{p_4}{p_4 + n_4} \log\left(\frac{p_4}{p_4 + n_4}\right) - \frac{n_4}{p_4 + n_4} \log\left(\frac{n_4}{p_4 + n_4}\right)$$

Now that we have this information, we can calculate the *expected entropy* after splitting on Feature A, by adding the four entropies just mentioned, each multiplied by its respective probability: $p(value_1) = \frac{k_1}{m}$, $p(value_2) = \frac{k_2}{m}$, $p(value_3) = \frac{k_3}{m}$, and $p(value_4) = \frac{k_4}{m}$.

Therefore, the expected entropy after splitting on Feature A would be:

Expected Entropy(Feature A)

$= p(value_1)\text{Entropy}(value_1) + p(value_2)\text{Entropy}(value_2)$

$+ p(value_3)\text{Entropy}(value_3) + p(value_4)\text{Entropy}(value_4)$

$= \frac{k_1}{m}\text{Entropy}(value_1) + \frac{k_2}{m}\text{Entropy}(value_2) + \frac{k_3}{m}\text{Entropy}(value_3) + \frac{k_4}{m}\text{Entropy}(value_4)$

So what would be the information gained from using Feature A to split on? It would be the difference between the entropy of the outcome feature without any information from Feature A and the expected entropy of Feature A. That is, we have a formula for *information gain* given that we decide to split on Feature A:

Information Gain=

$Entropy$(Outcome Feature) – Expected Entropy(Feature A) =

$-\frac{p}{p + n} \log\left(\frac{p}{p + n}\right) - \frac{n}{p + n} \log\left(\frac{n}{p + n}\right)$ – Expected Entropy(Feature A)

Now it is easy to go through each feature of the training data subset and calculate the information gain resulting from using that feature to split on. Ultimately, the decision tree algorithm decides to split on the feature with highest information gain. The algorithm does this recursively for each node and at each layer of the tree, until it runs out of features to split on or data instances. This is how we obtain our entropy-based decision tree.

Multi-class output.   It is not too difficult to generalize this logic to the case where we have a multiclass output, for example, a classification problem with three or more target labels. The classical Iris data set (*https://oreil.ly/LZ1V9*) from the UCI Machine Learning Repository (*https://oreil.ly/iOnAc*) is a great example with three target labels. This data set has four features for a given Iris flower: its sepal length and width, and its petal length and width. Note that each of these features is a continuous random variable, not discrete. So we have to devise a test to split on the values of each feature, *before* applying the previously mentioned logic. This is part of the feature engineering stage of a data science project. The engineering step here is: transform a continuous valued feature into a Boolean feature; for example, is the petal length > 2.45? We will not go over how to choose the number 2.45, but by now you probably can guess that there is an optimization process that should go on here as well.

Gini impurity

Each decision tree is characterized by its nodes, branches, and leaves. A node is considered *pure* if it only contains data instances from the training data subset that have the same target label (this means they belong to the same class). Note that a pure node is a desired node, since we know its class. Therefore, an algorithm would want to grow a tree in a way that minimizes the impurity of the nodes: if the data instances in a node do not all belong in the same class, then the node is *impure*. Gini impurity quantifies this impurity the following way.

Suppose that our classification problem has three classes, like the Iris data set (*https://oreil.ly/Gvr3J*). Suppose also that a certain node in a decision tree grown to fit this data set has n training instances, with n_1 of these belonging in the first class, n_2 in the second class, and n_3 in the third class (so $n_1 + n_2 + n_3 = n$). Then the Gini impurity of this node is given by:

$$\text{Gini impurity} = 1 - \left(\frac{n_1}{n}\right)^2 - \left(\frac{n_2}{n}\right)^2 - \left(\frac{n_3}{n}\right)^2$$

So for each node, the fraction of the data instances belonging to each class is calculated, squared, then the sum of those is subtracted from 1. Note that if all the data instances of a node belong in the same class, then that formula gives a Gini impurity equal to 0.

The decision tree growing algorithm now looks for the feature and split point in each feature that produce children nodes with the lowest Gini impurity, on average. This means the children of a node must on average be purer than the parent node. Thus, the algorithm tries to minimize a weighted average of the Gini impurities of two of the children nodes (of a binary tree). Each child's Gini impurity is weighted by its relative size, which is the ratio between its number of instances relative to the total number of instances in that tree layer (which is the same as the number of instances as its parent's). Thus, we end up having to search for the feature and the split point (for each feature) combination that solve the following minimization problem:

$$\min_{Feature, FeatureSplitValue} \frac{n_{left}}{n} Gini(\text{Left Node}) + \frac{n_{right}}{n} Gini(\text{Right Node})$$

where n_{left} and n_{right} are the number of data instances that end up being in the left and right children nodes, and n is the number of data instances that are in the parent node (note that n_{left} and n_{right} must add up to n).

Regression decision trees

It is important to point out that decision trees can be used for both regression and classification. A regression decision tree returns a predicted value rather than a class, but a similar process to a classification tree applies.

Instead of splitting a node by selecting a feature and a feature value (for example, is height > 3 feet?) that maximize information gain or minimize Gini impurity, we select a feature and a feature value that minimize a mean squared distance between the true labels and the average of the labels of all the instances in each of the left and right children nodes. That is, the algorithm chooses a feature and feature value to split on, then looks at the left and right children nodes resulting from that split, and calculates:

- The average value of all the labels of the training data instances in the left node. This average will be the left node value y_{left}, and is the value predicted by the decision tree if this node ends up being a leaf node.

- The average value of all the labels of the training data instances in the right node. This average will be the right node value y_{right}. Similarly, this is the value predicted by the decision tree if this node ends up being a leaf node.

- The sum of the squared distance between the left node value and the true label of each instance in the left node $\Sigma_{LeftNodeInstances} \left| y^i_{true} - y_{left} \right|^2$.

- The sum of the squared distance between the right node value and the true label of each instance in the right node $\Sigma_{RightNodeInstances} \left| y^i_{true} - y_{right} \right|^2$.

- A weighted average of the just-mentioned two sums, where each node is weighted by its size relative to the parent node, just like we did for the Gini impurity:

$$\frac{n_{left}}{n} \sum_{LeftNodeInstances} \left| y^i_{true} - y_{left} \right|^2 +$$

$$\frac{n_{right}}{n} \sum_{RightNodeInstances} \left| y^i_{true} - y_{right} \right|^2$$

That algorithm is *greedy* and computation-heavy, in the sense that it has to do this for *each feature and each possible feature split value*, then choose the feature and feature split that provide the smallest weighted squared error average between the left and right children nodes.

The famous CART (Classification and Regression Tree) algorithm is used by software packages, including Python's scikit-learn, which we use in the Jupyter notebooks supplementing this book. This algorithm produces trees with nodes that only have two children (binary trees), where the test at each node only has Yes or No answers. Other algorithms such as ID3 can produce trees with nodes that have two or more children.

Shortcomings of decision trees

Decision trees are very easy to interpret and are popular for many good reasons: they adapt to large data sets, different data types (discrete and continuous features, no scaling of data needed), and can perform both regression and classification tasks. However, they can be unstable, in the sense that adding just one instance to the data set can change the tree at its root and hence result in a very different decision tree.

They are also sensitive to rotations in the data, since their decision boundaries are usually horizontal and vertical (not slanted like support vector machines). This is because splits usually happen at specific feature values, so the decision boundaries end up parallel to the feature axes. One fix is to transform the data set to match its *principal axes*, using the *singular value decomposition method* presented in Chapter 6. Decision trees tend to overfit the data, so they need pruning. This is usually done using statistical tests. The greedy algorithms involved in constructing the trees, where the search happens over all features and their values, makes them computationally expensive and less accurate. Random forests, discussed next, address some of these shortcomings.

Random Forests

When I first learned about decision trees, the most perplexing aspects for me were:

- How do we start the tree, meaning how do we decide which data feature is the root feature?
- At what particular feature value do we decide to split a node?

- When do we stop?
- In essence, how do we grow a tree?

(Note that we answered some of these questions in the previous subsection.) It didn't make matters any easier that I would surf the internet looking for answers, only to encounter declarations that decision trees are so easy to build and understand, so it felt like I was the only one deeply confused by decision trees.

My puzzlement instantly disappeared when I learned about *random forests*. The amazing thing about random forests is that we can get incredibly good regression or classification results *without* answering any of my bewildering questions. Randomizing the whole process means building many decision trees while answering all my questions with two words: *choose randomly*. Then aggregating the predictions in an ensemble produces very good results, even better than one carefully crafted decision tree. It has been said that *randomization often produces reliability*!

Another very useful property of random forests is that they give a measure of *feature importance*, helping us pinpoint which features significantly affect our predictions, and aid in feature selection as well.

k-means Clustering

One common goal of data analysts is to partition data into *clusters*, each cluster highlighting certain common traits. *k-means clustering* is a common machine learning method that partitions n data points (vectors) into k clusters, where each data point gets assigned to the cluster with the nearest mean. The mean of each cluster, or its centroid, serves as the prototype of the cluster. Overall, k-means clustering minimizes the variance (the squared Euclidean distances to the mean) within each cluster.

The most common algorithm for k-means clustering is iterative:

1. Start with an initial set of k means. This means that we specify the number of clusters ahead of time, which raises the question: How to initialize it? How to select the locations of the first k centroids? There is literature on that.

2. Assign each data point to the cluster with the nearest mean in terms of squared Euclidean distance.

3. Recalculate the means of each cluster.

The algorithm converges when the data point assignments to each cluster do not change anymore.

Performance Measures for Classification Models

It is relatively easy to develop mathematical models that compute things and produce outputs. It is a completely different story to develop models that perform well for our desired tasks. Furthermore, models that perform well according to some metrics behave badly according to some other metrics. We need extra care in developing performance metrics and deciding which ones to rely on, depending on our specific use cases.

Measuring the performance of models that predict numerical values, such as regression models, is easier than classification models, since we have many ways to compute distances between numbers (good predictions and bad predictions). On the other hand, when our task is classification (we can use models such as logistic regression, softmax regression, support vector machines, decision trees, random forests, or neural networks), we have to put some extra thought into evaluating performance. Moreover, there are usually trade-offs. For example, if our task is to classify YouTube videos as being safe for kids (positive) or not safe for kids (negative), do we tweak our model so as to reduce the number of false positives or false negatives? It is obviously more problematic if a video is classified as safe while in reality it is unsafe (false positive) than the other way around, so our performance metric needs to reflect that.

The following are the performance measures commonly used for classification models. Do not worry about memorizing their names, as the way they are named does not make logical sense. Instead, spend your time understanding their meanings:

Accuracy
Percentage of times the prediction model got the classification right:

$$Accuracy = \frac{\text{true positives + true negatives}}{\text{all predicted positives + all predicted negatives}}$$

Confusion matrix
Counting all true positives, false positives, true negatives, and false negatives:

True negative	False positive
False negative	True positive

Precision score
Accuracy of the positive predictions:

$$Precision = \frac{\text{true positives}}{\text{all predicted positives}} = \frac{\text{true positives}}{\text{true positives + false positives}}$$

Recall score

Ratio of the positive instances that are correctly classified:

$$Recall = \frac{true\ positives}{all\ positive\ labels} = \frac{true\ positives}{true\ positives\ +\ false\ negatives}$$

Specificity

Ratio of the negative instances that are correctly classified:

$$Specificity = \frac{true\ negatives}{all\ negative\ labels} = \frac{true\ negatives}{true\ negatives\ +\ false\ positives}$$

F_1 *score*

This quantity is only high when both precision and recall scores are high:

$$F_1 = \frac{2}{\frac{1}{precision} + \frac{1}{recall}}$$

AUC (area under the curve) and ROC (receiver operating characteristics) curves

These curves provide a performance measure for a classification model at various threshold values. We can use these curves to measure how well a certain variable predicts a certain outcome; for example, how well does the GRE subject test score predict passing a graduate school's qualifying exam in the first year?

Andrew Ng's book, *Machine Learning Yearning* (*https://oreil.ly/pnZF8*) (self-published), provides an excellent guide for performance metrics best practices. Please read carefully before diving into real AI applications, since the book's recipes are based on many trials, successes, and failures.

Summary and Looking Ahead

In this chapter, we surveyed some of the most popular machine learning models, emphasizing a particular mathematical structure that appears throughout the book: training function, loss function, and optimization. We discussed linear, logistic, and softmax regression, then breezed over support vector machines, decision trees, ensembles, random forests, and k-means clustering.

Moreover, we made a decent case for studying these topics from mathematics:

Calculus

The minimum and maximum happen at the boundary or at points where one derivative is zero or does not exist.

Linear algebra
- Linearly combining features: $\omega_1 x_1 + \omega_2 x_2 + \cdots + \omega_n x_n$.
- Writing various mathematical expressions using matrix and vector notation.
- The scalar product of two vectors $\vec{a}^t \vec{b}$.
- The l^2 norm of a vector.
- Avoid working with ill-conditioned matrices. Get rid of linearly dependent features. This also has to do with feature selection.
- Avoid multiplying matrices by each other; this is too expensive. Multiply matrices by vectors instead.

Optimization
- For convex functions, we do not worry about getting stuck at local minima since local minima are also global minima. We do worry about narrow valleys (see Chapter 4).
- Gradient descent methods need only one derivative (see Chapter 4).
- Newton's methods need two derivatives or an approximation of two derivatives (inconvenient for large data).
- Quadratic programming, the dual problem, and coordinate descent (all appear in support vector machines).

Statistics
- Correlation matrix and scatterplots
- The F-test and mutual information test for feature selection
- Standardizing the data features (subtracting the mean and dividing by the standard deviation)

More steps we did not and will not go over (yet):

- Validating our models: tune the weight values and the hyperparameters so as not to overfit.
- Test the trained model on the testing subset of the data, which our model had not used (or seen) during the training and validation steps (we do this in the accompanying Jupyter notebook on the book's GitHub page (*https://github.com/halanelson/Essential-Math-For-AI*)).
- Deploy and monitor the finalized model.
- Never stop thinking on how to improve our models and how to better integrate them into the whole production pipeline.

In Chapter 4, we step into the new and exciting era of neural networks.

Optimization for Neural Networks

I have lived each and every day of my life optimizing....My first aha moment was when I learned that our brain, too, learns a model of the world.

—H.

Various artificial neural networks have fully connected layers in their architecture. In this chapter, we explain how the mathematics of a fully connected neural network works. We design and experiment with various training and loss functions. We also explain that the optimization and backpropagation steps used when training neural networks are similar to how learning happens in our brains. The brain learns by reinforcing neuron connections when faced with a concept it has seen before, and weakening connections if it learns new information that contradicts previously learned concepts. Machines only understand numbers. Mathematically, stronger connections correspond to larger numbers, and weaker connections correspond to smaller numbers.

Finally, we walk through various regularization techniques, explaining their advantages, disadvantages, and use cases.

The Brain Cortex and Artificial Neural Networks

Neural networks are modeled after the brain cortex, which involves billions of neurons arranged in a layered structure. Figure 4-1 shows an image of three vertical cross-sections of the brain neocortex, and Figure 4-2 shows a diagram of a fully connected artificial neural network.

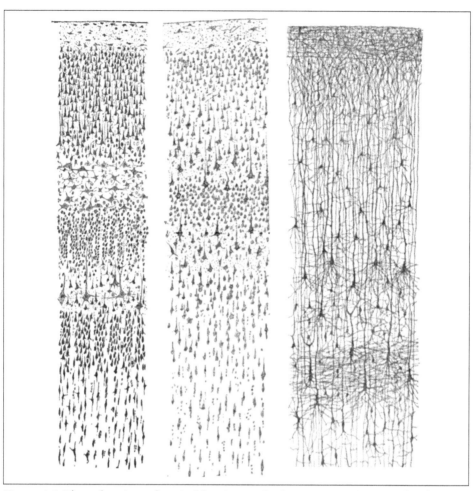

Figure 4-1. Three drawings of cortical lamination by Santiago Ramón y Cajal, taken from the book Comparative Study of the Sensory Areas of the Human Cortex *(Andesite Press) (image source: Wikipedia (https://oreil.ly/r5xJu))*

In Figure 4-1, each drawing shows a vertical cross-section of the cortex, with the surface (outermost side closest to the skull) of the cortex at the top. On the left is a Nissl-stained visual cortex of a human adult. In the middle is a Nissl-stained motor cortex of a human adult. On the right is a Golgi-stained cortex of a month-and-a half-old infant. The Nissl stain shows the cell bodies of neurons. The Golgi stain shows the dendrites and axons of a random subset of neurons. The layered structure of the neurons in the cortex is evident in all three cross-sections.

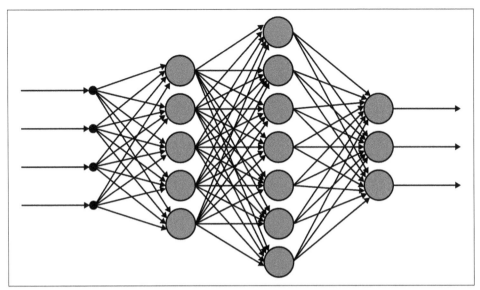

Figure 4-2. A fully connected or dense artificial neural network with four layers

Even though different regions of the cortex are responsible for different functions, such as vision, auditory perception, logical thinking, language, speech, etc., what actually determines the function of a specific region are its *connections*: which sensory and motor skills input and output regions it connects to. This means that if a cortical region is connected to a different sensory input/output region—for example, a vision locality instead of an auditory one—then it will perform vision tasks (computations), not auditory tasks. In a very simplified sense, the cortex performs one basic function at the neuron level. In an artificial neural network, the basic computation unit is *the perceptron*, and it functions in the same way across the whole network. The various connections, layers, and architecture of the neural network (both the brain cortex and artificial neural networks) are what allow these computational structures to do very impressive things.

Training Function: Fully Connected, or Dense, Feed Forward Neural Networks

In a *fully connected* or *dense* artificial neural network (see Figure 4-2), every neuron, represented by a node (the circles), in every layer is connected to all the neurons in the next layer. The first layer is the input layer, the last layer is the output layer, and the intermediate layers are called *hidden layers*. The neural network itself, whether fully connected or not (the networks that we will encounter in the new few chapters are *convolutional* and are not fully connected), is a computational graph representing the formula of the training function. Recall that we use this function to make predictions after training.

Training in the neural networks context means finding the parameter values, or weights, that enter into the formula of the training function via minimizing a loss function. This is similar to training linear regression, logistic regression, softmax regression, and support vector machine models, which we discussed in Chapter 3. The mathematical structure here remains the same:

1. Training function
2. Loss function
3. Optimization

The only difference is that for the simple models of Chapter 3, the formulas of the training functions are very uncomplicated. They linearly combine the data features, add a bias term (ω_0), and pass the result into at most one nonlinear function (for example, the logistic function in logistic regression). As a consequence, the results of these models are also simple: a linear (flat) function for linear regression, and a linear division boundary between different classes in logistic regression, softmax regression, and support vector machines. Even when we use these simple models to represent nonlinear data, such as in the cases of polynomial regression (fitting the data into polynomial functions of the features) or support vector machines with the kernel trick, we still end up with linear functions or division boundaries, but these will either be in higher dimensions (for polynomial regression, the dimensions are the feature and its powers) or in transformed dimensions (such as when we use the kernel trick with support vector machines).

For neural network models, on the other hand, the process of linearly combining the features, adding a bias term, then passing the result through a nonlinear function (now called *activation function*) is the computation that happens *only in one neuron*. This simple process happens over and over again in dozens, hundreds, thousands, or sometimes millions of neurons, arranged in layers, where the output of one layer acts as the input of the next layer. Similar to the brain cortex, the aggregation of simple and similar processes over many neurons and layers produces, or allows for, the representation of, much more complex functionalities. This is sort of miraculous. Thankfully, we are able to understand much more about artificial neural networks than our brain's neural networks, mainly because we design them, and after all, an artificial neural network is just one mathematical function. No black box remains dark once we dissect it under the lens of mathematics. That said, the mathematical analysis of artificial neural networks is a relatively new field. There are still many questions to be answered and a lot to be discovered.

A Neural Network Is a Computational Graph Representation of the Training Function

Even for a network with only five neurons, such as the one in Figure 4-3, it is pretty messy to write the formula of the training function. This justifies the use of computational graphs to represent neural networks in an organized and easy way. Graphs are characterized by two things: nodes and edges (congratulations, this was lesson one in graph theory). In a neural network, an edge connecting node i in layer m to node j in layer n is assigned a weight $\omega_{mn,ij}$. That is four indices for only one edge! At the risk of drowning in a deep ocean of indices, we must organize a neural network's weights in matrices.

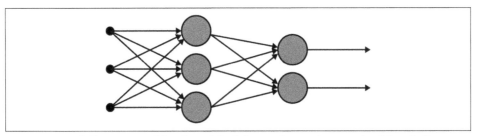

Figure 4-3. A fully connected (or dense) feed forward neural network with only five neurons arranged in three layers. The first layer (the three black dots on the very left) is the input layer, the second layer is the only hidden layer with three neurons, and the last layer is the output layer with two neurons.

Let's model the training function of a *feed forward* fully connected neural network. Feed forward means that the information flows forward through the computational graph representing the network's training function.

Linearly Combine, Add Bias, Then Activate

What kind of computations happen within a neuron when it receives input from other neurons? Linearly combine the input information using different weights, add a bias term, then use a nonlinear function to *activate* the neuron. We will go through this process one step at a time.

The weights

Let the matrix W^1 contain the weights of the edges *incident to* hidden layer 1, W^2 contain the weights of the edges incident to hidden layer 2, and so on, until we reach the output layer.

So for the small neural network represented in Figure 4-3, we only have $h = 1$ hidden layer, obtaining two matrices of weights:

$$W^1 = \begin{pmatrix} \omega_{11}^1 & \omega_{12}^1 & \omega_{13}^1 \\ \omega_{21}^1 & \omega_{22}^1 & \omega_{23}^1 \\ \omega_{31}^1 & \omega_{32}^1 & \omega_{33}^1 \end{pmatrix} \text{ and } W^{h+1} = W^2 = W^{output} = \begin{pmatrix} \omega_{11}^2 & \omega_{12}^2 & \omega_{13}^2 \\ \omega_{21}^2 & \omega_{22}^2 & \omega_{23}^2 \end{pmatrix},$$

where the superscripts indicate the layer to which the edges point. Note that if we only had one node at the output layer instead of two, then the last matrix of weights $W^{h+1} = W^{output}$ would only be a row vector:

$$W^{h+1} = W^2 = W^{output} = \begin{pmatrix} \omega_{11}^2 & \omega_{12}^2 & \omega_{13}^2 \end{pmatrix}$$

Now at one node of this neural network, two computations take place:

1. A linear combination plus bias
2. Passing the result through a nonlinear activation function (the composition operation from calculus)

We elaborate on these two, then ultimately construct the training function of the fully connected feed forward neural network represented in Figure 4-3.

A linear combination plus bias

At the first node in the first hidden layer (the only hidden layer for this small network), we linearly combine the inputs:

$$z_1^1 = \omega_{11}^1 x_1 + \omega_{12}^1 x_2 + \omega_{13}^1 x_3 + \omega_{01}^1$$

At the second node in the first hidden layer, we linearly combine the inputs using different weights than the previous linear combination:

$$z_2^1 = \omega_{21}^1 x_1 + \omega_{22}^1 x_2 + \omega_{23}^1 x_3 + \omega_{02}^1$$

At the third node in the first hidden layer, we linearly combine the inputs using different weights than the previous two linear combinations:

$$z_3^1 = \omega_{31}^1 x_1 + \omega_{32}^1 x_2 + \omega_{33}^1 x_3 + \omega_{03}^1$$

Let's express the three equations above using vector and matrix notation. This will be extremely convenient for our optimization task later, and of course it will preserve our sanity:

$$\begin{pmatrix} z_1^1 \\ z_2^1 \\ z_3^1 \end{pmatrix} = \begin{pmatrix} \omega_{11}^1 \\ \omega_{21}^1 \\ \omega_{31}^1 \end{pmatrix} x_1 + \begin{pmatrix} \omega_{12}^1 \\ \omega_{22}^1 \\ \omega_{32}^1 \end{pmatrix} x_2 + \begin{pmatrix} \omega_{13}^1 \\ \omega_{23}^1 \\ \omega_{33}^1 \end{pmatrix} x_3 + \begin{pmatrix} \omega_{01}^1 \\ \omega_{02}^1 \\ \omega_{03}^1 \end{pmatrix} = \begin{pmatrix} \omega_{11}^1 & \omega_{12}^1 & \omega_{13}^1 \\ \omega_{21}^1 & \omega_{22}^1 & \omega_{23}^1 \\ \omega_{31}^1 & \omega_{32}^1 & \omega_{33}^1 \end{pmatrix} \begin{pmatrix} x_1 \\ x_2 \\ x_3 \end{pmatrix} + \begin{pmatrix} \omega_{01}^1 \\ \omega_{02}^1 \\ \omega_{03}^1 \end{pmatrix}$$

We can now summarize the above expression compactly as:

$$\vec{z}^1 = W^1 \vec{x} + \vec{\omega}_0^1$$

Pass the result through a nonlinear activation function

Linearly combining the features and adding bias are not enough to pick up on more complex information in the data, and neural networks would have never been successful without this crucial but very simple step: compose with a *nonlinear* function at each node of the hidden layers.

Linear Combination of a Linear Combination Is Still a Linear Combination

If we skip the step of composing with a nonlinear function, and pass the information from the first layer to the next layer using only linear combinations, then our network will not learn anything new in the next layer. It will not be able to pick up on more complex features from one layer to the next. The math of why this is the case is straightforward. Suppose for simplicity that we only have two input features, the first hidden layer has only two nodes, and the second one has two nodes as well. Then the outputs of the first hidden layer without a nonlinear activation function would be:

$$z_1^1 = \omega_{11}^1 x_1 + \omega_{12}^1 x_2 + \omega_{01}^1$$
$$z_2^1 = \omega_{21}^1 x_1 + \omega_{22}^1 x_2 + \omega_{02}^1$$

At the second hidden layer, these will be linearly combined again, so the output of the first node of this layer would be:

$$\begin{aligned} z_1^2 &= \omega_{11}^2 z_1^1 + \omega_{12}^2 z_2^1 + \omega_{01}^2 \\ &= \omega_{11}^2 \left(\omega_{11}^1 x_1 + \omega_{12}^1 x_2 + \omega_{01}^1 \right) + \omega_{21}^2 \left(\omega_{21}^1 x_1 + \omega_{22}^1 x_2 + \omega_{02}^1 \right) + \omega_{01}^2 \\ &= \left(\omega_{11}^2 \omega_{11}^1 + \omega_{21}^2 \omega_{21}^1 \right) x_1 + \left(\omega_{11}^2 \omega_{12}^1 + \omega_{21}^2 \omega_{22}^1 \right) x_2 + \left(\omega_{11}^2 \omega_{01}^1 + \omega_{21}^2 \omega_{02}^1 + \omega_{01}^2 \right) \\ &= \omega_1 x_1 + \omega_2 x_2 + \omega_3 \end{aligned}$$

This output is nothing but a simple linear combination of the original features plus bias. Hence, adding a layer without any nonlinear activation contributes nothing new. In other words, the training function would remain linear and would lack the ability to pick up on any nonlinear relationships in the data.

We are the ones who decide on the formula for the nonlinear activation function, and different nodes can have different activation functions, even though it is rare to do this in practice. Let f be this activation function, then the output of the first hidden layer will be:

$$\vec{s}^1 = \vec{f}\left(\vec{z}^1\right) = \vec{f}\left(W^1\vec{x} + \vec{w}_0^1\right)$$

It is now straightforward to see that if we had more hidden layers, their outputs will be *chained* with those of previous layers, making writing the training function a bit tedious:

$$\vec{s}^2 = \vec{f}\left(\vec{z}^2\right) = \vec{f}\left(W^2\vec{s}^1 + \vec{w}_0^2\right) = \vec{f}\left(W^2\left(\vec{f}\left(W^1\vec{x} + \vec{w}_0^1\right)\right) + \vec{w}_0^2\right)$$

$$\vec{s}^3 = \vec{f}\left(\vec{z}^3\right) = \vec{f}\left(W^3\vec{s}^2 + \vec{w}_0^3\right) = \vec{f}\left(W^3\left(\vec{f}\left(W^2\left(\vec{f}\left(W^1\vec{x} + \vec{w}_0^1\right)\right) + \vec{w}_0^2\right)\right) + \vec{w}_0^3\right)$$

This chaining goes on until we reach the output layer. What happens at this very last layer depends on the task of the network. If the goal is regression (predict one numerical value) or binary classification (classify into two classes), then we only have one output node (see Figure 4-4).

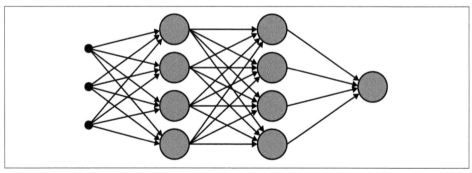

Figure 4-4. A fully connected (or dense) feed forward neural network with only nine neurons arranged in four layers. The first layer on the very left is the input layer, the second and third layers are the two hidden layers with four neurons each, and the last layer is the output layer with only one neuron (this network performs either a regression task or a binary classification task).

- If the task is regression, we linearly combine the outputs of the previous layer at the final output node, add bias, and go home (we *do not* pass the result through a nonlinear function in this case). Since the output layer only has one node, the output matrix is just a row vector $W^{output} = W^{h+1}$, and one bias ω_0^{h+1}. The prediction of the network will now be:

$$y_{predict} = W^{h+1} \overrightarrow{s^h} + \omega_0^{h+1}$$

where h is the total number of hidden layers in the network (this does not include the input and output layers).

- If, on the other hand, the task is binary classification, then again we have only one output node, where we linearly combine the outputs of the previous layer, add bias, then pass the result through the logistic function $\sigma(s) = \dfrac{1}{1+e^{-s}}$, resulting in the network's prediction:

$$y_{predict} = \sigma\left(W^{h+1} \overrightarrow{s^h} + \omega_0^{h+1} \right)$$

- If the task is to classify into multiple classes, say, five classes, then the output layer would include five nodes. At each of these nodes, we linearly combine the outputs of the previous layer, add bias, then pass the result through the softmax function:

$$\sigma\left(z^1\right) = \frac{e^{z^1}}{e^{z^1} + e^{z^2} + e^{z^3} + e^{z^4} + e^{z^5}}$$

$$\sigma\left(z^2\right) = \frac{e^{z^2}}{e^{z^1} + e^{z^2} + e^{z^3} + e^{z^4} + e^{z^5}}$$

$$\sigma\left(z^3\right) = \frac{e^{z^3}}{e^{z^1} + e^{z^2} + e^{z^3} + e^{z^4} + e^{z^5}}$$

$$\sigma\left(z^4\right) = \frac{e^{z^4}}{e^{z^1} + e^{z^2} + e^{z^3} + e^{z^4} + e^{z^5}}$$

$$\sigma\left(z^5\right) = \frac{e^{z^5}}{e^{z^1} + e^{z^2} + e^{z^3} + e^{z^4} + e^{z^5}}$$

Group those into a vector function $\vec{\sigma}$ that also takes vectors as input: $\vec{\sigma}\left(\vec{z}\right)$, then the final prediction of the neural network is a vector of five probability scores where a data instance belongs to each of the five classes:

$$\vec{y}_{predict} = \vec{\sigma}\left(\vec{z}\right)$$

$$= \vec{\sigma}\left(W^{output}\vec{s}^{h} + \vec{w_0}^{h+1}\right)$$

Notation overview

We will try to remain consistent with notation throughout our discussion of neural networks. The x's are the input features, the W's are the matrices or column vectors containing the weights that we use for linear combinations, the w_0's are the biases that are sometimes grouped into a vector, the z's are the results of linear combinations plus biases, and the s's are the results of passing the z's into the nonlinear activation functions.

Common Activation Functions

In theory, we can use any nonlinear function to *activate our nodes* (think of all the calculus functions we've ever encountered). In practice, there are some popular ones, listed next and graphed in Figure 4-5.

By far the Rectified Linear Unit function (ReLU) is the most commonly used in today's networks, and the success of AlexNet (*https://oreil.ly/c8RCQ*) in 2012 is partially attributed to the use of this activation function, as opposed to the hyperbolic tangent and logistic functions (sigmoid) that were commonly used in neural networks at the time (and are still in use).

The first four functions in the following list and in Figure 4-5 are all inspired by computational neuroscience, where they attempt to model a threshold for the activation (firing) of one neuron cell. Their graphs look similar to each other: some are smoother variants of others, some output only positive numbers, others output more balanced numbers between –1 and 1, or between $-\frac{\pi}{2}$ and $\frac{\pi}{2}$. They all *saturate* for small or large inputs, meaning their graphs become flat for inputs large in magnitude. This creates a problem for *learning*, since if these functions output the same numbers over and over again, there will not be much learning happening.

Mathematically, this phenomenon manifests itself as *the vanishing gradient problem*. The second set of activation functions attempts to rectify this saturation problem, which it does, as we see in the graphs of the second row in Figure 4-5. This, however, introduces another problem, called *the exploding gradient problem*, since these activation functions are unbounded and can now output big numbers, and if these numbers grow over multiple layers, we have a problem.

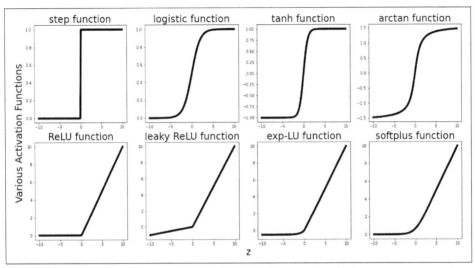

Figure 4-5. Various activation functions for neural networks. The first row consists of sigmoidal-type activation functions, shaped like the letter S. These saturate (become flat and output the same values) for inputs large in magnitude. The second row consists of ReLU-type activation functions, which do not saturate. An engineer pointed out the analogy of these activation functions to the physical function of a transistor.

Every new set of problems that gets introduced comes with its own set of techniques attempting to fix it, such as *gradient clipping*, normalizing the outputs after each layer, etc. The take-home lesson is that none of this is magic. A lot of it is trial and error, and new methods emerge to fix problems that other new methods introduced. We only need to understand the principles, the why and the how, and get a decent exposure to what is popular in the field, while keeping an open mind for improving things, or doing things entirely differently.

Let's state the formulas of common activation functions, as well as their derivatives. We need to calculate one derivative of the training function when we optimize the loss function in our search for the best weights of the neural network:

- Step function: $f(z) = \begin{cases} 0 \text{ if } z < 0 \\ 1 \text{ if } z \geq 0 \end{cases}$

  Its derivative: $f'(z) = \begin{cases} 0 \text{ if } z \neq 0 \\ undefined \text{ if } z = 0 \end{cases}$

- Logistic function: $\sigma(z) = \dfrac{1}{1 + e^{-z}}$.

    Its derivative: $\sigma'(z) = \dfrac{e^{-z}}{\left(1 + e^{-z}\right)^2} = \sigma(z)(1 - \sigma(z))$.

- Hyperbolic tangent function: $\tanh(z) = \dfrac{e^z - e^{-z}}{e^z + e^{-z}} = \dfrac{2}{1 + e^{-2z}} - 1$

    Its derivative: $\tanh'(z) = \dfrac{4}{\left(e^z + e^{-z}\right)^2} = 1 - f(z)^2$

- Inverse tangent function: $f(z) = \arctan(z)$.

    Its derivative: $f'(z) = \dfrac{1}{1 + z^2}$.

- Rectified Linear Unit function or ReLU(z): $f(z) = \begin{cases} 0 \text{ if } z < 0 \\ z \text{ if } z \geq 0 \end{cases}$

    Its derivative: $f'(z) = \begin{cases} 0 \text{ if } z < 0 \\ undefined \text{ if } z = 0 \\ 1 \text{ if } z > 0 \end{cases}$

- Leaky Rectified Linear Unit function (or parametric linear unit): $f(z) = \begin{cases} \alpha z \text{ if } z < 0 \\ z \text{ if } z \geq 0 \end{cases}$

    Its derivative: $f'(z) = \begin{cases} \alpha \text{ if } z < 0 \\ undefined \text{ if } z = 0 \\ 1 \text{ if } z > 0 \end{cases}$

- Exponential Linear Unit function: $f(z) = \begin{cases} \alpha(e^z - 1) \text{ if } z < 0 \\ z \text{ if } z \geq 0 \end{cases}$

    Its derivative: $f'(z) = \begin{cases} f(z) + \alpha \text{ if } z < 0 \\ 1 \text{ if } z \geq 0 \end{cases}$

- Softplus function: $f(z) = \ln(1 + e^z)$

    Its derivative: $f'(z) = \dfrac{1}{1 + e^{-z}} = \sigma(z)$

Note that all of these activations are rather elementary functions. This is a good thing, since both they and their derivatives are usually involved in massive computations with thousands of parameters (weights) and data instances during training, testing, and deployment of neural networks, so better keep them elementary. The other reason is that in theory, it doesn't really matter what activation function we end up choosing because of the *universal function approximation theorems*, discussed next. Careful here: operationally, it definitely matters what activation function we choose for our neural network nodes. As we mentioned earlier in this section, AlexNet's success in image classification tasks is partly due to its use of the Rectified Linear Unit function, ReLU(z). Theory and practice do not contradict each other in this case, even though it seems so on the surface. We explain this in the next subsection.

Universal Function Approximation

Approximation theorems, when available, are awesome because they tell us, with mathematical confidence and authority, that if we have a function that we do not know, or that we know but that is difficult to include in our computations, then we do not have to deal with this unknown or difficult function altogether. We can, instead, approximate it using known functions that are much easier to compute, to a great degree of precision. This means that under certain conditions on both the unknown or complicated function, and the known and simple (sometimes elementary) functions, we can use the simple functions and be confident that our computations are doing the right thing. These types of approximation theorems quantify how far off the true function is from its approximation, so we know exactly how much error we are committing when substituting the true function with this approximation.

The fact that neural networks, even sometimes *nondeep* neural networks with only one hidden layer, have proved so successful for accomplishing various tasks in vision, speech recognition, classification, regression, and others means that they have some universal approximation property going on for them. The training function that a neural network represents (built from elementary linear combinations, biases, and very simple activation functions) approximates the underlying unknown function that truly represents or generates the data rather well.

The natural questions that mathematicians must now answer with a theorem, or a bunch of theorems, are:

Given some function that we don't know but we really care about (because we think it is the true function underlying or generating our data), is there a neural network that can approximate it to a good degree of precision (without ever having to know this true function)?

> Practice using neural networks successfully suggests that the answer is yes, and universal approximation theorems for neural networks *prove* that the answer is yes for a certain class of functions and networks.

If there is a neural network that approximates this true and elusive data generating function, how do we construct it? How many layers should it have? How many nodes in each layer? What type of activation function should it include?

> In other words, what is the *architecture* of this network? Sadly, as of now, little is known on how to construct these networks, and experimentation with various architectures and activations is the only way forward until more mathematicians get on this.

Are there multiple neural network architecture that work well? Are there some that are better than others?

Experiments suggest that the answer is yes, given the comparable performance of various architectures on the same tasks and data sets.

Note that having definite answers for these questions is very useful. An affirmative answer to the first question tells us: hey, there is no magic here, neural networks do approximate a *wide class of functions* rather well! This *wide coverage*, or universality, is crucial, because recall that we do not know the underlying generating function of the data, but if the approximation theorem covers a wide class of functions, our unknown and elusive function might as well be included, hence the success of the neural network. Answering the second and third sets of questions is even more useful for practical applications, because if we know which architecture works best for each task type and data set, then we would be saved from so much experimentation, and we'd immediately choose an architecture that performs well.

Before stating the universal approximation theorems for neural networks and discussing their proofs, let's go over two examples where we already encountered approximation type theorems, even when we were in middle school. The same principle applies for all examples: we have an unruly quantity that for whatever reason is difficult to deal with or is unknown, and we want to approximate it using another quantity that is easier to deal with. If we want universal results, we need to specify three things:

1. What class or what kind of *space* does the unruly quantity or function belong to? Is it the set of real numbers $\mathbb{R}$? The set of irrational numbers? The space of continuous functions on an interval? The space of compactly supported functions on $\mathbb{R}$? The space of Lebesgue measurable functions (I did slide in some *measure theory* stuff in here, hoping that no one notices or runs away)? Etc.

2. What kind of easier quantities or functions are we using to approximate the unruly entities, and how does using these quantities instead of the true function benefit us? How do these approximations fare against *other approximations*, if there are already some other popular approximations?

3. In what sense is the approximation happening, meaning that when we say we can approximate f_{true} using $f_{approximate}$, how exactly are we measuring the distance between f_{true} and $f_{approximate}$? Recall that in math we can measure sizes of objects, including distances, in many ways. So exactly which way are we using for our particular approximations? This is where we hear about the Euclidean norm, uniform norm, supremum norm, L^2 norm, etc. What do norms (sizes) have to do with distances? A norm induces a distance. This is intuitive: if our space allows us to talk about sizes of objects, then it better allow us talk about distances as well.

Example 1: Approximating irrational numbers with rational numbers

Any irrational number can be approximated by a rational number, up to any precision that we desire. Rational numbers are so well-behaved and useful, since they are just pairs of whole numbers. Our minds can easily intuit about whole numbers and fractions. Irrational numbers are quite the opposite. Have you ever been asked, maybe in grade 6, to calculate $\sqrt{47} = 6.8556546...$ without a calculator, and stay at it until you had a definite answer? I have. Pretty mean! Even calculators and computers approximate irrational numbers with rationals. But I had to sit there thinking I could keep writing digits until I either found a pattern or the computation terminated. Of course neither happened, and around 30 digits later, I learned that some numbers are just irrational.

There is more than one way to write a mathematical statement quantifying this approximation. They are all equivalent and useful:

The approximating entity can be made arbitrarily close to the true quantity
This is the most intuitive way.

Given an irrational number s and any precision ϵ, no matter how small, we can find a rational number q within a distance ϵ from s:

$$|s - q| < \epsilon$$

This means that rational and irrational numbers live arbitrarily close to each other on the real line $\mathbb{R}$. This introduces the idea of denseness.

Denseness and closure
Approximating entities are *dense* in the space where the true quantities live.

This means that if we focus only on the space of approximating members, then add in all the limits of all their sequences, we get the whole space of the true members. Adding in all the limiting points of a certain space S is called *closing* the space, or taking its *closure*, $\bar{S}$. For example, when we add to the open interval (a,b) its limit points a and b, we get the *closed* interval [a,b]. Thus the *closure* of (a,b) is [a,b]. We write $\overline{(a,b)}$ =[a,b].

The set of rational numbers $\mathbb{Q}$ is *dense* in the real line $\mathbb{R}$. In other words, the *closure* of $\mathbb{Q}$ is $\mathbb{R}$. We write $\overline{\mathbb{Q}} = \mathbb{R}$.

Limits of sequences
The true quantity is the limit of a sequence of the approximating entities.

The idea of adding in the limit points in the previous bullet introduces approximation using the terminology of sequences and their limits.

In the context of rational numbers approximating irrational numbers, we can therefore write: for any irrational number s, there is a sequence q_n of rational numbers such that $\lim_{n \to \infty} q_n = s$. This gives us the chance to write as an example one of the favorite definitions of the most famous irrational number: $e = 2.71828182...$

$$\lim_{n \to \infty} \left(1 + \frac{1}{n}\right)^n = e$$

That is, the *irrational* number e is the limit of the sequence of *rational* numbers $\left(1 + \frac{1}{1}\right)^1, \left(1 + \frac{1}{2}\right)^2, \left(1 + \frac{1}{3}\right)^3, \cdots$ which is equivalent to $2, 2.25, 2.370370..., \cdots$.

Whether we approximate an irrational number with a rational number using the *arbitrarily close* concept, the *denseness and closure* concepts, or the *limits of sequences* concept, any distance involved in the mathematical statements is measured using the usual Euclidean norm: $d(s,q) = |s - q|$, which is the normal distance between two numbers.

Closeness Statements Need to Be Accompanied by a Specific Norm

We might wonder: what if we change the norm? Would the approximation property still hold? Can we still approximate irrationals using rationals if we measure the distance between them using some other definition of distance than the usual Euclidean norm? Welcome to *mathematical analysis*. In general, the answer is *no*. Quantities can be close to each other using some norm and very far using another norm. So in mathematics, when we say that quantities are close to each other, approximate others, or converge somewhere, we need to mention the accompanying norm in order to pinpoint in what sense these closeness statements are happening.

Example 2: Approximating continuous functions with polynomials

Continuous functions can be anything. A child can draw a wiggly line on a piece of paper and that would be a continuous function that no one knows the formula of. Polynomials, on the other hand, are a special type of continuous function that are extremely easy to evaluate, differentiate, integrate, explain, and do computations with. The only operations involved in polynomial functions are powers, scalar multiplication, addition, and subtraction. A polynomial of degree n has a simple formula:

$$p_n(x) = a_0 + a_1 x + a_2 x^2 + a_3 x^3 + \cdots + a_n x^n$$

where the a_i's are scalar numbers. Naturally, it is extremely desirable to be able to approximate nonpolynomial continuous functions using polynomial functions. The wonderful news is that we can, up to any precision ϵ. This is a classical result in mathematical analysis, called the Weierstrass Approximation Theorem:

> Suppose f is a continuous real-valued function defined on a real interval [a,b]. For any precision $\epsilon > 0$, there exists a polynomial p_n such that for all x in [a,b], we have $|f(x) - p_n(x)| < \epsilon$, or equivalently, the supremum norm $\| f - p_n \| < \epsilon$.

Note that the same principle as the one that we discussed for using rational numbers to approximate irrationals applies here. The theorem asserts that we can always find polynomials that are *arbitrarily close* to a continuous function, which means that the set of polynomials is *dense* in the space of continuous functions over the interval [a,b]; or equivalently, for any continuous function f, we can find a sequence of polynomial functions that converge to f (so f is the limit of a sequence of polynomials). In all of these variations of the same fact, the distances are measured with respect to the supremum norm. In Figure 4-6, we verify that the continuous function $\sin x$ is the limit of the sequence of the polynomial functions $\left\{ x, x - \frac{x^3}{3!}, x - \frac{x^3}{3!} + \frac{x^5}{5!}, \cdots \right\}$.

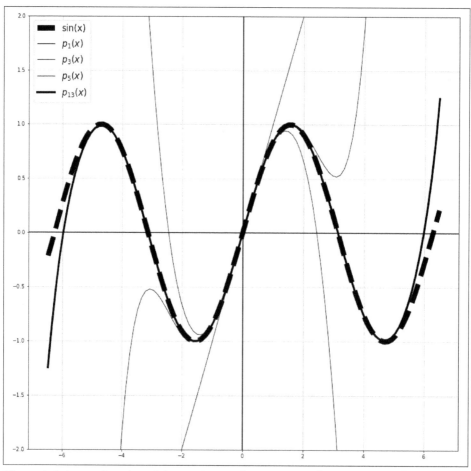

Figure 4-6. Approximation of the continuous function sin *x by a sequence of polynomials*

Statement of the universal approximation theorem for neural networks

Now that we understand the principles of approximation, let's state the most recent approximation theorems for neural networks.

Recall that a neural network is the representation of the training function as a computational graph. We want this training function to approximate the unknown function that generates the data well. This allows us to use the training function instead of the underlying true function, which we do not know and probably will never know, to make predictions. The following approximation theorems assert that neural networks can approximate the underlying functions up to any precision. When we compare the statements of these theorems to the two previous examples on irrational numbers and continuous functions, we notice that they are the same kind of mathematical statements.

The following result is from Hornik, Stinchombe, and White (1989) (*https://oreil.ly/ e0VWk*): let f be a continuous function on a compact set K (this is the true but unknown function underlying the data) whose outputs are in $\mathbb{R}^d$. Then:

Arbitrarily close
> There exists a feed forward neural network, having only a single hidden layer, which uniformly approximates f to within an arbitrary $\epsilon > 0$ on K.

Denseness
> The set of neural networks, with prescribed nonlinear activations and bounds on the number of neurons and layers depending on d, is dense in the uniform topology of $C\left(K, \mathbb{R}^d\right)$.

In both variations of the same fact, the distances are measured with respect to the supremum norm on continuous functions.

The proof needs mathematical concepts from *measure theory* and *functional analysis*. We will introduce measure theory in Chapter 11 on probability. For now we only list what is needed for the proof without any details: Borel and Radon measures, Hahn-Banach theorem, and Riesz representation theorem.

Approximation Theory for Deep Learning

We only motivated approximation theory and stated one of its main results for deep learning. For more information and a deeper discussion, we point to the state-of-the-art results such as the ability of neural networks to learn probability distributions (*https://oreil.ly/gZk03*), Barron's theorem, the neural tangent kernel, and others (*https://oreil.ly/pNRT0*).

Loss Functions

Even though in this chapter we transitioned from Chapter 3's traditional machine learning to the era of deep learning, the structure of the training function, loss function, and optimization is still exactly the same. The loss functions used for neural networks are not different from those discussed in Chapter 3, since the goal of a loss function has not changed: to capture the error between the ground truth and prediction made by the training function. In deep learning, the neural network represents the training function, and for feed forward neural networks, we saw that this is nothing more than a sequence of linear combinations followed by compositions with nonlinear activation functions.

The most popular loss functions used in deep learning are still the mean squared error for regression tasks and the cross-entropy function for classification tasks. Go back to Chapter 3 for a thorough explanation of these functions.

There are other loss functions that we sometimes come across in the field. When we encounter a new loss function, usually the designers of the model have a certain reason to prefer it over the other more popular ones, so make sure you go through their rationale for using that specific loss function for their particular setup. Ideally, a good loss function penalizes bad predictions, is not expensive to compute, and has one derivative that is easy to compute. We need this derivative so that our optimization method behaves well. As we discussed in Chapter 3, functions with one good derivative have smoother terrains than functions with discontinuous derivatives, and hence are easier to navigate during the optimization process\when searching for minimizers of the loss function.

Cross-Entropy Function, log-likelihood Function, and KL Divergence

Minimizing the cross-entropy loss function is the same as maximizing the log-likelihood function; KL divergence for probability distributions is closely related. Recall that the cross-entropy function is borrowed from information theory and statistical mechanics, and it quantifies the cross entropy between the true (empirical) distribution of the data and the distribution (of predictions) produced by the neural network's training function. The cross-entropy function has a negative sign and a log function in its formula. Minimizing the minus of a function is the same as maximizing the same function without the minus sign, so sometimes you encounter the following statement in the field: *maximizing the log likelihood function*, which for us is equivalent to *minimizing the cross-entropy loss function*. A closely related concept is the *Kullback-Leibler divergence*, also called *KL divergence*. Sometimes, as in the cases where we generate images or machine audio, we need to learn a probability distribution, not a deterministic function. Our loss function in this case should capture the *difference* (I will not say distance since its mathematical formula is not a distance metric) between the true probability distribution of the data and the learned probability distribution. KL divergence is an example of such a loss function that quantifies the amount of information lost when the learned distribution is used to approximate the true distribution, or the relative entropy of the true distribution with respect to the learned distribution.

Optimization

Faithful to our *training function*, *loss function*, and *optimization* mathematical structure, we now discuss the optimization step. Our goal is to perform an efficient search of the landscape of the loss function $L(\vec{\omega})$ to find the minimizing ω's. Note that when we previously explicitly wrote formulas for the training functions of the neural network, we bundled up the ω weights in matrices W, and the biases in vectors $\vec{\omega}_0$. In this section, for the sake of simplifying notation and to keep the focus on the mathematics, we put all the weights and biases in one very long vector $\vec{\omega}$. That is, we write the loss function as $L(\vec{\omega})$, while in reality, for a fully connected neural network with h hidden layers, it is:

$$\text{Loss function} = L\left(W^1, \vec{\omega}^1_0, W^2, \vec{\omega}^2_0, \cdots, W^{h+1}, \vec{\omega}^{h+1}_0 \right)$$

We only need that representation when we explicitly compute the derivative of the loss function using *backpropagation*, which we will cover later in this chapter.

For deep learning, the number of ω's in the vector $\vec{\omega}$ can be extremely high, as in tens of thousands, millions, or even billions. OpenAI's GPT-2 for natural language (*https://oreil.ly/u8S5L*) has 1.5 billion parameters, and was trained on a data set of eight million web pages. We need to solve for these many unknowns! Think parallel computing, or mathematical and algorithmic pipelining.

Using optimization methods, such as Newton-type methods that require computing matrices of second derivatives of the loss function in that many unknowns, is simply unfeasible even with our current powerful computational abilities. This is a great example where the mathematical theory of a numerical method works perfectly fine but is impractical for computational and real-life implementation. The sad part here is that numerical optimization methods that use the second derivative usually converge faster than those that use only the first derivative, because they take advantage of the extra knowledge about the concavity of the function (the shape of its bowl), as opposed to only using the information on whether the function is increasing or decreasing that the first derivative provides. Until we invent even more powerful computers, we have to satisfy ourselves with first-order methods that use only one derivative of the loss function with respect to the unknown ω's. These are the *gradient-descent-type* methods, and luckily, they perform extremely well for many real-life AI systems that are currently deployed for use in our everyday life, such as Amazon's Alexa.

Mathematics and the Mysterious Success of Neural Networks

It is worth pausing here to reflect on the success of neural networks, which in the context of this section translates to: our ability to locate a minimizer for the loss function $L\left(\vec{\omega}\right)$ that makes the training function generalize to new and unseen data really well. I do not have a North American accent, and Amazon's Alexa understands me perfectly fine. Mathematically, this success of neural networks is still puzzling for various reasons:

- The loss function's $\vec{\omega}$-domain $L\left(\vec{\omega}\right)$, where the search for the minimum is happening, is very high-dimensional (can reach billions of dimensions). We have billions or even trillions of options. How are we finding the right one?

- The landscape of the loss function itself is nonconvex, so it has a bunch of local minima and saddle points where optimization methods can get stuck or converge to the wrong local minimum. Again, how are we finding the right one?

- In some AI applications, such as computer vision, there are much more ω's than data points (images). Recall that for images, each pixel is a feature, so that is already a lot of ω's only at the input level. For such applications, there are much more unknowns (the ω's) than information required to determine them (the data points). Mathematically, this is an *under-determined* system, and such systems have infinitely many possible solutions! So exactly how is the optimization method for our network picking up on the good solutions? The ones that generalize well?

Some of this mysterious success is attributed to techniques that have become a staple during the training process, such as regularization (discussed later in this chapter), validation, testing, etc. However, deep learning still lacks a solid theoretical foundation. This is why a lot of mathematicians have recently converged to answer such questions. The efforts of the National Science Foundation (NSF) in this direction, and the quotes that we copy next from its announcements are quite informative and give great insight on how mathematics is intertwined with advancing AI:

> The NSF has recently established 11 new artificial intelligence research institutes (*https://www.nsf.gov/cise/ai.jsp*) to advance AI in various fields, such as human-AI interaction and collaboration, AI for advances in optimization, AI and advanced cyberinfrastructure, AI in computer and network systems, AI in dynamic systems, AI-augmented learning, and AI-driven innovation in agriculture and the food system. The NSF's ability to bring together numerous fields of scientific inquiry, including computer and information science and engineering, along with cognitive science and psychology, economics and game theory, engineering and control theory, ethics, linguistics, mathematics, and philosophy, uniquely positions the agency to lead the nation in expanding the frontiers of AI. NSF funding will help the U.S. capitalize on the full

potential of AI to strengthen the economy, advance job growth, and bring benefits to society for decades to come.

The following is quoted from the NSF's Mathematical and Scientific Foundations of Deep Learning (SCALE MoDL) webcast (*https://oreil.ly/ttUVi*).

Deep learning has met with impressive empirical success that has fueled fundamental scientific discoveries and transformed numerous application domains of artificial intelligence. Our incomplete theoretical understanding of the field, however, impedes accessibility to deep learning technology by a wider range of participants. Confronting our incomplete understanding of the mechanisms underlying the success of deep learning should serve to overcome its limitations and expand its applicability. The SCALE MoDL program will sponsor new research collaborations consisting of mathematicians, statisticians, electrical engineers, and computer scientists. Research activities should be focused on explicit topics involving some of the most challenging theoretical questions in the general area of Mathematical and Scientific Foundations of Deep Learning. Each collaboration should conduct training through research involvement of recent doctoral degree recipients, graduate students, and/or undergraduate students from across this multidisciplinary spectrum. A wide range of scientific themes on theoretical foundations of deep learning may be addressed in these proposals. Likely topics include but are not limited to geometric, topological, Bayesian, or game-theoretic formulations, to analysis approaches exploiting optimal transport theory, optimization theory, approximation theory, information theory, dynamical systems, partial differential equations, or mean field theory, to application-inspired viewpoints exploring efficient training with small data sets, adversarial learning, and closing the decision-action loop, not to mention foundational work on understanding success metrics, privacy safeguards, causal inference, and algorithmic fairness.

Gradient Descent $\vec{\omega}^{i+1} = \vec{\omega}^i - \eta \nabla L(\vec{\omega}^i)$

The widely used gradient descent method for optimization in deep learning is so simple that we could fit its formula in this subsection's title. This is how gradient descent searches the landscape of the loss function $L(\vec{\omega})$ for a local minimum:

Initialize somewhere at $\vec{\omega}^0$

Randomly pick starting numerical values for $\vec{\omega}^0 = (\omega_0, \omega_1, \cdots, \omega_n)$. This choice places us somewhere in the search space and at the landscape of $L(\vec{\omega})$. One big warning here: *where we start matters!* Do not initialize with all zeros or all equal numbers. This will diminish the network's ability to learn different features, since different nodes will output exactly the same numbers. We will discuss initialization shortly.

Move to a new point $\overrightarrow{\omega}^1$

The gradient descent moves in the direction opposite to the gradient vector of the loss function $-\nabla L\left(\overrightarrow{\omega}^0\right)$. This is guaranteed to decrease the gradient if the step size η, also called the *learning rate*, is not too large:

$$\overrightarrow{\omega}^1 = \overrightarrow{\omega}^0 - \eta \nabla L\left(\overrightarrow{\omega}^0\right)$$

Move to a new point $\overrightarrow{\omega}^2$

Again, the gradient descent moves in the direction opposite to the gradient vector of the loss function $-\nabla L\left(\overrightarrow{\omega}^1\right)$. This is guaranteed to decrease the gradient if the learning η is not too large:

$$\overrightarrow{\omega}^2 = \overrightarrow{\omega}^1 - \eta \nabla L\left(\overrightarrow{\omega}^1\right)$$

Keep going until the sequence of points $\left\{\overrightarrow{\omega}^0, \overrightarrow{\omega}^1, \overrightarrow{\omega}^2, \cdots\right\}$ converges

Note that in practice, we sometimes have to stop before it is clear that this sequence has converged—for example, when it becomes painfully slow due to a flattening landscape.

Figure 4-7 shows minimizing a certain loss function $L(\omega_1,\omega_2)$ using gradient descent. We as humans are limited to the three-dimensional space that we exist in, so we cannot visualize beyond three dimensions. This is a severe limitation for us in terms of visualization since our loss functions usually act on very high-dimensional spaces. They are functions of many ω's, but we can only visualize them accurately if they depend on at most two ω's. That is, we can visualize a loss function $L(\omega_1,\omega_2)$ depending on two ω's, but not a loss function $L(\omega_1,\omega_2,\omega_3)$ depending on three (or more) ω's. Even with this severe limitation on our capacity to visualize loss functions acting on high-dimensional spaces, Figure 4-7 gives an accurate picture of how the gradient descent method operates in general. In Figure 4-7, the search happens in the two-dimensional (ω_1,ω_2) plane (the flat ground in Figure 4-7), and we track the progress on the landscape of the function $L(\omega_1,\omega_2)$ that is embedded in $\mathbb{R}^3$. The search space always has one dimension less than the dimension of the space in which the landscape of the loss function is embedded. This makes the optimization process harder, since we are looking for a minimizer of a busy landscape in a flattened or squished version of its terrain (the ground level in Figure 4-7).

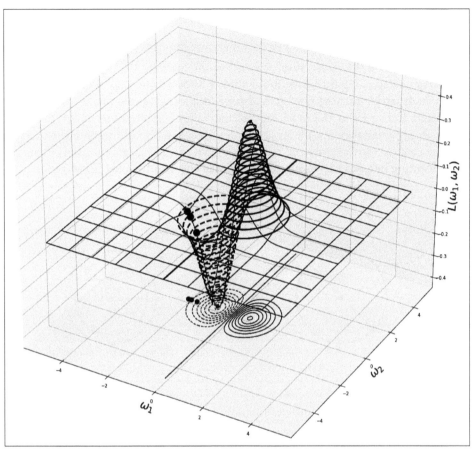

Figure 4-7. Two gradient descent steps. Note that if we start on the other side of the mountain, we wouldn't converge to the minimum. So when we are searching for the minimum of a nonconvex function, where we start or how we initiate the ω's matters a lot.

Explaining the Role of the Learning Rate Hyperparameter η

At each iteration, the gradient descent method $\overrightarrow{\omega}^{i+1} = \overrightarrow{\omega}^{i} - \eta \nabla L\left(\overrightarrow{\omega}^{i}\right)$ moves us from the point $\overrightarrow{\omega}^{i}$ in the search space to another point $\overrightarrow{\omega}^{i+1}$. The gradient descent adds $-\eta \nabla L\left(\overrightarrow{\omega}^{i}\right)$ to the current $\overrightarrow{\omega}^{i}$ to obtain $\overrightarrow{\omega}^{i+1}$. The quantity $-\eta \nabla L\left(\overrightarrow{\omega}^{i}\right)$ is made up of a scalar number η multiplied by the negative of the gradient vector $-\nabla L\left(\overrightarrow{\omega}^{i}\right)$, which points in the direction of the steepest decrease of the loss function from the point $\overrightarrow{\omega}^{i}$. Thus, the scaled $-\eta \nabla L\left(\overrightarrow{\omega}^{i}\right)$ tells us how far in the search space we are going to go along the steepest descent direction in order to choose the next point $\vec{w}^{i+1}$. In other words, the vector $-\nabla L\left(\vec{w}^i\right)$ specifies in which direction we will move away from our current point, and the scalar number η, called the *learning rate*, controls how far we are going to step along that direction. Figure 4-8 shows one step of gradient descent with two different learning rates η. Too large of a learning rate might overshoot the minimum and cross to the other side of the valley. On the other hand, too small of a learning rate takes a while to get to the minimum. So the trade-off is between choosing a large learning rate and risking overshooting the minimum, and choosing a small learning rate and increasing computational cost and time for convergence.

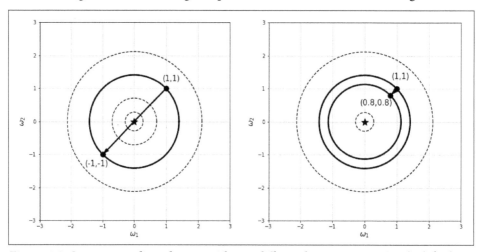

Figure 4-8. One-step gradient descent with two different learning rates. On the left, the learning rate is too large, so the gradient descent overshoots the minimum (the star point) and lands on the other side of the valley. On the right, the learning rate is small, however, it will take a while to get to the minimum (the star point). Note how the gradient vector at a point is perpendicular to the level set at that point.

The learning rate η is another example of a hyperparameter of a machine learning model. It is not one of the weights that goes into the formula of the training function. It is a parameter that is intrinsic to the algorithm that we employ to estimate the weights of the training function.

The scale of the features affects the performance of the gradient descent

This is one of the reasons to standardize the features ahead of time. Standardizing a feature means subtracting from each data instance the mean and dividing by the standard deviation. This forces all the data values to have the same scale, with mean zero and standard deviation one, as opposed to having vastly different scales, such as a feature measured in the millions and another measured in 0.001. But why does this affect the performance of the gradient descent method? Read on.

Recall that the values of the input features get multiplied by the weights in the training function, and the training function in turn enters into the formula of the loss function. Very different scales of the input features change the shape of the bowl of the loss function, making the minimization process harder. Figure 4-9 shows the level sets of the function $L(\omega_1,\omega_2) = \omega_1^2 + a\omega_2^2$ with different values of a, mimicking different scales of input features. Note how the level sets of loss function become much more narrow and elongated as the value of a increases. This means that the shape of the bowl of the loss function is a long, narrow valley.

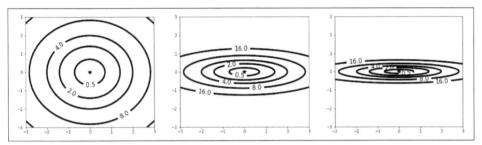

Figure 4-9. The level sets of the loss function $L(\omega_1,\omega_2) = \omega_1^2 + a\omega_2^2$ become much more narrow and elongated as the value of a increases from 1 to 20 to 40

When the gradient descent method tries to operate in such a narrow valley, its points hop from one side of the valley to the other, zigzagging as it tries to locate the minimum, and slowing down the convergence considerably. Imagine zigzagging along all the streets of Rome before arriving at the Vatican, as opposed to taking a helicopter straight to the Vatican.

But why does this zigzagging behavior happen? One hallmark of the gradient vector of a function is that it is perpendicular to the level sets of that function. So if the valley of the loss function is so long and narrow, its level sets almost look like lines that are parallel to each other, and with a large enough step size (learning rate), we can literally cross from one side of the valley to the other since it is so narrow. Google *gradient descent zigzag* and you will see many images illustrating this behavior.

One fix for zigzagging, even with a narrow, long valley (assuming we did not scale the input feature values ahead of time), is to choose a very small learning rate, preventing the gradient descent method from stepping from one side of the valley to the other. However, that slows down the arrival to the minimum in its own way, since the method will step only incrementally at each iteration. We will eventually arrive at the Vatican from Rome, but at a turtle's pace.

Near the minima (local and/or global), flat regions, or saddle points of the loss function's landscape, the gradient descent method crawls

The gradient descent method updates the current point $\overrightarrow{\omega}^i$ by adding the vector $-\eta \nabla L\left(\overrightarrow{\omega}^i\right)$. Therefore, the exact length of the step from the point $\overrightarrow{\omega}^i$ in the direction of the negative gradient vector is η multiplied by the length of the gradient vector $\nabla L\left(\overrightarrow{\omega}^i\right)$. At a minimum, maximum, saddle point, or any flat region of the landscape of the loss function, the gradient vector is zero, hence its length is zero as well. This means that near a minimum, maximum, saddle point, or any flat region, the step size of the gradient descent method becomes very small, and the method slows down significantly. If this happens near a minimum, then there is not much worry since this can be used as a stopping criterion, unless this minimum is a local minimum very far from the global minimum. If, on the other hand, it happens in a flat region or near a saddle point, then the method will get stuck there for a while, and that is undesirable. Some practitioners put the learning rate η on a schedule, changing its value during the optimization process. When we look into these, we notice that the goals are to avoid crawling, save computational time, and speed up convergence.

We will discuss stochastic (random) gradient descent later in this chapter. Due to the random nature of this method, the points hop around a lot, as opposed to following a more consistent route toward the minimum. This works to our advantage in situations where we are stuck, such as saddle points or local minima, since we might get randomly propelled out of the local minimum or away from the saddle point into a part of the landscape with a better route toward the minimum.

Convex Versus Nonconvex Landscapes

We cannot have an optimization chapter without discussing *convexity*. In fact, entire mathematical fields are dedicated solely to *convex optimization* (*https://oreil.ly/xq7bx*). It is equally important to immediately note that optimization for neural networks is in general, *nonconvex*.

When we use nonconvex activation functions, such as the sigmoid-type functions in the first row of Figure 4-5, the landscapes of the loss functions involved in the resulting neural networks are not convex. This is why we spend a good amount of time talking about getting stuck at local minima, flat regions, and saddle points, which we wouldn't worry about for convex landscapes. The contrast between convex and nonconvex landscapes is obvious in Figure 4-10, which shows a convex loss function and its level sets, and Figures 4-11 and 4-12, which show nonconvex functions and their level sets.

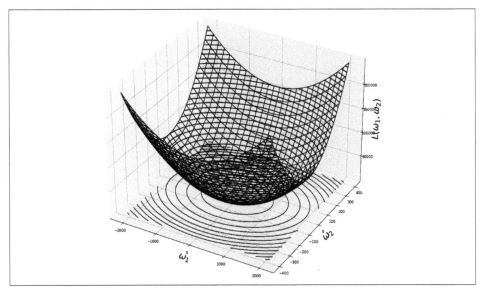

Figure 4-10. Plot of three-dimensional convex function and its level sets. Gradient vectors live in the same space ($\mathbb{R}^2$) as the level sets, not in $\mathbb{R}^3$.

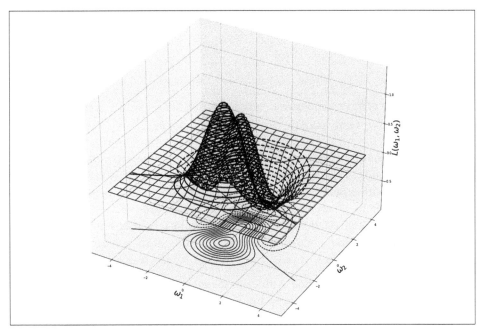

Figure 4-11. Plots of three-dimensional nonconvex functions and their level sets. Gradient vectors live in the same space ($\mathbb{R}^2$) as the level sets, not in $\mathbb{R}^3$.

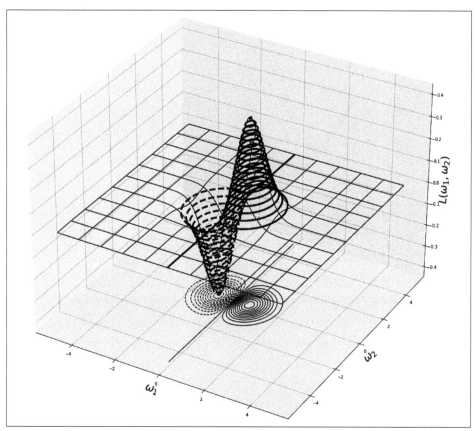

Figure 4-12. Plots of three-dimensional nonconvex functions and their level sets. Gradient vectors live in the same space ($\mathbb{R}^2$) as the level sets, not in $\mathbb{R}^3$.

When we use convex activation functions throughout the network, such as the ReLU-type functions in the second row of Figure 4-5, *and* convex loss functions, we can still end up with a *nonconvex* optimization problem, because the composition of two convex functions is not necessarily convex. If the loss function happens to be nondecreasing and convex, then its composition with a convex function is convex. The loss functions that are popular for neural networks, such as mean squared error, cross-entropy, and hinge loss, are all convex but not nondecreasing.

It is important to become familiar with central concepts from convex optimization. If you do not know where to start, keep in mind that convexity replaces linearity when linearity is too simplistic or unavailable, then learn everything about the following (which will be tied to AI, deep learning, and reinforcement learning when we discuss operations research in Chapter 10):

- Max of linear functions is convex
- Max-min and min-max
- Saddle points
- Two-player zero-sum games
- Duality

Since convex optimization is such a well-developed and understood field (at least more than the mathematical foundations for neural networks), and neural networks still have a long way to go mathematically, it would be nice if we could exploit our knowledge about convexity in order to gain a deeper understanding of neural networks. Research in this area is ongoing. For example, in a 2020 paper (*https:// oreil.ly/mHc7S*) titled "Convex Geometry of Two-Layer ReLU Networks: Implicit Autoencoding and Interpretable Models," Tolga Ergen and Mert Pilanci frame the problem of training two layered ReLU networks as a convex analytic optimization problem. The following is the abstract of the paper:

We develop a convex analytic framework for ReLU neural networks which elucidates the inner workings of hidden neurons and their function space characteristics. We show that rectified linear units in neural networks act as convex regularizers, where simple solutions are encouraged via extreme points of a certain convex set. For one dimensional regression and classification, we prove that finite two-layer ReLU networks with norm regularization yield linear spline interpolation. In the more general higher dimensional case, we show that the training problem for two-layer networks can be cast as a convex optimization problem with infinitely many constraints. We then provide a family of convex relaxations to approximate the solution, and a cutting-plane algorithm to improve the relaxations. We derive conditions for the exactness of the relaxations and provide simple closed form formulas for the optimal neural network weights in certain cases. Our results show that the hidden neurons of a ReLU network can be interpreted as convex autoencoders of the input layer. We also establish a connection to $l_0 - l_1$ equivalence for neural networks analogous to the minimal cardinality solutions in compressed sensing. Extensive experimental results show that the proposed approach yields interpretable and accurate models.

Stochastic Gradient Descent

So far, training a feed forward neural network has progressed as follows:

1. Fix an initial set of weights $\vec{\omega}^0$ for the training function.
2. Evaluate this training function at *all* the data points in the training subset.
3. Calculate the individual losses at *all* the data points in the training subset by comparing their true labels to the predictions made by the training function.
4. Do this for *all* the data in the training subset.

5. Average *all* these individual losses. This average is the loss function.

6. Evaluate the gradient of this loss function at this initial set of weights.

7. Choose the next set of weights according to the steepest descent rule.

8. Repeat until you converge somewhere, or stop after a certain number of iterations determined by the performance of the training function on the validation set.

The problem with this process is that when we have a large training subset with thousands of points, and a neural network with thousands of weights, it gets too expensive to evaluate the training function, the loss function, and the gradient of the loss function on *all* the data points in the training subset. The remedy is to randomize the process: randomly choose a very small portion of the training subset to evaluate the training function, loss function, and gradient of this loss function at each step. This slashes the computational cost dramatically.

Keep repeating this random selection (in principle with replacement but in practice without replacement) of small portions of the training subset until you converge somewhere, or stop after a certain number of iterations determined by the performance of the training function on the validation set. One pass through the whole training subset is called *one epoch*.

Stochastic gradient descent performs remarkably well, and it has become a staple in training neural networks.

Initializing the Weights $\vec{w}^0$ for the Optimization Process

We have already established that initializing with all zero weights or all the same weights is a really bad idea. The next logical step, and what was the traditional practice (before 2010), would be to choose the weights in the initial $\vec{w}^0$ randomly, sampled either from the uniform distribution over small intervals, such as [-1,1], [0,1], or [-0.3,0.3], or from the Gaussian distribution with a preselected mean and variance. Even though this has not been studied in depth, it seems from empirical evidence that it doesn't matter whether the initial weights are sampled from the uniform distribution or Gaussian distribution, but it does seem that the scale of the initial weights matters when it comes to both the progress of the optimization process and the ability of the network to generalize well to unseen data. It turns out that some choices are better than others in this respect. Currently, the two state-of-the-art choices depend on the choice of the activation function: whether it is sigmoid-type or ReLU-type.

Xavier Glorot initialization

Here, initial weights are sampled from uniform distribution over the interval $[-\frac{\sqrt{6}}{\sqrt{n+m}}, \frac{\sqrt{6}}{\sqrt{n+m}}]$, where n is the number of inputs to the node (e.g., the number of nodes in the previous layer), and m is the number of outputs from the layer (e.g., the number of nodes in the current layer).

Kaiming He initialization

Here, the initial weights are sampled from the Gaussian distribution with zero mean and variance $2/n$, where n is the number of inputs to the node.

Regularization Techniques

Regularization helps us arrive at a good choice for the weights of the training function while at the same time avoiding overfitting the data. We want our trained function to follow the signal in the data rather than the noise, so it can generalize well to unseen data. Here we include four simple yet popular regularization techniques that are used while training a neural network: dropout, early stopping, batch normalization, and weight decay (ridge, lasso, and elastic net) regularizations.

Dropout

Drop some randomly selected neurons from each layer during training. Usually, about 20% of the input layer's nodes and about half of each of the hidden layers' nodes are randomly dropped. No nodes from the output layer are dropped. Dropout is partially inspired by genetic reproduction, where half a parent's genes are dropped and there is a small random mutation. This has the effect of training different networks at once (with a different number of nodes at each layer) and averaging their results, which typically produces more reliable results.

One way to implement dropout is by introducing a hyperparameter p for each layer that specifies the probability at which each node in that layer will be dropped. Recall the basic operations that take place at each node: linearly combine the outputs of the nodes of the previous layer, then activate. With dropout, each output of the nodes of the previous layer (starting with the input layer), is multiplied by a random number r, which can be either 0 or 1 with probability p. Thus when a node's r takes the value 0, that node is essentially dropped from the network, which now forces the other retained nodes to pick up the slack when adjusting the weights in one gradient descent step. We will explain this further in "Backpropagation in Detail" on page 155, and this link (*https://oreil.ly/HGt8T*) provides a step-by-step route to implementing dropout.

For a deeper mathematical exploration, a 2015 paper (*https://oreil.ly/uvwNf*) connects dropout to Bayesian approximations of model uncertainty.

Early Stopping

As we update the weights during training, in particular during gradient descent, after each epoch, we evaluate the error made by the training function at the current weights on the validation subset of the data.

This error should be decreasing as the model learns the training data; however, after a certain number of epochs, this error will start increasing, indicating that the training function has now started overfitting the training data and is failing to generalize well to the validation data. Once we observe this increase in the model's prediction over the validation subset, we stop training and go back to the set of weights where that error was lowest, right before we started observing the increase.

Batch Normalization of Each Layer

The main idea here is to normalize the inputs to each layer of the network. This means that the inputs to each layer will have mean 0 and variance 1. This is usually accomplished by subtracting the mean and dividing by the variance for each of the layer's inputs. We will detail this in a moment. The reason this is good to do at each hidden layer is similar to why it is good at the original input layer.

Applying *batch normalization* often eliminates the need for dropout, and allows us to be less particular about initialization. It makes the training faster and safer from vanishing and exploding gradients. It also has the added advantage of regularization. The cost for all of these gains is not too high, as it usually involves training only two additional parameters, one for scaling and one for shifting, at each layer.

The 2015 paper by Ioffe and Szegedy (*https://oreil.ly/pZwV0*) introduced the method. The abstract of their paper describes the batch normalization process and the problems it addresses (the brackets are my own comments):

> Training Deep Neural Networks is complicated by the fact that the distribution of each layer's inputs changes during training, as the parameters of the previous layers change. This slows down the training by requiring lower learning rates and careful parameter initialization, and makes it notoriously hard to train models with saturating nonlinearities [such as the sigmoid type activation functions, in Figure 4-5, which become almost constant, outputting the same value when the input is large in magnitude. This renders the nonlinearity useless in the training process, and the network stops learning at subsequent layers]. We refer to this phenomenon [the change in the distribution of the inputs to each layer] as internal covariate shift, and address the problem by normalizing layer inputs. Our method draws its strength from making normalization a part of the model architecture and performing the normalization for each training mini-batch. Batch Normalization allows us to use much higher learning rates and be less careful about initialization, and in some cases eliminates the need for Dropout. Applied to a state-of-the-art image classification model, Batch Normalization achieves the same accuracy with 14 times fewer training steps, and beats the original model by a significant margin. Using an ensemble of batch-normalized networks, we improve

upon the best published result on ImageNet classification: reaching 4.82% top-5 test error, exceeding the accuracy of human raters.

Batch normalization is often implemented in the architecture of a network either in its own layer before the activation step, or after activation. The process, *during training*, usually follows these steps:

1. Choose a batch from the training data of size b. Each data point in this has feature vector $\vec{x_i}$, so the whole batch has feature vectors $\vec{x_1},\vec{x_2},\cdots,\vec{x_b}$.

2. Calculate the vector whose entries are the means of each feature in this particular batch: $\vec{\mu} = \dfrac{\vec{x_1} + \vec{x_2} + \cdots + \vec{x_b}}{b}$.

3. Calculate the variance across the batch: subtract $\vec{\mu}$ from each $\vec{x_1},\vec{x_2},\cdots,\vec{x_b}$, calculate the result's l^2 norm, add, and divide by b.

4. Normalize each of $\vec{x_1},\vec{x_2},\cdots,\vec{x_b}$ by subtracting the mean and dividing by the square root of the variance:

5. Scale and shift by trainable parameters that can be initialized and learned by gradient descent, the same way the weights of the training function are learned. This becomes the input to the first hidden layer.

6. Do the same for the input of each of the subsequent layers.

7. Repeat for the next batch.

During testing and prediction, there is no batch of data to train on, and the parameters at each layer are already learned. The batch normalization step, however, is already incorporated into the formula of the training function. During training, we were changing these *per batch* of training data. This in turn was changing the formula of the loss function slightly per batch. However, the point of normalization was partly not to change the formula of the loss function too much, because that in turn would change the locations of its minima, and that would cause us to forever chase a moving target. Alright, we fixed that with batch normalization during training, and now we want to validate, test, and predict. Which mean vector and variance do we use for a particular data point that we are testing/predicting at? Do we use the means and variances of the features of the original data set? We have to make such decisions.

Control the Size of the Weights by Penalizing Their Norm

Another way to regularize the training function to avoid overfitting the data is to introduce a *competing term* into the minimization problem. Instead of solving for the set of weights $\vec{\omega}$ that minimizes *only* the loss function:

$$\min_{\vec{\omega}} L\left(\vec{\omega}\right)$$

we introduce a new term $\alpha \parallel \vec{\omega} \parallel$ and solve for the set of weights $\vec{\omega}$ that minimizes:

$$\min_{\vec{\omega}} L\left(\vec{\omega}\right) + \alpha \parallel \vec{\omega} \parallel$$

For example, for the mean squared error loss function usually used for regression problems, the minimization problem looks like:

$$\min_{\vec{\omega}} \frac{1}{m}\Sigma_{i=1}^{m} \left| y_{predict}\left(\vec{\omega}\right) - y_{true}\right|^2 + \alpha \parallel \vec{\omega} \parallel$$

Recall that so far we have established two ways to solve that minimization problem:

The minimum happens at points where the derivative (gradient) is equal to zero

So the minimizing $\vec{\omega}$ must satisfy $\nabla L\left(\vec{\omega}\right) + \alpha\nabla(\parallel \vec{\omega} \parallel) = 0$. Then we solve this equation for $\vec{\omega}$ if we have the luxury to get a *closed form* for the solution. In the case of linear regression (which we can think about as an extremely simplified neural network, with only one layer and zero nonlinear activation function), we do have this luxury, and for this *regularized* case, the formula for the minimizing $\vec{\omega}$ is:

$$\vec{\omega} = \left(X^t X + \alpha B\right)^{-1} X^t \vec{y}_{true}$$

where the columns of X are the feature columns of the data augmented with a vector of ones, and B is the identity matrix (if we use ridge regression, discussed later). The closed form solution for the extremely simple linear regression problem with regularization helps us appreciate weight decay type regularization and see the important role it plays. Instead of inverting the matrix $\left(X^t X\right)$ in the *unregularized* solution and worrying about its ill-conditioning (for example, from highly correlated input features) and the resulting instabilities, we invert $\left(X^t X + \alpha B\right)$ in the *regularized* solution. Adding this αB term is equivalent to adding a small positive term to the denominator of a scalar number that helps us avoid division by zero. Instead of using $1/x$ where x runs the risk of being zero, we use $1/(x + \alpha)$ where α is a positive constant. Recall that matrix inversion is the analog of scalar number division.

Gradient descent

We use gradient descent or any of its variations, such as stochastic gradient descent, when we do not have the luxury of obtaining closed form solutions for the derivative equals zero equation, and when our problem is very large, that computing second-order derivatives is extremely expensive.

Commonly used weight decay regularizations

There are three popular regularizations that control the size of the weights that we are forever searching for in this book:

Ridge regression

Penalize the l^2 norm of $\vec{\omega}$. In this case, we add the term $\alpha\Sigma_{i=1}^{n}|\omega_i|^2$ to the loss function, then we minimize.

Lasso regression

Penalize the l^1 norm of the ω's. In this case, we add the term $\alpha\Sigma_{i=1}^{n}|\omega_i|$ to the loss function, then we minimize.

Elastic net

This is a middle-ground case between ridge and lasso regressions. We introduce one additional hyperparameter γ, which can take any value between zero and one, and add a term to the loss function that combines both ridge and lasso regressions through γ: $\gamma\alpha\Sigma_{i=1}^{n}|\omega_i|^2 + (1-\gamma)\alpha\Sigma_{i=1}^{n}|\omega_i|$. When $\gamma = 0$, this becomes lasso regression; when it is equal to one, it is ridge regression; and when it is between zero and one, it is some sort of a middle ground.

When do we use plain linear regression, ridge, lasso, or elastic net?

If you are already confused and slightly overwhelmed by the multitude of choices that are available for building machine learning models, join the club, but do not get frustrated. Until the mathematical analysis that tells us exactly which choices are better than others and under what circumstances becomes available (or catches us with mathematical computation and experimentation), think about the enormity of available choices the same way you think about a home renovation: we have to choose from many available materials, designs, and architectures to produce a final product. This is a home renovation, not a home decoration, so our decisions are fateful and more consequential than a mere home decoration. They do affect the *quality* and the *function* of our final product, but they are choices nevertheless. Rest easy, there is more than one way to skin AI:

- Some regularization is always good. Adding a term that controls the sizes of the weights and competes with minimizing the loss function is good in general.

- Ridge regression is usually a good choice because the l^2 norm is differentiable. Minimizing this is more stable than minimizing the l^1 norm.

- If we decide to go with the l^1 norm, even though it is not differentiable at 0, we can still define its *subdifferential* or *subgradient* at 0. For example, we can set that to be zero. Note that $f(x) = |x|$ is differentiable when $x \neq 0$: it has derivative 1 when $x > 0$ and -1 when $x < 0$; the only problematic point is $x=0$.

- If we suspect only a few features are useful, then it is good to use either lasso or elastic net as a data preprocessing step to kill off the less important features.

- Elastic net is usually preferred over lasso because lasso might behave badly when the number of features is greater than the number of training instances or when several features are strongly correlated.

Penalizing the l^2 Norm Versus Penalizing the l^1 Norm

Our goal is to find $\vec{w}$ that solves the minimization:

$$\min_{\vec{w}} L\left(\vec{w},\vec{w_0}\right) + \alpha \parallel \vec{w} \parallel$$

The first term wants to decrease the loss $L\left(\vec{w},\vec{w_0}\right)$. The other term wants to decrease the values of the coordinates of $\vec{w}$ all the way to zeros. The type of the norm that we choose for $\parallel \vec{w} \parallel$ determines the path $\vec{w}$ follows on its way to $\vec{0}$.

If we use the l^1 norm, the coordinates of $\vec{w}$ will decrease; however, a lot of them might encounter premature death, hitting zero before others. That is, the l^1 norm encourages sparsity: when a weight dies, it kills the contribution of the associated feature to the training function.

The plot on the right in Figure 4-13 shows the diamond-shaped level sets of $\parallel \vec{w} \parallel_{l^1} = |w_1| + |w_2|$ in two dimensions (if we only had two features), namely, $|w_1| + |w_2| = c$ for various values of c. If a minimization algorithm follows the path of steepest descent, such as the gradient descent, then we must travel in the direction perpendicular to the level sets, and as the arrow shows in the plot, w_2 becomes zero pretty fast since going perpendicular to the diamond-shaped level sets is bound to hit one of the coordinate axes, effectively killing the respective feature. w_1 then travels to zero along the horizontal axis.

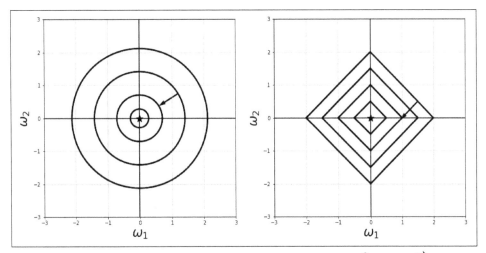

Figure 4-13. The plot on the left shows the circular level sets of the l^2 norm of $\vec{w}$, along with the direction the gradient descent follows toward the minimum at (0,0). The plot on the right shows the diamond-shaped level sets of the l^1 norm of $\vec{w}$, along with the direction the gradient descent follows toward the minimum at (0,0).

- If we use the l^2 norm, the weight sizes get smaller without necessarily killing them. The plot on the left in Figure 4-13 shows the circular-shaped level sets of $\| \vec{w} \|_{l^2}^2 = w_1^2 + w_2^2$ in two dimensions, namely, $w_1^2 + w_2^2 = c$ for various values of c. We see that following the path perpendicular to the circular level sets toward the minimum at (0,0) decreases the values of both w_1 and w_2 without either of them becoming zero before the other.

Which norm to choose depends on our use cases. Note that in all cases, we do not regularize the bias weights $\vec{w_0}$. This is why in this section we wrote them separately in the loss function $L\left(\vec{w},\vec{w_0}\right)$.

Explaining the Role of the Regularization Hyperparameter α

The minimization problem with weight decay regularization looks like:

$$\min_{\vec{w}} L\left(\vec{w}\right) + \alpha \| \vec{w} \|$$

To understand the role of the regularization hyperparameter α, we observe the following:

- There is a competition between the first term, where the loss function $L(\vec{\omega})$ chooses ω's that fit the training function to the training data, and the second term that just cares about making the ω values small. These two objectives are not necessarily in sync. The values of ω's that make the first term smaller might make the second term bigger and vice versa.

- If α is big, then the minimization process will compensate by making values of ω's very small, regardless of whether these small values of ω's will make the first term small as well. So the more we increase α, the more important minimizing the second term becomes than the first term, so our ultimate model might end up not fitting the data perfectly (high bias), but this is sometimes desired (low variance) so that it generalizes well to unseen data.

- If, on the other hand, α is small (say, close to zero), then we can choose larger ω values, and minimizing the first term becomes more important. Here, the minimization process will result in ω values that make the first term happy, so the data will fit into the model nicely (low bias) but the variance might be high. In this case, our model would work well on seen data (it is designed to fit it nicely through minimizing $L(\vec{\omega})$), but might not generalize well to unseen data.

- As $\alpha \to 0$, we can prove mathematically that the solution of the regularized problem converges to the solution of the unregularized problem.

Hyperparameter Examples That Appear in Machine Learning

We have now encountered many hyperparameters that enter machine learning models. It is good practice to list the ones that enter our particular model along with their values. Let's list the ones we have come across and recall that tuning these enhances the performance of our models. Most of the time, there are recommended values for us to use. These are usually implemented as default values in machine learning libraries and software packages. However, it is always good to experiment with different values during the validation stage of our modeling process, given that we have the available time and resources. The hyperparameters include:

- The learning rate in gradient descent.
- Weight decay coefficients, such as the ones that appear in ridge, lasso, and elastic net regularizations.
- The number of epochs before we stop training.
- The sizes of data split into training, validation, and testing subsets.

- The sizes of mini-batches during stochastic gradient descent and its variants.
- The acceleration coefficients in momentum methods.
- The architecture of a neural network: number of layers, number of neurons in each layer, what happens at each layer (batch normalization, type of activation function), type of regularization (dropout, ridge, lasso), type of network (feed forward, dense, convolutional, adversarial, recurrent), type of loss functions, etc.

Chain Rule and Backpropagation: Calculating $\nabla L\left(\vec{\omega}^{\,i}\right)$

It is time to get our hands dirty and compute something important: the gradient of the loss function, namely, $\nabla L\left(\vec{\omega}^{\,i}\right)$. Whether we decide to find our optimal weights using gradient descent, stochastic gradient descent, mini-batch gradient descent, or any other variant of gradient descent, there is no escape from calculating this quantity. Recall that the loss function includes in its formula the neural network's training function, which in turn is made up of subsequent linear combinations and compositions with activation functions. This means that we have to cleverly use the chain rule. Back in calculus, we only used the single variable chain rule for derivatives, but now we somehow have to transition to a chain rule of several variables: several, as in, sometimes billions.

It is the layered architecture of a neural network that forces us to pause and think: how exactly are we going to compute this one derivative of the loss function? The workhorse here is the *backpropagation* algorithm (also called *backward mode automatic differentiation*), and it is a powerful one.

Before writing formulas, let's summarize the steps that we follow as we train a neural network:

- The training function is a function of $\vec{\omega}$, so the outcome of the neural network after a data point passes through it, which is the same as evaluating the training function at the data point, is: $outcome = function\left(\vec{\omega}\right)$. This is made up of linear combinations of node outputs, followed by compositions with activation functions, repeated over all of the network's layers. The output layer might or might not have an activation function and could have one node or multiple nodes, depending on the ultimate task of the network.

- The loss function provides a measure of how badly the outcome of the training function diverged from what is true.

- We initialize our learning function with a *random* set of weights $\vec{\omega}^{\,0}$, according to preferred initialization rules prescribed in the previous sections. Then we compute the loss, or error, that we committed because of using these particular weight values. This is the forward pass of the data point through the net.

- We want to move to the next set of weights $\vec{w}^1$ that gives a lower error. We move in the direction opposite to the gradient vector of the loss function.

- But: the training function is built into the loss function, and given the layered structure of this function, which comes from the architecture of the neural network, along with its high dimensionality, how do we efficiently perform the multivariable chain rule to find the gradient and evaluate it at the current set of weights?

The answer is that we send the data point back through the network, computing the gradient *backward* from the output layer all the way to the input layer, evaluating along the way how each node contributed to the error. In essence, we compute $\frac{\partial L}{\partial \text{node functions}}$, then we tweak the weights accordingly, updating them from $\vec{w}^0$ to $\vec{w}^1$. The process continues as we pass more data points into the network, usually in batches. One *epoch* is then counted each time the network has seen the full training set.

Backpropagation Is Not Too Different from How Our Brain Learns

When we encounter a new math concept, the neurons in our brain make certain connections. The next time we see the same concept, the same neurons connect better. The analogy for our neural network is that the value w of the edge connecting the neurons increases. When we see the same concept again and again, it becomes part of our brain's model. This model will not change, unless we learn new information that undoes the previous information. In that case, the connection between the neurons weakens. For our neural network, the w value connecting the neurons decreases. Tweaking the w's via minimizing the loss function accomplishes exactly that: establishing the correct connections between the neurons.

The neuroscientist Donald Hebb mentions in his 1949 book *The Organization of Behavior: A Neuropsychological Theory* (paraphrased): *When a biological neuron triggers another neuron often, the connection between these two neurons grows stronger. In other words, cells that fire together, wire together.*

Similarly, a neural network's computational model takes into account the error made by the network when it produces an outcome. Since computers only understand numbers, the w of an edge increases if the node contributes to lowering the error, and decreases if the node contributes to increasing the error function. So a neural network's learning rule reinforces the connections that reduce the error by increasing the corresponding w's, and weakens the connections that increase the error by decreasing the corresponding w's.

Why Is It Better to Backpropagate?

Backpropagation computes the derivative of the training function with respect to each node, moving *backward* through the network. This measures the contribution of each node to both the training function and the loss function $L\left(\vec{\omega}\right)$.

The most important formula to recall here is: the chain rule from calculus. This calculates the derivatives of *chained* functions (or function compositions). The calculus chain rule mostly deals with functions depending only on one variable ω; for example, for three chained functions, the derivative with respect to ω is:

$$\frac{d}{d\omega}f_3(f_2(f_1(\omega))) = \left\{\frac{d}{d\omega}f_1(\omega)\right\}\left\{\frac{d}{df_1}f_2(f_1(\omega))\right\}\left\{\frac{d}{f_2}f_3(f_2(f_1(\omega)))\right\}$$

For neural networks, we must apply the chain rule to the loss function that depends on the matrices and vectors of variables W and $\vec{\omega_0}$. So we have to generalize the above rule to a *many variables chain rule*. The easiest way to do this is to follow the structure of the network computing the derivatives backward, from the outcome layer all the way back to the input layer.

If instead we decide to compute the derivatives forward through the network, we would not know whether these derivatives with respect to each variable will ultimately contribute to our final outcome, because we do not know if they will connect through the graph of the network. Even when the graph is fully connected, the weights for deeper layers are not present in earlier layers, so it is a big waste to compute for their derivatives in the early layers.

When we compute the derivatives backward through the network, we start with the output and follow the edges of the graph of the network back, computing the derivatives at each node. Each node's contribution is calculated only from the edges leading to it and edges going out of it. This is computationally much cheaper because now we are sure of how and when these nodes contribute to the network's outcome.

In linear algebra, it is much cheaper to compute the multiplication of a matrix with a vector than to compute the multiplication of two matrices together. We must always avoid multiplying two matrices with each other: computing $A(B\mathbf{v})$ is cheaper than computing $(AB)\mathbf{v}$, even though in theory, these two are exactly the same. Over large matrices and vectors, this simple observation provides enormous cost savings.

Backpropagation in Detail

Let's pause and be thankful that software packages exist so that we never have to implement the following computation ourselves. Let's also not forget to be grateful to the creators of these software packages. Now we compute.

For a neural network with h hidden layers, we can write the loss function as a function of the training function, which in turn is a function of all the weights that appear in the network:

$$L = L\left(g\left(W^1,\overrightarrow{\omega_0^1}, W^2,\overrightarrow{\omega_0^2}, \cdots, W^h,\overrightarrow{\omega_0^h}, W^{h+1},\overrightarrow{\omega_0^{h+1}}\right)\right)$$

We will compute the partial derivatives of L backward, starting with $\dfrac{\partial L}{\partial W^{h+1}}$ and $\dfrac{\partial L}{\partial \overrightarrow{\omega_0^{h+1}}}$, and working our way back to $\dfrac{\partial L}{\partial W^1}$ and $\dfrac{\partial L}{\partial \overrightarrow{\omega_0^1}}$. Their derivatives are taken with respect to each entry in the corresponding matrix or vector.

Suppose for simplicity, but without loss of generality, that the network is a regression network predicting a single numerical value, so that the training function g is scalar (not a vector). Suppose also that we use the same activation function f for each neuron throughout the network. The output neuron has no activation since this is a regression.

Let's find the derivatives with respect to the weights pointing to the output layer. The loss function is:

$$L = L\left(W^{h+1}\overrightarrow{s^h} + \omega_0^{h+1}\right)$$

so that

$$\frac{\partial L}{\partial \omega_0^{h+1}} = 1 \times L'\left(W^{h+1}\overrightarrow{s^h} + \omega_0^{h+1}\right)$$

and

$$\frac{\partial L}{\partial W^{h+1}} = \left(\overrightarrow{s^h}\right)^t L'\left(W^{h+1}\overrightarrow{s^h} + \omega_0^{h+1}\right)$$

Recall that $\overrightarrow{s^h}$ is the output of the last layer, so it depends on all the weights of the previous layers, namely, $\left(W^1,\overrightarrow{\omega_0^1}, W^2,\overrightarrow{\omega_0^2}, \cdots, W^h,\overrightarrow{\omega_0^h}\right)$.

To compute derivatives with respect to the weights pointing to the last hidden layer, we show them explicitly in the formula of the loss function:

$$L = L\left(W^{h+1}\left(f\left(W^h\overrightarrow{s}^{h-1} + \overrightarrow{\omega}_0^h\right)\right) + \omega_0^{h+1}\right)$$

so that

$$\frac{\partial L}{\partial \overrightarrow{\omega}_0^h} = \overrightarrow{1}\left(W^{h+1}f'\left(W^h\overrightarrow{s}^{h-1} + \overrightarrow{\omega}_0^h\right)\right)L'\left(W^{h+1}\left(f\left(W^h\overrightarrow{s}^{h-1} + \overrightarrow{\omega}_0^h\right)\right) + \omega_0^{h+1}\right)$$

and

$$\frac{\partial L}{\partial W^h} = \overrightarrow{s}^{h-1}\left(W^{h+1}f'\left(W^h\overrightarrow{s}^{h-1} + \overrightarrow{\omega}_0^h\right)\right)L'\left(W^{h+1}\left(f\left(W^h\overrightarrow{s}^{h-1} + \overrightarrow{\omega}_0^h\right)\right) + \omega_0^{h+1}\right)$$

Recall that $\overrightarrow{s}^{h-1}$ is the output of the hidden layer before the last hidden layer, so it depends on all the weights of the previous layers, namely, $\left(W^1, \overrightarrow{\omega}_0^1, W^2, \overrightarrow{\omega}_0^2, \cdots, W^{h-1}, \overrightarrow{\omega}_0^{h-1}\right)$.

We continue the process systematically until we reach the input layer.

Assessing the Significance of the Input Data Features

One goal of data analysts is to assess the significance of the input variables (data features) with respect to the output or target variable.

The main question to answer here is: if we tweak the value of a certain input variable, what is the relative change of the output?

For example, if we add one more bus on a given bus route, would that affect the overall bus ridership?

The math question that we are asking is a *derivative* question: find the partial derivative of the output with respect to the input variable in question.

We have plenty of literature in statistics on variable significance when the models are linear (sensitivity analysis). When the models are nonlinear, such as our neural network models, there isn't as much literature. We cannot make our predictions based on nonlinear models, then employ variable significance analysis that is built for linear models. Many data analysts who use built-in software packages for their analysis fall into this trap. This is another reason to seek to understand deeply the assumptions of the models on which we base our business decisions.

Summary and Looking Ahead

This chapter represents our official transition to the deep learning era in the AI field. While Chapter 3 presented traditional yet still very useful machine learning models, Chapter 4 adds neural networks to our arsenal of machine learning models. Both chapters built the models with the general mathematical structure of: training function, loss function, and optimization, where each was tailored to the particular task and model at hand.

By employing nonlinear activation functions at each neuron of a neural network, over multiple layers, our training function is able to pick up on complex features in the data that are otherwise hard to describe using an explicit formula of a nonlinear function. Mathematical analysis—in particular, universal approximation theorems for neural networks—back up this intuition and provide a theoretical background that justifies the wild success of neural networks. These theorems, however, still lack the ability to provide us with a map to construct special networks tailored to specific tasks and data sets, so we must experiment with various architectures, regularizations, and hyperparameters until we obtain a neural network model that performs well on new and unseen data.

Neural networks are well tailored to large problems with large data sets. The optimization task for such large problems requires efficient and computationally inexpensive methods, though all computations at that scale can be considered expensive. Stochastic gradient descent is the popular optimization method of choice, and the backpropagation algorithm is the workhorse of this method. More specifically, the backpropagation algorithm computes the gradient of the loss function (or the objective function when we add weight decay regularization) at the current weight choice. Understanding the landscape of the objective function remains central for any optimization task, and as a rule of thumb, convex problems are easier to optimize than nonconvex ones. Loss functions involved in neural network models are generally nonconvex.

Chapter 4 is the last foundational (and long) chapter in this book. We can finally discuss more specialized AI models, as well as deeper mathematics, when needed. The next chapters are independent from each other, so read them in the order that feels most relevant to your immediate application area.

Finally, let's summarize the mathematics that appeared in this chapter, which we must elaborate more on as we progress in the field:

Probability and measure
  This is needed to prove universal approximation-type theorems, and will be discussed in Chapter 11. It is also related to uncertainty analysis for dropout.

Statistics

Input standardizing steps during batch normalization at each layer of the neural network, and the resulting reshaping of the related distributions.

Optimization

Gradient descent, stochastic gradient descent, convex, and nonconvex landscapes.

Calculus on linear algebra

Backpropagation algorithm: this is the chain rule from calculus applied on functions of matrices of variables.

CHAPTER 5

Convolutional Neural Networks and Computer Vision

They. Can. See.

—H.

Convolutional neural networks have revolutionized the computer vision and the natural language processing fields. Application areas, irrespective of the ethical questions associated with them (such as surveillance, automated weapons, etc.), are limitless: self-driving cars, smart drones, facial recognition, speech recognition, medical imaging, generating audio, generating images, robotics, etc.

In this chapter, we start with the simple definitions and interpretations of *convolution* and *cross-correlation*, and highlight the fact that these two slightly different mathematical operations are conflated in machine learning terminology. We perpetrate the same sin and conflate them as well, but with a good reason.

We then apply the convolution operation to filtering grid-like signals, which it is perfectly suited for, such as time series data (one-dimensional), audio data (one-dimensional), and images (two-dimensional if the images are grayscale, and three-dimensional if they are color images, with the extra dimension corresponding to the red, green, and blue channels). When data is one-dimensional, we use one-dimensional convolutions, and when it is two-dimensional, we use two-dimensional convolutions (for the sake of simplicity and conciseness, we will not do three-dimensional convolutions in this chapter, corresponding to three-dimensional color images, called *tensors*). In other words, we adapt our network to the shape of the data, a process that has wildly contributed to the success of convolutional neural networks. This is in contrast to forcing the data to adapt to the shape of the network's input, such as flattening a two-dimensional image into one long vector to make it fit a network that only takes one-dimensional data as its input. In later chapters, we see that the same applies for the success of graph neural networks for graph-type data.

Next, we incorporate convolution into the architecture of a feed forward neural network. The convolution operation has the effect of making the network *locally connected* as opposed to fully connected. Mathematically, the matrix containing the weights for each layer is not *dense* (so most of the weights for convolutional layers are zero). Moreover, the weights have similar values (*weight sharing*), unlike the case of fully connected neural networks, where a different weight is assigned to each input. Therefore, the matrix containing the weights of a convolutional layer is mostly zeros, and the nonzero parts are localized and share similar values. For image data or audio data, this is great, since most of the information is contained locally. Moreover, this dramatically reduces the number of weights that we need to store and compute during the optimization step, rendering convolutional neural networks ideal for data with a massive amount of input features (recall that for images, each pixel is a feature).

We then discuss *pooling*, which is another layer that is common to the architecture of convolutional neural networks. As in the previous chapter, both the multilayer structure and the nonlinearity at each layer enable us to extract increasingly complex features from images, significantly enhancing computer vision tasks.

Once we comprehend the basic anatomy of a convolutional neural network, it is straightforward to apply the same mathematics to tasks involving natural language processing, such as sentiment analysis, speech recognition, audio generation, and others. The fact that the same mathematics works for both computer vision and natural language processing is akin to the impressive ability of our brain to physically change in response to circumstances, experiences, and thoughts (the brain's version of virtual simulations). Even when some portions of the brain are damaged, other parts can take over and perform new functions. For example, the parts of the brain devoted to sight can start performing hearing or remembering tasks when those are impaired. In neuroscience, this is called *neuroplasticity*. We are very far from having a complete understanding of the brain and how it works, but the simplest explanation to this phenomenon is that each neuron in the brain performs the same basic function, similar to how neurons in a neural network perform one basic mathematical calculation (actually two: linearly combine, then activate), and the various neural connections over multiple layers produce the observed complexity in perception and behavior.

Convolutional neural networks are in fact inspired by the brain's visual neocortex. The success in 2012 for image classification (AlexNet2012 (*https://oreil.ly/QxjXD*)) propelled AI back into the mainstream, inspiring many and bringing us here. If you happen to have some extra time, a good bedtime read accompanying this chapter would be about the function of the brain's visual neocortex, and its analogy to convolutional neural networks designed for computer vision.

Convolution and Cross-Correlation

Convolution and cross-correlation are slightly different operations and measure different things in a signal, which can be a digital image, a digital audio signal, or others. They are exactly the same if we use a symmetric function k, called a *filter* or a *kernel*. Briefly speaking, convolution *flips* the filter, then slides it across the function, while cross-correlation slides the filter across the function *without flipping* it. Naturally, if the filter happens to be symmetric, then convolution and cross-correlation are exactly the same. Flipping the kernel has the advantage of making the convolution operation commutative, which in turn is good for writing theoretical proofs. That said, from a neural networks perspective, commutativity is not important, for three reasons:

- First, the convolution operation does not usually appear alone in a neural network, but is composed with other nonlinear functions, so we lose commutativity irrespective of whether we flip the kernel or not.

- Second, a neural network usually *learns* the values of the entries in the kernel during the training process; thus, it learns the correct values in the correct locations, and flipping becomes immaterial.

- Third, and this is important for practical implementation, convolutional networks usually use *multichannel convolution*; for example, the input can be a color image with red-green-blue channels, or even a video, with red-green-blue space channels and one time channel. Furthermore, they use batch mode convolution, meaning they take input vectors, images, videos, or other data types, in batches and apply parallel convolution operations simultaneously. Even with kernel flipping, these operations are not guaranteed to be commutative unless each operation has the same number of output channels as input channels. This is usually not the case, since the outputs of multiple channels are usually summed together as a whole or partially, producing a different number of output channels than input channels.

For all these reasons, many machine learning libraries do not flip the kernel when implementing convolution, in essence implementing cross-correlation and calling it convolution. We do the same here.

The convolution operation between two real valued functions k (the filter) and f is defined as:

$$(k \star f)(t) = \int_{-\infty}^{\infty} f(s)k(-s+t)ds$$

$$= \int_{-\infty}^{\infty} f(-s+t)k(s)ds$$

and the discrete analog for discrete functions is:

$$(k \star f)(n) = \sum_{s=-\infty}^{\infty} f(s)k(-s+n)$$

$$= \sum_{s=-\infty}^{\infty} f(-s+n)k(s)$$

The cross-correlation operation between two real valued functions k (the filter) and f is defined as:

$$(k * f)(t) = \int_{-\infty}^{\infty} f(s)k(s+t)ds$$

$$= \int_{-\infty}^{\infty} f(s-t)k(s)ds$$

and the discrete analog for discrete functions is:

$$(k * f)(n) = \sum_{s=-\infty}^{\infty} f(s)k(s+n)$$

$$= \sum_{s=-\infty}^{\infty} f(s-n)k(s)$$

Note that the formulas defining convolution and cross-correlation look exactly the same, except that for convolution we use -s instead of s. This corresponds to flipping the involved function (before shifting). Note also that the indices involved in the convolution integral and sum add up to t or n, which is not the case for cross-correlation. This makes convolution commutative, in the sense that $(f \star k)(n) = (k \star f)(n)$, and cross-correlation not necessarily commutative. We comment further on commutativity later.

Every time we encounter a new mathematical object, it is good to pause and ask ourselves a few questions before diving in. This way we can build a solid mathematical foundation while managing to avoid the vast and dark ocean of technical and complicated mathematics.

What kind of mathematical object do we have on our hands?

For convolution and cross-correlation, we are looking at infinite sums for the discrete case, and integrals over infinite domains for the continuum case. This guides our next question.

Since we are summing over infinite domains, what kind of functions can we allow without our computations blowing up to infinity?

In other words, for what kind of functions do these infinite sums and integrals exist and are well-defined? Here, start with the easy answers, such as when f and k are compactly supported (are zeros except in a finite portion of the domain), or when f and k decay rapidly enough to allow the infinite sum or integral to converge. Most of the time, these simple cases are enough for our applications, such as image and audio data, and the filters that we accompany them with. Only when we find that our simple answer does not apply to our particular use case do we seek more general answers. These come in the form of theorems and proofs. Do not seek the most general answers first, as these are usually built on top of a large mathematical theory that took centuries, and countless questions and searches for answers, to take form and materialize. Taking the most general road first is overwhelming, counterintuitive, counter-historical, and time- and resource-draining. It is not how mathematics and analysis naturally evolve. Moreover, if we happen to encounter someone talking in the most general and technical language without providing any context or motivation for why this level of generality and complexity is needed, we just tune them out and move on peacefully with our lives; otherwise they might confuse us beyond repair.

What is this mathematical object used for?

The convolution operation has far-reaching applications that range from purely theoretical mathematics to applied sciences to the engineering and systems design fields. In mathematics, it appears in many fields, such as differential equations, measure theory, probability, statistics, analysis, and numerical linear algebra. In the applied sciences and engineering, it is used in acoustics, spectroscopy, image processing and computer vision, and in the design and implementation of finite impulse response filters in signal processing.

How is this mathematical object useful in my particular field of study and how does it apply to my particular interest or use case?

For our AI purposes, we will use the convolution operation to construct convolutional neural networks for both one-dimensional text and audio data and two-dimensional image data. The same ideas generalize to any type of high-dimensional data where most of the information is contained locally. We use convolutional neural networks in two contexts: *understanding* image, text, and audio data; and *generating* image, text, and audio data.

Additionally, in the context of data and distributions of data, we use the following result that has to do with the probability distribution of the sum of two independent random variables. If μ and ν are probability measures on the topological group $(\mathbb{R}, +)$, and if X and Y are two independent random variables whose respective distributions are μ and ν, then the convolution $\mu \star \nu$ is the probability distribution of the sum random variable $X + Y$. We elaborate on this in Chapter 11 on probability.

How did this mathematical object come to be?

It is worth investing a tad bit of extra time to learn some of the history and the chronological order of when, how, and why the object that we care about first appeared, along with the main results associated with it. In other words, rather than studying our mathematical object through a series of dry lemmas, propositions, and theorems, which are usually deprived of all context, we learn through its own story, and through the ups and downs that mathematicians encountered while attempting to develop it. One of the most valuable insights we learn here is that mathematics develops organically with our quest to answer various questions, establish connections, and gain a deeper understanding of something that we need to use. Modern mathematical analysis developed while attempting to answer very simple questions related to Fourier sine and cosine series (decomposing a function into its component frequencies), which turned out to be not so simple for many types of functions. For example: when can we interchange the integral and the infinite sum, what is an integral, and what is dx anyway?

What comes as a surprise to many, especially to people who feel intimidated or scared of mathematics, is that during the quest to gain understanding, some of the biggest names in mathematics, including the fathers and mothers of certain fields, made multiple mistakes along the way and corrected them at later times, or were corrected by others until the theory finally took shape.

One of the earliest uses of the convolution integral appeared in 1754, in d'Alembert's derivation of Taylor's theorem. Later, between 1797 and 1800, it was used by Sylvestre François Lacroix in his book, *Treatise on Differences and Series*, which is part of his encyclopedic series, *An Elementary Treatise on Differential Calculus and Integral Calculus*. Soon thereafter, convolution operations appear in the works of very famous names in mathematics, such as Laplace, Fourier, Poisson, and Volterra. What is common here is that all of these studies have to do with integrals, derivatives, and series of functions. In other words, calculus, and again, decomposing functions into their component frequencies (Fourier series and transform).

What are the most important operations, manipulations, and/or theorems related to this mathematical object that we must be aware of before diving in?

Things do not spring into fame or viral success without being immensely beneficial to many, many people. The convolution operation is so simple yet so useful and generalizes neatly to more involved mathematical entities, such as measures and distributions. It is commutative, associative, distributive over addition and scalar multiplication; its integral becomes a product, and its derivative transforms to differentiating only one of its component functions.

Translation Invariance and Translation Equivariance

These properties of convolutional neural networks enable us to detect similar features in different parts of an image. That is, a pattern that occurs in one location of an image is easily recognized in other locations of the image. The main reason is that at a convolutional layer of a neural network, we convolute with the same filter (also called a kernel, or template) throughout the image, picking up on the same patterns (such as edges, or horizontal, vertical, and diagonal orientations). This filter has a fixed set of weights. Recall that this is not the case for fully connected neural networks, where we would have to use different weights for different pixels of an image. Since we use convolution instead of matrix multiplication to perform image filtering, we have the benefit of *translation invariance*, since usually all we care for is the presence of a pattern, irrespective of its location. Mathematically, translation invariance looks like:

$$trans_a(k) \star f = k \star trans_a(f) = trans_a(k \star f)(t)$$

where $trans_a$ is the translation of a function by a. For our AI purposes, this implies that given a filter designed to pick up on a certain feature in an image, convolving it with a translated image (in the horizontal or vertical directions), is the same as filtering the image *then* translating it. This property is sometimes called *translation equivariance*, and translation invariance is instead attributed to the pooling layer that is often built into the architecture of convolutional neural networks. We discuss this later in this chapter. In any case, the fact that we use one filter (one set of weights) throughout the whole image at each layer means that we detect one pattern at various locations of the image, when present. The same applies to audio data or any other type of grid-like data.

Convolution in Usual Space Is a Product in Frequency Space

The Fourier transform of the convolution of two functions is the product of the Fourier transform of each function, up to a scaling. In short, the Fourier transform resolves a function into its frequency components (we elaborate on the Fourier transform in Chapter 13). Therefore, the convolution operation does not create new frequencies, and the frequencies present in the convolution function are simply the product of the frequencies of the component functions.

In mathematics, we are forever updating our arsenal of useful tools for tackling different problems, so depending on our domain of expertise, it is at this point that we branch out and study related but more involved results, such as *circular convolutions for periodic functions*, preferred algorithms for computing convolutions, and others. The best route for us to dive into convolution in a way that is useful for our AI purposes is through signal and system design, which is the topic of the next section.

Convolution from a Systems Design Perspective

We are surrounded by systems, each interacting with its environment and designed to accomplish a certain task. Examples include HVAC systems in our buildings, adaptive cruise control systems in our cars, city transportation systems, irrigation systems, various communication systems, security systems, navigation systems, data centers, etc. Some systems interact with each other, others do not. Some are very large, others are as small and simple as a single device, such as a thermostat, receiving signals from its environment through sensors, processing them, and outputting other signals, for example, to actuators.

The convolution operation appears when designing and analyzing such simple systems, which process an input signal and produce an output signal, if we impose two special constraints: *linearity* and *time invariance*. Linearity and time invariance become linearity and *translation or shift invariance* if we are dealing with *space-dependent signals*, such as images, instead of *time-dependent signals*, such as electrical signals or audio signals. Note that a video is both space- (two or three spatial dimensions) and time-dependent. Linearity in this context has to do with the output of a scaled signal (amplified or reduced), and the output of two superimposed signals. Time and translation invariance have to do with the output of a delayed (time-dependent) signal, or the output of a translated or shifted (space-dependent) signal. We will detail these in the next section.

Together, linearity and time or translation invariance are very powerful. They allow us to find the system's output of *any signal* only if we know the output of a simple *impulse signal*, called the system's *impulse response*. The system's output to *any input signal* is obtained by merely convolving the signal with the system's impulse response. Therefore, imposing the conditions of linearity and time or translation invariance dramatically simplifies the analysis of signal processing systems.

The nagging question is then: how realistic are these wonder-making conditions? In other words, how prevalent are linear and time/translation invariant systems, or even only approximately linear and approximately time/translation invariant systems?

Aren't most realistic systems nonlinear and complex? Thankfully, we have control over system designs, so we can just decide to design systems with these properties. One example is any electrical circuit consisting of capacitors, resistors, inductors, and linear amplifiers. This is in fact mathematically equivalent to an ideal mechanical spring, mass, and damper system. Other examples that are relevant for us include processing and filtering various types of signals and images. We discuss these in the next few sections.

Convolution and Impulse Response for Linear and Translation Invariant Systems

Let's formalize the concepts of a linear system and a time/translation invariant system, then understand how the convolution operation naturally arises when attempting to quantify the response of a system possessing these properties to *any* signal. From a math perspective, a system is a function H that takes an input signal x and produces an output signal y. The signals x and y can depend on time, space (single or multiple dimensions), or both. If we enforce linearity on such a function, then we are claiming two things:

1. Output of a scaled input signal is nothing but a scaling of the original output: $H(ax) = aH(x) = ay$.

2. Output of two superimposed signals is nothing but the superposition of the two original outputs: $H(x_1 + x_2) = H(x_1) + H(x_2) = y_1 + y_2$.

If we enforce time/translation invariance, then we are claiming that the output of a delayed/translated/shifted signal is nothing but a delayed/translated/shifted original output: $H(x(t - t_0)) = y(t - t_0)$.

We can leverage these conditions when we think of any arbitrary signal, whether discrete or continuous, in terms of a superposition of a bunch of *impulse signals* of various amplitudes. This way, if we are able to measure the system's output to a single impulse signal, called the system's *impulse response*, then that is enough information to measure the system's response to any other signal. This becomes the basis of a very rich theory. We only walk through a discrete case in this chapter, since the signals we care for in AI (for example, for natural language processing, human machine interaction and computer vision), whether one-dimensional audio signals, or two- or three-dimensional images, are discrete. The continuous case is analogous except that we consider infinitesimal steps instead of discrete steps, we use integrals instead of sums, and we enforce continuity conditions (or whatever conditions we need to make the involved integrals well-defined). Actually, the continuous case brings a little extra complication: having to properly define an *impulse* in a mathematically sound way, for it is not a function in the usual sense. Thankfully, there are multiple mathematical ways to make it well-defined, such as the theory of distributions, or defining it as an operator acting on usual functions, or as a measure and using the help of Lebesgue integrals. However, I have been avoiding measures, Lebesgue integrals, and any deep mathematical theory until we truly need them with their added functionality, so for now, the discrete case is very sufficient.

We define a unit impulse $\delta(k)$ to be zero for each nonzero k, and one when $k = 0$, and define its response as $H(\delta(k)) = h(k)$. Then $\delta(n - k)$ would be zero for each k, and 1 for $k = n$. This represents a unit impulse located at $k = n$. Therefore, $x(k)\delta(n - k)$ is an impulse of amplitude $x(k)$ located at $k = n$. Now we can write the input signal $x(n)$ as:

$$x(n) = \Sigma_{k = -\infty}^{\infty} x(k)\delta(n - k)$$

The above sum might seem like such a convoluted way to write a signal, and it is, but it is very helpful, since it says that any discrete signal can be expressed as an infinite sum of unit impulses that are scaled correctly at the right locations. Now, using the linearity and translation invariance assumptions on H, it is straightforward to see that the system's response to the signal $x(n)$ is:

$$H(x(n)) = H\left(\sum_{k = -\infty}^{\infty} x(k)\delta(n - k) \right)$$

$$= \sum_{k = -\infty}^{\infty} x(k)H(\delta(n - k))$$

$$= \sum_{k = -\infty}^{\infty} x(k)h(n - k)$$

$$= (x \star h)(n)$$

$$= y(n)$$

Therefore, a linear and translation invariant system is completely described by its impulse response $h(n)$. But there is another way to look at this, which is independent of linear and translation invariant systems, and which is very useful for our purposes in the next few sections. The statement:

$$y(n) = (x \star h)(n)$$

says that the signal $x(n)$ can be transformed into the signal $y(n)$ after we convolve it with the filter $h(n)$. Hence, designing the filter $h(n)$ carefully can produce a $y(n)$ with some desired characteristics, or can *extract certain features* from the signal $x(n)$, for example, all the edges. Moreover, using different filters $h(n)$ extracts different features from the same signal $x(n)$. We elaborate more on these concepts in the next few sections; meanwhile, keep the following in mind: as information, such as signals or images, flows through a convolutional neural network, different features get extracted (mapped) at each convolutional layer.

Before leaving linear and time/translation invariant systems, we must mention that such systems have a very simple response for sinusoidal inputs. If the input to the system is a sine wave with a given frequency, then the output is also a sine wave with the same frequency, but possibly with a different amplitude and phase. Moreover, knowing the impulse response of the system allows us to compute its *frequency response*, which is the system's response for sine waves at all frequencies, and vice versa. That is, determining the system's impulse response allows us to compute its frequency response, and determining the frequency response allows us to compute its impulse response, which in turn completely determines the system's response to any arbitrary signal. This connection is very useful from both theoretical and applications perspectives, and is closely connected to Fourier transforms and frequency domain representations of signals. In short, the frequency response of a linear and translation invariant system is simply the Fourier transform of its impulse response. We do not go over the computational details here, since these concepts are not essential for the rest of this book; nevertheless, it is important to be aware of these connections and understand how things from seemingly different fields come together and relate to each other.

Convolution and One-Dimensional Discrete Signals

Let's dissect the convolution operation and understand how it creates a new signal from an input signal by sliding a filter (kernel) against it. We will not flip the kernel here, as we established the fact that flipping the kernel is irrelevant from a neural network perspective. We start with a one-dimensional discrete signal $x(n)$, then we convolve it with a filter (kernel) $k(n)$, and produce a new signal $z(n)$. We show how we obtain $z(n)$ one entry at a time by sliding $k(n)$ along $x(n)$. For simplicity, let $x(n) = (x_0, x_1, x_2, x_3, x_4)$ and $k(n) = (k_0, k_1, k_2)$. In this example the input signal $x(n)$ only has five entries, and the kernel has three entries. In practice, such as in signal processing, image filtering, or AI's neural networks, the input signal $x(n)$ is orders of magnitude larger than the filter $k(n)$. We will see this very soon, but the example here is only for illustration. Recall the formula for discrete cross-correlation, which is the convolution without flipping the kernel:

$$(k*x)(n) = \sum_{s=-\infty}^{\infty} x(s)k(s+n)$$
$$= \cdots + x(-1)k(-1+n) + x(0)k(n) + x(1)k(1+n) + x(2)k(2+n) + \cdots$$

Since neither $x(n)$ nor $k(n)$ has infinitely many entries, and the sum is infinite, we pretend that the entries are zero whenever the indices are not defined. The new filtered signal resulting from the convolution will only have nontrivial entries with indices: -4, -3, -2, -1, 0, 1, 2. Let's write each of these entries:

$$(k*x)(-4) = x_4 k_0$$
$$(k*x)(-3) = x_3 k_0 + x_4 k_1$$
$$(k*x)(-2) = x_2 k_0 + x_3 k_1 + x_4 k_2$$
$$(k*x)(-1) = x_1 k_0 + x_2 k_1 + x_3 k_2$$
$$(k*x)(0) = x_0 k_0 + x_1 k_1 + x_2 k_2$$
$$(k*x)(1) = x_0 k_1 + x_1 k_2$$
$$(k*x)(2) = x_0 k_2$$

That operation is easier to understand with a mental picture: fix the signal $x(n) = (x_0, x_1, x_2, x_3, x_4)$ and slide the filter $k(n) = (k_0, k_1, k_2)$ against it from right to left. This can also be summarized concisely using linear algebra notation, where we multiply the input vector $x(n)$ by a special kind of matrix, called the *Toeplitz matrix*, containing the filter weights. We will elaborate on this later in this chapter.

When we slide a filter across the signal, the output signal will peak at the indices where $x(n)$ and $k(n)$ match, so we can design the filter in a way that picks up on certain patterns in $x(n)$. This way, convolution (unflipped) provides a measure of similarity between the signal and the kernel.

We can also view cross-correlation, or unflipped convolution, in a different way: each entry of the output signal is a *weighted average* of the entries of the input signal. This way, we emphasize both the linearity of this transformation and the fact that we can choose the weights in the kernel in way that emphasizes certain features over others. We see this more clearly with image processing, discussed next, but for that we have to write a formula for convolution (unflipped) in two dimensions. In linear algebra notation, discrete two-dimensional convolution amounts to multiplication of the two-dimensional signal by another special kind of matrix, called the *doubly block circulant matrix*, also appearing later in this chapter. Keep in mind that multiplication by a matrix is a linear transformation.

Convolution and Two-Dimensional Discrete Signals

The convolution (unflipped) operation in two dimensions looks like:

$$(k*x)(m,n) = \sum_{q=-\infty}^{\infty} \sum_{s=-\infty}^{\infty} x(m+q, n+s) k(q,s)$$

For example, the (2,1) entry of the convolution (unflipped) between the following 4×4 matrix A and the 3×3 kernel K:

$$A * K = \begin{pmatrix} a_{00} & a_{01} & a_{02} & a_{03} \\ a_{10} & a_{11} & a_{12} & a_{13} \\ a_{20} & \boxed{a_{21}} & a_{22} & a_{23} \\ a_{30} & a_{31} & a_{32} & a_{33} \end{pmatrix} * \begin{pmatrix} k_{00} & k_{01} & k_{02} \\ k_{10} & \boxed{k_{11}} & k_{12} \\ k_{20} & k_{21} & k_{22} \end{pmatrix}$$

is $z_{21} = a_{10}k_{00} + a_{11}k_{01} + a_{12}k_{02} + a_{20}k_{10} + a_{21}k_{11} + a_{22}k_{12} + a_{30}k_{20} + a_{31}k_{21} + a_{32}k_{22}$. To see this, imagine placing the kernel K exactly on top of the matrix A with its center k_{11} on top of a_{21}, meaning the entry of A with the required index, then multiply the entries that are on top of each other and add all the results together. Note that here we only computed one entry of the output signal, meaning if we were working with images, it would be the value of only one pixel of the filtered image. We need all the others! For this, we have to know which indices are valid convolutions. By valid we mean *full*, as in they take all of the kernel entries into account during the computation. The word *valid* is a bit misleading since *all* entries are valid if we allow ourselves to *pad the boundaries of matrix A by zeros*. To find the indices that take the full kernel into account, recall our mental image of placing K exactly on top of A, with K's center at the index that we want to compute. With this placement, the rest of K should not exceed the boundaries of A, so for our example, the good indices will be (1,1), (1,2), (2,1), and (2,2), producing the output:

$$Z = \begin{pmatrix} z_{11} & z_{12} \\ z_{21} & z_{22} \end{pmatrix}$$

This means that if we were working with images, the filtered image Z would be smaller in size than the original image A. If we want to produce an image with the same size as the original image, then we must pad the original image with zeros before applying the filter. For our example, we would need to pad with one layer of zeros around the full boundary of A, but if K was larger, then we would need more layers of zeros. The following is A padded with one layer of zeros:

$$A_{padded} = \begin{pmatrix} 0 & 0 & 0 & 0 & 0 & 0 \\ 0 & a_{00} & a_{01} & a_{02} & a_{03} & 0 \\ 0 & a_{10} & a_{11} & a_{12} & a_{13} & 0 \\ 0 & a_{20} & a_{21} & a_{22} & a_{23} & 0 \\ 0 & a_{30} & a_{31} & a_{32} & a_{33} & 0 \\ 0 & 0 & 0 & 0 & 0 & 0 \end{pmatrix}$$

There are other ways to retain the size of the output image than zero padding, even though zero padding is the simplest and most popular approach. These include:

Reflection
> Instead of adding a layer or multiple layers of zeros, add a layer or multiple layers of the same values as the boundary pixels of the image, and the ones below them, and so on if we need more. That is, instead of turning off the pixels outside an image boundary, extend the image using the same pixels that are already on or near the boundary.

Wraparound
> This is used for periodic signals. Mathematically, this is cyclic convolution, circulant matrix, discrete Fourier transform producing the eigenvalues of the circulant matrix, and the Fourier matrix containing its eigenvectors as columns. We will not dive into that here, but recall that periodicity usually simplifies things, and makes us think of periodic waves, which in turn makes us think of Fourier stuff.

Multiple channels
> Apply more than one *channel* of independent filters (weight matrices), each sampling the original input, then combine their outputs.

Filtering Images

Figure 5-1 shows an example of how convolving the same image with various kernels extracts various features of the image.

For example, the third kernel in the table has 8 at its center and the rest of its entries are –1. This means that this kernel makes the current pixel 8 times as intense, then subtracts from that the values of all the pixels surrounding it. If we are in a uniform region of the image, meaning all the pixels are equal or very close in value, then this process will give zero, returning a black or turned-off pixel. If, on the other hand, this pixel lays on an edge boundary, for example, the boundary of the eye or the boundary of the face of the deer, then the output of the convolution will have a nonzero value, so it will be a bright pixel. When we apply this process to the whole image, the result is a new image with many edges traced with bright pixels and the rest of the image dark.

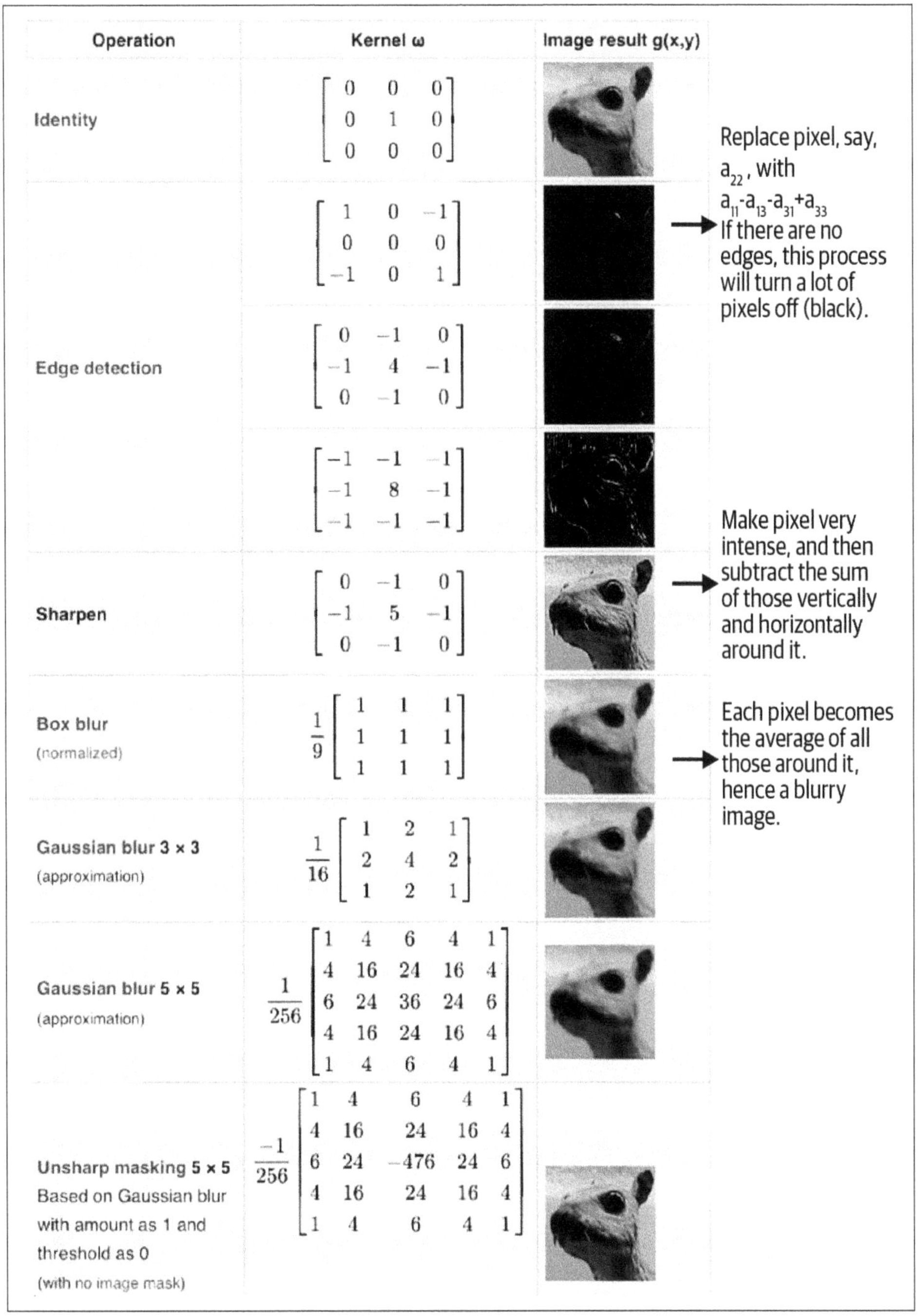

Operation	Kernel ω	Image result g(x,y)	
Identity	$\begin{bmatrix} 0 & 0 & 0 \\ 0 & 1 & 0 \\ 0 & 0 & 0 \end{bmatrix}$		
	$\begin{bmatrix} 1 & 0 & -1 \\ 0 & 0 & 0 \\ -1 & 0 & 1 \end{bmatrix}$		Replace pixel, say, a_{22}, with a_{11}-a_{13}-a_{31}+a_{33} → If there are no edges, this process will turn a lot of pixels off (black).
Edge detection	$\begin{bmatrix} 0 & -1 & 0 \\ -1 & 4 & -1 \\ 0 & -1 & 0 \end{bmatrix}$		
	$\begin{bmatrix} -1 & -1 & -1 \\ -1 & 8 & -1 \\ -1 & -1 & -1 \end{bmatrix}$		Make pixel very intense, and then subtract the sum of those vertically and horizontally around it.
Sharpen	$\begin{bmatrix} 0 & -1 & 0 \\ -1 & 5 & -1 \\ 0 & -1 & 0 \end{bmatrix}$		
Box blur (normalized)	$\frac{1}{9}\begin{bmatrix} 1 & 1 & 1 \\ 1 & 1 & 1 \\ 1 & 1 & 1 \end{bmatrix}$		Each pixel becomes the average of all those around it, hence a blurry image.
Gaussian blur 3 × 3 (approximation)	$\frac{1}{16}\begin{bmatrix} 1 & 2 & 1 \\ 2 & 4 & 2 \\ 1 & 2 & 1 \end{bmatrix}$		
Gaussian blur 5 × 5 (approximation)	$\frac{1}{256}\begin{bmatrix} 1 & 4 & 6 & 4 & 1 \\ 4 & 16 & 24 & 16 & 4 \\ 6 & 24 & 36 & 24 & 6 \\ 4 & 16 & 24 & 16 & 4 \\ 1 & 4 & 6 & 4 & 1 \end{bmatrix}$		
Unsharp masking 5 × 5 Based on Gaussian blur with amount as 1 and threshold as 0 (with no image mask)	$\frac{-1}{256}\begin{bmatrix} 1 & 4 & 6 & 4 & 1 \\ 4 & 16 & 24 & 16 & 4 \\ 6 & 24 & -476 & 24 & 6 \\ 4 & 16 & 24 & 16 & 4 \\ 1 & 4 & 6 & 4 & 1 \end{bmatrix}$		

Figure 5-1. Applying various filters to an image (image source (https://oreil.ly/S2Nfu))

As we see in the table, the choice of the kernel that is able to detect edges, blur, etc., is not unique. The same table includes two-dimensional discrete Gaussian filters for blurring. When we discretize a one-dimensional Gaussian function, we do not lose its symmetry, but when we discretize a two-dimensional Gaussian function, we lose its radial symmetry since we have to approximate its natural circular or elliptical shape by a square matrix. Note that a Gaussian peaks at the center and decays as it spreads away from the center. Moreover, the area under its curve (surface for two dimensions) is one. This has the overall effect of averaging and smoothing (removing noise) when we convolve this with another signal. The price we pay is removing sharp edges, which is exactly what blurring is (think that a sharp edge is replaced by a smooth decaying average of itself and all the surrounding pixels within a distance of a few standard deviations from the center). The smaller the standard deviation (or the variance, which is the square of the standard deviation), the more detail we can retain from the image.

Another great example is shown in Figure 5-2. Here, each image on the right is the result of the convolution between the image on the left and the filter (kernel) portrayed in the middle of each row. These filters are called Gabor filters (*https:// oreil.ly/LRVLi*). They are designed to pick up on certain patterns/textures/features in an image, and they work in a similar way to the filters found in the human visual system.

Our eyes detect the parts of an image where things change, i.e., contrasts. They pick up on edges (horizontal, vertical, diagonal) and gradients (measuring the steepness of a change). We design filters that do the same things mathematically, sliding them via convolution across a signal. These produce smooth and uneventful results (zeros or numbers close to each other in value) when nothing is changing in the signal, and spiking when an edge or a gradient that lines up with the kernel is detected.

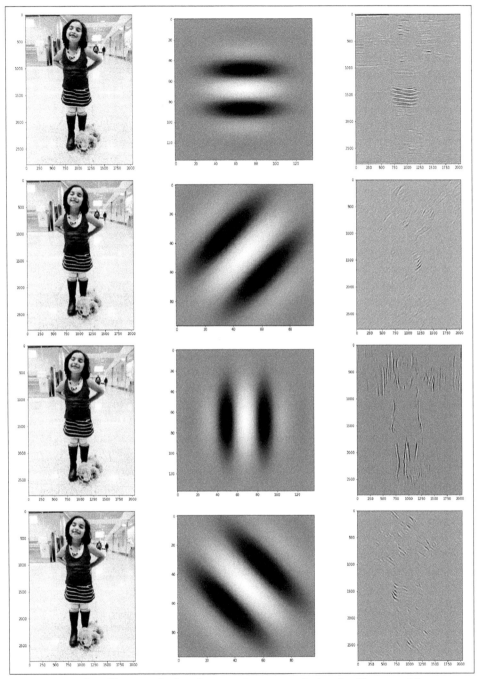

Figure 5-2. Gabor filters applied to the same image; we design the filters to pick up on different textures and orientations in the image. Visit the book's GitHub page (https://github.com/halanelson/Essential-Math-For-AI) to reproduce these images.

Feature Maps

A convolutional neural network *learns* the kernels from the data by optimizing a loss function in the same way a fully connected network does. The unknown weights that enter the formula of the training function are the entries of each kernel at each convolutional layer, the biases, and the weights related to any fully connected layer involved in the network's architecture. The output of a convolutional layer (which includes the nonlinear activation function) is called a *feature map*, and the learned kernels are feature detectors. A common observation in neural networks is that earlier layers in the network (close to the input layer) learn low-level features such as edges, and later layers learn higher-level features such as shapes. This is naturally expected since at each new layer we compose with a nonlinear activation function, so complexity increases over multiple layers and so does the network's ability to express more elaborate features.

How do we plot feature maps?

Feature maps help us open the black box and directly observe what a trained network detects at each of its convolutional layers. If the network is still in the training process, feature maps allow us to pinpoint the sources of error, then tweak the model accordingly. Suppose we input an image to a trained convolutional neural network. At the first layer, using the convolution operation, a kernel slides through the whole image and produces a new filtered image. This filtered image then gets passed through a nonlinear activation function, producing yet another image. Finally, this gets passed through a pooling layer, explained soon, and produces the final output of the convolutional layer. This output is a different image than the one we started with, and possibly of different dimensions, but if the network was trained well, the output image would highlight some important features of the original image, such as edges, texture, etc. This output image is usually a matrix of numbers or a tensor of numbers (three-dimensional for color images, or four-dimensional if we are working in batches of images or with video data, where there is one extra dimension for the time series). It is easy to visualize these as feature maps using the matplotlib library in Python, where each entry in the matrix or tensor is mapped to the intensity of a pixel in the same location as the matrix entry. Figure 5-3 shows various feature maps at various convolutional layers of a convolutional neural network.

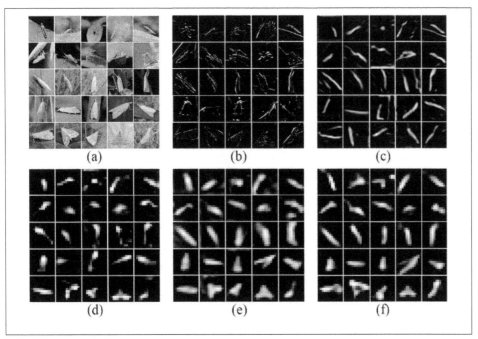

Figure 5-3. Feature maps at various convolutional layers of a convolutional neural network (image source (https://oreil.ly/WvY1l))

Linear Algebra Notation

We now know that the most fundamental operation inside a convolutional neural network is the convolution. Given a filter k (could be one-dimensional, two-dimensional, three-dimensional, or higher), a convolution operation takes an input signal and applies k to it by sliding it across the signal. This operation is linear, meaning each output is a linear combination (by the weights in k) of the input components, thus it can be efficiently represented as a matrix multiplication. We need this efficiency, since we must still write the training function representing the convolutional neural network in a way that is easy to evaluate and differentiate. The mathematical structure of the problem is still the same as for many machine learning models, which we are very familiar with now:

Training function

For convolutional neural networks, this usually includes linear combinations of components of the input, composed with activation functions, then a pooling function (discussed soon), over multiple layers of various sizes and connections, and finally topped with a logistic function, a support vector machine function, or other functions depending on the ultimate purpose of the network (classification, image segmentation, data generation, etc.).

The difference from fully connected neural networks is that the linear combination happens now with the weights associated with the filter, which means they are not all different from each other (unless we are only doing locally connected layers instead of convolutional layers).

Moreover, the sizes of the filters are usually orders of magnitude smaller than the input signals, so when we express the convolution operation in matrix notation, most of the weights in the matrix will actually be zero. Recall that there is a weight per each input feature at each layer of a network. For example, if the input is a color image, there is a different weight for each pixel in each channel, unless we decide to implement a convolutional layer–then there would only be a few unique weights, and many, many zeros. This simplifies storage requirements and computation time tremendously, while at the same time capturing the important local interactions.

That said, it is common in the architecture of a convolutional neural network to have fully connected layers (with a different weight assigned to each connection) near the output layer. One can rationalize this as a distillation of features: after important and locally dependent features, which increase in complexity with each layer, are captured, these are combined to make a prediction. That is, when information arrives to a fully connected layer, it would have been distilled to its most important components, which in turn act as unique features contributing to a final prediction.

Loss function
This is similar to all the loss functions we went over in the previous chapters, always providing a measure of the error made by the network's predictions and ground truths.

Optimization
This again uses stochastic gradient descent, due to the enormity of the size of the problem and the number of variables that go into it. Here, as usual, we need to evaluate one derivative of the loss function, which includes one derivative of the training function. We compute the derivative with respect to all the unknown weights involved in the filters of all the network's layers and channels, and the associated biases. Computationally, the backpropagation algorithm is still the workhorse of the differentiation process.

Linear algebra and computational linear algebra have all the tools we need to produce trainable convolutional neural networks. In general, the worst kind of matrix to be involved in our computations is a dense (mostly nonzero entries) matrix with no obvious structure or pattern to its entries (even worse if it is nondiagonalizable, etc.). But when we have a sparse matrix (mostly zeros), or a matrix with a certain pattern to its entries (such as diagonal, tridiagonal, circular, etc.), or both, then we are in a computationally friendly world, given that we learn how to exploit the special matrix structure to our advantage. Researchers who study large matrix computations and algorithms are the kings and queens of such necessary exploitation, and without their work we'd be left with theory that is very hard to implement in practice and at scale.

In one dimension, the convolution operation can be represented using a special kind of matrix, called the *Toeplitz* matrix; and in two dimensions, it can be represented using another special kind of matrix, called a *doubly block circulant* matrix. Let's only focus on these two, but with the take-home lesson that matrix notation, in general, is the best way to go, and we would be fools not to discover and make the most use of the inherent structure within our matrices. In other words, attacking the most general cases first might just be a sad waste of time, which happens to be a rare commodity in this life. A good compromise between working with the most general matrix and the most specific matrix is to accompany results with complexity analysis, such as computing the order of a method ($O(n^3)$ or $O(n \log n)$, etc.), so that stakeholders are made aware of the trade-offs between implementing certain methods versus others.

We borrow a simple example from the free ebook *Deep Learning* (Chapter 9, page 334) (*https://www.deeplearningbook.org*) by Ian Goodfellow et al. (MIT Press), on the efficiency of using either convolution or exploiting the many zeros in a matrix versus using the usual matrix multiplication to detect vertical edges within a certain image. Convolution is an extremely efficient way of describing transformations that apply the same linear transformation of a small local region across the entire input.

The image on the right of Figure 5-4 was formed by taking each pixel in the original image and subtracting the value of its neighboring pixel on the left. This shows the strength of all the vertically oriented edges in the input image, which can be a useful operation for object detection.

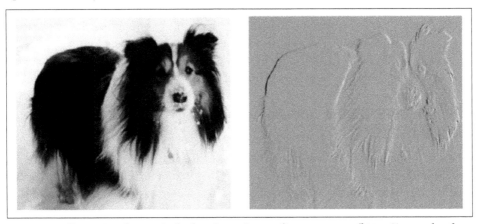

Figure 5-4. Detecting the vertical edges in an image (image source (https://www.deeplear ningbook.org))

Both images are 280 pixels tall. The input image is 320 pixels wide, while the output image is 319 pixels wide. This transformation can be described by a convolution kernel containing two elements, and requires $319 \times 280 \times 3 = 267{,}960$ floating-point operations (two multiplications and one addition per output pixel) to compute using convolution.

To describe the same transformation with a matrix multiplication would take $320 \times 280 \times 319 \times 280$, or over eight billion entries in the matrix, making convolution four billion times more efficient for representing this transformation. The straightforward matrix multiplication algorithm performs over 16 billion floating-point operations, making convolution roughly 60,000 times more efficient computationally. Of course, most of the entries of the matrix would be zero. If we stored only the nonzero entries of the matrix, then both matrix multiplication and convolution would require the same number of floating-point operations to compute. The matrix would still need to contain $2 \times 319 \times 280 = 178{,}640$ entries.

—*Deep Learning*, by Ian Goodfellow et al.

The One-Dimensional Case: Multiplication by a Toeplitz Matrix

A *banded* Toeplitz matrix looks like:

$$Toeplitz = \begin{pmatrix} k_0 & k_1 & k_2 & 0 & 0 & 0 & 0 \\ 0 & k_0 & k_1 & k_2 & 0 & 0 & 0 \\ 0 & 0 & k_0 & k_1 & k_2 & 0 & 0 \\ 0 & 0 & 0 & k_0 & k_1 & k_2 & 0 \\ 0 & 0 & 0 & 0 & k_0 & k_1 & k_2 \end{pmatrix}$$

Multiplying this Toeplitz matrix by a one-dimensional signal $x = (x_0, x_1, x_2, x_3, x_4, x_5, x_6, x_7)$ yields the exact result of the convolution of a one-dimensional filter $k = (k_1, k_2, k_3)$ with the signal x, namely, $(Toeplitz)x^t = k * x$. Performing the multiplication, we see the *sliding effect* of the filter across the signal.

The Two-Dimensional Case: Multiplication by a Doubly Block Circulant Matrix

The two-dimensional analog involves the two-dimensional convolution operation and filtering images. Here, instead of multiplying with a Toeplitz matrix, we end up multiplying with a *doubly block circulant matrix*, where each row is a circular shift of a given vector. It is a nice exercise in linear algebra to write this matrix down along with its equivalence to two-dimensional convolution. In deep learning, we end up learning the weights, which are the entries of these matrices. This linear algebra notation (in terms of Toeplitz or circulant matrices) helps us find compact formulas for the derivatives of the loss function with respect to these weights.

Pooling

One step common to almost all convolutional neural networks is *pooling*. This is typically implemented after the input gets filtered via convolution, then passed through the nonlinear activation function. There is more than one type of pooling, but the idea is the same: replace the current output at a certain location with a summary statistic of the nearby outputs. An example for images is to replace four pixels with one pixel containing the maximum value of the original four (*max pooling*), or with their average value, or a weighted average, or the square root of the sum of their squares, etc.

Figure 5-5 shows how max pooling works.

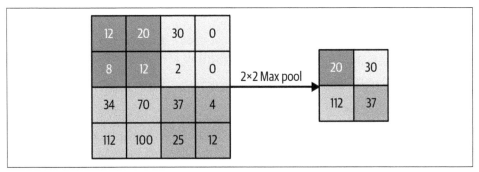

Figure 5-5. Max pooling (image source (https://oreil.ly/y9x9P))

In effect, this reduces the dimension and summarizes whole neighborhoods of outputs, at the expense of sacrificing fine detail. So pooling is not very good for use cases where fine detail is essential for making predictions. Nevertheless, pooling has many advantages:

- It provides approximate invariance to small spatial translations of the input. This is useful if we care more about whether some feature is present than its exact location.
- It can greatly improve the statistical efficiency of the network.
- It improves the computational efficiency and memory requirements of the network because it reduces the number of inputs to the next layer.
- It helps with handling inputs of varying sizes, because we can control the size of the pooled neighborhoods and the size of the output after pooling.

A Convolutional Neural Network for Image Classification

It is impossible to have an exhaustive list of all the different architectures and varia-tions that are involved in neural networks without diverting from the main purpose of the book: understanding the mathematics that underlies the different models. It is possible, however, to go over the essential components and how they all come together to accomplish an AI task, such as image classification for computer vision. During the training process, the steps of Chapter 4 still apply:

1. Initialize random weights (according to the initialization processes we described in Chapter 4).

2. Forward pass a batch of images through the convolutional network and output a class for the image.

3. Evaluate the loss function for this particular choice of weights.

4. Backpropagate the error through the network.

5. Adjust the weights that contributed to the error (stochastic gradient descent).

6. Repeat for a certain number of iterations, or until you converge.

Thankfully, we do not have to do any of this on our own. Python's Keras library has many pre-trained models, which means that their weights have already been fixed, and all we have to do is evaluate the trained model on our particular data set.

What we can and should do is observe and learn the architecture of successful and winning networks. Figure 5-6 shows the simple architecture of LeNet1 (*https://oreil.ly/kKgqL*) by LeCun et al. (1989), and Figure 5-7 shows AlexNet's (2012) architecture.

A good exercise is to count the number of the weights that go into the training function of LeNet1 and AlexNet. Note that more units in each layer (feature maps) means more weights. When I tried to count the weights involved in LeNet1 based on the architecture in figure Figure 5-6, I ended up with 9,484 weights, but the original paper mentions 9,760 weights, so I do not know where the rest of the weights are. If you find them please let me know. Either way, the point is we need to solve an optimization problem in $\mathbb{R}^{9760}$. Now do the same computation for AlexNet in Figure 5-7: we have around 62.3 million weights, so the optimization problem ends up in $\mathbb{R}^{62.3 million}$. Another startling number: we need 1.1 billion computation units for one forward pass through the net.

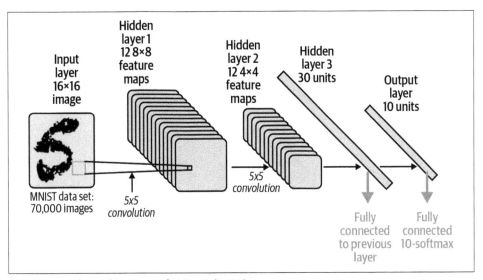

Figure 5-6. The architecture of LeNet1 (1989)

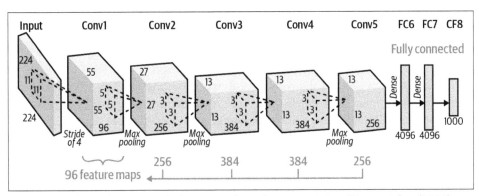

Figure 5-7. The architecture of AlexNet with a whopping 62.3 million weights (2012); adapted from https://oreil.ly/eEWgJ

Figure 5-8 shows a wonderful illustration of an image of the handwritten digit 8 passing through a pre-trained LeNet1 and ultimately getting correctly classified as 8.

Finally, if the choice of architecture seems arbitrary to you, meaning that if you are wondering whether we could accomplish similar performance with simpler architecture, join the club. The whole community is wondering the same thing.

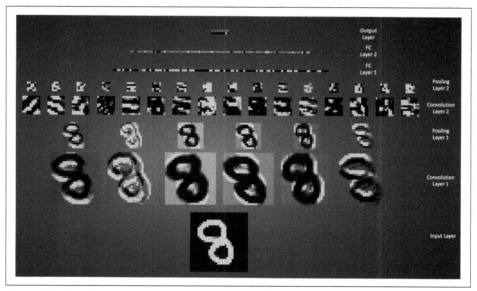

Figure 5-8. Passing an image of a handwritten 8 through a pretrained LeNet1 (image source (https://oreil.ly/6vgOc))

Summary and Looking Ahead

In this chapter, we defined the convolution operation: the most significant component of convolutional neural networks. Convolutional neural networks are essential for computer vision, machine audio processing, and other AI applications.

We presented convolution from a systems design perspective, then through filtering one-dimensional and two-dimensional signals. We highlighted the linear algebra equivalent of the convolution operation (multiplying by matrices of special structure), and ended with an example of image classification.

We will encounter convolutional neural networks frequently in this book, as they have become a staple in many AI systems that include vision and/or natural language.

Singular Value Decomposition: Image Processing, Natural Language Processing, and Social Media

Show me the essential, and only the essential.

—H.

The singular value decomposition is a mathematical operation from linear algebra that is widely applicable in the fields of data science, machine learning, and artificial intelligence. It is the mathematics behind principal component analysis (in data analysis) and latent semantic analysis (in natural language processing). This operation transforms a dense matrix into a diagonal matrix. In linear algebra, diagonal matrices are very special and highly desirable. They behave like scalar numbers when we multiply by them, only stretching or squeezing in certain directions.

When computing the singular value decomposition of a matrix, we get the extra bonus of revealing and quantifying the action of the matrix on space itself: rotating, reflecting, stretching, and/or squeezing. There is no warping (bending) of space, since this operation is linear (after all, it is called linear algebra). Extreme stretching or squeezing in one direction versus the others affects the stability of any computations involving our matrix, so having a measure of that allows us direct control over the sensitivity of our computations to various perturbations, for example, noisy measurements.

The power of the singular value decomposition lies in the fact that it can be applied to any matrix. That and its wide use in the field of AI earns it its own chapter in this book. In the following sections, we explore the singular value decomposition, focusing on the big picture rather than the tiny details, and on applications to image processing, natural language processing, and social media.

Given a matrix C (an image, a data matrix, etc.), we omit the details of computing its singular value decomposition. Most linear algebra books do that, presenting a theoretical method based on computing the eigenvectors and eigenvalues of the symmetric matrices $C^t C$ and CC^t, which for us are the covariance matrices of the data (if the data is centered). While it is still very important to understand the theory, the method it provides for computing the singular value decomposition is not useful for efficient computations, and is especially impossible for the large matrices involved in many realistic problems. Moreover, we live in an era where software packages help us compute it so easily. In Python, all we have to do is call the numpy.linalg.svd method from the *numpy* library. We peek briefly into the numerical algorithms that go into these software packages later in this chapter. However, our main focus is on understanding how the singular value decomposition works and why this decomposition is important for reducing the storage and computational requirements of a given problem without losing its essential information. We will also understand the role it plays in clustering data.

Matrix Factorization

We can factorize a scalar number in multiple ways; for instance, we can write the number $12 = 4 \times 3$, $12 = 2 \times 2 \times 3$, or $12 = 0.5 \times 24$. Which factorization is better depends on our use case. The same can be done for matrices of numbers. Linear algebra provides us with a variety of useful matrix factorizations. The idea is that we want to break down an object into its smaller components, and these components give us insight about the function and the action of the object itself. This breakdown also gives us a good idea about which components contain the most information, and thus are more important than others. In this case, we might benefit from throwing away the less important components, and building a smaller object with a similar function. The smaller object might not be as detailed as the object that we started with, as that contained all of its components; however, it contains enough significant information from the original object that using it with its smaller size provides benefits. The singular value decomposition is a matrix factorization that does exactly that. Its formula looks like:

$$C_{m \times n} = U_{m \times m} \Sigma_{m \times n} V^t_{n \times n},$$

where we break down the matrix C into three component matrices: U, Σ, and V^t. U and V are square matrices that have orthonormal rows and columns. Σ is a diagonal matrix that has the same shape as C (see Figure 6-1).

Let's start with matrix multiplication. Suppose A is a matrix with 3 rows and 3 columns:

$$A = \begin{pmatrix} 1 & 2 & 3 \\ 4 & 5 & 6 \\ 7 & 8 & 9 \end{pmatrix}_{3 \times 3}$$

and B is a matrix with 3 rows and 2 columns:

$$B = \begin{pmatrix} 1 & 3 \\ 4 & -2 \\ 0 & 1 \end{pmatrix}_{3 \times 2}$$

Then $C = AB$ is a matrix with 3 rows and 2 columns:

$$C_{3 \times 2} = A_{3 \times 3} B_{3 \times 2} = \begin{pmatrix} 1 & 2 & 3 \\ 4 & 5 & 6 \\ 7 & 8 & 9 \end{pmatrix}_{3 \times 3} \begin{pmatrix} 1 & 3 \\ 4 & -2 \\ 0 & 1 \end{pmatrix}_{3 \times 2} = \begin{pmatrix} 9 & 2 \\ 24 & 8 \\ 39 & 14 \end{pmatrix}_{3 \times 2}$$

We can think of C as being factorized into the product of A and B, in the same way as the number $12 = 4 \times 3$. The previous factorization of C has no significance, since neither A nor B is a special type of matrix. A very significant factorization of C is its singular value decomposition. Any matrix has a singular value decomposition. We calculate it using Python (see the associated Jupyter notebook for the code):

$$C_{3 \times 2} = U_{3 \times 3} \Sigma_{3 \times 2} V^t_{2 \times 2} =$$
$$\begin{pmatrix} -0.1853757 & 0.8938507 & 0.4082482 \\ -0.5120459 & 0.2667251 & -0.8164965 \\ -0.8387161 & -0.3604005 & 0.4082482 \end{pmatrix} \begin{pmatrix} 49.402266 & 0 \\ 0 & 1.189980 \\ 0 & 0 \end{pmatrix} \begin{pmatrix} -0.9446411 & -0.3281052 \\ 0.3281052 & -0.9446411 \end{pmatrix}$$

Observe the following in this decomposition: the rows of V^t are the *right singular vectors* (these are exactly the columns of V), the columns of U are the *left singular vectors*, and the diagonal entries of Σ are the *singular values*. The singular values are always positive and always arranged in decreasing order along the diagonal of Σ. The ratio of largest singular value to the smallest singular value is the *condition number κ* of a matrix. In our case there are only two singular values, and $\kappa = \frac{49.402266}{1.189980} = 41.515207$. This number plays an important role in the stability of computations involving our matrix. Well-conditioned matrices are those with condition numbers that are not very large.

The left singular vectors are orthonormal (orthogonal to each other and have length 1). Similarly, the right singular vectors are also orthonormal.

For qualitative properties, images are faster to assess than endless arrays of numbers. It is easy to visualize matrices as images using Python (and vice versa, images are stored as matrices of numbers): the value of an entry of the matrix corresponds to the intensity of the corresponding pixel. The higher the number, the brighter the pixel. Smaller numbers in the matrix show up as darker pixels, and larger numbers show up as brighter pixels. Figure 6-1 shows the previously mentioned singular value decomposition. We observe that the diagonal matrix Σ has the same shape as C, and its diagonal entries are arranged in decreasing order, with the brightest pixel, corresponding to the largest *singular value*, at the top-left corner.

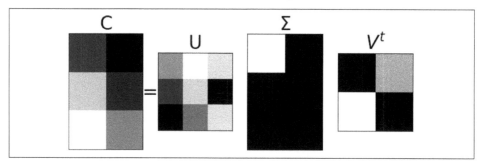

Figure 6-1. Visualizing the singular value decomposition

Figures 6-2 and 6-3 visualize the singular value decompositions of two rectangular matrices A and B, where A is wide and B is tall:

$$A = \begin{pmatrix} -1 & 3 & -5 & 4 & 18 \\ 1 & -2 & 4 & 0 & -7 \\ 2 & 0 & 4 & -3 & -8 \end{pmatrix}_{3 \times 5} = U_{3 \times 3} \Sigma_{3 \times 5} V^t_{5 \times 5}$$

$$B = \begin{pmatrix} 5 & 4 \\ 4 & 0 \\ 7 & 10 \\ -1 & 8 \end{pmatrix}_{4 \times 2} = U_{4 \times 4} \Sigma_{4 \times 2} V^t_{2 \times 2}$$

In Figure 6-2, we note that the last two columns of Σ are all zeros (black pixels), hence we can economize in storage and throw away these two columns along with the last two rows of V^t (see the next section for multiplying by a diagonal matrix from the left). Similarly, in Figure 6-3, we note that the last two rows of Σ are all zeros (black pixels), hence we can economize in storage and throw away these two rows along with the last two columns of U (see the next section for multiplying by a diagonal matrix from the right). The singular value decomposition is already saving us some space (note that we usually only store the diagonal entries of Σ as opposed to the whole matrix with all its zeros).

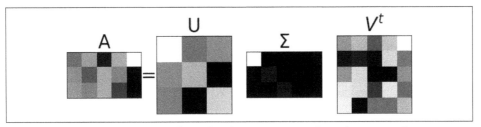

Figure 6-2. Visualizing the singular value decomposition of a wide rectangular matrix. The last two columns of Σ are all zeros (black pixels), allowing for storage reduction: throw away the last two columns of Σ along with the last two rows of V^t.

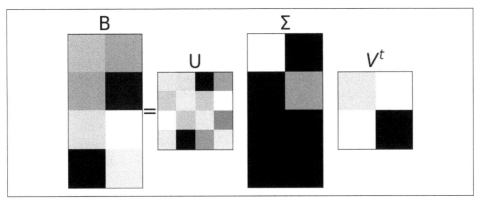

Figure 6-3. Visualizing the singular value decomposition of a tall rectangular matrix. The last two rows of Σ are all zeros (black pixels), allowing for storage reduction: throw away the last two rows of Σ along with the last two columns of U.

Diagonal Matrices

When we multiply a vector by a scalar number, say 3, we obtain a new vector along the same direction with the same orientation, but whose length is stretched three times. When we multiply the same vector by another scalar number, say –0.5, we get another vector, again along the same direction, but this time its length is halved and its orientation is flipped. Multiplying by a scalar number is such a simple operation, and it would be nice if we had matrices that behaved equally easily when we applied them to (in other words, multiplied them by) vectors. If our life was one-dimensional we would only have to deal with scalar numbers, but since our life and applications of interest are higher dimensional, then we have to satisfy ourselves with diagonal matrices (Figure 6-4). These are the good ones.

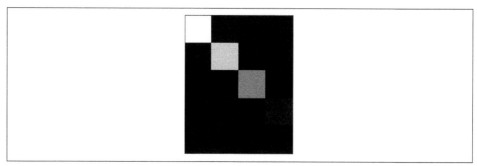

Figure 6-4. An image of a 5 × 4 diagonal matrix with diagonal entries: 10 (brightest pixel), 6, 3, and 1 (darkest pixel other than the zeros)

Multiplying by a diagonal matrix corresponds to stretching or squeezing in certain directions in space, with orientation flipping corresponding to any negative numbers on the diagonal. As we very well know, most matrices are very far from being diagonal. The power of the singular value decomposition is that it provides us with the directions in space along which the matrix *behaves like* (albeit in a broad sense) a diagonal matrix. A diagonal matrix usually stretches/squeezes in the same directions as those for the vector coordinates. If, on the other hand, the matrix is not diagonal, it generally does not stretch/squeeze in the same directions as the coordinates. It does so in *other* directions, *after* a change of coordinates. The singular value decomposition gives us the required coordinate change (*right singular vectors*), the directions along which vectors will be stretched/squeezed (*left singular vectors*), as well as the magnitude of the stretch/squeeze (*singular values*). We detail this in the next section, but we first clarify multiplication by diagonal matrices from the left and from the right.

If we multiply a matrix A by a diagonal matrix Σ from the right, $A\Sigma$, then we scale the columns of A by the σ's, for example:

$$A\Sigma = \begin{pmatrix} a_{11} & a_{12} \\ a_{21} & a_{22} \\ a_{31} & a_{32} \end{pmatrix} \begin{pmatrix} \sigma_1 & 0 \\ 0 & \sigma_2 \end{pmatrix} = \begin{pmatrix} \sigma_1 a_{11} & \sigma_2 a_{12} \\ \sigma_1 a_{21} & \sigma_2 a_{22} \\ \sigma_1 a_{31} & \sigma_2 a_{32} \end{pmatrix}$$

If we multiply A by Σ from the left ΣA, then we scale the rows of A by the σ's, for example:

$$\Sigma A = \begin{pmatrix} \sigma_1 & 0 & 0 \\ 0 & \sigma_2 & 0 \\ 0 & 0 & \sigma_3 \end{pmatrix} \begin{pmatrix} a_{11} & a_{12} \\ a_{21} & a_{22} \\ a_{31} & a_{32} \end{pmatrix} = \begin{pmatrix} \sigma_1 a_{11} & \sigma_1 a_{12} \\ \sigma_2 a_{21} & \sigma_2 a_{22} \\ \sigma_3 a_{31} & \sigma_3 a_{32} \end{pmatrix}$$

Matrices as Linear Transformations Acting on Space

One way we can view matrices is as linear transformations (no warping) that act on vectors in space, and on space itself. If no warping is allowed because it would render an operation nonlinear, then what actions are allowed? The answers are rotation, reflection, stretching, and/or squeezing, which are all nonwarping operations. The singular value decomposition $A = U\Sigma\ V^t$ captures this concept. When A acts on a vector $\vec{v}$, let's go over the multiplication $A\vec{v} = U\Sigma\ V^t\vec{v}$ step-by-step:

1. First $\vec{v}$ gets rotated/reflected because of the orthogonal matrix V^t.
2. Then it gets stretched/squeezed along special directions because of the diagonal matrix Σ.
3. Finally, it gets rotated/reflected again because of the other orthogonal matrix U.

Reflections and rotations do not really change space, as they preserve size and symmetries (think of rotating an object or looking at its reflection in a mirror). The amount of stretch and/or squeeze encoded in the diagonal matrix Σ (via its singular values on the diagonal) is very informative regarding the action of A.

Orthogonal Matrix

An orthogonal matrix has orthonormal rows and orthonormal columns. It never stretches or squeezes, it only rotates and/or reflects, meaning that it does not change the size and shape of objects when acting on them, only their direction and/or orientation. As with many things in mathematics, these names are confusing. It is called an *orthogonal* matrix even though its rows and columns are an *orthonormal*, which means orthogonal *and* of length equal to one. One more useful fact: if C is an orthogonal matrix, then $CC^t = C^tC = I$, that is, the inverse of this matrix is its transpose. Computing the inverse of a matrix is usually a very costly operation, but for orthogonal matrices, all we have to do is exchange its rows for its columns.

We illustrate these concepts using two-dimensional matrices since they are easy to visualize. In the following subsections, we explore:

- The action of a matrix A on the right singular vectors, which are the columns $\vec{v}_1$ and $\vec{v}_2$ of the matrix V. These get sent to multiples of the left singular vectors $\vec{u}_1$ and $\vec{u}_2$, which are the columns of U.

- The action of A on the standard unit vectors $\vec{e}_1$ and $\vec{e}_2$. We also notice that the unit square gets transformed to a parallelogram.

- The action of A on a general vector $\vec{x}$. This will help us understand the matrices U and V as rotations or reflections in space.

- The action of A on the unit circle. We see that A transforms the unit circle to an ellipse, with its principal axes along the left singular vectors (the $\vec{u}$'s), and the lengths of its principal axes are the singular values (the σ's). Since the singular values are ordered from largest to smallest, then $\vec{u}_1$ defines the direction with the most variation, and $\vec{u}_2$ defines the direction with the second most variation, and so on.

Action of A on the Right Singular Vectors

Let A be the 2×2 matrix:

$$A = \begin{pmatrix} 1 & 5 \\ -1 & 2 \end{pmatrix}$$

Its singular value decomposition $A = U\Sigma V^t$ is given by:

$$A = \begin{pmatrix} 0.93788501 & 0.34694625 \\ 0.34694625 & -0.93788501 \end{pmatrix} \begin{pmatrix} 5.41565478 & 0 \\ 0 & 1.29254915 \end{pmatrix} \begin{pmatrix} 0.10911677 & 0.99402894 \\ 0.99402894 & -0.10911677 \end{pmatrix}$$

The expression $A = U\Sigma V^t$ is equivalent to:

$$AV = U\Sigma$$

since all we have to do is multiply $A = U\Sigma V^t$ by V from the right and exploit the fact that $V^t V = I$ due to the orthogonality of V.

We can think of AV as the matrix A acting on each column of the matrix V. Since $AV = U\Sigma$, then the action of A on the orthonormal columns of V is the same as stretching/squeezing the columns of U by the singular values. That is:

$$A\vec{v}_1 = \sigma_1 \vec{u}_1$$

and

$$A\vec{v}_2 = \sigma_2 \vec{u}_2$$

This is demonstrated in Figure 6-5.

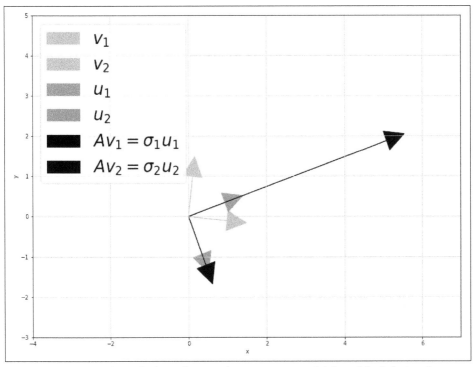

Figure 6-5. Matrix A sends the right singular vectors to multiples of the left singular vectors: $Av_1 = \sigma_1 u_1$ and $Av_2 = \sigma_2 u_2$

Action of A on the Standard Unit Vectors and the Unit Square Determined by Them

The matrix A sends the standard unit vectors to its own columns and transforms the unit square into a parallelogram. There is no warping (bending) of space. Figure 6-6 shows this transformation.

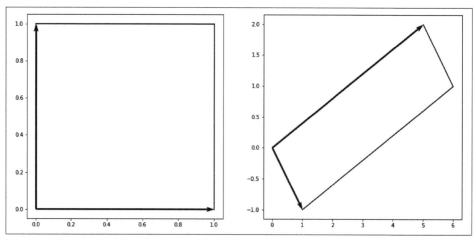

Figure 6-6. Transforming the standard unit vectors

Action of A on the Unit Circle

Figure 6-7 shows that the matrix A sends the unit circle to an ellipse. The principal axes along the u's and lengths of the principal axes are equal to the σ's. Again, since matrices represent linear transformations, there is reflection/rotation and stretching/squeezing of space, but no warping.

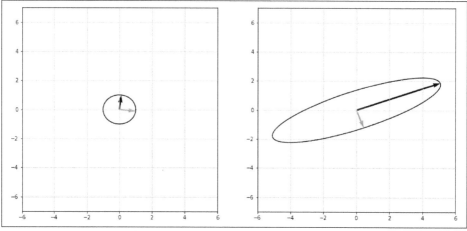

Figure 6-7. Matrix A sends the unit circle to an ellipse with principal axes along the left singular vectors and lengths of principal axes equal to the singular values

We can easily see the described action from the singular value decomposition.

The polar decomposition:

$$A = QS$$

is a very easy way that geometrically shows how a circle gets transformed into an ellipse.

Breaking Down the Circle-to-Ellipse Transformation According to the Singular Value Decomposition

Figure 6-8 shows four subplots that break down the steps of the circle-to-ellipse transformation illustrated previously:

1. First we multiply the unit circle and the vectors $\vec{v}_1$ and $\vec{v}_2$ by V^t. Since $V^t V = I$, we have $V^t \vec{v}_1 = \vec{e}_1$ and $V^t \vec{v}_2 = \vec{e}_2$. So, in the beginning, the right singular vectors get *straightened out*, aligning correctly with the standard unit vectors.

2. Then we multiply by Σ. All that happens here is stretching/squeezing the standard unit vectors by σ_1 and σ_2 (the stretch or squeeze depend on whether the magnitude of the singular value is greater or smaller than one).

3. Finally we multiply by U. This either reflects the ellipse across a line or rotates it a certain amount clockwise or counterclockwise. The next subsection explains this in detail.

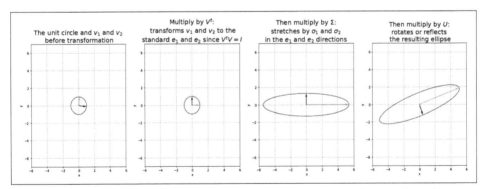

Figure 6-8. Steps of the unit circle-to-ellipse transformation using the singular value decomposition

Rotation and Reflection Matrices

The matrices U and V^t that appear in the singular value decomposition $A = U\Sigma V^t$ are orthogonal matrices. Their rows and columns are orthonormal, and their inverse is the same as their transpose. In two dimensions, the U and V could either be rotation or reflection (about a line) matrices.

Rotation matrix

A matrix that rotates clockwise by an angle θ is given by:

$$\begin{pmatrix} \cos\theta & \sin\theta \\ -\sin\theta & \cos\theta \end{pmatrix}$$

The transpose of a rotation matrix is a rotation in the opposite direction. So if a matrix rotates clockwise by an angle θ, then its transpose rotates counterclockwise by θ and is given by:

$$\begin{pmatrix} \cos\theta & -\sin\theta \\ \sin\theta & \cos\theta \end{pmatrix}$$

Reflection matrix

A reflection matrix about a line L making an angle θ with the x-axis is:

$$\begin{pmatrix} \cos 2\theta & \sin 2\theta \\ \sin 2\theta & -\cos 2\theta \end{pmatrix}$$

The slope of the straight line L is $\tan\theta$ and it passes through the origin, so its equation is $y = (\tan\theta)x$. This line acts like a mirror for the reflection operation. Figure 6-9 shows the two straight lines about which the matrices V^t and U reflect, together with a vector $\vec{x}$ and its subsequent transformation.

The determinant of a rotation matrix is 1, and the determinant of the reflection matrix is -1.

In higher dimensions, reflection and rotation matrices look different. Always make sure you understand the object you are dealing with. If we have a rotation in a three-dimensional space, then about what axis? If we have a reflection, then about what plane? If you want to dive deeper, this is a good time to read about orthogonal matrices and their properties.

Action of A on a General Vector $\vec{x}$

We have explored the action of A on the right singular vectors (they get mapped to the left singular vectors), the standard unit vectors (they get mapped to the columns of A), the unit square (it gets mapped to a parallelogram), and the unit circle (it gets mapped to an ellipse with principal axes along the left singular vectors and whose lengths are equal to the singular values). Finally, we explore the action of A on a general, nonspecial, vector $\vec{x}$. This gets mapped to another nonspecial vector $A\vec{x}$. However, breaking down this transformation into steps using the singular value decomposition is informative.

Recall our matrix A and its singular value decomposition:

$$
\begin{aligned}
A &= \begin{pmatrix} 1 & 5 \\ -1 & 2 \end{pmatrix} \\
&= U\Sigma V^t \\
&= \begin{pmatrix} 0.93788501 & 0.34694625 \\ 0.34694625 & -0.93788501 \end{pmatrix} \begin{pmatrix} 5.41565478 & 0 \\ 0 & 1.29254915 \end{pmatrix} \begin{pmatrix} 0.10911677 & 0.99402894 \\ 0.99402894 & -0.10911677 \end{pmatrix}
\end{aligned}
$$

Both U and V^t in this singular value decomposition happen to be reflection matrices. The straight lines L_U and L_{V^t} that act as mirrors for these reflections are plotted in Figure 6-9, and their equations are easy to find from their respective matrices: $\cos(2\theta)$ and $\sin(2\theta)$ are on the first row, so we can use those to find the slope $\tan(\theta)$. The equation of the line along which V^t reflects is then $y = (\tan\theta_{V^t})x = 0.8962347008436108x$, and that of the line along which U reflects is $y = (\tan\theta_U)x = 0.17903345403184898x$. Since $A\vec{x} = U\Sigma V^t\vec{x}$, first $\vec{x}$ gets reflected across the line L_{V^t}, arriving at $V^t\vec{x}$. Then, when we multiply by Σ from the left, the first coordinate of $V^t\vec{x}$ gets stretched horizontally by the first singular value, and the second coordinate gets stretched by the second singular value, obtaining $\Sigma V^t\vec{x}$. Finally, when we multiply by U, the vector $\Sigma V^t\vec{x}$ gets reflected across the line L_U, arriving at $A\vec{x} = U\Sigma V^t\vec{x}$. Figure 6-9 illustrates this process.

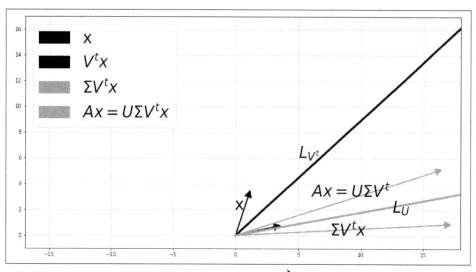

Figure 6-9. Action of a matrix A on a general vector $\vec{x}$. The transformation is done in steps using the singular value decomposition.

Three Ways to Multiply Matrices

Efficient algorithms for matrix multiplication are so desirable in the age of big data. In theory, there are three ways to multiply two matrices $A_{m \times n}$ and $B_{n \times s}$:

Row-column approach
> Produce one entry $(ab)_{ij}$ at a time by taking the dot product of the *i*th row of A with the *j*th column of B:

$$(ab)_{ij} = A_{row_i}B_{col_j} = \Sigma_{k=1}^{n} a_{ik}b_{kj}$$

Column-columns approach
> Produce one column $(AB)_{col_i}$ at a time by linearly combining the columns of A using the entries of the *i*th column of B:

$$(AB)_{col_i} = b_{1i}A_{col_1} + b_{2i}A_{col_2} + \cdots + b_{ni}A_{col_n}$$

Column-row approach
> Produce *rank one pieces* of the product, one piece at a time, by multiplying the first column of A with the first row of B, the second column of A with the second row of B, and so on. Then add all these rank one matrices together to get the final product AB:

$$AB = A_{col_1}B_{row_1} + A_{col_2}B_{row_2} + \cdots + A_{col_n}B_{row_n}$$

How does this help us understand the usefulness of the singular value decomposition? We can expand the product $A = U\Sigma V^t$ of the singular value decomposition as a sum of rank one matrices, using the *column-row* approach for matrix multiplication. Here, we multiply the matrix $U\Sigma$ (which scales each column U_{col_i} of U by σ_i) with V^t:

$$A = U\Sigma V^t = \sigma_1 U_{col_1} V^t_{row_1} + \sigma_2 U_{col_2} V^t_{row_2} + \cdots + \sigma_r U_{col_r} V^t_{row_r}$$

where r is the number of nonzero singular values of A (also called the *rank of A*).

The great thing about this expression is that it splits A into a sum of rank one matrices arranged according to their order of importance, since the σ's are arranged in decreasing order. Moreover, it provides a straightforward way to approximate A by lower rank matrices: throw away lower singular values. The *Eckart–Young–Mirsky theorem* (*https://oreil.ly/5eYKY*) asserts that this is in fact the *best way* to find low rank approximation of A, when the *closeness* of approximation is measured using the *Frobenius norm* (*https://oreil.ly/Yev1c*) (which is the square root of the sum of squares of the singular values) for matrices. Later in this chapter, we take advantage of this rank one decomposition of A for digital image compression.

Algorithms for Matrix Multiplication

Finding efficient algorithms for matrix multiplication is an essential, yet surprisingly difficult, goal. In matrix multiplication algorithms, saving on even one multiplication operation is worthy (saving on addition is not as much of a big deal). Recently, DeepMind developed AlphaTensor (2022) (*https://oreil.ly/HZPbd*) to automatically discover more efficient algorithms for matrix multiplication. This is a milestone because matrix multiplication is a fundamental part of a vast array of technologies, including neural networks, computer graphics, and scientific computing.

The Big Picture

So far we have focused on the singular value decomposition of a matrix $A = U\Sigma V^t$ in terms of A's action on space and in terms of approximating A using lower rank matrices. Before moving to applications relevant to AI, let's have an eagle-eye perspective and address the big picture.

Given a matrix of real numbers, we want to understand the following, depending on our use case:

- If the matrix represents data that we care about, like images or tabular data, what are the most important components of this matrix (data)?
- Along what important directions is the data mostly spread (directions with most variation in the data)?
- If I think of a matrix $A_{m \times n}$ as a transformation from the initial space $\mathbb{R}^n$ to the target space $\mathbb{R}^m$, what is the effect of this matrix on vectors in $\mathbb{R}^n$? To which vectors do they get sent in $\mathbb{R}^m$?
- What is the effect of this matrix on space itself? Since this is a linear transformation, we know there is no space warping, but there is space stretching, squeezing, rotating, and reflecting.
- Many physical systems can be represented as a system of linear equations $A\vec{x} = \vec{b}$. How can we solve this system (find $\vec{x}$)? What is the most efficient way to go about this, depending on the properties of A? If there is no solution, is there an approximate solution that satisfies our purposes? Note that here we are looking for the unknown vector $\vec{x}$ that gets transformed to $\vec{b}$ when A acts on it.

The singular value decomposition can be used to answer all these questions. The first two are intrinsic to the matrix itself, while the second two have to do with the effect of multiplying the matrix with vectors (the matrix acts on space and the vectors in this space). The last question has to do with the very important problem of solving systems of linear equations and appears in all kinds of applications.

Therefore, we can investigate a matrix of numbers in two ways:

- What are its intrinsic properties?
- What are its properties when viewed as a transformation?

These two are related because the matrix's intrinsic properties affect how it acts on vectors and on space.

The following are properties to keep in mind:

- A sends the orthonormal vectors v_i (right singular vectors) of its initial space to scalar multiples of the orthonormal vectors u_i (left singular vectors) of its target space:

$$Av_i = \sigma_i u_i$$

- If our matrix is square, then the absolute value of its determinant is equal to the product of all its singular values: $\sigma_1\sigma_2\cdots\sigma_r$.

- The condition number of the matrix, with respect to the l^2 norm, which is the usual distance in Euclidean space, is the ratio of the largest singular value to the smallest singular value:

$$\kappa = \frac{\sigma_1}{\sigma_r}$$

The Condition Number and Computational Stability

The condition number is very important for computational stability:

- The condition number measures how much A stretches space. If the condition number is too large, then it stretches space too much in one direction relative to another direction, and it could be dangerous to do computations in such an extremely stretched space. Solving $A\vec{x} = \vec{b}$ when A has a large condition number makes the solution $\vec{x}$ unstable in the sense that it is extremely sensitive to perturbations in $\vec{b}$. A small error in $\vec{b}$ will result in a solution $\vec{x}$ that is wildly different from the solution without the error in $\vec{b}$. It is easy to envision this instability geometrically.

- Numerically solving $A\vec{x} = \vec{b}$ (say, by Gaussian elimination) and iterative methods works fine when the involved matrices have reasonable (not very large) condition numbers.

- One thing about a matrix with an especially large condition number: it stretches space so much that it almost collapses into a space of lower dimension. The interesting part is that if we decide to throw away that very small singular value and hence work in the collapsed space of lower dimension, our computations become perfectly fine. So at the boundaries of extremeness lies normalcy, except that this normalcy now lies in a lower dimension.

- Many iterative numerical methods, including the very useful gradient descent, have matrices involved in their analysis. If the condition number of these matrices is too large, then the iterative method might not converge to a solution. The condition number controls how fast these iterative methods converge.

The Ingredients of the Singular Value Decomposition

In this chapter we have been dissecting only one formula: $A = U\Sigma V^t$. We used Python to compute the entries of U, Σ, and V, but what exactly are these entries? The answer is short, if we happen to know what eigenvectors and eigenvalues are, which we clarify in the next section. For now, we list the ingredients of U, Σ, and V:

- The columns of V (the right singular vectors) are the orthonormal eigenvectors of the symmetric matrix $A^t A$.

- The columns of U (the left singular vectors) are the orthonormal eigenvectors of the symmetric matrix AA^t.

- The singular values $\sigma_1, \sigma_2, \cdots \sigma_r$ are the square roots of the eigenvalues of $A^t A$ or AA^t. The singular values are nonnegative and arranged in decreasing order. The singular values can be zero.

- $Av_i = \sigma_i u_i$.

 Every real symmetric positive semi-definite (nonnegative eigenvalues) matrix is diagonalizable $S = PDP^{-1}$, which means that it is similar to a diagonal matrix D when viewed in a different set of coordinates (the columns of P). $A^t A$ and AA^t both happen to be symmetric positive semi-definite, so they are diagonalizable.

Singular Value Decomposition Versus the Eigenvalue Decomposition

It is important to learn more about symmetric matrices if we want to understand the ingredients of the singular value decomposition. This will also help us discern the difference between the singular value decomposition $A = U\Sigma V^t$ and the eigenvalue decomposition $A = PDP^{-1}$ when the latter exists.

 The singular value decomposition (SVD) always exists, but the eigenvalue decomposition exists only for special matrices, called diagonalizable. Rectangular matrices are never diagonalizable. Square matrices may or may not be diagonalizable. When the square matrix is diagonalizable, the SVD and the eigenvalue decomposition are *not equal*, unless the matrix is symmetric and has nonnegative eigenvalues.

We can think of a hierarchy in terms of the desirability of matrices:

1. The best and easiest matrices are square diagonal matrices with the same number along the diagonal.

2. The second best ones are square diagonal matrices D that don't necessarily have the same numbers along the diagonal.

3. The third best matrices are symmetric matrices. These have real eigenvalues and orthogonal eigenvectors. They are the next closest type of matrices to diagonal matrices, in the sense that they are diagonalizable $S = PDP^{-1}$, or similar to a diagonal matrix after a change in basis. The columns of P (eigenvectors) are orthogonal.

4. The fourth best matrices are square matrices that are diagonalizable $A = PDP^{-1}$. These are similar to a diagonal matrix after a change of basis; however, the columns of P (eigenvectors) need not be orthogonal.

5. The fifth best matrices are all the rest. These are not diagonalizable, meaning there is no change of basis that can turn them diagonal; however, there is the next closest approach to making them similar to a diagonal matrix via the singular value decomposition $A = U\Sigma V^t$. Here U and V are different from each other, and they have orthonormal columns and rows. Their inverse is very easy, since it is the same as their transpose. The singular value decomposition works for both square and nonsquare matrices.

Given a matrix A, both $A^t A$ and $A A^t$ happen to be symmetric and positive semi-definite (meaning their eigenvalues are nonnegative); thus, they are diagonalizable with two bases of orthogonal eigenvectors. When we divide by the norm of these orthogonal eigenvectors, they become orthonormal. These are the columns of V and of U, respectively.

$A^t A$ and $A A^t$ have exactly the same nonnegative eigenvalues, $\lambda_i = \sigma_i^2$. Arrange the square root of these in decreasing order (keeping the corresponding eigenvector order in U and V), and we get the diagonal matrix Σ in the singular value decomposition.

What if the matrix we start with is symmetric? How is its singular value decomposition $A = U\Sigma V^t$ related to its diagonalization $A = PDP^{-1}$? The columns of P, which are the eigenvectors of symmetric A, are orthogonal. When we divide by their lengths, they become orthonormal. Stack these orthonormal eigenvectors in a matrix in the order corresponding to the decreasing absolute value of the eigenvalues and we get both the U and the V for the singular value decomposition. Now if all the eigenvalues of symmetric A happen to be nonnegative, the singular value decomposition of this positive semi-definite symmetric matrix will be the same as its eigenvalue decomposition, provided you normalize the orthogonal eigenvectors in P, ordering them with respect to the nonnegative eigenvalues in decreasing order.

So $U = V$ in this case. What if some (or all) of the eigenvalues are negative? Then $\sigma_i = |\lambda_i| = -\lambda_i$, but now we have to be careful with the corresponding eigenvectors $Av_i = -\lambda_i v_i = \lambda_i(-v_i) = \sigma_i u_i$. This makes U and V in the singular value decomposition unequal. So the singular value decomposition of a symmetric matrix that has some negative eigenvalues can be easily extracted from its eigenvalue decomposition, but it is not exactly the same.

What if the matrix we start with is not symmetric but diagonalizable? How is its singular value decomposition $A = U\Sigma V^t$ related to its diagonalization $A = PDP^{-1}$? In this case, the eigenvectors of A, which are the columns of P, are in general not orthogonal, so the singular value decomposition and the eigenvalue decomposition of such a matrix are not related.

Computation of the Singular Value Decomposition

How do Python and others numerically calculate the singular value decomposition of a matrix? What numerical algorithms lie under the hood? The fast answer is: *QR decomposition*, *Householder reflections*, and *iterative algorithms* for eigenvalues and eigenvectors.

In theory, calculating the singular value decomposition for a general matrix, or the eigenvalues and the eigenvectors for a square matrix, requires setting a polynomial equal to 0 to solve for the eigenvalues, then setting up a linear system of equations to solve for the eigenvectors. This is far from being practical for applications. The problem of finding the zeros of a polynomial is very sensitive to any variations in the coefficients of the polynomials, so the computational problem becomes prone to roundoff errors that are present in the coefficients. We need stable numerical methods that find the eigenvectors and eigenvalues without having to numerically compute the zeros of a polynomial. Moreover, we need to make sure that the matrices involved in linear systems of equations are well conditioned, otherwise popular methods like Gaussian elimination (the LU decomposition) do not work.

Most numerical implementations of the singular value decomposition try to avoid computing AA^t and A^tA. This is consistent with one of the themes of this book: avoid multiplying matrices; instead, multiply a matrix with vectors. The popular numerical method for the singular value decomposition uses an algorithm called *Householder reflections* to transform the matrix to a bidiagonal matrix (sometimes preceded by a QR decomposition), then uses iterative algorithms to find the eigenvalues and eigenvectors. The field of numerical linear algebra develops such methods and adapts them to the types and sizes of matrices that appear in applications. In the next subsection, we present an iterative method to compute one eigenvalue and its corresponding eigenvector for a given matrix.

Computing an Eigenvector Numerically

An eigenvector of a square matrix A is a nonzero vector that does not change its direction when multiplied by A; instead, it only gets scaled by an eigenvalue λ:

$$Av = \lambda v$$

The following iterative algorithm is an easy numerical method that finds an eigenvector of a matrix corresponding to its largest eigenvalue:

1. Start at a random unit vector (of length 1) v_0.

2. Multiply by A: $v_{i+1} = Av_i$.

3. Divide by the length of v_{i+1} to avoid the size of our vectors growing too large.

4. Stop when you converge.

This iterative method is very simple but has a drawback: it only finds one eigenvector of the matrix—the eigenvector corresponding to its largest eigenvalue. So it finds the direction that gets stretched the most when we apply A.

For example, consider the matrix $A = \begin{pmatrix} 1 & 2 \\ 2 & -3 \end{pmatrix}$. We start with the vector $\vec{v}_0 = \begin{pmatrix} 1 \\ 0 \end{pmatrix}$ and apply the above algorithm. We note the algorithm after 28 iterations to the vector $\vec{v} = \begin{pmatrix} -0.38268343 \\ 0.92387953 \end{pmatrix}$. The code is in the linked Jupyter notebook and the output is shown here:

```
[1, 0]
[0.4472136  0.89442719]
[ 0.78086881 -0.62469505]
[-0.1351132   0.99083017]
[ 0.49483862 -0.86898489]
[-0.3266748   0.9451368]
[ 0.40898444 -0.91254136]
[-0.37000749  0.92902877]
[ 0.38871252 -0.92135909]
[-0.37979817  0.92506937]
[ 0.3840601  -0.9233081]
[-0.38202565  0.92415172]
[ 0.38299752 -0.92374937]
[-0.38253341  0.92394166]
[ 0.38275508 -0.92384985]
[-0.38264921  0.92389371]
[ 0.38269977 -0.92387276]
[-0.38267563  0.92388277]
[ 0.38268716 -0.92387799]
[-0.38268165  0.92388027]
[ 0.38268428 -0.92387918]
```

```
[-0.38268303  0.9238797 ]
[ 0.38268363 -0.92387945]
[-0.38268334  0.92387957]
[ 0.38268348 -0.92387951]
[-0.38268341  0.92387954]
[ 0.38268344 -0.92387953]
[-0.38268343  0.92387953]
[ 0.38268343 -0.92387953]


 v= [-0.38268343  0.92387953]
Av= [ 1.46507563 -3.53700546]
$\lambda=$ -3.828427140993716
```

Figure 6-10 shows this iteration. Note that all the vectors have length 1 and that the direction of the vector does not change when the algorithm converges, hence capturing an eigenvector of A. For the last few iterations, the sign keeps oscillating, so the vector keeps flipping orientation, and the eigenvalue must be negative. Indeed, we find it to be $\lambda = -3.828427140993716$.

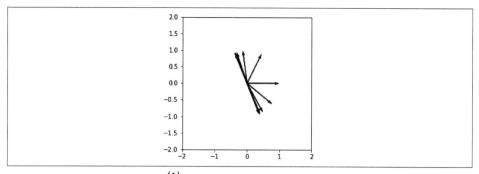

Figure 6-10. We start at $\overrightarrow{v}_0 = \begin{pmatrix} 1 \\ 0 \end{pmatrix}$, then multiply by A and normalize until we converge to an eigenvector

The Pseudoinverse

Many physical systems can be represented by (or approximated by) a linear system of equations $A\overrightarrow{x} = \overrightarrow{b}$. If $\overrightarrow{x}$ is an unknown vector that we care for, then we need to divide by matrix A in order to find $\overrightarrow{x}$. The matrix equivalent of division is finding the inverse A^{-1}, so that the solution $\overrightarrow{x} = A^{-1}\overrightarrow{b}$. Matrices that have an inverse are called *invertible*. These are square matrices with a nonzero determinant (the determinant is the product of the eigenvalues; the product of the singular values and the determinant will have the same absolute value). But what about all the systems whose matrices are rectangular? How about those with noninvertible matrices? And those whose matrices are square and invertible, but are almost noninvertible (their determinant is very close to zero)? We still care about finding solutions to such systems. The power of the singular value decomposition is that it exists for any matrix, including those mentioned above, and it can help us invert any matrix.

Given any matrix and its singular value decomposition $A = U\Sigma V^t$, we can define its *pseudoinverse* as:

$$A^+ = V\Sigma^+ U^t$$

where Σ^+ is obtained from Σ by inverting all its diagonal entries except for the ones that are zero (or very close to zero if the matrix happens to be ill-conditioned).

This allows us to find solutions to any system of linear equation $Ax = b$, namely $x = A^+ b$.

The pseudoinverse of a matrix coincides with its inverse when the latter exists.

Applying the Singular Value Decomposition to Images

We are finally ready for real-world applications of the singular value decomposition. We start with image compression. Digital images are stored as matrices of numbers, where each number corresponds to the intensity of a pixel. We will use the singular value decomposition to reduce the storage requirements of an image without losing its most essential information. All we have to do is throw away the insignificant singular values, along with the corresponding columns of U and rows of V^t. The mathematical expression that helps us here is:

$$A = U\Sigma V^t = \sigma_1 U_{col_1} V^t_{row_1} + \sigma_2 U_{col_2} V^t_{row_2} + \cdots + \sigma_r U_{col_r} V^t_{row_r}$$

Recall that the σ's are arranged from the largest value to the smallest value, so the idea is that we can keep the first few large σ's and throw away the rest of the σ's, which are small anyway.

Let's work with the image in Figure 6-11. The code and details are in the book's GitHub page (*https://github.com/halanelson/Essential-Math-For-AI*). Each color image has three channels: red, green, and blue (see Figures 6-12 and 6-13). Each channel is a matrix of numbers, just like the ones we have been working with in this chapter.

Each channel of the image in Figure 6-11 is a $size = 960 \times 714$ matrix, so to store the full image, we need $size = 960 \times 714 \times 3 = 2{,}056{,}320$ numbers. Imagine the storage requirements for a streaming video, which contains many image frames. We need a compression mechanism, so as not to run out of memory. We compute the singular value decomposition for each channel (see Figure 6-14 for an image representation of the singular value decomposition for the red channel). We then perform a massive reduction, retaining for each channel only the first 25 singular values (out of 714), 25 columns of U (out of 960), and 25 rows of V^t (out of 714). The storage reduction for each channel is substantial: U is now 960×25, V^t is 25×714, and we only need to store 25 singular values (no need to store the zeros of the diagonal matrix Σ). This adds up 41,875 numbers for each channel, so for all 3 channels, we need to store $41{,}875 \times 3 = 125{,}625$ numbers, a whopping 93% storage requirement reduction.

We put the image back together, one channel at a time, by multiplying the reduced U, reduced Σ, and reduced V^t together:

$$Channel_{reduced} = U_{960 \times 25} \Sigma_{25 \times 25} \left(V^t \right)_{25 \times 714}$$

Figure 6-15 shows the result of this multiplication for each of the red, green, and blue channels.

Finally, we layer the reduced channels to produce the reduced image (Figure 6-16). It is obvious that we lost a lot of detail in the process, but that is a trade-off we have to live with.

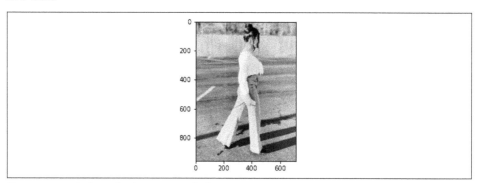

Figure 6-11. A digital color image of $size = 960 \times 714 \times 3$

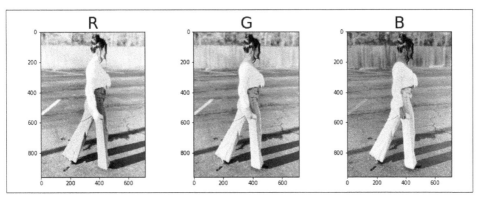

Figure 6-12. The red, green, and blue channels of the digital image. Each has $size = 960 \times 714$.

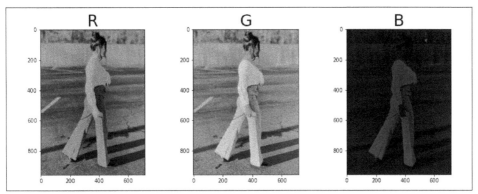

Figure 6-13. Showing the red, green, and blue tints for the three channels of the digital image. Each has $size = 960 \times 714 \times 3$. See the color image on GitHub (https://oreil.ly/ xhF8A).

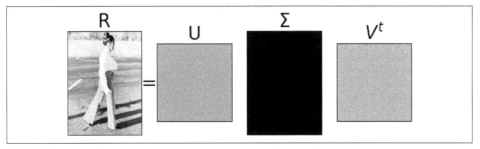

Figure 6-14. The singular value decomposition of the red channel. We have 714 nonzero singular values, but few significant ones. Even though the diagonal matrix Σ appears all black, it had nonzero singular on its diagonal. The pixels are not bright enough to appear clearly at this resolution level.

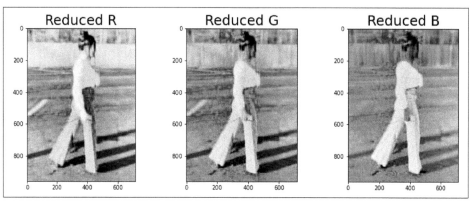

Figure 6-15. The red, green, and blue channels after rank reduction. For each channel we have retained only the first 25 singular values, the first 25 columns of U, and the first 25 rows of V^t.

Figure 6-16. Original image with 714 singular values versus reduced rank image with only 25 singular values. Both still have $size = 960 \times 714 \times 3$ but require different storage space.

For advanced image compression techniques, check this article (*https://oreil.ly/7dsZ3*).

Principal Component Analysis and Dimension Reduction

Principal component analysis is widely popular for data analysis. It is used for dimension reduction and clustering in unsupervised machine learning. In a nutshell, it is the singular value decomposition performed on the data matrix X, after *centering* the data, which means subtracting the average value of each feature from each feature column (each column of X). The principal components are then the right singular vectors, which are the rows of V^t in the now familiar decomposition $X = U\Sigma V^t$.

Statisticians like to describe the principal component analysis (*https://oreil.ly/HdOrY*) using the language of variance, or variation in the data, and *uncorrelating* the data. They end up working with the eigenvectors of the *covariance matrix* of the data. This is a familiar description of principal component analysis in statistics:

> It is a method that reduces the dimensionality of a dataset, while preserving as much variability, or statistical information, as possible. Preserving as much variability as possible translates into finding new features that are linear combinations of those of the dataset, that successively maximize variance and that are uncorrelated with each other.

The two descriptions (the right singular vectors of the centered data, and the eigenvectors of the covariance matrix) are exactly the same, since the rows of V^t are the eigenvectors of $X^t_{centered}X_{centered}$, which in turn is the covariance matrix of the data. Moreover, the term *uncorrelating* in statistics corresponds to *diagonalizing* in mathematics and linear algebra, and the singular value decomposition says that any matrix *acts like* a diagonal matrix, namely Σ, when written in a new set of coordinates, namely the rows of V^t, which are the columns of V.

Let's explain this in detail. Suppose X is a centered data matrix and its singular value decomposition is $X = U\Sigma V^t$. This is the same as $XV = U\Sigma$, or, when we resolve the expression by column, $XV_{col_i} = \sigma_i U_{col_i}$. Note that XV_{col_i} is just a linear combination of the features of the data using the entries of that particular column of V. Now, faithful to what we have been doing all along in this chapter, we can throw away the less significant components, meaning the columns of V and U corresponding to the lower singular values.

Suppose now that our data has 200 features, but only 2 singular values are significant, so we decide to only keep the first 2 columns of V and the first 2 columns of U. Thus, we have reduced the dimension of the features from 200 to 2. The first new feature is a linear combination of all the original 200 features using the entries of the first column of V, but that is exactly $\sigma_1 U_{col_1}$, and the second new feature is a linear combination of all the original 200 features using the entries of the second column of V, but that is exactly $\sigma_2 U_{col_2}$.

Now let's think of individual data points. A data point in the data matrix X has 200 features. This means that we need 200 axes to plot this data point. However, taking the dimension reduction we performed previously using only the first two principal components, this data point will now have only two coordinates, which are the corresponding entries of $\sigma_1 U_{col_1}$ and $\sigma_2 U_{col_2}$. So if this was the third data point in the data set, then its new coordinates will be the third entry of $\sigma_1 U_{col_1}$ and the third entry of $\sigma_2 U_{col_2}$. Now it is easy to plot this data point in a two-dimensional space, as opposed to plotting it in the original 200-dimensional space.

We choose how many singular values (and thus principal components) to retain. The more we keep, the more faithful to the original data set we would be, but of course the dimension will be higher. This *truncation* decision (finding the optimal threshold for singular value truncation) is the subject of ongoing research. The common method is determining the desired rank ahead of time, or keeping a certain amount of variance in the original data. Other techniques plot all the singular values, observe an obvious change in the graph, and decide to truncate at that location, hopefully separating the essential patterns in the data from the noise.

It is important not to only center the data, but to also standardize it: subtract the mean of each feature and divide by the standard deviation. The reason is that the singular value decomposition is sensitive to the scale of the feature measurements. When we standardize the data, we end up working with the correlation matrix instead of the covariance matrix. To not confuse ourselves, the main point to keep in mind is that we perform the singular value decomposition on the standardized data set, then the principal components are the columns of V, and the new coordinates of the data points are the entries of $\sigma_i U_{col_i}$.

Principal Component Analysis and Clustering

We saw in the previous section how we can use principal component analysis to reduce the number of features of the data, providing a new set of features in hierarchical order in terms of variation in the data. This is incredibly useful for visualizing data, since we can only visualize in two or three dimensions. It is important to be able to visualize patterns and correlations in high-dimensional data, for example, in genetic data. Sometimes, in the reduced dimensional space determined by the principal components, there is an inherent clustering of the data by category. For example, if the data set contains both patients with cancer and patients without cancer, along with their genetic expression (usually in the thousands), we might notice that plotting the data in the first three principal components space, patients with cancer cluster separately from patients without cancer.

A Social Media Application

In the same essence of principal component analysis and clustering, a recent publication (Dec 2020) (*https://oreil.ly/H1sLr*) by Dan Vilenchik presents a wonderful application from social media: an unsupervised approach to characterizing users in online social media platforms. Here's the abstract from a talk he gave on the subject, along with the abstract from his publication:

Making sense of data that is automatically collected from online platforms such as online social media or e-learning platforms is a challenging task: the data is massive, multidimensional, noisy, and heterogeneous (composed of differently behaving individuals). In this talk we focus on a central task common to all on-line social platforms and that is the task of user characterization. For example, automatically identify a spammer or a bot on Twitter, or a disengaged student in an e-learning platform.

Online social media channels play a central role in our lives. Characterizing users in social networks is a long-standing question, dating back to the 50's when Katz and Lazarsfeld studied influence in "Mass Communication". In the era of Machine Learning, this task is typically cast as a supervised learning problem, where a target variable is to be predicted: age, gender, political incline, income, etc. In this talk we explore what can be achieved in an unsupervised manner. Specifically, we harness principal component analysis to understand what underlying patterns and structures are inherent to some social media platforms, but not to others, and why. We arrive at a Simpson-like paradox that may give us a deeper understanding of the data-driven process of user characterization in such platforms.

The idea of principal component analysis for creating clusters with maximal variance in the data will appear multiple times throughout this book.

Latent Semantic Analysis

Latent semantic analysis for natural language data (documents) is similar to principal component analysis for numerical data.

Here, we want to analyze the relationships between a set of documents and the words they contain. The distributional hypothesis for latent semantic analysis states that words that have similar meaning occur in similar pieces of text, and hence in similar documents. Computers only understand numbers, so we have to come up with a numerical representation of our word documents before doing any analysis on them. One such representation is the *word count* matrix X: the columns represent unique words (such as apple, orange, dog, city, intelligence, etc.) and the rows represent each document. Such a matrix is very large but very sparse (has many zeros). There are too many words (these are the features), so we need to reduce the dimension of the features while preserving the similarity structure among the documents (the data points). By now we know what to do: perform the singular value decomposition on the word count matrix, $X = U\Sigma V^t$, then throw away the smaller singular values along with the corresponding columns from U and rows from V^t. We can now represent each document in the lower-dimensional space (of linear combinations of words) in exactly the same way principal component analysis allows for data representation in lower-dimensional feature space.

Once we have reduced the dimension, we can finally compare the documents using *cosine similarity*: compute the cosine of the angle between the two vectors representing the documents. If the cosine is close to 1, then the documents point in the same direction in *word space* and hence represent very similar documents. If the cosine is close to 0, then the vectors representing the documents are orthogonal to each other and hence are very different from each other.

In its early days, Google search was more like an index, then it evolved to accept more natural language searches. The same is true for smartphone autocomplete. Latent semantic analysis compresses the meaning of a sentence or a document into a vector, and when this is integrated into a search engine, it dramatically improves the quality of the engine, retrieving the exact documents we are searching for.

Randomized Singular Value Decomposition

In this chapter, we have avoided computing the singular value decomposition on purpose because it is expensive. We did mention, however, that common algorithms use a matrix decomposition called QR decomposition (which obtains an orthonormal basis for the columns of the data matrix), then Householder reflections to transform to a bidiagonal matrix, and finally iterative methods to compute the required eigen-vectors and eigenvalues. Sadly, for the ever-growing data sets, the matrices involved are too large even for these efficient algorithms. Our only salvation is through *randomized linear algebra*. This field provides extremely efficient methods for matrix decomposition, relying on *the theory of random sampling*. Randomized numerical methods work wonders, providing accurate matrix decompositions while at the same time being much cheaper than deterministic methods. Randomized singular value decomposition samples the column space of the large data matrix X, computes the QR decomposition of the sampled (much smaller) matrix, projects X onto the smaller space ($Y = Q^t X$, so $X \approx QY$), then computes the singular value decomposition of Y ($Y = U \Sigma V^t$). The matrix Q is orthonormal and approximates the column space of X, so the matrices Σ and V are the same for X and Y. To find the U for X, we can compute it from the U for Y and Q $U_X = QU_Y$.

Like all randomized methods, they have to be accompanied by error bounds, in terms of the expectation of how far off the original matrix X is from the sampled QY. We do have such error bounds, but we postpone them until Chapter 7, which discusses large random matrices.

Summary and Looking Ahead

The star of the show in this chapter was one formula:

$$X = U \Sigma V^t = \sigma_1 U_{col_1} V^t_{row_1} + \sigma_2 U_{col_2} V^t_{row_2} + \cdots + \sigma_r U_{col_r} V^t_{row_r}$$

This is equivalent to $XV = U\Sigma$ and $XV_{col_i} = \sigma_i U_{col_i}$.

The power of the singular value decomposition is that it allows for rank reduction without losing essential information. This enables us to compress images, reduce the dimension of the feature space of a data set, and compute document similarity in natural language processing.

We discussed principal component analysis, latent semantic analysis, and the clustering structure inherent in the principal component space. We discussed examples where principal component analysis was used as an unsupervised clustering technique for cancer patients according to their gene expression, as well as characterizing social media users.

We ended the chapter with randomized singular value decomposition, highlighting a recurring theme for this book: when things are too large, sample them. *Randomness is pretty much dependable!*

If you are interested in diving deeper, you can read about tensor decompositions and N-way data arrays, and the importance of data alignment for the singular value decomposition to work properly. I learned this stuff from the book *Deep Learning* (*https://www.deeplearningbook.org*). If you are interested in other popular examples, you can read about eigenfaces from a modern perspective.

CHAPTER 7

Natural Language and Finance AI: Vectorization and Time Series

They. Can. Read.

—H.

One of the hallmarks of human intelligence is our mastery of language at a very early age: comprehension of written and spoken language, written and spoken expression of thoughts, conversation between two or more people, translation from one language to another, and the use of language to express empathy, convey emotions, and process visual and audio data perceived from our surroundings. Leaving the philosophical question of consciousness aside, if machines acquire the ability to perform these language tasks, deciphering the intent of words, at a level similar to humans, or above humans, then it is a major propeller toward general artificial intelligence. These tasks fall under the umbrellas of *natural language processing, computational linguistics, machine learning,* and/or *probabilistic language modeling.* These fields are vast and it is easy to find interested people wandering aimlessly in a haze of various models with big promises. We should not get lost. The aim of this chapter is to lay out the natural processing field all at once so we can have a bird's-eye view without getting into the weeds.

The following questions guide us at all times:

- What type of task is at hand? In other words, what is our goal?
- What type of data is at hand? What type of data do we need to collect?
- What state-of-the-art models are out there that deal with similar tasks and similar types of data? If there are none, then we have to come up with the models ourselves.

- How do we train these models? In what formats do they consume their data? In what formats do they produce their outputs? Do they have a training function, loss function (or objective function), and optimization structure?

- What are the advantages and disadvantages of various models versus others?

- Are there Python packages or libraries available for model implementation? Luckily, nowadays, most models come out accompanied with Python implementations and very simple APIs. Even better, there are many pre-trained models available to download and ready for use in applications.

- How much computational infrastructure do we need to train and/or deploy these models?

- Can we do better? There is always room for improvement.

We also need to *extract the math from the best-performing models*. Thankfully, this is the easy part, since similar mathematics underlies many models, even when relating to different types of tasks or from dissimilar application areas, such as predicting the next word in a sentence or predicting stock market behavior.

The state-of-the-art models that we intend to cover in this chapter are:

- Transformers or attention models (since 2017). The important math here is extremely simple: the dot product between two vectors.

- Recurrent long short-term memory neural networks (since 1995). The important math here is *backpropagation in time*. We covered backpropagation in Chapter 4, but for recurrent nets, we take the derivatives with respect to time.

- Convolutional neural networks (since 1989) for time series data. The important math is the *convolution* operation, which we covered in Chapter 5.

These models are very well suited for *time series data*, that is, data that appears sequentially with time. Examples of time series data are movies, audio files such as music and voice recordings, financial markets data, climate data, dynamic systems data, documents, and books.

We might wonder why documents and books can be considered as time dependent, even though they have already been written and are just *there*. How come an image is not time dependent but a book and, in general, reading and writing are? The answer is simple:

- When we read a book, we comprehend what we read one word at a time, then one phrase at a time, then one sentence at a time, then one paragraph at a time, and so on. This is how we grasp the concepts and topics of the book.

- The same is true when we write a document, outputting one word at time, even though the whole idea we are trying to express is already there, *encoded*, before we output the words, *sequentially*, on paper.

- When we caption an image, the image itself is not time dependent, but our captioning (the output) is.

- When we summarize an article, answer a question, or translate from one language to another, the output text is time dependent. The input text could be time dependent if processed using a recurrent neural network, or stationary if processed all at once using a transformer or a convolutional model.

Until 2017, the most popular machine learning models to process time series data were based either on *convolutional neural networks* or on *recurrent neural networks with long short-term memory*. In 2017, *transformers* took over, abandoning recurrence altogether in certain application areas. The question of whether recurrent neural networks are obsolete is out there, but with things changing every day in the AI field, who knows which models will die and which will survive the test of time. Moreover, recurrent neural networks power many AI engines and are still subjects of active research.

In this chapter, we answer the following questions:

- How do we transform natural language text to numerical quantities that retain meaning? Our machines only understand numbers, and we need to process natural language using these machines. We must *vectorize* our samples of text data, or *embed* them into finite dimensional vector spaces.

- How do we lower the dimension of the vectors from the enormous ones initially required to represent natural language? For example, the French language has around 135,000 distinct words, so how do we get around having to one-hot code words in a French sentence using vectors of 135,000 entries each?

- Does the model at hand consider (as its input and/or output) our natural language data as a time dependent sequence fed into it one term at a time, or a stationary vector consumed all at once?

- How exactly do various models for natural language processing work?

- Why is there finance in this chapter as well?

Along the way, we discuss the types of natural language and finance applications that our models are well suited for. We keep the focus on the mathematics and not the programming, since such models (especially for language applications) require substantive computational infrastructures. For example, the DeepL Translator (*https://oreil.ly/9NZ6b*) generates its translations using a supercomputer operated with hydropower from Iceland, which reaches 5.1 petaflops. We also note that the AI-specialized chip industry is booming, led by NVIDIA, Google's Tensor Processing Unit, AWS

Inferentia, AMD's Instinct GPU, and startups like Cerebras and Graphcore. While conventional chips have struggled to keep pace with Moore's law, which predicted a doubling of processing power every 18 months, AI-specialized chips have outpaced this law by a wide margin.

Even though we do not write code for this chapter, we note that most programming can be accomplished using Python's TensorFlow and Keras libraries.

Throughout our discussion, we have to be mindful of whether we are in the *training* phase of a model or in the *prediction* phase (using the pre-trained model to do tasks). Moreover, it is important to differentiate whether our model needs labeled data to be trained, such as English sentences along with their French translations as labels, or can learn from unlabeled data, such as computing the meanings of words from their contexts.

Natural Language AI

Natural language processing applications are ubiquitous. This technology has been integrated into so many aspects of our lives that we just take it for granted: when using apps on our smartphones, digital calendars, digital home assistants, Siri, Alexa, and others. The following list is partially adapted from the excellent book *Natural Language Processing in Action* by Hobson Lane, Hannes Hapke, and Cole Howard (Manning 2019), demonstrating how indispensable natural language processing has become:

- Search and information retrieval: web, documents, autocomplete, chatbots
- Email: spam filter, email classification, email prioritization
- Editing: spelling check, grammar check, style recommendation
- Sentiment analysis: product reviews, customer care, monitoring of community morale
- Dialog: chatbots, digital assistants such as Amazon's Alexa, scheduling
- Writing: indexing, concordance, table of contents
- Text mining: summarization, knowledge extraction such as mining election campaigns' finance and natural language data (finding connections between political donors), résumé-to-job matching, medical diagnosis
- Law: legal inference, precedent search, subpoena classification
- News: event detection, fact checking, headline composition
- Attribution: plagiarism detection, literary forensics, style coaching
- Behavior prediction: finance applications, election forecasting, marketing

- Creative writing: movie scripts, poetry, song lyrics, bot-powered financial and sports news stories
- Captioning: computer vision combined with natural language processing
- Translation: Google Translate and DeepL Translate

Even though the past decade has brought impressive feats, machines are still nowhere close to mastering natural language. The processes involved are tedious, requiring attentive statistical bookkeeping and substantive memory, the same way humans require memory to master languages. The point here is: there is plenty of room for new innovations and contributions to the field.

Language models have recently shifted from handcoded to data driven. They do not implement hardcoded logical and grammar rules. Instead, they rely on detecting the statistical relationships between words. Even though there is a school of thought in linguistics that asserts grammar is an innate property for humans, or in other words, is *hardcoded* into our brains, humans have a striking ability to master new languages without ever encountering any grammatical rules for these languages. From personal experience, attempting to learn the grammar of a new language seems to impede the learning process, but do not quote me on that.

One major challenge is that data for natural language is extremely high-dimensional. There are millions of words across thousands of languages. There are huge corpuses of documents, such as entire collections of authors' works, billions of tweets, Wikipedia articles, news articles, Facebook comments, movie reviews, etc. A first goal is then to reduce the number of dimensions for efficient storage, processing, and computation, while at the same time avoiding the loss of essential information. This has been a common theme in the AI field, and one cannot help but wonder how many mathematical innovations would have never seen the light of the day had we possessed unlimited storage and computational infrastructures.

Preparing Natural Language Data for Machine Processing

For a machine to process any natural language task, the first thing it must do is to break down text and organize it into building blocks that retain meaning, intent, context, topics, information, and sentiments. To this end, it must establish a correspondence between words and number tags, using processes called *tokenizing*, *stemming* (such as giving singular words and their plural variation the same token), *lemmatization* (associating several words of similar meaning together), *case normalization* (such as giving capitalized and lowercase words of the same spelling the same tokens), and others. This correspondence is not for individual characters that make up words, but for full words, pairs or more of words (2-grams or n-grams), punctuations, significant capitalizations, etc., that carry meaning. This creates a *vocabulary* or a *lexicon* of numerical tokens corresponding to a given corpus of natural language documents. A vocabulary or a lexicon in this sense is similar to a Python dictionary: each individual natural language building block object has a unique token.

An *n-gram* is a sequence of *n* words that carry a meaning when kept ordered together that is different from the meaning of each word on its own. For example, a 2-gram is a couple of words together whose meaning would change if we unpair them, such as ice cream or was not, so the whole 2-gram gets one numerical token, retaining the meaning of the two words within their correct context. Similarly, a 3-gram is a triplet of ordered words, such as John F. Kennedy, and so on. A *parser* for natural language is the same as a compiler for computers. Do not worry if these new terms confuse you. For our mathematical purposes, all we need are the numerical tokens associated with unique words, n-grams, emojis, punctuation, etc., and the resulting vocabulary for a corpus of natural language documents. These are saved in a dictionary like objects, allowing us to flip back and forth easily between text and numerical tokens.

We leave the actual details of tokenizing, stemming, lemmatization, parsing, and other natural language data preparations for computer scientists and their collaborations with linguists. In fact, collaboration with linguists has become less important as the models mature in their ability to detect patterns directly from the data, thus, the need for coding handcrafted linguistic rules into natural language models has diminished. Note also that not all natural language pipelines include stemming and lemmatization. They all, however, involve tokenizing. The quality of tokenizing text data is crucial for the performance of our natural language pipeline. It is the first step containing fundamental building blocks representing the data that we feed into our models. The quality of both the data and the way it is tokenized affects the outputs of the entire natural language processing pipeline. For your production applications, use the *spaCy* parser, which does sentence segmentation, tokenization, and multiple other things in one pass.

After tokenizing and building a healthy vocabulary (the collection of numerical tokens and the entities they correspond to in the natural language text), we need to represent entire natural language documents using vectors of numbers. These documents can range from very long, such as a book series, to very short, such as a Twitter tweet or a simple search query for Google Search or DuckDuckGo. We can then express a corpus of one million documents as a collection of one million numerical vectors, or a matrix with one million columns. These columns will be as long as our chosen vocabulary, or shorter if we decide to *compress* these documents further. In linear algebra language, the length of these vectors is the dimension of our *vector space* that our documents are *embedded in*.

The whole point of this process is to obtain numerical vector representations of our documents so that we can do math on them: now comes linear algebra with its arsenal of linear combinations, projections, dot products, and singular value decompositions. There is, however, one caveat: for natural language applications, the lengths of the vectors representing our documents, or the size of our vocabulary, are prohibitively enormous to do any useful computations with. The curse of dimensionality becomes a real thing.

The Curse of Dimensionality

Vectors become exponentially farther apart in terms of Euclidean distance as the number of dimensions increases. One natural language example is sorting documents based on their *distance* from another document, such as a search query. This simple operation becomes impractical when we go above 20 dimensions or so if we use the Euclidean distance to measure the *closeness* of documents (see Wikipedia's "curse of dimensionality" (*https://oreil.ly/575IJ*) for more details). Thus, for natural language applications, we must use another measure for distance between documents. We will discuss *cosine similarity* shortly, which measures the *angle* between two document vectors, as opposed to their Euclidean distance.

Therefore, a main driver for natural language processing models is to represent these documents using shorter vectors that convey the main topics and retain meaning. Think how many unique tokens or combinations of tokens we have to use to represent this book while at the same time preserving its most important information.

To summarize, our natural language processing pipeline proceeds as follows:

1. From text to numerical tokens, then to an acceptable vocabulary for an entire corpus of documents.

2. From documents of tokens to high-dimensional vectors of numbers.

3. From high-dimensional vectors of numbers to lower-dimensional vectors of topics using techniques like *direct projection onto a smaller subset of the vocabulary space* (just dropping part of the vocabulary, making the corresponding entries zero), *latent semantic analysis* (projecting onto special vectors determined by special linear combinations of the document vectors), *word2vec, Doc2Vec, thought vectors, Dirichlet allocation*, and others. We discuss these shortly.

As is usually the case in mathematical modeling, there is more than one way to represent a given document as a vector of numbers. We decide on the vector space that our documents inhabit, or get *embedded in*. Each vector representation has advantages and disadvantages, depending on the goal of our natural language task. Some are simpler than others too.

Statistical Models and the log Function

When representing a document as a vector of numbers starts with counting the number of times certain terms appear in the document, then our document vectorizing model is statistical, since it is frequency based.

When we deal with term frequencies, it is better to apply the *log* function to our counts as opposed to using raw counts. The log function is advantageous when we deal with quantities that could get extremely large, extremely small, or could have extreme variations in scale. Viewing these extreme counts or variations within a logarithmic scale brings them back to the normal realm.

For example, the number 10^{23} is huge, but $\log\left(10^{23}\right) = 23\log(10)$ is not. Similarly, if the term "shark" appears in 2 documents of a corpus of 20 million documents (20 million/2 = 10 million), and the term "whale" appears in 20 documents of this corpus (20 million/20 = 1 million), then that is a 9 million difference, which seems excessive for terms that appeared in 2 and 20 documents, respectively. Computing the same quantities but on a *log* scale, we get 7log(10) and 6log(10), respectively (it doesn't matter which log base we use), which doesn't seem excessive anymore, and more in line with the terms' occurrence in the corpus.

The need for using the log function when dealing with word counts in particular is reinforced by *Zipf's law*. This law says that term counts in a corpus of natural language naturally follow a power law, so it is best to temper that with a log function, transforming differences in term frequencies into a linear scale. We discuss this next.

Zipf's Law for Term Counts

Zipf's law (https://oreil.ly/hF3Ab) for natural language has to do with word counts. It is very interesting and so surprising that I am tempted to try and see if it applies to my own book. It is hard to imagine that as I write each word in this book, my unique word counts are actually following some law. Are we, along with the way we word our ideas and thoughts, *that predictable*? It turns out that Zipf's law extends to counting many things around us, not only words in documents and corpuses.

Zipf's law reads as follows: *for a corpus of natural language where the terms have been ordered according to their frequencies, the frequency of the first item is twice that of the second item, three times the third item, and so on.* That is, the frequency with which an item appears in a corpus is related to its ranking: $f_1 = 2f_2 = 3f_3 = \ldots$

We can verify if Zipf's law applies by plotting the frequency of the terms against their respective ranks and verifying the power law: $f_r = f(r) = f_1 r^{-1}$. To verify power laws, it is easier to make a *log-log* plot, plotting $\log\left(f_r\right)$ against $\log\left(r\right)$. If we obtain a straight line in the log-log plot, then $f_r = f(r) = f_1 r^\alpha$, where α is the slope of the straight line.

Various Vector Representations for Natural Language Documents

Let's list the most common document vector representations for state-of-the-art natural language processing models. The first two, *term frequency* and *term frequency-inverse document frequency* (TF-IDF), are statistical representations since they are frequency based, relying on counting word appearances in documents. They are slightly more involved than a simple binary representation detecting the presence or nonpresence of certain words, nevertheless, they are still shallow, merely counting words. Even with this shallowness, they are very useful for applications such as spam filtering and sentiment analysis.

Term Frequency Vector Representation of a Document or Bag of Words

Here, we represent a document using a *bag of words*, discarding the order in which words appear in the document. Even though word order encodes important information about a document's content, ignoring it is usually an OK approximation for short sentences and phrases.

Suppose we want to embed our given document in a *vocabulary space* of 10,000 tokens. Then the vector representing this document will have 10,000 entries, with each entry counting how many times each particular token appears in the document. For obvious reasons, this is called the *term frequency* or *bag-of-words* vector representation of the document, where each entry is a nonnegative integer (a whole number).

For example, the Google Search query "What's the weather tomorrow?" will be vectorized as zeros everywhere except for ones at the tokens representing the words "what," "the," "weather," and "tomorrow," if they exist in the vocabulary. We then *normalize* this vector, dividing each entry by the total number of terms in the document so that the length of the document doesn't skew our analysis. That is, if a document has 50,000 terms and the term "cat" gets mentioned a hundred times, and another document has a hundred terms only and the term "cat" gets mentioned 10 times, then obviously the word "cat" is more important for the second document than for the first, and a mere word count without normalizing would not be able to capture that.

Finally, some natural language processing classes take the log of each term in the document vector for the reasons mentioned in the previous two sections.

Term Frequency-Inverse Document Frequency Vector Representation of a Document

Here, for each entry of the vector representing the document, we still count the number of times the token appears in the document, *but then we divide by the number of documents in our corpus in which the token occurs.*

The idea is that if a term appears many times in one document and not as much in the others, then this term must be important for this one document, getting a higher score in the corresponding entry of the vector representing this document.

To avoid division by zero, if a term does not appear in any document, it is common practice to add one to the denominator. For example, the inverse document frequency of the token *cat* is:

$$\text{IDF for cat} = \frac{\text{number of documents in corpus}}{\text{number of documents containing cat} + 1}$$

Obviously, using TF-IDF representation, the entries of the document vectors will be nonnegative rational numbers, each providing a measure of the *importance* of that particular token to the document. Finally, we take the log of each entry in this vector, for the same reasons stated in the previous section.

There are many alternative TF-IDF approaches relevant to information retrieval systems, such as Okapi BM25 (*https://oreil.ly/xrntI*), and Molino 2017 (*https://oreil.ly/jT56u*).

Topic Vector Representation of a Document Determined by Latent Semantic Analysis

TF-IDF vectors are very high-dimensional (as many dimensions as tokens in the corpus, so it could be in the millions), sparse, and have no special meaning when added or subtracted from each other. We need more compact vectors, in the hundreds of dimensions or less, which is a big squeeze from millions of dimensions. In addition to the dimension reduction advantage, these vectors capture some meaning, not only word counts and statistics. We call them *topic vectors*. Instead of focusing on the *statistics of words in documents*, we focus on the *statistics of connections between words in documents and across corpuses*. The topics produced here will be linear combinations of word counts.

First, we process the whole TF-IDF matrix X of our corpus, producing our *topic space*. Processing in this case means that we compute the *singular value decomposition* of the TF-IDF matrix from linear algebra, namely, $X = U \Sigma V^t$. Chapter 6 is dedicated to singular value decomposition, so for now we will only explain how it is used for producing our topic space for a corpus. Singular value decomposition from linear algebra is called *latent semantic analysis* (*https://oreil.ly/Ul0nb*) in natural language processing. We will use both terms synonymously.

We have to pay attention to whether the columns of the corpus's TF-IDF matrix X represent the word tokens or the documents. Different authors and software packages use one or the other, so we must be careful and process either the matrix or its transpose to produce our topic space. In this section we follow the representation that the rows are all the words (tokens for words, n-grams, etc.) of the entire corpus, and the columns are the TF-IDF vector representations for each document in the corpus. This is slightly divergent from the usual representation of a data matrix, where the features (the words within each document) are in the columns, and the instances (the documents) are in the rows. The reason for this switch will be apparent shortly. However, this is not divergent from our representation for documents as column vectors.

Next, given a new document with its TF-IDF vector representation, we convert it to a much more compact *topic vector* by *projecting it onto* the topic space produced by the singular value decomposition of the corpus's TF-IDF matrix. *Projecting* in linear algebra is merely computing the *dot product* between the appropriate vectors and saving the resulting scalar numbers into the entries of a new *projected* vector:

- We have a TF-IDF vector of a document that has as many entries as the number of tokens in the entire corpus.
- We have *topic weight vectors*, which are the columns of the matrix U produced by the singular value decomposition of the TF-IDF matrix $X = U \Sigma V^t$. Again, each topic weight vector has as many entries as tokens in our corpus. Initially, we also have as many topic weight vectors as tokens in our entire corpus (columns of U). The *weights* in the column of U tell us how much a certain token contributes to the topic, with a big contribution if it is a positive number close to 1, an ambivalent contribution if it is close to 0, and even a negative contribution if it is a negative number close to -1. Note that the entries of U are always numbers between -1 and 1, so we interpret them as weighing factors for our corpus's tokens.

Topic selection and dimension reduction

You might be wondering, if we have as many topic weight vectors as tokens in our corpus, each having as many entries as tokens as well, then where are the savings, and when will compression or dimension reduction happen? Keep reading.

Goal 1

Compute how much of a certain topic our document contains. This is simply the dot product between the document's TF-IDF vector and the column of U corresponding to the topic that we care for. Record this as the first scalar number.

Goal 2

Compute how much of *another* topic our document contains. This is the dot product between the document's TF-IDF vector and the column of U corresponding to this other topic that we care for. Record this as the second scalar number.

Goal 3

Repeat this for as many (as there are columns of U, which is the same as the total number of tokens in the corpus) or as few (as one) topics as we like, recording the scalar number from each dot product that we compute. It is clear now that "a topic" in this context means a column vector containing weights between -1 and 1 assigned to each token in the corpus.

Goal 4

Reduce the dimension by keeping only the topics that matter. That is, if we decide to keep only two topics, then the *compressed vector representation* of our document will be the *two-dimensional* vector containing the two scalar numbers produced using the two dot products between the document's TF-IDF vector and the two topics' weight vectors. This way, we would have reduced the dimension of our document from possibly *millions* to just two. Pretty cool stuff.

Goal 5

Choose the right topics to represent our documents. This is where the singular value decomposition works its magic. The columns of U are organized in order, from the most important topic across the corpus to the least important. In the language of statistics, the columns are organized from the topic with the *most variance* across the corpus and hence encodes more information, to the one with the *least variance* and hence encodes little information. We explain how variance and singular value decomposition are related in Chapter 10. Thus, if we decide to project our high-dimensional document onto the first few column vectors of U only, we are guaranteed that we are not missing much in terms of capturing enough variation of possible topics across the corpus, and assessing how much of these our document contains.

Goal 6

Understand that *this is still a statistical method for capturing topics in a document*. We started with the TF-IDF matrix of a corpus, simply counting token occurrences in documents. In this sense, a topic is captured based only on the premise that documents that refer to similar things use similar words. This is different from capturing topics based on the *meanings* of the words they use. That is, if we have two documents discussing the same topic but using entirely different vocabulary, they will be far apart in topic space. The remedy to this would be to store words together with other words of similar meaning, which is the word2vec approach, discussed later in this chapter.

Question 1

What happens if we add another document to our corpus? Luckily, we do not have to reprocess the whole corpus to produce the document's topic vector, we just project it onto the corpus's existing topic space. This of course breaks down if we add a new document that has nothing in common with our corpus, such as an article on pure mathematics added to a corpus on Shakespeare's love sonnets. Our math article in this case will be represented by a bunch of zeros or close-to-zero entries, which does not capture the ideas in the article adequately.

Now we have the following questions:

Question 2

What about the matrix V^t in the singular value decomposition $X = U\Sigma V^t$, what does it mean in the context of natural language processing of our corpus? The matrix V^t has the same number of rows and columns as the number of documents in our corpus. It is the document-document matrix and gives the shared meaning between documents.

Question 3

When we move to a lower-dimensional topic space using latent semantic analysis, are large distances between documents preserved? Yes, since the singular value decomposition focuses on *maximizing* the variance across the corpus's documents.

Question 4

Are small distances preserved, meaning does latent semantic analysis preserve the *fine structure* of a document that separates it from *not so different* other documents? No. Latent Dirichlet allocation, discussed soon, does a better job here.

Question 5

Can we improve latent semantic analysis to also keep close document vectors together in the lower-dimensional topic space? Yes, we can *steer* the vectors by taking advantage of extra information, or *metadata*, of the documents, such as messages having the same sender, or by penalizing using a cost function so that the method spills out topic vectors that preserve *closeness* as well.

To summarize, latent semantic analysis chooses the topics in an optimal way that maximizes the diversity in the topics across the corpus. The matrix U from the singular value decomposition of the TF-IDF matrix is very important for us. It returns the directions along which the variance is maximal. We usually get rid of the topics that have the least amount of variance between the documents in the corpus, throwing away the last columns of U. This is similar to manually getting rid of stop words (and, a, the, etc.) during text preparation, but latent semantic analysis does that for us in an optimized way. The matrix U has the same number of rows and columns as our vocabulary. It is the cross-correlation between words and topics based on word co-occurrence in the same document. When we multiply a new document by U (project it onto the columns of U), we would get the amount of each topic in the document. We can truncate U as we wish and throw away less important topics, reducing the dimension to as few topics as we want.

Shortcomings of latent semantic analysis

The topic spaces it produces, or the columns of U, are mere *linear combinations* of tokens that are thrown together in a way that captures as much variance in the usage across the vocabulary's tokens as possible. This doesn't necessarily translate into word combinations that are in any way meaningful to humans. Bummer. Word2vec, discussed later, addresses these shortcomings.

Finally, the topic vectors produced via latent semantic analysis are just linear transformations performed on the TF-IDF vectors. They should be the first choice for semantic searches, clustering documents, and content-based recommendation engines. All of this can be accomplished by measuring distances between these topic vectors, which we explain later in this chapter.

Topic Vector Representation of a Document Determined by Latent Dirichlet Allocation

Unlike topic vectors using latent semantic analysis, with *latent Dirichlet allocation* (*https://oreil.ly/QsSPp*) (LDA) we do have to reprocess the entire corpus if we add a new document to the corpus to produce its topic vector. Moreover, we use a nonlinear statistical approach to bundle words into topics: we assume a Dirichlet distribution of word frequencies. This makes the method more precise than latent semantic analysis in terms of the statistics of allocating words to topics. Thus, the method is explainable: the way words are allocated to topics, based on how often they occurred together in a document, and the way topics are allocated to documents, tend to make sense to us as humans.

This nonlinear method takes longer to train than the linear latent semantic analysis. For this reason it is impractical for applications involving corpuses of documents, even though it is explainable. We can use it instead for summarizing single documents, where each sentence in the document becomes its own *document*, and the mother document becomes the corpus.

LDA was invented in 2000 by geneticists for the purpose of inferring population structure, and adopted in 2003 for natural language processing. The following are its assumptions:

- We start with raw word counts (rather than normalized TF-IDF vectors), but there is still no sequencing of words to make sense of them. Instead, we still rely on modeling the statistics of words for each document, except this time we incorporate the word distribution explicitly into the model.

- A document is a linear combination of an arbitrary number of topics (specify this number ahead of time so that the method allocates the document's tokens to this number of topics).

- We can represent each topic by a certain distribution of words based on their term frequencies.

- The probability of occurrence of a certain topic in a document follows a Dirichlet probability distribution.

- The probability of a certain word being assigned to a topic also follows a Dirichlet probability distribution.

As a result, topic vectors obtained using Dirichlet allocation are sparse, indicating clean separation between the topics in the sense of which words they contain, which makes them explainable.

With Dirichlet allocation, words that occur frequently together are assigned to the same topics. So this method keeps tokens that were close together, close together, when we move to the lower-dimensional topic space. Latent semantic analysis, on the other hand, keeps tokens that were spread apart, spread apart, when we move to the lower-dimensional topic space, so this is better for classification problems where the separation between the classes is maintained even as we move to the lower-dimensional space.

Topic Vector Representation of a Document Determined by Latent Discriminant Analysis

Unlike latent semantic analysis and latent Dirichlet allocation, which break down a document into as many topics as we choose, latent discriminant analysis breaks down a document into *only one topic*, such as spamness, sentiment, etc. This is good for binary classification, such as classifying messages as spam or nonspam, or classifying reviews as positive or negative. Rather than what latent semantic analysis does, maximizing the separation between all the vectors in the new topic space, latent discriminant analysis maximizes the separation only between the centroids of the vectors belonging to each class.

But how do we determine the vector representing this one topic? Given the TF-IDF vectors of labeled spam and nonspam documents, we compute the centroid of each class, then our vector is along the line connecting the two centroids (see Figure 7-1).

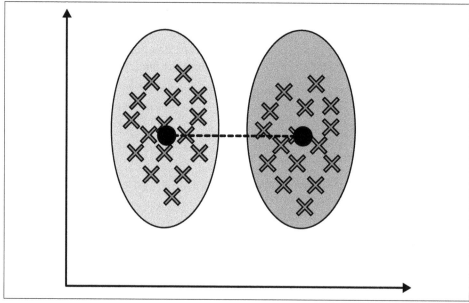

Figure 7-1. Latent discriminant analysis

Each new document can now be projected onto this one dimension. The coordinate of our document along that line is the dot product between its TF-IDF and the direction vector of the centroids line. The whole document (with millions of dimensions) is now squashed into one number along one dimension (one axis) that carries the two centroids along with their midpoint. We can then classify the document as belonging to one class or the other depending on its distance from each centroid along that one line. Note that the decision boundary for separating classes using this method is linear.

Meaning Vector Representations of Words and of Documents Determined by Neural Network Embeddings

The previous models for vectorizing documents of natural language text only considered linear relationships between words, or in latent Dirichlet allocation, we had to use human judgment to select the model's parameters and extract features. We now know that the power of neural networks lies in their ability to capture nonlinear relationships, extract features, and find appropriate model parameters automatically. We will now use neural networks to create vectors that represent individual words and terms, and we will employ similar methods to create vectors representing the meanings of entire paragraphs. Since these vectors encode the meaning and the logical and contextual usage of each term, we can reason with them simply by doing the usual vector additions and subtractions.

Word2vec vector representation of individual terms by incorporating continuous-*ness* attributes

By using TF vectors or TF-IDF vectors as a starting point for our topic vector models, we have ignored the nearby context of words and the effect that has on their meanings. Word vectors solve this problem. A word vector is a numerical vector representation of a word's meaning, so every single term in the corpus becomes a vector of semantics. This vector representation with floating-point number entries of single words enables semantic queries and logical reasoning.

Word vector representations are learned using a neural network. They usually have 100 to 500 dimensions encoding how much of each meaning dimension a word carries within it. When training a word vector model, the text data is unlabeled. Once trained, two terms can be determined to be close in meaning or far apart by comparing their vectors via some closeness metrics. Cosine similarity, discussed next, is the go-to method.

In 2013, Google created a word-to-vector model, word2vec, that it trained on the Google News feed containing 100 billion words. The resulting pre-trained word2vec model contains 300 dimensional vectors for 3 million words and phrases. It is freely available to download at the Google Code Archive page for the word2vec project (*https://oreil.ly/qvGQo*).

The vector that word2vec builds up captures much more of a word's meaning than the topic vectors discussed earlier in this chapter. The abstract of the paper "Efficient Estimation of Word Representations in Vector Space" (Mikolov et al. 2013) (*https://oreil.ly/lGT6i*) is informative:

> We propose two novel model architectures for computing continuous vector representations of words from very large data sets. The quality of these representations is measured in a word similarity task, and the results are compared to the previously best performing techniques based on different types of neural networks. We observe large improvements in accuracy at much lower computational cost, i.e. it takes less than a day to learn high quality word vectors from a 1.6 billion words data set. Furthermore, we show that these vectors provide state-of-the-art performance on our test set for measuring syntactic and semantic word similarities.

A month later, the paper "Distributed Representations of Words and Phrases and their Compositionality" (Mikolov et al. 2013) (*https://oreil.ly/RyijL*) addressed the representation of word phrases that mean something different than their individual components, such as "Air Canada":

The recently introduced continuous Skip-gram model is an efficient method for learning high quality distributed vector representations that capture a large number of precise syntactic and semantic word relationships. In this paper we present several extensions that improve both the quality of the vectors and the training speed. By subsampling of the frequent words we obtain significant speedup and also learn more regular word representations. We also describe a simple alternative to the hierarchical softmax called negative sampling. An inherent limitation of word representations is their indifference to word order and their inability to represent idiomatic phrases. For example, the meanings of "Canada" and "Air" cannot be easily combined to obtain "Air Canada". Motivated by this example, we present a simple method for finding phrases in text, and show that learning good vector representations for millions of phrases is possible.

The publication that introduced word2vec representations, "Linguistic Regularities in Continuous Space Word Representations" (Mikolov et al. 2013) (*https://oreil.ly/ vKzgZ*), demonstrates how these meaning vectors for words encode logical regularities and how this enables us to answer regular analogy questions:

Continuous space language models have recently demonstrated outstanding results across a variety of tasks. In this paper, we examine the vector space word representations that are implicitly learned by the input layer weights. We find that these representations are surprisingly good at capturing syntactic and semantic regularities in language, and that each relationship is characterized by a relation specific vector offset. This allows vector oriented reasoning based on the offsets between words. For example, the male/female relationship is automatically learned, and with the induced vector representations, "King - Man + Woman" results in a vector very close to "Queen." We demonstrate that the word vectors capture syntactic regularities by means of syntactic analogy questions (provided with this paper), and are able to correctly answer almost 40% of the questions. We demonstrate that the word vectors capture semantic regularities by using the vector offset method to answer SemEval-2012 Task 2 questions. Remarkably, this method outperforms the best previous systems.

The performance of word2vec has improved dramatically since 2013 by training it on much larger corpuses.

Word2vec takes one word and assigns to it a vector of attributes, such as place-ness, animal-ness, city-ness, positivity (sentiment), brightness, gender, etc. Each attribute is a dimension, capturing how much of the attribute the meaning of the word contains.

These word meaning vectors and the attributes are not encoded manually, but during training, where the model learns the meaning of a word from the company it keeps: the five or so nearby words in the same sentence. This is different from latent semantic analysis, where the topics are learned only from words occurring in the same document, not necessarily close to each other. For applications involving short documents and statements, word2vec embeddings have actually replaced topic vectors obtained through latent semantic analysis. We can also use word vectors to derive word clusters from huge data sets by performing k-means clustering on top of the word vector representations. See the Google Code Archive page for the word2vec project (*https://oreil.ly/7tqd8*) for more information.

The advantage of representing words through vectors that mean something (rather than count something) is that we can reason with them. For example, as mentioned earlier, if we subtract the vector representing "man" from the vector representing "king" and add the vector representing "woman," then we get a vector very close to the vector representing the word "queen." Another example is capturing the relationship between singular and plural words. If we subtract vectors representing the singular form of words from vectors representing their plural forms, we obtain vectors that are roughly the same for all words.

The next questions are: how do we compute word2vec embeddings? That is, how do we train a word2vec model? What are training data, the neural network's architecture, and its input and output? The neural networks that train word2vec models are shallow with only one hidden layer. The input is a large corpus of text, and the outputs are vectors of several hundred dimensions, one for each unique term in the corpus. Words that share common linguistic contexts end up with vectors that are *close* to each other.

There are two learning algorithms for word2vec, which we will not detail here. However, by now we have a very good idea of how neural networks work, especially shallow ones with only one hidden layer. The two learning algorithms are:

Continuous bag-of-words
    This predicts the current word from a window of surrounding context words; the order of the context words does not influence the prediction.

Continuous skip-gram
    This uses the current word to predict the surrounding window of context words; the algorithm weighs nearby context words more heavily than more distant context words.

Both algorithms learn the vector representation of a term that is useful for prediction of other terms in a sentence. Continuous bag-of-words is apparently faster than continuous skip-gram, while skip-gram is better for infrequent words.

For more details, refer to the tutorial "The Amazing Power of Word Vectors" (Colyer 2016) (*https://oreil.ly/mDBZf*), the Wikipedia page on word2vec (*https://oreil.ly/rwL93*), and the three original papers on the subject: "Efficient Estimation of Word Representations in Vector Space" (Mikolov et al. 2013) (*https://oreil.ly/1PeLI*), "Distributed Representations of Words and Phrases and their Compositionality" (Mikolov et al. 2013) (*https://oreil.ly/XDZzr*), and "Linguistic Regularities in Continuous Space Word Representations" (Mikolov et al. 2013) (*https://oreil.ly/vKzgZ*).

The trained Google News word2vec model has 3 million words, each represented with a vector of 300 dimensions. To download this, you would need 3 GB of available memory. There are ways around downloading the whole pre-trained model if we have limited memory or if we only care about a fraction of the words.

How to visualize vectors representing words

Word vectors are very high-dimensional (100–500 dimensions), but humans can only visualize two- and three-dimensional vectors, so we need to project our high-dimensional vectors onto these drastically lower-dimensional spaces and still retain their most essential characteristics. By now we know that the singular value decomposition (principal component analysis) accomplishes that for us, giving us the vectors along which to project in decreasing order of importance, or the directions along which a given collection of word vectors varies the most. That is, the singular value decomposition ensures that this projection gives the best possible view of the word vectors, keeping them as far apart as possible.

There are many nice examples on the web. In the publication "Word Embedding-Topic Distribution Vectors for MOOC (Massive Open Online Courses) video lectures dataset" (Kastrati et al. 2020) (*https://oreil.ly/ktDdy*), the authors use a data set from the education domain with the transcripts of 12,032 video lectures from 200 courses collected from Coursera (*https://www.coursera.org*) to generate two things: word vectors using the word2vec model, and document topic vectors using latent Dirichlet allocation. The data set has 878,000 sentences and more than 79,000,000 tokens. The vocabulary size is over 68,000 unique words. The individual video transcripts are of different lengths, varying from 228 to 32,767 tokens, with an average of 6,622 tokens per video transcript. The authors use word2vec and latent Dirichlet allocation implementations in the Gensim package in Python. Figure 7-2 shows the publication's three-dimensional visualization of a subset of the word vectors using principal component analysis.

Note that word vectors and document topic vectors are not an end unto themselves; instead they are a means to an end, which is usually a natural language processing task, such as: classification within specific domains (such as the massive open online courses in the example), benchmarking and performance analysis of existing and new models, transfer learning, recommendation systems, contextual analysis, short text enrichment with topics, personalized learning, and organizing content that is easy to search and for maximum visibility. We visit such tasks shortly.

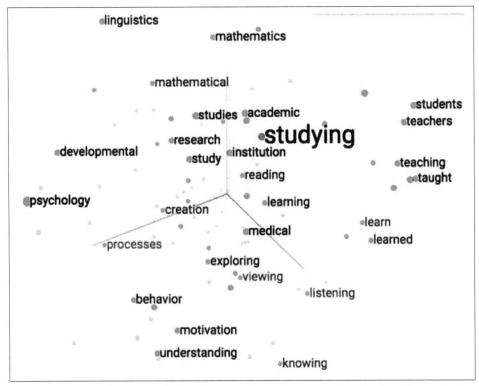

Figure 7-2. Three-dimensional visualization of word vectors using the first three principal components. This example highlights the vector representing the word "studying" and its neighbors: academic, studies, institution, reading, etc. (image source (https://oreil.ly/Pj0GF)).

Facebook's fastText vector representation of individual n-character grams

Facebook's fastText is similar to word2vec, but instead of representing full words or n-grams as vectors, it is trained to output a vector representation for every *n-character* gram. This enables fastText to handle rare, misspelled, and even partial words, such as the ones frequently appearing in social media posts. During training, word2vec's skip-gram algorithm learns to predict the surrounding context of a given word. Similarly, fastText's n-character gram algorithm learns to predict a word's surrounding n-character grams, providing more granularity and flexibility. For example, instead of only representing the full word "lovely" as a vector, it will represent the 2- and 3-grams as vectors as well: lo, lov, ov, ove, ve, vel, el, ely, and ly.

Facebook released its pre-trained fastText models for 294 languages, trained on available Wikipedia collections for these languages. These range from Abkhazian to Zulu, and include rare languages spoken by only a handful of people. Of course, the accuracy of the released models varies across languages and depends on the availability and quality of the training data.

Doc2vec or par2vec vector representation of a document

How about representing documents semantically? In previous sections we were able to represent entire documents as topic vectors, but word2vec only represents individual words or phrases as vectors. Can we then extend the word2vec model to represent entire documents as vectors carrying meaning? The paper "Distributed Representations of Sentences and Documents" (Le et al. 2014) (*https://oreil.ly/Y1NzF*) does exactly that, with an unsupervised algorithm that learns fixed-length dense vectors from variable-length pieces of texts, such as sentences, paragraphs, and documents. The tutorial "Doc2Vec tutorial using Gensim" (Klintberg 2015) (*https://oreil.ly/tWRcj*) walks through the Python implementation process, producing a fixed-size vector for each full document in a given corpus.

Global vector or vector representation of words

There are other ways to produce vectors representing meanings of words. Global vector or GloVe (2014) (*https://oreil.ly/8P32U*) is a model that obtains such vectors using the singular value decomposition. It is trained only on the nonzero entries of a global word-word co-occurrence matrix, which tabulates how frequently words co-occur with one another across an entire corpus.

GloVe is essentially a *log-bilinear model* with a weighted least-squares objective. The log-bilinear model is perhaps the simplest neural language model. Given the preceding *n-1* words, the log-bilinear model computes an initial vector representation for the next word simply by linearly combining the vector representations of these preceding *n-1* words. Then the probability of the occurrence of the next word given those *n-1* preceding words is computed based on computing the similarity (dot product) between the linear combination vector representation and the representations of all words in the vocabulary:

$$
\begin{aligned}
&Prob\left(w_n = w \mid w_1, w_2, \cdots, w_{n-1}\right) \\
&= \frac{exp\left(w^t_{vocab_i} w\right)}{exp\left(w^t_{vocab_1} w\right) + exp\left(w^t_{vocab_2} w\right) + \cdots + exp\left(w^t_{vocab_{vocab_{size}}} w\right)}
\end{aligned}
$$

The main intuition underlying the Global Vector model is the simple observation that ratios of word-word co-occurrence probabilities potentially encode some form of meaning. The example on the GloVe project website considers the co-occurrence probabilities for the target words "ice" and "steam" with various probe words from the vocabulary. Table 7-1 shows the actual probabilities from a six-billion-word corpus.

Table 7-1. Probability table showing the occurrence of the words "ice" and "steam" with the words "solid," "gas," "water," and "fashion"

Probability and ratio	k = solid	k = gas	k = water	k = fashion
P (k|ice)	1.9×10^{-4}	6.6×10^{-5}	3.0×10^{-3}	1.7×10^{-5}
P (k|steam)	2.2×10^{-5}	7.8×10^{-4}	2.2×10^{-3}	1.8×10^{-5}
P (k|ice)/P (k|steam)	8.9	8.5×10^{-2}	1.36	0.96

Observing the table in Table 7-1, we notice that, as expected, the word *ice* co-occurs more frequently with the word *solid* than it does with the word *gas*, whereas the word *steam* co-occurs more frequently with the word *gas* than it does with the word *solid*. Both *ice* and *steam* co-occur with their shared property *water* frequently, and both co-occur with the unrelated word *fashion* infrequently. Calculating the ratio of probabilities cancels out the noise from nondiscriminative words like *water*, so that large values, much greater than 1, correlate well with properties specific to *ice*, and small values, much less than 1, correlate well with properties specific to *steam*. This way, the ratio of probabilities encodes some crude form of meaning associated with the abstract concept of the thermodynamic phase.

The training objective of GloVe is to learn word vectors such that their dot product equals the logarithm of the probability of co-occurrence of words. Since the logarithm of a ratio is equal to the difference of logarithms, this objective considers vector differences in the word vector space. Because these ratios can encode some form of meaning, this information gets encoded as vector differences as well. For this reason, the resulting word vectors perform very well on word analogy tasks, such as those discussed in the word2vec package.

Since singular value decomposition algorithms have been optimized for decades, GloVe has an advantage in training over word2vec, which is a neural network and relies on gradient descent and backpropagation to perform its error minimization. If training our own word vectors from a certain corpus that we care about, we are probably better off using a Global Vector model than word2vec, even though word2vec is the first to accomplish semantic and logical reasoning with words, since Global Vector trains faster, has better RAM and CPU efficiency, and gives more accurate results than word2vec, even on smaller corpuses.

Cosine Similarity

So far in this chapter we have worked toward one goal only: convert a document of natural language text into a vector of numbers. Our document can be one word, one sentence, a paragraph, multiple paragraphs, or longer. We discovered multiple ways to get our vectors—some are more semantically representative of our documents than others.

Once we have a document's vector representation, we can feed it into machine learning models, such as classification algorithms, clustering algorithms, or others. One example is to cluster the document vectors of a corpus with some clustering algorithm such as *k-means* to create a document classifier. We can also determine how semantically similar our document is to other documents, for search engines, information retrieval systems, and other applications.

We have established that due to the curse of dimensionality, measuring the Euclidean distance between two very high-dimensional document vectors is useless, since they would come out extremely far apart, only because of the vastness of the space they inhabit. So how do we determine whether vectors representing documents are *close* or *far*, or *similar* or *different*? One successful way is to use *cosine similarity*, measuring the cosine of the angle between the two document vectors. This is given by the dot product of the vectors, each normalized by its length (had we normalized the document vectors ahead of time, then their lengths would have already been one):

$$\cos\left(\text{angle between } \overrightarrow{doc_1} \text{ and } \overrightarrow{doc_2}\right) = \frac{\overrightarrow{doc_1}^t \; \overrightarrow{doc_2}}{length\left(\overrightarrow{doc_1}\right) length\left(\overrightarrow{doc_2}\right)}$$

Figure 7-3 shows three documents represented in a two-dimensional vector space. We care about the angles between them.

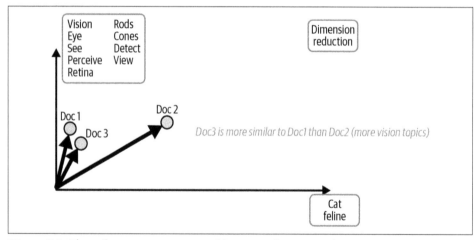

Figure 7-3. Three documents represented in a two-dimensional vector space

The cosine of an angle is always a number between -1 and 1. When two document vectors are perfectly aligned and pointing in the same direction along all the dimensions, their cosine similarity is 1; when they are perfect opposites of each other regarding every single dimension, their cosine similarity is -1; and when they are orthogonal to each other, their cosine similarity is 0.

Natural Language Processing Applications

The bulk of this chapter has been about converting a given document of natural language text to a vector of numbers. We have established that there are multiple ways to get our document vectors, all leading to varying representations (and hence conclusions), or emphasizing certain aspects of the given natural language data over others. For people entering the natural language processing subfield of AI, this is one of the hardest barriers to overcome, especially if they are from a quantitative background, where the entities they work with are inherently numerical, ripe for mathematical modeling and analysis. Now that we have overcome this barrier, equipped with concrete vector representations for natural language data, we can think mathematically about popular applications. It is important to be aware that there are multiple ways to accomplish each of the following. Traditional approaches are hardcoded rules, assigning scores to words, punctuations, emojis, etc., then relying on the existence of these in a data sample to produce a result. Modern approaches rely on various machine learning models, which in turn rely on (mostly) labeled training data sets. To excel in this field, we must set time aside and try different models on the same task, compare performance, and gain an in-depth understanding of each model along with its strengths, weaknesses, and mathematical justifications of its successes and failures.

Sentiment Analysis

The following are common approaches for extracting sentiment from natural language text:

Hardcoded rules
A successful algorithm is VADER, or Valence Aware Dictionary for sEntiment Reasoning. A tokenizer here needs to handle punctuation and emojis properly, since these convey a lot of sentiment. We also have to manually compile thousands of words along with their sentiment score, as opposed to having the machine accomplish this automatically.

Naive Bayes classifier (https://oreil.ly/uzJDd)
This is a set of classifying algorithms based on the Bayes' Theorem from probability, the decision rule for classification on maximum likelihood. This will be discussed in Chapter 11.

Latent discriminant analysis
In the previous section, we learned how to classify documents into two classes using latent discriminant analysis. To recap, we start with the data labeled into

two classes, then we compute the centroid of each class and find the direction connecting them. We project each new data instance along that direction, and classify it according to which centroid it falls closer to.

Using latent semantic analysis

Clusters of document vectors formed using latent semantic analysis can be used for classification. Ideally, positive reviews cluster away from negative reviews in latent semantic analysis topic spaces. Given a bunch of reviews labeled positive or negative, we first compute their topic vectors using latent semantic analysis. Now, to classify a new review, we can compute its topic vector, then that topic vector's cosine similarity with the positive and negative topic vectors. Finally, we classify the review as positive if it is more similar to positive topic vectors, and negative if more similar to negative topic vectors.

Transformers, convolutional neural network, recurrent long short-term memory neural network

All of these modern machine learning methods require passing our document in vector form into a neural network with a certain architecture. We will spend time on these state-of-the-art methods shortly.

Spam Filter

Mathematically, spam filtering is a similar classification problem to sentiment analysis discussed previously, when the sentiment of a document is either positive or negative. Thus, the same methods for sentiment classification apply for spam filtering. In all cases, it doesn't matter how we create our document vectors–we can use them to predict whether a social post is spam or not spam, predict how likely it is to get likes, etc.

Search and Information Retrieval

Again, no matter how we create the numerical vectors representing documents, we can use them for search and information retrieval tasks. The search can be index based or semantic based.

Full text search

When we search for a document based on a word or a partial word that it contains. Search engines break documents into words that can be indexed, similar to the indexes we find at the end of textbooks. Of course, spelling errors and typos require a lot of tracking and sometimes guessing. Indices, when available, work pretty well.

Semantic search

Here, our search for documents takes into account the meaning of the words in both our query and in the documents within which we are searching.

The following are common approaches for search and information retrieval:

Based on cosine similarity between the TF-IDF of documents
This is good for corpuses containing billions of documents. Any search engine with a millisecond response time employs an underlying TF-IDF matrix.

Based on semantics
Cosine similarity between topic vectors of documents obtained through latent semantic analysis (for corpuses containing millions of documents) or latent Dirichlet allocation (for much smaller corpuses). This is similar to how we classified whether a message is spam or not spam using latent semantic analysis, except that now we compute the cosine similarity between the new document's topic vector and *all* the topic vectors of our database, returning the ones that are most similar to our document.

Based on eigenvector iteration
This has to do with ranking algorithms in search results, such as the PageRank algorithm (which we go over in "Example: PageRank Algorithm" on page 302). The following is a useful excerpt from the paper "Role of Ranking Algorithms for Information Retrieval" (Choudhary and Burdak 2012) (*https://oreil.ly/jEbMX*):

> There are three important components in a search engine. They are the crawler, the indexer, and the ranking mechanism. The crawler, also called a robot or spider, traverses the web and downloads the web pages. These downloaded pages are sent to an indexing module that parses the web pages and builds the index based on the keywords in those pages. An index is generally maintained using keywords. When a user types a query using keywords into a search engine's interface, the query processor component matches the query keywords with the index and returns URLs to the user. But before showing the pages to the user, a ranking mechanism is used by the search engine to show the most relevant pages at the top and less relevant ones at the bottom.

Semantic search and queries using word vectors (word2vec or GloVe)
Consider a search like this one, adapted from the book *Natural Language Processing in Action*: "She invented something to do with physics in Europe in the early 20th century." When we enter our search sentence into Google or Bing, we may not get the direct answer *Marie Curie*. Google Search will most likely only give us links to lists of famous physicists, both men and women. After searching several pages we find *Marie Curie*, our answer. Google will take note of that, and refine our results next time we search. Now using word vectors, we can do simple arithmetic on the word vectors representing woman+Europe+physics+scientist+famous, then we would obtain a new vector, close in cosine similarity to the vector representing *Marie Curie*, and voilà! We have our answer. We can even subtract gender bias in the natural sciences from word vectors by simply subtracting the vector representing the token *man*, *male*, etc., so we can search for the word vector closest to: woman+Europe+physics+scientist-male-2*man.

Search based on analogy questions

To compute a search such as *They are to music what Marie Curie is to science*, all we have to do is simple vector arithmetic of the word vectors representing Marie Curie-science+music.

The following paragraph, regarding indexing and semantic searches is paraphrased from the book *Natural Language Processing in Action*:

Traditional indexing approaches work with binary word occurrence vectors, discrete vectors (bag-of-word vectors), sparse floating-point number vectors (TF-IDF vectors), and low-dimensional floating-point number vectors (such as three-dimensional geographic information system data). But high-dimensional floating-point number vectors, such as topic vectors from latent semantic analysis or latent Dirichlet allocation, are challenging. Inverted indexes work for discrete vectors or binary vectors, because the index only needs to maintain an entry for each nonzero discrete dimension. That value of that dimension is either present or not present in the referenced vector or document. Because TF-IDF vectors are sparse, mostly zero, we do not need an entry in our index for most dimensions for most documents. Latent semantic analysis and latent Dirichlet allocation produce topic vectors that are high dimensional, continuous, and dense, where zeros are rare. Moreover, the semantic analysis algorithm does not produce an efficient index for scalable search. This is exacerbated by the curse of dimensionality, which makes an exact index impossible. One solution to the challenge of high-dimensional vectors is to index them with a locality-sensitive hash, like a zip code, that designates a region of hyperspace. Such a hash is similar to a regular hash: it is discrete and only depends on the values in the vector. But even this doesn't work perfectly once we exceed about 12 dimensions. An exact semantic search wouldn't work for a large corpus, such as a Google search or even a Wikipedia semantic search. The key is to settle for *good enough* rather than striving for a perfect index or a latent hashing algorithm for our high-dimensional vectors. There are now several open source implementations of some efficient and accurate approximate nearest neighbors algorithms that use latent semantic hashing to efficiently implement semantic search. Technically, these indexing or hashing solutions cannot guarantee that we will find all the best matches for our semantic search query. But they can get a good list of close matches almost as fast as with a conventional reverse index on a TF-IDF vector or bag-of-words vector if we are willing to give up a little precision. Neural network models fine-tune the concepts of topic vectors so that the vectors associated with words are more precise and useful, hence enhancing searches.

Machine Translation

The goal here is to translate a sequence of tokens of any length (such as a sentence or a paragraph) to a sequence of any length in a different language. The encoder-decoder architecture, discussed in the context of transformers and recurrent neural networks, has proven successful for translation tasks. The encoder-decoder architecture is different than the auto-encoder architecture.

Image Captioning

This combines computer vision with natural language processing.

Chatbots

This is the ultimate application of natural language processing. A chatbot requires more than one kind of processing: parse language, search, analyze, generate responses, respond to requests, and execute them. Moreover, it requires a database to maintain a memory of past statements and responses.

Other Applications

Other applications include *named-entity recognition (https://oreil.ly/5Mg9k)*, *conceptual focus (https://oreil.ly/QYDOn)*, relevant information extraction from text (such as dates), and language generation, which we visit in Chapter 8.

Transformers and Attention Models

Transformers and attention models are the state-of-the-art for natural language processing applications such as machine translation, question answering, language generation, named-entity recognition, image captioning, and chatbots (as of 2022). Currently, they underlie large language models such as Google's BERT (Bidirectional Encoder Representations from Transformers) (*https://oreil.ly/sP7uM*) and OpenAI's GPT-2 (Generative Pre-trained Transformer) (*https://oreil.ly/9rJQo*) and GPT-3 (*https://oreil.ly/NzDKo*).

Transformers bypass both recurrence and convolution architectures, which were the go-to architectures for natural language processing applications up until 2017, when the paper "Attention Is All You Need" (Vaswani et al. 2017) (*https://oreil.ly/AUAl9*) introduced the first transformer model.

The dethroned recurrent and convolutional neural network architectures are still in use (and work well) for certain natural language processing applications, as well as other applications such as finance. We elaborate on these models later in this chapter. However, the reasons that led to abandoning them for natural language are:

- For short input sequences of natural language tokens, the attention layers that are involved in transformer models are faster than recurrent layers. Even for long sequences, we can modify attention layers to focus on only certain neighborhoods within the input.

- The number of sequential operations required by a recurrent layer depends on the length of the input sequence. This number stays constant for an attention layer.

- In convolutional neural networks, the width of the kernel directly affects the long-term dependencies between pairs of inputs and corresponding outputs. Tracking long-term dependencies then requires large kernels, or stacks of convolutional layers, all increasing the computational cost of the natural language model employing them.

The Transformer Architecture

Transformers are an integral part of enormous language models, such as GPT-2, GPT-3, ChatGPT, Google's BERT (which trains the language model by looking at the sequential text data from both left to right and right to left) and Wu Dao's transformer (*https://oreil.ly/Submn*). These models are massive: GPT-2 has around 1.5 billion parameters trained on millions of documents, drawn from 8 million websites from all around the internet. GPT-3 has 175 billion parameters trained on an even larger data set. Wu Dao's transformer has a whopping 1.75 trillion parameters, consuming a ton of computational resources for training and inference.

Transformers were originally designed for language translation tasks, so they have an *encoder-decoder* structure. Figure 7-4 illustrates the architecture of the transformer model originally introduced by the paper "Attention Is All You Need" (Vaswani et al. 2017) (*https://oreil.ly/S3vEz*). However, each encoder and decoder is its own module, so they can be used separately to perform various tasks. For example, we can use the encoder alone to perform a classification task such as a part of speech tagging, meaning we input the sentence: *I love cooking in my kitchen*, and the output will be a class for each word: *I*: pronoun; *love*: verb; *cooking*: noun, etc.

The input to the full transformer model (with both the encoder and decoder included) is a sequence of natural language tokens of any length, such as a question to a chatbot, a paragraph in English that requires translation to French, or a summarization into a headline. The output is another sequence of natural language tokens, also of any length, such as the chatbot's answer, the translated paragraph in French, or the headline.

Do not confuse the training phase with the inference phase of a model:

During training
    The model is fed both the data and the labels, such as an English sentence (input data sample) along with its French translation (label), and the model learns a mapping from the input to the target label that hopefully generalizes well to the entire vocabularies and grammars of both languages.

During inference
    The model is fed only the English sentence, and outputs its French translation. Transformers output the French sentence one new token at a time.

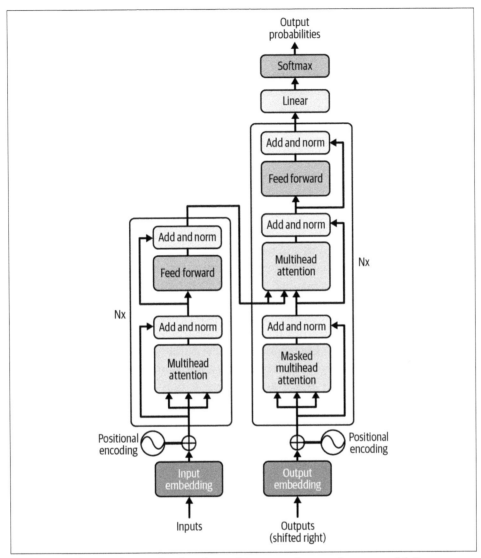

Figure 7-4. The simple encoder-decoder architecture of a transformer model (image source (https://oreil.ly/q1FOp))

The encoder, on the left half of the transformer architecture in Figure 7-4, receives an input of tokens, such as an English sentence, *How was your day?*, and produces multiple numerical vector representations for each token of this sentence—along with their linear transformations into new directions that capture alignment with other word vectors—encoding the token's contextual information from within the sentence. The linear transformations happen (multiplying word vectors by weight matrices) because the encoder wants to learn the important directions in which certain word vectors of a sentence align with others (contextual alignment).

The decoder, on the right half of the architecture in Figure 7-4, receives these vectors as part of its input, together with its own output from the previous time step. Ultimately, it generates an output of tokens, such as the French translation of the input sentence, *Comme se passe ta journée* (see Figure 7-5). What the decoder actually computes is a *probability* for each word in the French vocabulary (say, 50,000 tokens) using a softmax function, then produces the token with the highest probability. In fact, since computing a softmax for such a high-dimensional vocabulary is expensive, the decoder uses a *sampled softmax*, which computes the probability for each token in a random sample of the French vocabulary at each step. During training, it has to include the target token in this sample, but during inference, there is no target token.

Transformers use a process called *attention* to capture long-term dependencies in sequences of tokens. The word *sequence* is confusing here, especially for mathematicians, who have clear distinctions among the terms *sequence, series, vector,* and *list.* Sequences are usually processed one term at a time, meaning one term is processed, then the next, then the next, and so on, until the whole input is consumed. Transformers do not process input tokens sequentially. They process them all together, in parallel. This is different from the way recurrent neural networks process input tokens, which have to be fed sequentially, in effect prohibiting parallel computation. If it was up to us to correct this terminology, we should call a natural language sentence a *vector* if we are processing it using a transformer model, or a *matrix* since each word in the sentence is represented as its own vector, or a *tensor* if we process a batch of sentences at a time, which the architecture of the transformer allows. If we want to process the same exact sentence using a recurrent neural network model, then we should call it a *sequence*, since this model consumes its input data sequentially, one token at a time. If we process it using a convolutional neural network, then we would call it a vector (or matrix) again, since the network consumes it as a whole, not broken down into one token at a time.

It is an advantage when a model does not need to consume the input sequentially, because such architectures allow for parallel processing. That said, even though parallelization makes transformers computationally efficient, they cannot take full advantage of the inherent sequential nature of the natural language input and the information encoded within this sequentiality. Think of how humans process text. There are new transformer models that try to leverage this.

The transformer model runs as follows:

1. Represent each word from the input sequence as a d-dimensional vector.

2. Incorporate the order of words into the model by adding to the word vector information about its position (*positional encoding*). Introduce positional information into the input by accompanying each vector of each word with a positional encoding vector of the same length. The positional encoding vectors have the same dimension as the word vector embeddings (this allows the two vectors to be added together). There are many choices of positional encodings: some are learned during training, others are fixed. Discretized sine and cosine functions with varying frequencies are common.

3. Next, feed the positionally encoded word vectors to the encoder block. The encoder attends to all words in the input sequence, irrespective of if they precede or succeed the word under consideration, thus the transformer encoder is bidirectional.

4. The decoder receives as input its own predicted output word at time step $t-1$, along with the output vectors of the encoder.

5. The input to the decoder is also augmented by positional encoding.

6. The augmented decoder input is fed into the three sublayers. The decoder cannot attend to succeeding words, so we apply masking in its first sublayer. At the second sublayer, the decoder also receives the output of the encoder, which now allows the decoder to attend to all of the words in the input sequence.

7. The output of the decoder finally passes through a fully connected layer, followed by a softmax layer, to generate a prediction for the next word of the output sequence.

The Attention Mechanism

The transformer's magic is largely due to built-in *attention mechanisms*. An attention mechanism comes with bonuses:

Explainability
Pointing out which parts of the input sentence (or document) the model paid attention to when producing a particular output (see Figure 7-5).

Leveraging pre-trained attention models
We can adapt pre-trained models to domain-specific tasks. That is, we can further tweak their parameter values with extra training on domain-specific data.

More accurate modeling of longer sentences
Another value of attention mechanisms is that they allow the modeling of dependencies in sequences of natural language tokens without regard to how far apart related tokens occur in these sequences.

Figure 7-5 illustrates attention for a translation task from English to French.

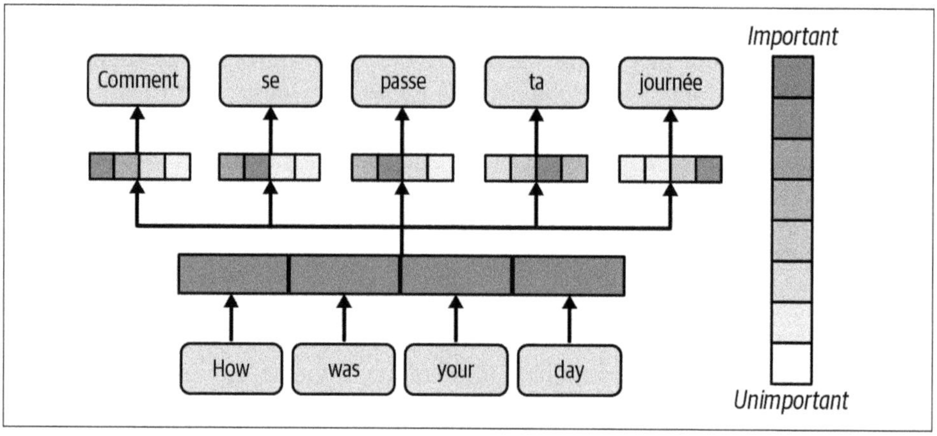

Figure 7-5. Illustrating attention via a translation task: how weights assigned to input tokens show which ones the model paid more attention to in order to produce each output token (image source (https://oreil.ly/2SZCW))

There is no hardcore mathematics involved in an attention mechanism: we only have to compute a scaled dot product. The main goal of attention is to highlight the most relevant parts of the input sequence, how strongly they relate to each other within the input itself, and how strongly they contribute to certain parts of the output. We compute these word alignment scores in multiple directions because words can relate to each other in multiple ways.

Self attention is when a sequence of vectors computes alignment within its own members. We are now familiar with the fact that the dot product measures the compatibility between two vectors. We can compute the simplest possible self attention weights by finding the dot products between all the members of the sequence of vectors. For example, for the sentence *I love cooking in my kitchen*, we would compute all the dot products between the word vectors representing the words *I*, *love*, *cooking*, *in*, *my*, and *kitchen*. We would expect the dot product between *I* and *my* to be high, similarly between *cooking* and *kitchen*. However, the dot product will be highest between *I* and *I*, *love* and *love*, etc., because these vectors are perfectly aligned with themselves, but there is no valuable information gleaned there.

The transformer's solution to avoiding this waste is multifold:

1. Apply three different linear transformations to each vector of the input sequence (each word of the sentence), multiplying them by three different weight matrices. We then obtain three different sets of vectors corresponding to each input word vector $\vec{w}$:

- The *query* vector $\overrightarrow{query} = W_q \vec{w}$, the vector *attended from*.

- The *key* vector $\overrightarrow{key} = W_k \vec{w}$, the vector *attended to*.

- The *value* vector $\overrightarrow{value} = W_v \vec{w}$, to capture the context that is being generated.

2. Obtain alignment scores between the query and key vectors for all words in the sentence by computing their dot product scaled by the inverse of the square root of the length of these vectors $\sqrt{l}$. We apply this scaling for numerical stability to keep the dot products from becoming large. (These dot products will soon be passed into a softmax function. Since the softmax function has a very small gradient when its input has a large magnitude, we offset this effect by dividing each dot product by $\sqrt{l}$.) Moreover, alignment of two vectors is independent from the lengths of these vectors. Therefore, the alignment score between *cooking* and *kitchen* in our sentence will be:

$$alignment_{cooking,kitchen} = \frac{1}{\sqrt{l}} \overrightarrow{query}_{cooking}^{t} \overrightarrow{key}_{kitchen}$$

Note that this will be different than the alignment score between *kitchen* and *cooking*, since the query and key vectors for each are different. Thus, the resulting alignment matrix is not symmetric.

3. Transform each alignment score between each two words in the sentence into a probability by passing the score into the *softmax* function. For example:

$$
\begin{aligned}
\omega_{cooking,kitchen} &= \\
softmax\left(alignment_{cooking,kitchen}\right) &= \\
\frac{exp\left(alignment_{cooking,kitchen}\right)}{\{exp\left(alignment_{cooking,I}\right) +} \\
exp\left(alignment_{cooking,love}\right) + \\
exp\left(alignment_{cooking,cooking}\right) + \\
exp\left(alignment_{cooking,in}\right) + \\
exp\left(alignment_{cooking,my}\right) + \\
exp\left(alignment_{cooking,kitchen}\right)\}
\end{aligned}
$$

4. Finally, encode the context of each word by linearly combining the value vectors using the alignment probabilities as weights for the linear combination. For example:

$$context_{cooking} = \omega_{cooking,I}\overrightarrow{value}_I + \omega_{cooking,love}\overrightarrow{value}_{love}$$
$$+ \omega_{cooking,cooking}\overrightarrow{value}_{cooking} + \omega_{cooking,in}\overrightarrow{value}_{in} + \omega_{cooking,my}\overrightarrow{value}_{my}$$
$$+ \omega_{cooking,kitchen}\overrightarrow{value}_{kitchen}$$

We have thus managed to capture in one vector the context of each word in the given sentence, with high worth assigned to those words in the sentence it mostly aligns with.

The good news here is that we can compute the context vector for all the words of a sentence (data sample) simultaneously, since we can pack the vectors previously mentioned in matrices and use efficient and parallel matrix computations to get the contexts for all the terms at once.

We implement all of the above in one *attention head*. That is, one attention head produces one context vector for each token in the data sample. We would benefit from producing multiple context vectors for the same token, since some information gets lost with all the averaging happening on the way to a context vector. The idea here is to be able to extract information using different representations of the terms of a sentence (data sample), as opposed to a single representation corresponding to a single attention head. So we implement a *multihead attention*, choosing for each head new transformation matrices W_q, W_k, and W_v.

Note that during the training process, the entries of the transformation matrices are model parameters that have to be learned from the training data samples. Imagine then how fast the number of the model's parameters will balloon.

Figure 7-6 illustrates the *multihead attention mechanism*, implementing h heads that receive different linearly transformed versions of the queries, keys, and values to produce h context vectors for each token, which are then concatenated to produce the output of the *multihead attention* part of the model's structure.

The decoder uses a similar self attention mechanism, but here each word can only attend to the words before it, since text is generated from left to right. Moreover, the decoder has an extra attention mechanism attending to the outputs it receives from the encoder.

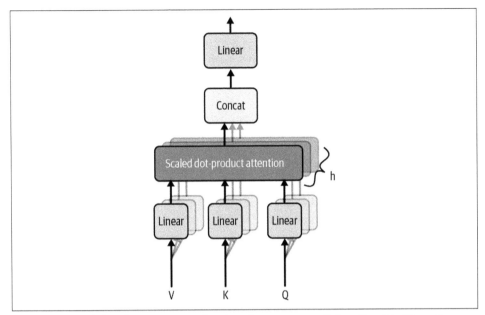

Figure 7-6. Multihead attention mechanism (image source (https://oreil.ly/j9ypn))

Transformers Are Far from Perfect

Even though transformer models have revolutionized the natural language processing field, they are far from perfect. Language models, in general, are *mindless mimics*. They understand neither their inputs nor their outputs. Critical articles, such as this (*https://oreil.ly/aaj80*) and this (*https://oreil.ly/ToNil*) by the *MIT Technology Review*, among others, detail their shortcomings, such as lack of comprehension of language, repetition when used to generate long passages of text, and so on. That said, the transformer model brought about a tidal wave for natural language, and it is making its way to other AI fields, such as biomedicine, computer vision, and image generation.

Convolutional Neural Networks for Time Series Data

The term *time series* in the natural language processing and finance domains should be *time sequence* instead. *Series* in mathematics refers to adding up the terms of an infinite *sequence*. So when our data is *not summed*, which is the case for all natural language data and most finance data, we actually have *sequences*, not *series*, of numbers, vectors, etc. Oh well, vocabulary collisions are unavoidable, even across different fields that heavily rely on one another.

Other than the definitions of words in a dictionary, their meanings are mostly correlated to the way they occur relative to each other. This is conveyed through the way words are *ordered* in sentences, as well as their context and proximity to other words in sentences.

We first emphasize the two ways in which we can explore the meanings behind words and terms in documents:

Spatially
    Exploring a sentence all at once as one vector of tokens, whichever way these tokens are represented mathematically.

Temporarily
    Exploring a sentence sequentially, one token at a time.

Convolutional neural networks, discussed in Chapter 5, explore sentences spatially, by sliding a fixed-width window (kernel or filter) along the tokens of the sentence. When using convolutional neural networks to analyze text data, the network expects an input of fixed dimensions. On the other hand, when using recurrent neural networks (discussed next) to analyze text data, the network expects tokens sequentially, hence, the input does not need to be of fixed length.

In Chapter 5, we were sliding two-dimensional windows (kernels or filters) over images, and in this chapter, we will slide one-dimensional kernels over text tokens. We now know that each token is represented as a vector of numbers. We can either use one-hot encoding or word vectors of the word2vec model. One-hot encoded tokens are represented with a very long vector that has a 0 for every possible vocabulary word that we want to include from the corpus, and a 1 in the position of the token we are encoding. Alternatively, we can use trained word vectors produced via word2vec. Thus, an input data sample to the convolutional neural network is a matrix made up of column vectors, one column for each token in the data sample. If we use word2vec to represent tokens, then each column vector would have 100 to 500 entries, depending on the particular word2vec model used. Recall that for a convolutional neural network, each data sample has to have the exact same number of tokens.

Therefore, one data sample (a sentence or a paragraph) is represented with a two-dimensional matrix, where the number of rows is the full length of the word vector. In this context, saying that we are sliding a *one-dimensional* kernel over our data sample is slightly misleading, but here is the explanation. The vector representation of the sample's tokens extends *downward*; however, the filter covers the whole length of that dimension all at once. That is, if the filter is three tokens wide, it would be a matrix of weights with three columns and as many rows as the vector representation of our tokens. Thus, one-dimensional convolution here refers to convolving only *horizontally*. This is different than two-dimensional convolution for images, where the two-dimensional filter travels across the image both horizontally and vertically.

As in Chapter 5, during a forward pass, the weight values in the filters are the same for one data sample. This means that we can parallelize the process, which is why convolutional neural networks are efficient to train.

Recall that convolutional neural networks can also process more than one channel of input at the same time, that is, three-dimensional tensors of input, not only two-dimensional matrices of numbers. For images this was processing the red, green, and blue channels of an input image all at once. For natural language, one input sample is a bunch of words represented as column vectors lined up next to each other. We now know that there are multiple ways to represent the same word as a vector of numbers, each perhaps capturing different semantics of the same word. These different vector representations of the same word are not necessarily of the same length. If we restrict them to be of the same length, then each representation can be a word's *channel*, and the convolutional neural network can process all the channels of the same data sample at once.

As in Chapter 5, convolutional neural networks are efficient due to weight sharing, pooling layers, dropout, and small filter sizes. We can run the model with multiple size filters, then concatenate the output of each size filter into a longer thought vector before passing it into the fully connected last layer. Of course, the last layer of the network accomplishes the desired task, such as sentiment classification, spam filtering, text generation, and others. We went over this in Chapters 5 and 6.

Recurrent Neural Networks for Time Series Data

Consider the following three sentences:

- She bought tickets to watch the movie.
- She, having free time, bought tickets to watch the movie.
- She, having heard about it nonstop for two weeks in a row, finally bought tickets to watch the movie.

In all three sentences, the predicate *bought tickets to watch the movie* corresponds to the sentence's subject *She*. A natural language model will be able to learn this if it is designed to handle long-term dependencies. Let's explore how different models handle such long-term dependencies:

Convolutional neural networks and long-term dependencies
A convolutional neural network, with its narrow filtering window ranging three to five tokens scanning the sentence, will be able to learn from the first sentence easily, and maybe the second sentence, given that the predicate's position changed only a little bit (pooling layers help with the network's resistance to small variations). The third sentence will be tough, unless we use larger filters (which increases the computation cost and makes the network more like a fully

connected network than a convolutional network), or if we deepen the network, stacking convolutional layers on top of each other so that the coverage widens as the sentence makes its way deeper into the network.

Recurrent neural networks with memory units

A completely different approach is to feed the sentence into the network sequentially, one token at a time, and maintain a *state* and a *memory* that hold on to important information for a certain amount of time. The network produces an outcome when all the tokens in the sentence have passed through it. If this is during training, only the outcome produced after the last token has been processed gets compared to the sentence's label, then the error *backpropagates through time*, to adjust the weights. Compare this to the way we hold on to information when reading a long sentence or paragraph. Recurrent neural networks with long short-term memory units are designed this way.

Transformer models and long-term dependencies

Transformer models, which we discussed earlier, abolish both convolution and recurrence, relying only on attention to capture the relationship between the subject of the sentence *She*, and the predicate *bought tickets to watch the movie*.

One more thing differentiates recurrence models from convolutional and transformer models: does the model expect its input to be of the same length for all data samples? Can we only input sentences of the same length? The answer for transformers and convolutional networks is that they expect only fixed-length data samples, so we have to preprocess our samples and make them all the same length. On the other hand, recurrent neural networks handle variable length inputs really well, since after all they only take them one token at a time.

The main idea for a recurrent neural network is that it holds on to past information as it processes new information. How does this holding happen? In a feed forward network, the output of a neuron leaves it and never gets back to it. In a recurrent network, the output loops back into the neuron, along with new input, in essence creating a *memory*. Such algorithms are great for autocompletion and grammar check. They have been integrated into Gmail's Smart Compose since 2018.

How Do Recurrent Neural Networks Work?

Here are the steps for how a recurrent neural network gets trained on a set of labeled data samples. Each data sample is made up of a bunch of tokens and a label. As always, the goal of the network is to learn the general features and patterns within the data that end up producing a certain label (or output) versus others. When tokens of each sample are input sequentially, our goal is then to detect the features, across all the data samples, that emerge when certain tokens appear in patterns relative to each other:

1. Grab one tokenized and labeled data sample from your data set (such as a movie review labeled positive, or a tweet labeled fake news).

2. Pass the first token $token_0$ of your sample into the network. Remember that tokens are vectorized, so you are really passing a vector of numbers into the network. In mathematical terms, we are evaluating a function at that token's vector and producing another vector. So far, our network has calculated $f(token_0)$.

3. Now pass the second token $token_1$ of your sample into the network, *along with the output of the first token,* $f(token_0)$. The network now will evaluate $f(token_1 + f(token_0))$. This is the recurrence step, and this is how the network does not forget $token_0$ as it processes $token_1$.

4. Now pass the third token $token_2$ of your sample into the network, *along with the output of the previous step,* $f(token_1 + f(token_0))$. The network now will evaluate $f(token_2 + f(token_1 + f(token_0)))$.

5. Keep going until you finish all the tokens of your one sample. Suppose this sample only had five tokens, then our recurrent network will output $f(token_4 + f(token_3 + f(token_2 + f(token_1 + f(token_0)))))$. Note that this output looks very similar to the output of the feed forward fully connected networks that we discussed in Chapter 4, except that this output *unfolds through time* as we input a sample's tokens one at a time *into one recurrent neuron,* while in Chapter 4, the network's output *unfolds through space,* as one data sample moves from one layer of the neural network to the next. Mathematically, when we write the formulas of each, they are the same, so we don't need more math beyond what we learned in the past three chapters. That's why we love math.

6. When training the network to produce the right thing, it is the final output of the sample, $f(token_4 + f(token_3 + f(token_2 + f(token_1 + f(token_0)))))$, that gets compared against the sample's true label via evaluating a loss function, exactly as we did in Chapters 3, 4, and 5.

7. Now pass the next data sample one token at a time into the network and do the same thing again.

8. We update the network's weights in exactly the same way we updated them in Chapter 4, by minimizing the loss function via a gradient descent-based algorithm, where we calculate the required gradient (the derivatives with respect to all the network's weights) via backpropagation. As we just said, this is exactly the same backpropagation mathematics we learned in Chapter 4, except now, of course, we get to say *we're backpropagating through time.*

In finance, dynamics, and feedback control, this process is called an autoregressive moving average (ARMA) model (*https://oreil.ly/GEwhG*).

Training a recurrent neural net can be expensive, especially for data samples of any significant length, say 10 tokens or more, since the number of weights to learn is directly related to the number of tokens in data samples: the more tokens, the more *depth in time* the recurrent network has. Other than the computational cost, this depth comes with all the troubles encountered by regular feed forward networks with many layers: vanishing or exploding gradients, especially with samples of data with hundreds of tokens, which will be the mathematical equivalent of a fully connected feed forward neural network with hundreds of layers! The same remedies for exploding and vanishing gradients for feed forward networks work here.

Gated Recurrent Units and Long Short-Term Memory Units

Recurrent neurons in recurrent networks are not enough to capture long-term dependencies in a sentence. A token's effect gets diluted and stepped on by the new information as more tokens pass through the recurrent neuron. In fact, a token's information is almost completely gone only after two tokens have passed. This problem can be addressed if we add memory units, called *long short-term memory units*, to the architecture of the network. These help learning dependencies stretching across a whole data sample.

Long short-term memory units contain neural networks, and they can be trained to find only the new information that needs to be retained for the upcoming input, and to forget, or reset to zero, information that is no longer relevant to learning. Therefore, long short-term memory units learn which information to hold on to, while the rest of the network learns to predict the target label.

There is no new mathematics beyond what we learned in Chapter 4 here, so we will not go into the weeds digging into the specific architecture of a long short-term memory unit, or a *gated unit*. In summary, the input token for each time step passes through the forget and update gates (functions), gets multiplied by weights and masks, then gets stored in a memory cell. The network's next output depends on a combination of the input token and the memory unit's current state. Moreover, long short-term memory units share the weights they learned across samples, so they do not have to relearn basic information about language as they go through each sample's tokens.

Humans are able to process language on a subconscious level, and long short-term memory units are a step into modeling that. They are able to detect patterns in language that allow us to address more complex tasks than mere classification, such as language generation. We can generate novel text from learned probability distributions. This is the topic of Chapter 8.

An Example of Natural Language Data

When faced with narratives about different models, it always makes things easier when we have specific examples in mind with real data, along with the models' hyperparameters. We can find the IMDb (*https://oreil.ly/HbLZX*) movie review data set at the Stanford AI website (*https://oreil.ly/by2UQ*). Each data sample is labeled with a 0 (negative review) or a 1 (positive review). We can start with the raw text data if we want to practice preprocessing natural language text. Then we can tokenize it and vectorize it using one-hot encoding over a chosen vocabulary, the Google word2vec model, or some other model. Do not forget to split the data into the training and test sets. Then choose the hyperparameters, for example, length of word vectors around 300, number of tokens per data sample around 400, mini-batches of 32, and number of epochs 2. We can play around with these to get a feel for the models' performance.

Finance AI

AI models have a wide use for the finance field. By now, we know the underlying structure of most AI models (except for graphs, which have a different mathematical structure; we discuss graph networks in Chapter 9). At this point, only mentioning an application area from finance is enough for us to have a very good idea about how to go about modeling it using AI. Moreover, many finance applications are naturally intertwined with natural language processing applications, such as marketing decisions based on customer reviews, or a natural language processing system used to predict economic trends and trigger large financial transactions based only on the models' outputs.

The following are only two AI applications in finance, among many. Think of ways we can put what we have learned so far to good use modeling these problems:

- Stock market time series prediction. A recurrent neural network can take a sequence of inputs and produce a sequence of outputs. This is useful for the time series prediction required for stock prices. We input the prices over the past *n* days, and the network outputs the prices from the past *n-1* days *along with tomorrow's price.*

- Autoregressive moving average (ARMA) model in finance, dynamics, and feedback control.

The stock market appears multiple times in this book. Keep an eye out for it when we are discussing stochastic processes in Chapter 11 on probability.

Summary and Looking Ahead

There was almost no new math in this chapter; however, it was one of the hardest to write. The goal was to summarize the most important ideas in the whole natural language processing field. Moving from words to relatively low-dimensional vectors of numbers that carry meaning was the main barrier to overcome. Once we learned multiple ways to do this, whether vectorizing one word at a time or the main topics in a long document or an entire corpus, feeding those vectors to different machine learning models with different architectures and purposes was just business as usual.

Calculus
> The *log* scale for term frequencies and inverse document frequencies

Statistics
- Zipf's law for word counts
- The Dirichlet probability distribution for assigning words to topics and topics to documents

Linear algebra
- Vectorizing documents of natural language
- The dot product of two vectors and how it provides a measure of similarity or compatibility between the entities that the vectors represent
- Cosine similarity
- Singular value decomposition, i.e., latent semantic analysis

Probability
- Conditional probabilities
- Bilinear log model

Time series data
- What it means
- How it is fed into machine learning models (as one bulk, or one item at a time)

AI model of the day
- The transformer

CHAPTER 8

Probabilistic Generative Models

AI ties up all the math that I know together, and I have been getting to know math for years.

—H.

If machines are ever to be endowed with an understanding of the world around them, and an ability to recreate it, like we do when we imagine, dream, draw, create songs, watch movies, or write books, then generative models are one significant step in that direction. We need to get these models right if we are ever going to achieve general artificial intelligence.

Generative models are built on the assumption that we can only interpret input data correctly if our model has learned the underlying statistical structure of this data. This is loosely analogous to our dreaming process, which points to the possibility that our brain has learned a model that is able to virtually recreate our environment.

In this chapter, we still have the mathematical structure of training function, loss function, and optimization presented throughout the book. However, unlike in the first few chapters, we aim to learn probability distributions, instead of deterministic functions. The overarching theme is that there is training data, and we want to come up with a mathematical model that generates new data similar to it.

There are two quantities of interest:

- The true (and unknown) joint probability distribution of the features of the input data $p_{data}\left(\overrightarrow{x}\right)$.
- The model joint probability distribution of the features of the data along with the parameters of the model: $p_{model}\left(\overrightarrow{x}; \overrightarrow{\theta}\right)$.

Ideally, we want these two as close as possible. In practice, we settle for parameter values $\vec{\theta}$ that allow $p_{model}\left(\vec{x}\,;\vec{\theta}\right)$ to work well for our particular use cases.

Throughout the chapter, we make use of three rules for probability distributions:

1. The product rule that decomposes the multivariable joint probability distribution into a product of single variable conditional probability distributions.
2. Bayes' Rule, which allows us to flip between variables seamlessly.
3. Independence or conditional independence assumptions on the features or on latent (hidden) variables, which allow us to simplify the product of single variable conditional probabilities even further.

In previous chapters we were minimizing the *loss function*. In this chapter the analogous function is the *log likelihood function*, and the optimization process always attempts to *maximize* this log likelihood (careful, we are not minimizing a loss function, we are maximizing an objective function instead). More on this soon.

Before we dive in, let's make a note that puts our previous deterministic machine learning models into probability language. Our previous models learned a training function that mapped the features of the input data $\vec{x}$ to an output y (target or label), or $f\left(\vec{x}\,;\vec{\theta}\right) = y$. When our goal was classification, f returned the label y that had the highest probability. That is, a classifier learns a direct map from input data $\vec{x}$ to class labels y; in other words, they model the posterior probability $p\left(y\,\middle|\,\vec{x}\right)$ directly. We will elaborate on this later in the chapter.

What Are Generative Models Useful For?

Generative models have made it possible to blur the lines between true and computer-generated data. They have been improving and are achieving impressive successes: machine-generated images, including those of humans, are increasingly more realistic. It is hard to tell whether an image of a model in the fashion industry is that of a real person or the output of a generative machine learning model.

The goal of a generative model is to use a machine to generate novel data, such as audio waveforms containing speech, images, videos, or natural language text. Generative models sample data from a learned probability distribution, where the samples mimic reality as much as possible. The assumption here is that there is some unknown probability distribution underlying the real-life data that we want to mimic (otherwise our whole reality will be some random chaotic noise, lacking any coherence or structure), and the model's goal is to learn an approximation of this probability distribution using the training data.

After collecting a large amount of data from a specific domain, we train a generative model to generate data similar to the collected data. The collected data can be millions of images or videos, thousands of audio recordings, or entire corpuses of natural language.

Generative models are useful for many applications, including augmenting data when data is scarce and more of it is needed, inputting missing values for higher-resolution images, and simulating new data for reinforcement learning or for semi-supervised learning when only few labels are available. Another application is image-to-image translation, such as converting aerial images into maps or converting hand-drawn sketches to images. More applications include image denoising, inpainting, super-resolution, and image editing, such as making smiles wider, cheekbones higher, and faces slimmer.

Moreover, generative models are built to generate more than one acceptable output by drawing multiple samples from the desired probability distribution. This is different than our deterministic models that average over the output with different features during training using a mean squared error loss function or some other averaging loss function. The downside here is that a generative model can draw some bad samples as well.

One type of generative model, namely *generative adversarial networks* (*https://oreil.ly/ pTJZN*) (invented in 2014 by Ian Goodfellow et al.), are incredibly promising and have a wide range of applications, from augmenting data sets to completing masked human faces to astrophysics and high energy physics, such as simulating data sets similar to those produced at the CERN Large Hadron Collider, or simulating distribution of dark matter and predicting gravitational lensing. Generative adversarial models set up two neural networks that compete against each other in a zero-sum game (think game theory in mathematics) until the machine itself cannot tell the difference between a real image and a computer-generated one. This is why their outputs seem very close to reality.

Chapter 7, which was heavily geared toward natural language processing, flirted with generative models without explicitly pointing them out. Most applications of natural language processing, which are not simple classification models (spam or not spam, positive sentiment or negative sentiment, and part of speech tagging), include language generation. Such examples are autocomplete on our smartphones or email, machine translation, text summarization, chatbots, and image captioning.

The Typical Mathematics of Generative Models

Generative models perceive and represent the world through probability distributions. That is, a color image is one sample from the joint probability distribution of pixels that together form a meaningful image (try to count the dimensions of such a joint probability distribution with all the red, green, and blue channels included), an audio wave is one sample from the joint probability distribution of audio signals that together make up meaningful sounds (these are also extremely high dimensional), and a sentence is one sample from the joint probability distribution of words or characters that together represent coherent sentences.

The glaring question is then: how do we compute these amazingly representative joint probability distributions that are able to capture the complexity of the world around us, but sadly happen to be extremely high dimensional?

The machine learning answer is predictable at this point. Start with an easy probability distribution that we know of, such as the Gaussian distribution, then find a way to mold it into another distribution that well approximates the empirical distribution of the data at hand. But how do we mold one distribution into another? We can apply a deterministic function to its probability density. So we must understand the following:

How do we apply a deterministic function to a probability distribution, and what is the probability distribution of the resulting random variable?
We use the following transformation formula:

$$p_x(\vec{x}) = p_z\left(g^{-1}(\vec{x})\right)\left| \det\left(\frac{\partial g^{-1}\vec{x}}{\partial \vec{x}}\right)\right|$$

This is very well documented in many probability books and we will extract what we need from there shortly.

What is the correct function that we must apply?
One way is to train our model to *learn* it. We now know that neural networks have the capacity to represent a wide range of functions, so we can pass the simple probability distribution that we start with through a neural network (the neural network would be the formula of the deterministic function that we are looking for), then we learn the network's parameters by minimizing the error between the empirical distribution of the given data and the distribution output by the network.

How do we measure errors between probability distributions?
Probability theory provides us with some measures of how two probability distributions diverge from each other, such as the Kullback-Leibler (KL) divergence. This is also related to cross-entropy from information theory.

Do all generative models work this way?
Yes and no. *Yes* in the sense that they are all trying to learn the joint probability distribution that presumably generated the training data. In other words, generative models attempt to learn the formula and the parameters of a joint probability

distribution that maximizes the likelihood of the training data (or maximizes the probability that the model assigns to the training data). *No* in the sense that we only outlined an *explicit* way to approximate our desired joint probability distribution. This is one school of thought. In general, a model that defines an explicit and tractable probability density function allows us to operate directly on the log-likelihood of the training data, compute its gradient, and apply available optimization algorithms to search for the maximum. There are other models that provide an explicit but intractable probability density function, in which case we must use approximations to maximize the likelihood. How do we solve an optimization problem *approximately*? We can either use a deterministic approximation, relying on *variational methods* (variational autoencoder models), or use a stochastic approximation, relying on Markov chain Monte Carlo methods. Finally, there are *implicit* ways to approximate our desired joint probability distribution. Implicit models learn to sample from the unknown distribution without ever explicitly defining a formula for it. Generative adversarial networks fall into this category.

Nowadays, the three most popular approaches to generative modeling are:

Generative adversarial networks
These are implicit density models.

Variational models that provide an explicit but intractable probability density function
We approximate the solution of the optimization problem within the framework of probabilistic graphical models, where we maximize a lower bound on the log-likelihood of the data, since immediately maximizing the log-likelihood of the data is intractable.

Fully visible belief networks
These provide explicit and tractable probability density functions, such as Pixel Convolutional Neural Networks (PixelCNN) 2016 (*https://oreil.ly/DJFM0*) and WaveNet (2016) (*https://oreil.ly/YODqz*). These models learn the joint probability distribution by decomposing it into a product of one-dimensional probability distributions for each individual dimension, conditioned on those that preceded it, and learning each of these distributions one at a time. This decomposition is thanks to the product rule or chain rule for probabilities. For example, PixelCNN trains a network that learns the conditional probability distribution of every individual pixel in an image given previous pixels (to the left and to the top of it), and WaveNet trains a network that learns the conditional probability distribution of every individual audio signal in a sound wave conditioned on those that preceded it. The drawbacks here are that these models generate the samples only one entry at a time, and they disallow parallelization. This slows down the generation process considerably. For example, it takes WaveNet two minutes of

computation time to generate one second of audio, so we cannot use it for live back-and-forth conversations.

There are other generative models that fall into the above categories but are less popular, due to expensive computational requirements or difficulties in selecting the density function and/or its transformations. These include models that require a change of variables, such as nonlinear independent component estimation (explicit and tractable density model), Boltzmann machine models (explicit and intractable density model, with a stochastic Markov chain approximation to the solution of the maximization problem), and generative stochastic network models (implicit density model, again depending on a Markov chain to arrive at its approximate maximum likelihood). We survey these models briefly toward the end of this chapter. In practice and away from mathematical theory and analysis, Markov chain approaches are out of favor due to their computational cost and reluctance to converge rapidly.

Shifting Our Brain from Deterministic Thinking to Probabilistic Thinking

In this chapter, we are slowly shifting our brains from deterministic thinking to probabilistic thinking. So far in this book, we have only used deterministic functions to make our predictions. The training functions were linear combinations of data features, sometimes composed with nonlinear activators, the loss functions were deterministic discriminators between the true values and the predicted ones, and the optimization methods were based on deterministic gradient descent methods. Stochasticity, or randomness, was only introduced when we needed to make the computations of the deterministic components of our model less expensive, such as stochastic gradient descent or stochastic singular value decomposition; when we split our data sets into training, validation, and test subsets; when we selected our minibatches; when we traversed some hyperparameter spaces; or when we passed the scores of data samples into the softmax function, which is a deterministic function, and interpreted the resulting values as probabilities. In all of these settings, stochasticity and the associated probability distributions related only to specific components of the model, serving only as a means to an end: enabling the practical implementation and computation of the deterministic model. They never constituted a model's core makeup.

Generative models are different than the models that we have seen in previous chapters in the sense that they are probabilistic at their core. Nevertheless, we still have the training, loss, and optimization structure, except that now the model learns a probability distribution (explicitly or implicitly) as opposed to learning a deterministic function. Our loss function then measures the error between the true and the predicted probability distributions (at least for the explicit density models), so we must understand how to define and compute some sort of error function between probabilities instead of deterministic values. We must also learn how to optimize and take derivatives in this probabilistic setting.

In mathematics, it is a much easier problem to evaluate a given function (forward problem) than to find its inverse (inverse problem), let alone when we only have access to a few observations of the function values, such as our data samples. In our probabilistic setting, the forward problem looks like this: given a certain probability distribution, sample some data. The inverse problem is the one we care about: given this finite number of realizations (data samples) of a probability distribution that we do not know, find the probability distribution that most likely generated them. One difficulty that comes to mind is the issue of uniqueness: there could be more than one distribution that fits our data. Moreover, the inverse problem is usually much harder because in essence we have to act backward and undo the process that the forward function followed to arrive at the given observations. The issue is that most processes cannot be undone, and this is somehow bigger than us, embedded in the laws of nature: the universe tends to increase entropy. On top of the hardship inherent in solving inverse problems, the probability distributions that we usually try to estimate for AI applications are high dimensional, with many variables, and we are not even sure that our probabilistic model has accounted for all the variables (but that is problematic for deterministic models as well). These difficulties should not deter us. Representing and manipulating high-dimensional probability distributions is important for many math, science, finance, engineering, and other applications. We must dive into generative models.

Throughout the rest of this chapter, we will differentiate the case when our estimated probability distribution is given with an explicit formula, and when we do not have a formula but instead we numerically generate new data samples from an implicit distribution. Note that in the previous chapters, with all of our deterministic models, we always had explicit formulas for our training functions, including the ones given by decision trees, fully connected neural networks, and convolutional neural networks. Back then, once we estimated these deterministic functions from the data, we could answer questions like: what is the predicted value of the target variable? In probabilistic models, we answer a different question: what is the probability that the target variable assumes a certain value, or lies in a certain interval? The difference is that we do not know how our model combined the variables to produce our result, as in the deterministic case. What we try to estimate in probabilistic models is the probability that the model's variables occur together with the target variable (their joint probability), ideally for all ranges of all variables. This will give us the probability distribution of the target variable, without having to explicitly formulate how the model's variables interact to produce this result. This purely depends on observing the data.

Maximum Likelihood Estimation

Many generative models either directly or indirectly rely on the *maximum likelihood principle*. For probabilistic models, the goal is to learn a probability distribution that approximates the true probability distribution of the observed data. One way to do this is to specify an explicit probability distribution $p_{model}\left(\vec{x}; \vec{\theta}\right)$ with some unknown parameters $\vec{\theta}$, then solve for the parameters $\vec{\theta}$ that make the training data set as likely to be observed as possible. That is, we need to find the $\vec{\theta}$ that *maximizes the likelihood* of the training data, assigning a high probability for these samples. If there are *m* training data points, we assume that they are sampled independently, so that the probability of observing them together is just the product of the probabilities of all the individual samples. So we have:

$$\vec{\theta}_{optimal} = \arg\max_{\vec{\theta}} \, p_{model}\left(\vec{x}^1; \vec{\theta}\right) p_{model}\left(\vec{x}^2; \vec{\theta}\right) \cdots p_{model}\left(\vec{x}^m; \vec{\theta}\right)$$

Recall that each probability is a number between zero and one. If we multiply all of these probabilities together, we would obtain numbers extremely small in magnitude, which introduces numerical instabilities and runs the risk of underflow (when the machine stores a very small number as zero, essentially removing all significant digits). The *log* function always solves this problem, transforming all numbers whose magnitude is extremely large or extremely small back to the reasonable magnitude realm. The good news is that the log transformation for our probabilities does not affect the values of the optimal $\vec{\theta}$, since the log function is an increasing function. That is, if $f\left(\vec{\theta}_{optimal}\right) \geq f\left(\vec{\theta}\right)$ for all $\vec{\theta}$, then $\log\left(f\left(\vec{\theta}_{optimal}\right)\right) \geq \log\left(f\left(\vec{\theta}\right)\right)$ for all $\vec{\theta}$ as well. Composing with increasing functions does not change the inequality sign. The point is that the maximum likelihood solution becomes equivalent to the maximum log-likelihood solution. Now recall that the log function transforms products to sums, so we have:

$$\vec{\theta}_{optimal} = \arg\max_{\vec{\theta}} \, \log\left(p_{model}\left(\vec{x}^1; \vec{\theta}\right)\right) + \log\left(p_{model}\left(\vec{x}^2; \vec{\theta}\right)\right) + \cdots +$$
$$\log\left(p_{model}\left(\vec{x}^m; \vec{\theta}\right)\right)$$

Note that this expression wants to increase each of $p_{model}\left(\vec{x}, \vec{\theta}\right)$ for each data sample. That is, it prefers that the values of $\vec{\theta}$ *push up* the graph of $p_{model}\left(\vec{x}, \vec{\theta}\right)$ above each data point $\vec{x}^i$. However, we cannot push up indefinitely. There must be a downward compensation since the hyper-area of the region under the graph has to add up to 1, knowing that $p_{model}\left(\vec{x},\vec{\theta}\right)$ is a probability distribution.

We can reformulate our expression in terms of expectation and conditional probabilities:

$$\vec{\theta}_{optimal} = \text{arg } \max_{\vec{\theta}} \mathbb{E}_{x \sim p_{data}} \log \left(p_{model}\left(\vec{x}\,|\,\vec{\theta}\right)\right)$$

The deterministic models that we discussed in the previous chapters find the models' parameters (or weights) by minimizing a loss function that measures the error between the models' predictions and the true values provided by the data labels, or in other words, between y_{model} and y_{data}. In this chapter, we care about finding the parameters that maximize the log-likelihood of the data. It would be nice if there was a formulation of log-likelihood maximization that is analogous to minimizing a quantity that measures an error between the probability distributions p_{model} and p_{data}, so that the analogy between this chapter and the previous chapters is obvious. Luckily, there is. The maximum likelihood estimation is the same as minimizing the *Kullback-Leibler (KL) divergence* (*https://oreil.ly/Dk4NY*) between the probability distribution that generated the data and the model's probability distribution:

$$\vec{\theta}_{optimal} = \text{arg } \min_{\vec{\theta}} Divergence_{KL}\left(p_{data}\left(\vec{x}\right)\|p_{model}\left(\vec{x}\,;\vec{\theta}\right)\right)$$

If p_{data} happens to be a member of the family of distributions $p_{model}\left(\vec{x}\,;\vec{\theta}\right)$, and if we were able to perform the minimization precisely, then we would recover the exact distribution that generated the data, namely p_{data}. However, in practice, we do not have access to the data-generating the distribution; in fact, it is the distribution that we are trying to approximate. We only have access to m samples from p_{data}. These samples define the empirical distribution $\hat{p}_{data}$ that places mass only on exactly these m samples. Now maximizing the log-likelihood of the training set is exactly equivalent to minimizing the KL divergence between $\hat{p}_{data}$ and $p_{model}\left(\vec{x}\,;\vec{\theta}\right)$:

$$\vec{\theta}_{optimal} = \text{arg } \min_{\vec{\theta}} Divergence_{KL}\left(\hat{p}_{data}\left(\vec{x}\right)\|p_{model}\left(\vec{x}\,;\vec{\theta}\right)\right)$$

At this point we might be confused between three optimization problems that are in fact mathematically equivalent, they just happen to come from different subdisciplines and subcultures of mathematics, statistics, natural sciences, and computer science:

- Maximizing the log-likelihood of the training data
- Minimizing the KL divergence between the empirical distribution of the training data and the model's distribution
- Minimizing the cross-entropy loss function between the training data labels and the model outputs, when we are classifying into multiple classes using composition with the softmax function.

Do not be confused. The parameters that minimize the KL divergence are the same as the parameters that minimize the cross-entropy and the negative log-likelihood.

Explicit and Implicit Density Models

The goal of maximum log-likelihood estimation (or minimum KL divergence) is to find a probability distribution $p_{model}\left(\vec{x};\vec{\theta}\right)$ that best explains the observed data. Generative models use this learned $p_{model}\left(\vec{x};\vec{\theta}\right)$ to generate new data. There are two approaches here, one explicit and the other implicit:

Explicit density models

Define the formula for the probability distribution *explicitly* in terms of $\vec{x}$ and $\vec{\theta}$, then find the values of $\vec{\theta}$ that maximize the log-likelihood of the training data samples by following the gradient vector (the partial derivatives with respect to the components of $\vec{\theta}$) uphill. One glaring difficulty here is coming up with a formula for the probability density that is able to capture the complexity in the data, while at the same time staying amiable to computing the log-likelihood with its gradient.

Implicit density models

Sample directly from $p_{model}\left(\vec{x};\vec{\theta}\right)$ without ever writing a formula for this distribution. Generative stochastic networks do this based on a Markov chain framework, which is slow to converge and thus unpopular for practical applications. Using this approach, the model stochastically transforms an existing sample to obtain another sample from the same distribution. Generative adversarial networks interact indirectly with the model's probability distribution without explicitly defining it. They set up a zero-sum game between two networks, where one network generates a sample and the other network acts like a classifier determining whether the generated sample is from the correct distribution or not.

Explicit Density-Tractable: Fully Visible Belief Networks

These models admit an explicit probability density function with tractable log-likelihood optimization. They rely on the chain rule of probability (*https://oreil.ly/gmC1m*) to decompose the joint probability distribution $p_{model}\!\left(\overrightarrow{x}\right)$ into a product of one-dimensional probability distributions:

$$p_{model}\!\left(\overrightarrow{x}\right) = \Pi_{i=1}^{n}\, p_{model}(x_i \,|\, x_1, x_2, \cdots, x_{i-1})$$

The main drawback here is that samples must be generated one component at a time (one pixel of an image, or one character of a word, or one entry of a discrete audio wave), therefore, the cost of generating one sample is *O(n)*.

Example: Generating Images via PixelCNN and Machine Audio via WaveNet

PixelCNN (*https://oreil.ly/y1lHU*) trains a convolutional neural network that models the conditional distribution of every individual pixel, given previous pixels (to the left and to the top of the target pixel). Figure 8-1 illustrates this.

WaveNet trains a convolutional neural network that models the conditional distribution of each entry of an audiowave, given the previous entries. We will only elaborate on WaveNet. It is the one-dimensional analog of PixelCNN and captures the essential ideas.

The goal of WaveNet is to generate wideband raw audio waveforms. So we must learn the joint probability distribution of an audio waveform $\overrightarrow{x} = (x_1, x_2, \cdots, x_T)$ from a certain genre.

We use the product rule to decompose the joint distribution into a product of single variable distributions where we condition each entry of the audio waveform on those that preceded it:

$$p_{model}\!\left(\overrightarrow{x}\right) = \Pi_{t=1}^{T}\, p_{model}(x_t \,|\, x_1, x_2, \cdots, x_{t-1})$$

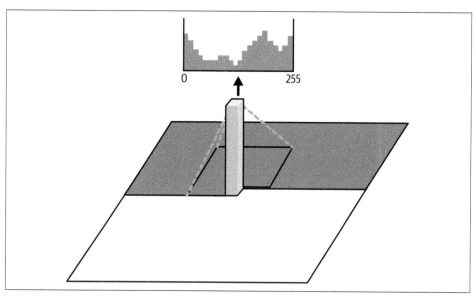

Figure 8-1. PixelCNN learning the conditional distribution of the nth pixel conditioned on the previous n-1 pixels (image source (https://oreil.ly/aTZkg))

One difficulty is that audio waveforms have very high temporal resolution, with at least 16,000 entries per 1 second of audio (so one data sample that is a minute long is a vector with T = 960,000 entries). Each entry represents one time step of discretized raw audio, and is usually stored as a 16-bit integer. That is, each entry can assume any value between 0 and 65,535. If we keep this range, the network has to learn the probability for each entry, so the softmax function at the output level has to output 65,536 probability scores for every single entry. The total number of entries we have to do this for, along with the computational complexity of the network itself, become very expensive. To make this more tractable, we must quantize, which in electronics means approximate a continuously varying signal by one whose amplitude is restricted to a prescribed set of values. WaveNet transforms the raw data to restrict the entries' values to 256 options each, ranging from 0 to 255, similar to the pixel range for digital images. Now, during training the network must learn the probability distribution of each entry over these 256 values, given the preceding entries, and during audio generation it samples from these learned distributions one entry at a time.

The last complication is that if the audio signal represents anything meaningful, then the vector representing it has long-range dependencies over multiple time scales. To capture these long-range dependencies, WaveNet uses dilated convolutions. These are one-dimensional kernels or filters that skip some entries to cover a wider range without increasing the number of parameters (see Figure 8-2 for an illustration).

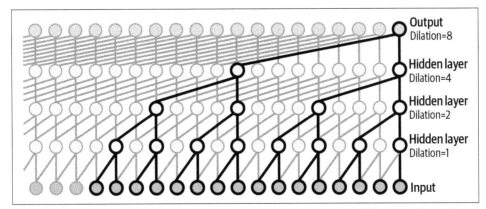

Figure 8-2. Dilated convolution with kernel size equals two. At each layer the kernel has only two parameters but it skips entries for larger coverage (image source that has a nice animation (https://oreil.ly/WiHpx)).

Note also that the network cannot peek into the future, so the filters at each layer cannot use entries from the training sample that are ahead of the target entry. In one dimension we just stop filtering earlier at each convolutional layer, so it is a simple time shift. In two dimensions we use *masked* filters, which have zeros to the right and to the bottom of the central entry.

WaveNet learns a total of T probability distributions, one for each entry of the audio waveform conditioned on those entries that preceded it: $p_{model}(x_1), p_{model}(x_2|x_1), p_{model}(x_3|x_1,x_2), \cdots$ and $p_{model}(x_T|x_1,x_2,\cdots,x_{T-1})$. During training, these distributions can be computed in parallel.

Now suppose we need to learn the probability distribution of the 100th entry, given the previous 99 entries. We input batches of audio samples from the training data, and the convolutional network uses only the first 99 entries of each sample, computing linear combinations (the filters linearly combine), passing through non-linear activation functions from one layer to the next to the next using some skip connections and residual layers to battle vanishing gradients, and finally passing the result through a softmax function and outputting a vector of length 256 containing probability scores for the value of the 100th entry. This is the probability distribution for the 100th entry output by the model. After comparing this output distribution with the empirical distribution of the data for the 100th entry from the training batch, the parameters of the network get adjusted to decrease the error (lower the cross-entropy or increase the likelihood). As more batches of data and more epochs pass through the network, the probability distribution for the 100th entry, given the previous 99, will approach the empirical distribution from the training data. What we save within the network after training are the values of the parameters. Now we can use the trained network to generate machine audio, one entry at a time:

1. Sample a value x_1 from the probability distribution $p_{model}(x_1)$.

2. Augment (x_1) with zeros to establish the required length for the network's input and pass the vector through the network. We will get as an output $p_{model}(x_2 | x_1)$, from which we can sample x_2.

3. Augment (x_1, x_2) with zeros and pass the vector through the network. We will get as an output $p_{model}(x_3 | x_1, x_2)$, from which we can sample x_3.

4. Keep going.

We can condition WaveNet on a certain speaker identity, so we can generate different voices using one model.

The fact that we can train WaveNet in parallel but use it to generate audio only sequentially is a major shortcoming. This has been rectified with Parallel WaveNet (*https://oreil.ly/BJCHO*), which is deployed online by Google Assistant, including serving multiple English and Japanese voices.

To summarize and place this discussion in the same mathematical context as this chapter, PixelCNN and WaveNet are models that aim to learn the joint probability distribution of image data or audio data from certain genres. They do so by decomposing the joint distribution into a product of one-dimensional probability distributions for each entry of their data, conditioned on all the preceding entries. To find these one-dimensional conditional distributions, they use a convolutional network to learn the way the observed entries interact together to produce a distribution of the next entry. This way, the input to the network is deterministic, and its output is a probability mass function. The network itself is also a deterministic function. We can view the network together with its output as a probability distribution with parameters that we tweak. As the training evolves, the output gets adjusted until it reaches an acceptable agreement with the empirical distribution of the training data. Therefore, we are not applying a deterministic function to a probability distribution and tweaking the function's parameters until we agree with the distribution of the training data. We are instead starting with an explicit formula for a probability distribution with many parameters (the network's parameters), then tweaking the parameters until this explicit probability distribution reasonably agrees with training data. We do this for each conditional probability distribution corresponding to each entry.

Explicit Density-Tractable: Change of Variables Nonlinear Independent Component Analysis

The main idea here is that we have the random variable representing the observed training data $\vec{x}$ and we want to learn the source random variable $\vec{s}$ that generated

---

it. We assume that there is a deterministic transformation $g(\vec{s}) = \vec{x}$ that is invertible and differentiable that transforms the unknown $\vec{s}$ to the observed $\vec{x}$. That is, $\vec{s} = g^{-1}(\vec{x})$. Now we need to find an appropriate g to find the probability distribution of $\vec{s}$. Moreover, we assume that $\vec{s}$ has independent entries, or components, so that its probability distribution is nothing but the product of the distributions of its components.

The formula that relates the probability distribution of a random variable with the probability distribution of a deterministic transformation of it is:

$$p_s(\vec{s}) = p_x(\vec{x}) \times determinant(Jacobian)$$

$$= p_x\left(g(\vec{s})\right) \left| \det\left(\frac{\partial g(\vec{s})}{\partial \vec{s}}\right) \right|$$

Multiplying by the determinant of the Jacobian of the transformation accounts for the change in volume in space due to the transformation.

Nonlinear independent component estimation (*https://oreil.ly/0nVvD*) models the joint probability distribution as the nonlinear transformation of the data $\vec{s} = g^{-1}(\vec{x})$. The transformation g is learned such that g^{-1} maps the data to a latent space where it conforms to a factorized distribution; that is, the mapping results in independent latent variables. The transformation g^{-1} is parameterized to allow for easy computation of the determinant of the Jacobian and the inverse Jacobian. g^{-1} is based on a deep neural network and its parameters are learned by optimizing the log-likelihood, which is tractable.

Note that the requirement that the transformation g must be invertible means that the latent variables $\vec{s}$ must have the same dimension as the data features (length of $\vec{x}$). This imposes restrictions on the choice of the function g and is a disadvantage of nonlinear independent component analysis models.

In comparison, generative adversarial networks impose very few requirements on g, and, in particular, allow $\vec{s}$ to have more dimensions than $\vec{x}$.

Explicit Density-Intractable: Variational Autoencoders Approximation via Variational Methods

Deterministic autoencoders are composed of an encoder that maps the data from x space to latent z space of lower dimension, and a decoder that in turn maps the data from z space to $\hat{x}$ space, with the objective of not losing much information, or reducing the reconstruction error, which means keeping x and $\hat{x}$ close, for example, in the Euclidean distance sense. In this sense, we can view *principal component analysis*, which is based on the singular value decomposition $X = U\Sigma V^t$, as a linear encoder, where the decoder is simply the transpose of the encoding matrix. Encoding and decoding functions can be nonlinear and/or neural networks.

For deterministic autoencoders, we cannot use the decoder as a data generator. At least, if we do, then we have to pick some z from latent z space and apply the decoder function to it. We are unlikely to get any $\hat{x}$ that is close to how the desired data x looks, unless we picked a z that corresponds to a coded x due to overfitting. We need a regularization that provides us with some control over z space, giving us the benefit of avoiding overfitting and using autoencoders as a data generator. We accomplish this by shifting from deterministic autoencoding to probabilistic autoencoding.

Variational autoencoders are probabilistic autoencoders: the encoder outputs probability distributions over the latent space z instead of single points. Moreover, during training, the loss function includes an extra regularization term that controls the distribution over the latent space. Therefore, the loss function for variational autoencoders contains a reconstruction term (such as mean squared distance) and a regularization term to control the probability distribution output by the encoder. The regularization term can be a KL divergence from a Gaussian distribution, since the underlying assumption is that simple probabilistic models best describe the training data. In other words, complex relationships can be probabilistically simple. We have to be careful here, since this introduces a bias: the simple assumption on the data distribution in the latent variable can be a drawback if it is too weak. That is, when the assumption on the prior distribution or the assumption on the approximate posterior distribution is too weak, even with a perfect optimization algorithm and infinite training data, the gap between the estimate $\mathscr{L}$ and the true log-likelihood can lead to p_{model} learning a completely different distribution than the true p_{data}.

Mathematically, we maximize a lower bound $\mathscr{L}$ on the log-likelihood of the data. In science, variational methods define lower bounds on an energy functional that we want to maximize, or upper bounds on an energy functional that we want to minimize. These bounds are usually easier to obtain and have tractable optimization algorithms, even when the log-likelihood does not. At the same time, they provide good estimates for the optimal values that we are searching for:

$$\mathscr{L}\left(\vec{x},\vec{\theta}\right) \leq \log p_{model}\left(\vec{x},\vec{\theta}\right)$$

Variational methods often achieve very good likelihood, but subjective evaluation of samples regard their generated samples as having lower quality. They are also considered more difficult to optimize than fully visible belief networks. Moreover, people find their mathematics more difficult than that of fully visible belief networks and of generative adversarial networks (discussed soon).

Explicit Density-Intractable: Boltzman Machine Approximation via Markov Chain

Boltzmann machines (originating in the 1980s) are a family of generative models that rely on Markov chains to train generative models. This is a sampling technique that happens to be more expensive than the simple sampling of a mini-batch from a data set to estimate a loss function. We will discuss Markov chains in the context of reinforcement learning in Chapter 11. In the context of data generation, they have many disadvantages that caused them to fall out of favor: high computational cost, impractical and less efficient to extend to higher dimensions, slow to converge, and no clear way to know whether the model has converged or not, even when the theory says it must converge. Markov chain methods have not scaled to problems like ImageNet generation.

A Markov chain has a transition operator q that encodes the probability of transitioning from one state of the system to another. This transition operator q needs to be explicitly defined. We can generate data samples by repeatedly drawing a sample $x' \sim q(x'|x)$, updating x' sequentially according to the transition operator q. This sequential nature of generation is another disadvantage compared to single step generation. Markov chain methods can sometimes guarantee that x' will eventually converge to a sample from $p_{model}(x)$, even though the convergence might be slow.

Some models, such as deep Boltzman machines, employ both Markov chain and variational approximations.

Implicit Density-Markov Chain: Generative Stochastic Network

Generative stochastic networks (Bengio et al. 2014 (*https://oreil.ly/DJI78*)) do not explicitly define a density function, and instead use a Markov chain transition operator that interacts indirectly with $p_{model}(x)$ by sampling from the training data. This Markov chain operator must be run several times to obtain a sample from $p_{model}(x)$. These methods still suffer from the shortcomings of Markov chain methods mentioned in the previous section.

Implicit Density-Direct: Generative Adversarial Networks

Currently the most popular generative models are:

- Fully visible deep belief networks, such as PixelCNN, WaveNet, and their variations.
- Variational autoencoders, consisting of a probabilistic encoder-decoder architecture.
- Generative adversarial networks, which have received a lot of attention from the scientific community due to the simplicity of their concept and the good quality of their generated samples. We discuss them now.

Generative adversarial networks (https://oreil.ly/2GKF3) were introduced in 2014 by Ian Goodfellow et al. The mathematics involved is a beautiful mixture between probability and game theory. Generative adversarial networks avoid some disadvantages associated with other generative models:

- Generating samples all at once, in parallel, as opposed to feeding a new pixel back into the network to predict the one, such as in PixelCNN.
- The generator function has few restrictions. This is an advantage relative to Boltzmann machines, for which few probability distributions admit tractable Markov chain sampling, and relative to nonlinear independent component analysis, for which the generator must be invertible and the latent variables z must have the same dimension as the samples x.
- Generative adversarial networks do not need Markov chains. This is an advantage relative to Boltzmann machines and to generative stochastic networks.
- While variational autoencoders might never converge to the true data generating distribution if they assume prior or posterior distributions that are too weak, generative adversarial networks converge to the true p_{data}, given that we have infinite training data and a large enough model. Moreover, generative adversarial networks do not need variational bounds, and the specific model families used within the generative adversarial network framework are already known to be universal approximators. Thus, generative adversarial networks are already known to be asymptotically consistent. On the other hand, some variational autoencoders are conjectured to be asymptotically consistent, but this still needs to be proven.

The disadvantage of generative adversarial networks is that training them requires spotting the Nash equilibrium of a game, which is more difficult than just optimizing an objective function. Moreover, the solution tends to be numerically unstable. This was improved in 2015 by Alec Radford et al. in their paper

"Unsupervised Representation Learning with Deep Convolutional Generative Adversarial Networks" (*https://oreil.ly/ARUUD*). This approach led to more stable models.

During training, generative adversarial networks formulate a game between two separate networks: a generator network and a discriminator network that tries to classify generator samples as either coming from the true distribution $p_{data}(x)$ or from the model $p_{model}(x)$. The loss functions of the two networks are related, so that the discriminator communicates the discrepancy between the two distributions, and the generator adjusts its parameters accordingly until it exactly reproduces the true data distribution (in theory) so that the discriminator's classifications are no better than random guesses.

The generator network wants to maximize the probability that the discriminator assigns the wrong label in its classification, whether the sample is from the training data or from the model, while the discriminator network wants to minimize that probability. This is a two-player zero-sum game, where one player's gain is another's loss. We end up solving a minimax problem instead of a purely maximizing or minimizing problem. A unique solution exists.

How Do Generative Adversarial Networks Work?

Keeping the goal of learning the generator's probability distribution $p_g\left(\vec{x}\,;\vec{\theta}\right)$ over the data, here's how the learning progresses for generative adversarial networks:

1. Start with a random sample $\vec{z}$ from a prior probability distribution $p_z\left(\vec{z}\right)$, which could be just uniform random noise for each component of $\vec{z}$.

2. Start also with a random sample $\vec{x}$ from the training data, so it is a sample from the probability distribution $p_{data}\left(\vec{x}\right)$ that the generator is trying to learn.

3. Apply to $\vec{z}$ the deterministic function $G\left(\vec{z},\vec{\theta}_g\right)$ representing the generative neural network. The parameters $\vec{\theta}_g$ are the ones we need to tweak via backpropagation until the output $G\left(\vec{z},\vec{\theta}_g\right)$ looks similar to samples from the training data set.

4. Pass the output $G\left(\vec{z},\vec{\theta}_g\right)$ into another deterministic function D representing the discriminative neural network. Now we have the new output $D\left(G\left(\vec{z},\vec{\theta}_g\right),\vec{\theta}_d\right)$ that is just a number closer to one or zero, signifying whether this sample came from the generator or from the training data. Thus, for this input from the

generator, $D\left(G\left(\vec{z},\vec{\theta}_g\right),\vec{\theta}_d\right)$ must return a number close to one. The parameters $\vec{\theta}_d$ are the ones we need to tweak via backpropagation until D returns the wrong classification around half of the time.

5. Pass also the sample $\vec{x}$ from the training data to D, so we evaluate $D\left(\vec{x},\vec{\theta}_d\right)$. For this input, $D\left(\vec{x},\vec{\theta}_d\right)$ must return a number close to zero.

6. What is the loss function for these two networks, that has in its formula both sets of parameters $\vec{\theta}_g$ and $\vec{\theta}_d$, along with the sampled vectors $\vec{x}$ and $\vec{z}$? The discriminator function D wants to get it right for both types of inputs, $\vec{x}$ and $G\left(\vec{z},\vec{\theta}_g\right)$. So its parameters $\vec{\theta}_d$ must be selected so that a number close to 1 is assigned a large score when the input is $G\left(\vec{z},\vec{\theta}_g\right)$, and a number close to 0 is assigned a large value when the input is $\vec{x}$. In both cases, we can use the negative of the log function since that is a function that is large near 0 and small near 1. Therefore, D needs the parameters $\vec{\theta}_d$ that maximize:

$$\mathbb{E}_{\vec{x} \sim p_{data}(\vec{x})}\left[\log D\left(\vec{x},\vec{\theta}_d\right)\right] + \mathbb{E}_{\vec{z} \sim p_z(\vec{z})}\left[\log\left(1 - D\left(G\left(\vec{z},\vec{\theta}_g\right),\vec{\theta}_d\right)\right)\right]$$

At the same time, G needs the parameters $\vec{\theta}_g$ that minimize $\log\left(1 - D\left(G\left(\vec{z},\vec{\theta}_g\right),\vec{\theta}_d\right)\right)$. Combined, D and G engage in a two-player minimax game with value function $V(D, G)$:

$$\min_{\vec{\theta}_g} \max_{\vec{\theta}_d} \mathbb{E}_{\vec{x} \sim p_{data}(\vec{x})}\left[\log D\left(\vec{x},\vec{\theta}_d\right)\right]$$
$$+ \mathbb{E}_{\vec{z} \sim p_z(\vec{z})}\left[\log\left(1 - D\left(G\left(\vec{z},\vec{\theta}_g\right),\vec{\theta}_d\right)\right)\right]$$

This is a very simple mathematical structure, where setting up a discriminator network allows us to get closer to the true data distribution without ever explicitly defining it or assuming anything about it.

Finally, we note that generative adversarial networks are highly promising for many applications. One example is the dramatic enhancement they have for semi-supervised learning, where the "NIPS 2016 Tutorial: Generative Adversarial Networks" (Goodfellow 2016) (https://oreil.ly/isXsx) reports:

We introduce an approach for semi-supervised learning with generative adversarial networks that involves the discriminator producing an additional output indicating the label of the input. This approach allows us to obtain state of the art results on MNIST, SVHN, and CIFAR-10 in settings with very few labeled examples. On MNIST, for example, we achieve 99.14% accuracy with only 10 labeled examples per class with a fully connected neural network—a result that's very close to the best known results with fully supervised approaches using all 60,000 labeled examples. This is very promising because labeled examples can be quite expensive to obtain in practice.

Another far-reaching application of generative adversarial networks (and machine learning in general) is simulating data for high energy physics. We discuss this next.

Example: Machine Learning and Generative Networks for High Energy Physics

The following discussion is inspired by and borrows from the Machine Learning for Jet Physics Workshop 2020 (*https://oreil.ly/Uy3Ah*) and the two articles "Deep Learning and Its Application to LHC Physics" (Guest et al. 2018) (*https://oreil.ly/4nlAj*) and "Graph Generative Adversarial Networks for Sparse Data Generation in High Energy Physics" (Kansal et al. 2021) (*https://oreil.ly/EHGrK*).

Before the deep learning revolution began in 2012, the field of high energy physics traditionally relied in its analyses and computations on physical considerations and human intuition, boosted decision trees, handcrafted data feature engineering and dimensionality reduction, and traditional statistical analysis. These techniques, while insightful, are naturally far from optimal and hard to automate or extend to higher dimensions. Several studies have demonstrated that traditional shallow networks based on physics-inspired engineered *high-level* features are outperformed by deep networks based on the higher-dimensional *lower-level* features that receive less preprocessing. Many areas of the Large Hadron Collider (*https://oreil.ly/WKudv*) data analysis have suffered from long-standing suboptimal feature engineering, and deserve reexamination. Thus, the high energy physics field is a breeding ground ripe for machine learning applications. A lot of progress is taking place on this front. The field is employing several machine learning techniques, including artificial neural networks, kernel density estimation, support vector machines, genetic algorithms, boosted decision trees, random forests, and generative networks.

The experimental program of the Large Hadron Collider probes the most fundamental questions in modern physics: the nature of mass, the dimensionality of space, the unification of the fundamental forces, the particle nature of dark matter, and the fine-tuning of the Standard Model (*https://oreil.ly/SMljD*). One driving goal is to understand the most fundamental structure of matter. Part of that entails searching for and studying exotic particles, such as the top quark and Higgs boson, produced in collisions at accelerators such as the Large Hadron Collider. Specific benchmarks and challenges include mass reconstruction, jet substructure, and jet-flavor classification. For example, one can identify jets from heavy *(c, b, t)* or light *(u, d, s)* quarks, gluons, and *W*, *Z*, and *H* bosons.

Running high energy particle experiments and collecting the resulting data is extremely expensive. The data collected is enormous in terms of the number of collisions and the complexity of each collision. In addition, the bulk of accelerator events does not produce interesting particles (signal particles versus background particles). Signal particles are rare, so high data rates are necessary. For example, the Large Hadron Collider detectors have $O(108)$ sensors used to record the large number of particles produced after each collision. It is thus of paramount importance to extract maximal information from experimental data (think regression and classification models), to accurately select and identify events for effective measurements, and to produce reliable methods for simulating new data similar to data produced by experiments (think generative models). High energy physics data is characterized by its high dimensionality, along with the complex topologies of many signal events.

This discussion ties into our chapter through the nature of collisions and the interaction of their products with Large Hadron Collider detectors. They are quantum mechanical, and therefore the observations resulting from a particular interaction are fundamentally probabilistic. The resulting data analysis must then be framed in statistical and probabilistic terms.

In our chapter, our aim is to learn the probability distribution $p\left(\vec{\theta}\,\middle|\,\vec{x}\right)$ of the model's parameters given the observed data. If the data is fairly low dimensional, such as less than five dimensions, the problem of estimating the unknown statistical model from the simulated samples would not be difficult, using histograms or kernel-based density estimates. However, we cannot easily extend these simple methods to higher dimensions, due to the curse of dimensionality. In a single dimension, we would need N samples to estimate the source probability density function, but in d dimensions, we would need $O\left(N^d\right)$. The consequence is that if the dimension of the data is greater than 10 or so, it is impractical or even impossible to use naive methods to estimate the probability distribution, requiring a prohibitive amount of computational resources.

High energy physicists have traditionally dealt with the curse of dimensionality by reducing the dimension of the data through a series of steps that operate both on individual collision events and collections of events. These established approaches were based on specific, hand-engineered features in the data to a number small enough to allow the estimation of the unknown probability distribution $p(x \mid \theta)$ using samples generated by simulation tools. Obviously, due to complexity of the data and the rarity of potential new physics, along with its subtle signatures, this traditional approach is probably suboptimal. Machine learning eliminates the need for hand-engineering features and manual dimensionality reduction that can miss crucial information in the lower-level higher-dimensional data. Moreover, the structure of lower-level data obtained directly from the sensors fits very well with well-established neural network models, such as convolutional neural networks and graph neural networks; for example, the projective tower structure of calorimeters present in nearly all modern high energy physics detectors is similar to the pixels of an image.

Note, however, that while the image-based approach has been successful, the actual detector geometry is not perfectly regular, thus some data preprocessing is required to represent jet images. In addition, jet images are typically very sparse. Both irregular geometry and sparsity can be addressed using *graph-based convolutional networks* instead of the usual convolutional networks for our particle data modeling. Graph convolutional networks extend the application of convolutional neural networks to irregularly sampled data. They are able to handle sparse, permutation invariant data with complex geometries. We will discuss graph networks in Chapter 9. They always come with nodes, edges, and a matrix encoding the relationships in the graph, called an *adjacency matrix*. In the context of high energy physics, the particles of a jet represent the nodes of the graph, and the edges encode how close the particles are in a learned adjacency matrix. In high energy physics, graph-based networks have been successfully applied to classification, reconstruction, and generation tasks.

The subject of our chapter is generative models, or generating data similar to a given data set. Generating or simulating data faithful to the experimental data collected in high energy physics is of great importance. In "Graph Generative Adversarial Networks for Sparse Data Generation in High Energy Physics" (*https://oreil.ly/eVkOP*), the authors develop graph-based generative models, using a generative adversarial network framework, for simulating sparse data sets like those produced at the CERN (*https://home.cern*) Large Hadron Collider.

The authors illustrate their approach by training on and generating sparse representations of MNIST handwritten digit images and jets of particles in proton-proton collisions like those at the Large Hadron Collider. The model successfully generates sparse MNIST digits and particle jet data. The authors use two metrics to quantify agreement between real and generated data: a graph-based Fréchet inception distance (*https://oreil.ly/LHron*) and the particle and jet feature-level 1-Wasserstein distance (*https://oreil.ly/V2UVS*).

Other Generative Models

We have discussed state-of-the-art generative models (as of 2022), but this chapter will be incomplete if we do not go over Naive Bayes, Gaussian mixture, and Boltzmann machine models. There are many others. That being said, Yann LeCun (VP and chief AI scientist at Meta) offers his perspective (*https://oreil.ly/fSGCz*) on some of these models:

Researchers in speech recognition, computer vision, and natural language processing in the 2000s were obsessed with accurate representations of uncertainty. This led to a flurry of work on probabilistic generative models such as Hidden Markov Models in speech, Markov random fields and constellation models in vision, and probabilistic topic models in NLP, e.g., with latent Dirichlet analysis. There were debates at computer vision workshops about generative models vs discriminative models. There were heroic-yet-futile attempts to build object recognition systems with non-parametric Bayesian methods. Much of this was riding on previous work on Bayesian networks, factor graphs and other graphical models. That's how one learned about exponential family, belief propagation, loopy belief propagation, variational inference, etc. Chinese restaurant process, Indian buffet process, etc. But almost none of this work was concerned with the problem of learning representations. Features were assumed to be given. The structure of the graphical model, with its latent variables, was assumed to be given. All one had to do was to compute some sort of log-likelihood by linearly combining features, and then use one of the above mentioned sophisticated inference methods to produce marginal distributions over the unknown variables, one of which being the answer, e.g., a category. In fact, exponential family pretty much means shallow: the log-likelihood can be expressed as a linearly parameterized function of features (or simple combinations thereof). Learning the parameters of the model was seen as just another variational inference problem. It's interesting to observe that almost none of this is relevant to today's top speech, vision, and NLP systems. As it turned out, solving the problem of learning hierarchical representations and complex functional dependencies was a much more important issue than being able to perform accurate probabilistic inference with shallow models. This is not to say that accurate probabilistic inference is not useful.

In the same vein, he continues:

> Generative Adversarial Networks are nice for producing pretty pictures (though they are being edged out by diffusion models, or "multistep denoising auto-encoders" as I like to call them), but for recognition and representation learning, GANs have been a big disappointment.

Nevertheless, there is a lot of math to be learned from all these models. In my experience, we understand and retain math at a much deeper level when we see it developed and utilized for specific purposes, as opposed to only train the neurons of the brain. Many mathematicians claim to experience pleasure while proving theories that have yet to find applications. I was never one of those.

Naive Bayes Classification Model

The Naive Bayes model is a very simple classification model that we can also use as a generative model, since it ends up computing a joint probability distribution $p(\vec{x}, y_k)$ for the data to determine its classification. The training data has features $\vec{x}$ and labels y_k. Therefore, we can use the Naive Bayes model to generate new data points, together with labels, by sampling from this joint probability distribution.

The goal of a Naive Bayes model is to compute the probability of the class y_k given the data features $\vec{x}$, which is the conditional probability $p\left(y_k \middle| \vec{x}\right)$. For data with many features (high-dimensional $\vec{x}$), this is expensive to compute, so we use Bayes' Rule and exploit the reverse conditional probability, which in turn leads to the joint probability distribution. That is:

$$p\left(y_k \middle| \vec{x}\right) = \frac{p(y_k)p\left(\vec{x} \middle| y_k\right)}{p\left(\vec{x}\right)} = \frac{p\left(\vec{x},y_k\right)}{p\left(\vec{x}\right)}$$

The Naive Bayes model makes the very strong and naive assumption, which in practice works better than one might expect, that the data features are mutually independent when conditioned on the class label y_k. This assumption helps simplify the joint probability distribution in the numerator tremendously, especially when we expand it as a product of single variable conditional probabilities. The feature independence assumptions conditional on the class label y_k means:

$$p(x_i | x_{i+1}, x_{i+2}, \cdots, x_n, y_k) = p(x_i | y_k)$$

Thus, the joint probability distribution factors into:

$$\begin{aligned} p\left(\vec{x}, y_k\right) &= p(x_1 | x_2, \cdots, x_n, y_k)p(x_2 | x_3, \cdots, x_n, y_k) \cdots p(x_n | y_k)p(y_k) \\ &= p(x_1 | y_k)p(x_2 | y_k) \cdots p(x_n | y_k)p(y_k) \end{aligned}$$

We can now estimate these single feature probabilities conditioned on each category of the data easily from the training data. We can similarly estimate the probability of each class $p(y_k)$ from the training data, or we can assume the classes are equally likely, so that $p(y_k) = \frac{1}{\text{number of classes}}$.

Note that in general, generative models find the joint probability distribution $p\left(\vec{x}, \vec{y_k}\right)$ between labels y_k and data $\vec{x}$. Classification models, on the other hand, calculate the conditional probabilities $p\left(y_k \middle| \vec{x}\right)$. They focus on calculating the decision boundaries between different classes in the data by returning the class y_k with the highest probability. So for the Naive Bayes classifier, it returns the label y_* with the highest value for $p\left(y_k \middle| \vec{x}\right)$, which is the same as the highest value for $p(x_1 | y_k)p(x_2 | y_k) \cdots p(x_n | y_k)p(y_k)$.

Gaussian Mixture Model

In a Gaussian mixture model, we assume that all the data points are generated from a mixture of a finite number of Gaussian distributions with unknown parameters (means and covariance matrices). We can think of mixture models as being similar to k-means clustering, but here we include information about the centers of the clusters (means of our Gaussians) along with the shape of the spread of the data in each cluster (determined by the covariance of the Gaussians). To determine the number of clusters in the data, Gaussian mixture models sometimes implement the Bayesian information criterion (*https://oreil.ly/YjP4F*). We can also restrict our model to control the covariance of the different Gaussians in the mixture: full, tied, diagonal, tied diagonal, and spherical (see Figure 8-3 for an illustration).

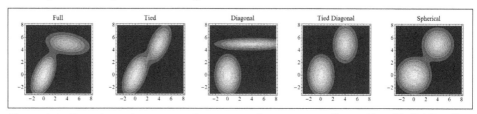

Figure 8-3. Gaussian mixture covariance types (image source (https://oreil.ly/BMF7a))

We finally need to maximize the likelihood of the data to estimate unknown parameters of the mixture (means and entries of the covariance matrices).

Maximum likelihood becomes intractable when there are latent or hidden variables in the data (variables that are not directly measured or observed). The way around this is to use an expectation maximization (EM) algorithm (*https://oreil.ly/QpWUo*) to estimate the maximum likelihood. The expectation maximization algorithm works as follows:

1. Estimate the values for the latent variables by creating a function for the expectation of the log-likelihood using the current estimate for the unknown parameters.

2. Optimize: compute new parameters that maximize the expected log-likelihood evaluated in step 1.

3. Repeat steps 1 and 2 until convergence.

We can see how Gaussian mixture models can be used as clustering, generative, or classification models. For clustering, this is the main part of the model buildup. For generation, sample new data points from the mixture after computing the unknown parameters via expectation maximization. For classification, given a new data point, the model assigns it to the Gaussian to which it most probably belongs.

The Evolution of Generative Models

In this section, we tell the story that contributed to ending the winter of neural networks, and ultimately led to modern probabilistic deep learning models, such as variational autoencoders, fully visible deep belief networks, and generative adversarial networks. We encounter the progression from Hopfield nets to Boltzmann machines to restricted Boltzmann machines. I have a special affinity for these models: in addition to their historical value, and learning the joint probability distribution of the data features by assembling a network of basic computational units, they employ the mathematical machinery of the extremely neat and well-developed field of *statistical mechanics*, my initial area of research.

In statistical mechanics, we define probability distributions in terms of energy functions. The probability of us finding a system in a certain state $\vec{x}$ depends on its energy $E\left(\vec{x}\right)$ at that state. More precisely, high energy states are less probable, which manifests itself in the negative sign in the exponential in the following formula:

$$p\left(\vec{x}\right) = \frac{exp\left(-E\left(\vec{x}\right)\right)}{Z}$$

The exponential function guarantees that p is positive, and the *partition function Z* in the denominator ensures that the sum (or integral if x is continuous) of $p\left(\vec{x}\right)$ overall states $\vec{x}$ is 1, making p a valid probability distribution. Machine learning models that define joint probability distributions this way are called *energy-based models*, for obvious reasons. They differ in how they assign the energy at each state, meaning in the specific formula they use for $E\left(\vec{x}\right)$, which in turn affects the formula for the partition function Z. The formula for $E\left(\vec{x}\right)$ contains the parameters of the model $\vec{\theta}$, which we need to compute from the data using maximum likelihood estimation. In fact, it is better if we have the dependence of p, E, and Z on $\vec{\theta}$ explicit in the joint probability distribution formula:

$$p\left(\vec{x},\vec{\theta}\right) = \frac{exp\left(-E\left(\vec{x},\vec{\theta}\right)\right)}{Z\left(\vec{\theta}\right)}$$

In most cases, it is not possible to compute a closed formula for the partition function Z, rendering the maximum likelihood estimation intractable. More precisely, when we maximize the log-likelihood, we need to compute its gradient, which includes computing its gradient with respect to the parameter $\vec{\theta}$, which in turn forces us to compute the gradient of the partition function Z with respect to $\vec{\theta}$. The following quantity appears frequently in these computations:

$$\nabla_{\vec{\theta}} \log Z\left(\vec{\theta}\right) = \mathbb{E}_{\vec{x} \sim p(\vec{x})}\left(\nabla_{\vec{\theta}} \log \left(numerator\left(\vec{x},\vec{\theta}\right)\right)\right)$$

where in our case the numerator in the formula of the energy-based joint probability distribution is $exp\left(-E\left(\vec{x},\vec{\theta}\right)\right)$, but this can also differ among models.

The cases where the partition function is intractable urges us to resort to approximation methods such as stochastic maximum likelihood and contrastive divergence. Other methods sidestep approximating the partition function and compute conditional probabilities without knowledge of the partition function. They take advantage of the ratio definition of conditional probabilities, along with the ratio in the definition of an energy-based joint probability distribution, effectively canceling out the partition function. These methods include score matching, ratio matching, and denoising score matching.

Other methods, such as noise contrastive estimation, annealed importance sampling, bridge sampling, or a combination of these relying on the strengths of each, approximate the partition function directly, not the log of its gradient.

We will not discuss any of these methods here. Instead we refer interested readers to *Deep Learning* by Ian Goodfellow et al. (2016) (*https://oreil.ly/ViEzk*).

Back to Hopfield nets and Boltzmann machines. These are the stepping stones to deep neural networks that are trained through backpropagation that the recent deterministic and probabilistic deep learning models rely on. These methods form the original *connectionist* (of neurons) approach to learning arbitrary probability distributions, initially only over binary vectors of zeros and ones, and later over vectors with arbitrary real number values.

Hopfield Nets

Hopfield nets take advantage of the elegant mathematics of statistical mechanics by identifying the states of neurons of an artificial neural network with the states of elements in a physical system. Even though Hopfield nets eventually proved to be computationally expensive and of limited practical use, they are the founding fathers of the modern era of neural networks, and are worth exploring if only to gauge the historical evolution of the AI field. Hopfield nets have no hidden units, and all their (visible) units are connected to each other. Each unit can be found in an *on* or *off* state (one or zero), and collectively they encode information about the whole network (or the system).

Boltzmann Machine

A Boltzmann machine is a Hopfield net, but with the addition of hidden units. We are already familiar with the structure of input units and hidden units in neural networks, so no need to explain them, but this is where they started. Similar to the Hopfield net, both input and hidden units are binary, where the states are either 0 or 1 (modern versions implement units that take real number values, not only binary values).

All Boltzmann machines have an intractable partition function, so we approximate the maximum likelihood gradient using the techniques surveyed at the introduction of this section.

Boltzmann machines rely only on the computationally intensive *Gibbs sampling* for their training. Gibbs is a name that appears repetitively in the statistical mechanics field. Gibbs sampling provides unbiased estimates of the weights of the network, but these estimates have high variance. In general, there is a trade-off between bias and variance, and this trade-off highlights the advantages and disadvantages of methods relying on each.

Restricted Boltzmann Machine (Explicit Density and Intractable)

Boltzmann machines have a very slow learning rate due to the many interconnections within visible layers and within hidden layers (think a very messy backpropagation). This makes their training very slow and prohibits their application to practical problems. Restricted Boltzmann machines, which restrict connections only to those between different layers, solve this problem. That is, there are no connections within each layer of a restricted Boltzmann machine, allowing all of the units in each layer to be updated simultaneously. Therefore, for two connected layers, we can collect co-occurrence statistics by alternately updating all of the units in each layer. In practice, there are larger savings because of minimal sampling procedures, such as contrastive divergence.

Conditional independence

The lack of connections within each layer means that the states of all units in the hidden layer do not depend on each other, but they do depend on the states of units in the previous layer. In other words, given the states of the previous layer's units, the state of each hidden unit is independent of the states of the other units in the hidden layer. This conditional independence allows us to factorize the joint probability of the state of a hidden layer $p\left(\vec{h} \mid \overrightarrow{h_{previous}}\right)$ as the product of the conditional probabilities of the states of individual hidden units. For example, if we have three units in a hidden layer, $p\left(\vec{h} \mid \overrightarrow{h_{previous}}\right) = p\left(h_1 \mid \overrightarrow{h_{previous}}\right)p\left(h_2 \mid \overrightarrow{h_{previous}}\right)p\left(h_3 \mid \overrightarrow{h_{previous}}\right)$. The other way is also true, the states of the units of a previous layer are conditionally independent of each other given the states of the current layer. This conditional independence means that we can sample unit states instead of iteratively updating them for long periods of time.

Universal approximation

The restricted connections in restricted Boltzmann machines allow for their stacking, that is, having a series of multiple hidden layers that are able to extract more complex features. We can see now how the architecture of the modern multilayer artificial neural network slowly emerged. Recall that in Chapter 4, we discussed the universal approximation of neural networks for a wide range of deterministic functions. In this chapter, we would like our networks to represent (or learn) joint probability distributions instead of deterministic function. In 2008, Le Roux and Bengio proved that Boltzmann machines can approximate any discrete probability distribution to an arbitrary accuracy. This result also applies to restricted Boltzmann machines. Moreover, under certain mild conditions, each additional hidden layer increases the value of the log-likelihood function, thus allowing the model distribution to be closer to the true joint probability distribution of the training set.

In 2015, Eldan and Shamir verified empirically that increasing the number of layers of the neural networks is exponentially more valuable than increasing the width of network layers by the number of units in each layer (depth versus width). We also know from practice (without proofs) that it is possible to train a network with hundreds of hidden layers, where deeper layers represent higher-order features. Historically, the problem of vanishing gradients had to be overcome in order to train deep networks.

The Original Autoencoder

The autoencoder architecture aims to compress the information of the input into its lower-dimensional hidden layers. The hidden layers should retain the same amount of information as the input layers, even when they have fewer units than the input layers. We have discussed modern variational autoencoders, which provide an efficient method for training autoencoder networks. During training, each vector should be mapped to itself (unsupervised), and the network tries to learn the best *encoding*. The input and the output layers must then have the same number of units. A Boltzmann machine set up with a certain number of input units, fewer number of hidden units, and an output layer with the same number of units as the input layer describes an original network autoencoder architecture. From a historical perspective this is significant: the autoencoder is one of the first examples of a network successfully learning a code, implicit in the states of the hidden units, to represent its inputs. This makes it possible to force a network to compress its input into a hidden layer with minimal loss of information. This is now an integral part of neural networks that we take for granted. The autoencoder architecture, with and without Boltzmann machines (with and without energy-based joint probability distribution), is still very influential in the deep learning world.

Earlier in this chapter, we discussed variational autoencoders. From a historical point of view, these synthesize the ideas of Boltzmann machine autoencoders, deep autoencoder networks, *denoising autoencoders*, and the information bottleneck (Tishby et al. 2000), which have their roots in the idea of analysis by synthesis (Selfridge 1958). Variational autoencoders use fast variational methods for their learning. In the context of bias-variance trade-off, variational methods provide biased estimates for the network's weights that have low variance.

Probabilistic Language Modeling

A natural connection between this chapter and Chapter 7, which focused almost exclusively on natural language processing and the various ways to extract meaning from natural language data, is to survey the fundamentals behind probabilistic language models, then highlight the models from Chapter 7 that adhere to these fundamentals.

This chapter started with maximum likelihood estimation. One of the reasons this appears everywhere when we need to estimate probability distributions is that the probability distribution attained via maximum likelihood estimation is supported by mathematical theory, under a couple conditions: maximum likelihood estimation does converge to the true distribution $p_{data}\left(\overrightarrow{x}\right)$ that generated the data, in the limit as the number of data samples goes to infinity (that is, assuming we have a ton of data), and provided that the model probability distribution $p_{model}\left(\overrightarrow{x},\overrightarrow{\theta}\right)$ already includes the true probability distribution. That is, in the limit as the number of samples goes to infinity, the model parameters $\overrightarrow{\theta}^{*}$ that will maximize the likelihood of the data satisfy

$$p_{model}\left(\overrightarrow{x},\overrightarrow{\theta}^{*}\right) = p_{data}\left(\overrightarrow{x}\right).$$

In language models, the training data is samples of text from some corpus and/or genre, and we would like to learn its probability distribution so that we can generate similar text. It is important to keep in mind that the true data distribution is most likely *not* included in the family of distributions provided by $p_{model}\left(\overrightarrow{x},\overrightarrow{\theta}\right)$, so the theoretical result in the previous paragraph might never hold in practice; however, this doesn't deter us and we usually settle for models that are useful enough for our purposes. Our goal is to build a model that assigns probabilities to pieces of language. If we randomly assemble some pieces of language, we most likely end up with gibberish. What we actually want is to find the distribution of those sentences that mean something. A good language model is one that assigns high probabilities to sentences that are meaningful, even when these sentences are not among the training data. People usually compute the *perplexity* of a language model on the training data set to evaluate its performance.

Language models are based on the assumption that the probability distribution of the next word depends on the *n-1* words that preceded it, for some fixed *n*, so we care about calculating $p_{model}(x_n | x_1, x_2, \cdots, x_{n-1})$. If we are using a word2vec model that embeds the meaning of each word in vector, then each of these x's is represented by a vector. Words that mean things or are frequently used in similar contexts tend to have similar vector values. We can use the transformer model from Chapter 7 to predict the next word vector based on the preceding word vectors.

Frequency-based language models construct conditional probability tables by counting the number of times words appear together in the training corpus. For example, we can estimate the conditional probability *p(morning|good)* of the word *morning* appearing after the word *good*, by counting the number of times *good morning* appears in the corpus divided by the number of times *good* appears in the corpus. That is:

$$p(morning \,|\, good) = \frac{p(good, morning)}{p(good)}$$

This breaks down for very large corpuses or for unstructured text data such as tweets, Facebook comments, or SMS messages where the usual rules of grammar/spelling, etc. are not totally adhered to.

We can formalize the notion of a probabilistic language model this way:

1. Specify the vocabulary *V* of your language. This could be a set of characters, spaces, punctuations, symbols, unique words, and/or n-grams. Mathematically, it is a finite discrete set that includes a stopping symbol signifying the end of a thought or sentence, like a period in English (even though a period does not always mean the end of a sentence in English, such as when it is used for abbreviations).

2. Define a sentence (which could be meaningful or not) as a finite sequence of symbols $\vec{x} = (x_1, x_2, \cdots, x_m)$ from the vocabulary *V* ending in the stop symbol. Each x_i can assume any value from the vocabulary *V*. We can specify *m* as a maximum length for our sentences.

3. Define our language space $l^m = \{(x_1, x_2, \cdots, x_m), x_i \in V\}$ as the set of all sentences of length less than or equal to *m*. The overwhelming majority of these sentences will mean nothing, and we need to define a language model that only captures

the sentences that mean something: high probabilities for meaningful sentences and low probabilities for nonmeaningful ones.

4. Let $\mathscr{L}$ be the collection of all subsets of l^m. This accounts for collections of all meaningful and meaningless sentences of maximal length m.

5. In rigorous probability theory, we usually start with probability triples: a space, a *sigma algebra* containing some subsets of that space, and a probability measure assigned to each member of the chosen sigma algebra (do not worry about these details in this chapter). A language model, in this context, is the probability triple: the language space l^m, the sigma algebra made up of all the subsets of the language space $\mathscr{L}$, and a probability measure P that we need to assign to each member of $\mathscr{L}$. Since our language space is discrete and finite, it is easier to assign a probability p to each member of l^m instead, that is, a probability to each sentence $\overrightarrow{x} = (x_1, x_2, \cdots, x_m)$ (since this will in turn induce a probability measure P on the collection of all subsets $\mathscr{L}$, so we will never worry about this for language models). It is this p that we need to learn from the training data. The usual approach is to select a full family of probability distributions $p\left(\overrightarrow{x}; \overrightarrow{\theta}\right)$ parameterized by $\overrightarrow{\theta}$.

6. Finally, we need to estimate the parameter $\overrightarrow{\theta}$ by maximizing the likelihood of the training data set that contains many sentence samples from l^m. Since the probabilities of meaningful sentences are very small numbers, we use the logarithm of these probabilities instead to avoid the risk of underflow.

For consistency, it is a nice exercise to check log-linear models and log-bilinear models (GloVe) and latent Dirichlet allocation from Chapter 7 in the context of this section.

Summary and Looking Ahead

This was another foundational chapter in our journey to pinpoint the mathematics that is required for the state-of-the-art AI models. We shifted from learning deterministic functions in earlier chapters to learning joint probability distributions of data features. The goal is to use those to generate new data similar to the training data. We learned, still without formalizing, a lot of properties and rules for probability distributions. We surveyed the most relevant models, along with some historical evolution that led us here. We made the distinction between models that provide explicit formulas for their joint distributions and models that interact indirectly with the underlying distribution, without explicitly writing down formulas. For models with explicit formulas, computing the log-likelihood and its gradients can be tractable or intractable, each of which requires its own methods. The goal is always the same: capture the underlying true joint probability distribution of the data by finding a model that maximizes its log-likelihood.

None of this would have been necessary if our data was low dimensional, with one or two features. Histograms and kernel density estimators do a good job of estimating probability distributions for low-dimensional data. One of the best accomplishments in machine learning is the ability to model high-dimensional joint probability distributions from a big volume of data.

All of the approaches that we presented in this chapter have their pros and cons. For example, variational autoencoders allow us to perform both learning and efficient Bayesian inference in probabilistic graphical models with hidden (latent) variables. However, they generate lower-quality samples. Generative adversarial networks generate better samples, but they are more difficult to optimize due to their unstable training dynamics. They search for an unstable saddle point instead of a stable maximum or minimum. Deep belief networks such as PixelCNN and WaveNet have a stable training process, optimizing the softmax loss function. However, they are inefficient during sampling and don't organically embed data into lower dimensions, as autoencoders do.

Two-player zero-sum games from game theory appeared naturally in this chapter due to the setup of generative adversarial networks.

Looking ahead into the next chapter on *graphical modeling*, we note that the connections in the graph of a neural network dictate the way we can write conditional probabilities, easily pointing out the various dependencies and conditional independences. We saw this while discussing restricted Boltzmann machines in this chapter. In the next chapter, we focus exclusively on graphical modeling, which we have managed to avoid for a good three-quarters of the book.

CHAPTER 9

Graph Models

Now this is something we all want to learn.

—H.

Graphs, diagrams, and networks are all around us: cities and roadmaps, airports and connecting flights, electrical networks, the power grid, the World Wide Web, molecular networks, biological networks such as our nervous system, social networks, terrorist organization networks, schematic representations of mathematical models, artificial neural networks, and many, many others. They are easily recognizable, with distinct nodes representing some entities that we care for, which are then connected by directed or undirected edges indicating the presence of some relationship between the connected nodes.

Data that has a natural graph structure is better understood by a mechanism that exploits and preserves that structure, building functions that operate directly on graphs (however they are mathematically represented), as opposed to feeding graph data into machine learning models that artificially reshape it before analyzing it. This inevitably leads to loss of valuable information. This is the same reason convolutional neural networks are successful with image data, recurrent neural networks are successful with sequential data, and so on.

Graph-based models are very attractive for data scientists and engineers. Graph structures offer a flexibility that is not afforded in spaces with a fixed underlying coordinate system, such as in Euclidean spaces or in relational databases, where the data along with its features is forced to adhere to a rigid and predetermined form. Moreover, graphs are the natural setting that allows us to investigate the relationships between the points in a data set. So far, our machine learning models consumed data represented as isolated data points. Graph models, on the other hand, consume isolated data points, *along with the connections between them*, allowing for deeper understanding and more expressive models.

The human brain naturally internalizes graphical structures: it is able to model entities and their connections. It is also flexible enough to generate new networks, or expand and enhance existing ones, for example, when city planning, project planning, or when continuously updating transit networks. Moreover, humans can transition from natural language text to graph models and vice versa seamlessly. When we read something new, we find it natural to formulate a graphical representation to better comprehend it or illustrate it to other people. Conversely, when we see graph schematics, we are able to describe them via natural language. There are currently models that generate natural language text based on knowledge graphs and vice versa. This is called reasoning over knowledge graphs.

At this point we are pretty comfortable with the building blocks of neural networks, along with the types of data and tasks they are usually suited for:

- Multilayer perceptron or fully connected neural network (Chapter 4)
- Convolutional layers (Chapter 5)
- Recurrent layers (Chapter 7)
- Encoder-decoder components (Chapter 7)
- Adversarial components and two-player zero-sum games (Chapter 8)
- Variational components (Chapter 8)

The main tasks are mostly classification, regression, clustering, coding and decoding, or new data generation, where the model learns the joint probability distribution of the data features.

We are also familiar with the fact that we can mix and match some of the components of neural networks to construct new models that are geared toward specific tasks. The good news is that graph neural networks use the same exact ingredients, so we do not need to go over any new machine learning concepts in this chapter. Once we understand how to mathematically represent graph data along with its features in a way that can be fed into a neural network, either for analysis or for new network (graph) data generation, we are good to go. We will therefore avoid going down a maze of surveys for all the graph neural networks out there. Instead, we will focus on the simple mathematical formulation, popular applications, common tasks for graph models, available data sets, and model evaluation methods. Our goal is to develop a strong intuition for the workings of the subject. The main challenge is, yet again, lowering the dimensionality of the problem in a way that makes it amenable to computation and analysis, while preserving the most amount of information. In other words, for a network with millions of users, we cannot expect our models to take as input vectors or matrices with millions of dimensions. We need efficient representation methods for graph data.

If you want to dive deeper and fast track into graph neural networks, the survey paper "A Comprehensive Survey on Graph Neural Networks" (*https://oreil.ly/938pf*) (Wu et al. 2019) is an excellent place to start (of course, only after carefully reading this chapter).

Graphs: Nodes, Edges, and Features for Each

Graphs are naturally well suited to model any problem where the goal is to understand a discrete collection of objects (with emphasis on discrete and not continuous) through the relationships among them. Graph theory is a relatively young discipline in discrete mathematics and computer science with virtually unlimited applications. This field is in need of more brains to tackle its many unsolved problems.

A graph (see Figure 9-1) is made up of:

Nodes or vertices
> Bundled together in a set as $Nodes = \{node_1, node_2, \cdots, node_n\}$. This can be as little as a handful of nodes (or even one node), or as massive as billions of nodes.

Edges
> Connecting any two nodes (this can include an edge from a node to itself, or multiple edges connecting the same two nodes) in a directed (pointing from one node to the other) or undirected way (the edge has no direction from either node to the other). The set of edges is $Edges = \{edge_{ij} = (node_i, node_j)$, such that there is an edge pointing from $node_i$ to $node_j\}$.

Node features
> We can assign to each $node_i$ a list of, say, d features (such as the age, gender, and income level of a social media user) bundled together in a vector $\overrightarrow{features}_{node_i}$. We can then bundle all the feature vectors of all the n nodes of the graph in a matrix $Features_{Nodes}$ of size $d \times n$.

Edge features
> Similarly, we can assign to each $edge_{ij}$ a list of, say, c features (such as the length of a road, its speed limit, and whether it is a toll road or not) bundled together in a vector $\overrightarrow{features}_{edge_{ij}}$. We can then bundle all the feature vectors of all the m edges of the graph in a matrix $Features_{Edges}$ of size $c \times m$.

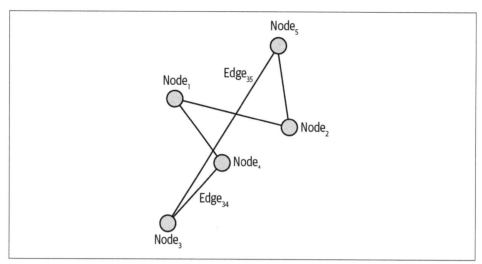

Figure 9-1. A graph is made up of nodes and directed or undirected edges connecting the nodes

Graph models are powerful because they are flexible and not necessarily forced to adhere to a rigid grid-like structure. We can think of their nodes as *floating through space* with no coordinates whatsoever. They are only held together by the edges that connect them. However, we need a way to represent their intrinsic structure. There are software packages that visualize graphs given their sets of nodes and edges, but we cannot do analysis and computations on these pretty (and informative) pictures. There are two popular graph representations that we can use as inputs to machine learning models: a graph's *adjacency matrix* and its *incidence matrix*.

There are other representations that are useful for graph theoretic algorithms, such as *edge listing, two linear arrays*, and *successor listing*. All of these representations convey the same information but differ in their storage requirements and the efficiency of graph retrieval, search, and manipulation. Most graph neural networks take as input the adjacency matrix along with the feature matrices for the nodes and the edges. Many times, they must do a dimension reduction (called graph representation or graph embedding) before feeding the graph data into a model. Other times, the dimension reduction step is part of the model itself.

Adjacency matrix

One algebraic way to store the structure of a graph on a machine and study its properties is through an adjacency matrix, which is an $n \times n$ whose entries $adjacency_{ij} = 1$ if there is an edge from $node_i$ $node_j$, and $adjacency_{ij} = 0$ if there is no edge from $node_i$ $node_j$. Note that this definition is able to accommodate a self edge, which is an edge from a vertex to itself, but not multiple edges between two distinct nodes, unless we decide to include the numbers 2, 3, etc.

as entries in the adjacency matrix. This, however, can mess up some results that graph theorists have established using the adjacency matrix.

Incidence matrix

This is another algebraic way to store the structure of the graph and retain its full information. Here, we list both the nodes and the edges, then formulate a matrix whose rows correspond to the vertices and whose columns correspond to the edges. An entry $incidence_{ij}$ of the matrix is 1 if $edge_j$ connects $node_i$ to some other node, and zero otherwise. Note that this definition is able to accommodate multiple edges between two distinct nodes, but not a self edge from a node to itself. Since many graphs have more edges than vertices, this matrix tends to be very wide and larger in size than the adjacency matrix.

The *Laplacian matrix* is another matrix that is associated with an undirected graph. It is an $n \times n$ symmetric matrix where each node has a corresponding row and column. The diagonal entries of the Laplacian matrix are equal to the degree of each node, and the off-diagonal entries are zero if there is no edge between nodes corresponding to that entry, and -1 if there is an edge between them. This is the discrete analog of the continuous Laplace operator from calculus and partial differential equations where the discretization happens at the nodes of the graph.

The Laplacian matrix takes into account the second derivatives of a continuous (and twice differentiable) function, which measure the concavity of a function, or how much its value at a point differs from its value at the surrounding points. Similar to the continuous Laplacian operator, the Laplacian matrix provides a measure of the extent a graph differs at one node from its values at nearby nodes. The Laplacian matrix of a graph appears when we investigate random walks on graphs and when we study electrical networks and resistances. We will see these later in this chapter.

We can easily infer simple node and edge statistics from the adjacency and incidence matrices, such as the degrees of nodes (the degree of a node is the number of edges connected to this node). The degree distribution $P(k)$ reflects the variability in the degrees of all the nodes. $P(k)$ is the empirical probability that a node has exactly k edges. This is of interest for many networks, such as web connectivity and biological networks.

For example, if the distribution of nodes of degree k in a graph follows a power law of the form $P(k) = k^{-\alpha}$, then such graphs have few nodes of high connectivity, or hubs, which are central to the network topology, holding it together, along with many nodes with low connectivity, which connect to the hubs.

We can also add time dependency, and think of dynamic graphs whose properties change as time evolves. Currently, there are models that add time dependency to the node and/or edge feature vectors (so each entry of these vectors becomes time dependent). For example, for a GPS system that predicts travel routes, the edge features connecting one point on the map to another change with time depending on the traffic situation.

Now that we have a mathematical framework for graph objects, along with their node and edge features, we can feed these representative vectors and matrices (and labels for supervised models) into machine learning models and do business as usual. Most of the time, half of the story is having a good representation for the objects at hand. The other half of the story is the expressive power of machine learning models in general, where we can get good results without encoding (or even having to learn) the rules that lead to these results. For the purposes of this chapter, this means that we can jump straight into graph neural networks *before* learning proper graph theory.

Directed Graphs

For directed graphs, on the one hand, we are interested in the same properties as undirected graphs, such as their spanning trees, fundamental circuits, cut sets, planarity, thickness, and others. On the other hand, directed graphs have their own unique properties that are different than undirected graphs, such as strong connectedness, arborescence (a directed form of rooted tree), decyclization, and others.

Example: PageRank Algorithm

PageRank (*https://oreil.ly/0yqGu*) is a retired algorithm (expired in 2019) that Google used to rank web pages in its search engine results. It provides a measure for the importance of a web page based on how many other pages link to it. In graph language, the nodes are the web pages, and the directed edges are the links pointing from one page to another. According to PageRank, node is important when it has many other web pages pointing to it, that is, when its incoming degree is large (see Figure 9-2).

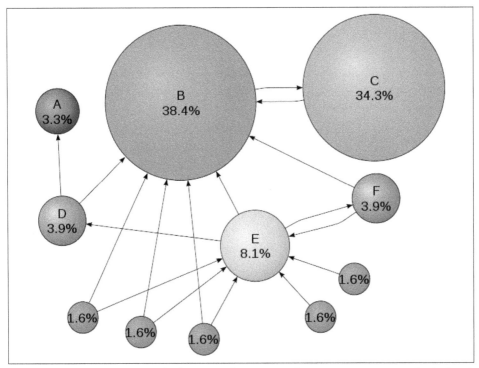

Figure 9-2. PageRank gives a higher score for pages with more pages pointing (or linking) to them (image source (https://oreil.ly/wMg3p))

As a concrete example involving graphs, adjacency matrix, linear algebra, and the web, let's walk through the PageRank algorithm for an absurdly simplified World Wide Web consisting of only four indexed web pages, such as in Figure 9-3, as opposed to billions.

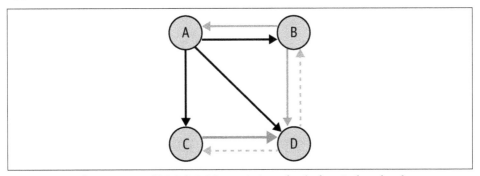

Figure 9-3. A fictitious World Wide Web consisting of only four indexed web pages (adapted from Coursera: Mathematics for Machine Learning (https://oreil.ly/gHqYs)).

In the graph of Figure 9-3, only B links to A; A and D link to B; A and D link to C; A, B, and C link to D; A links to B, C, and D; B links to A and D; C links to D; and D links to B and C.

Let's think of a web surfer who starts at some page then randomly clicks on a link from that page, then a link from this new page, and so on. This surfer simulates a *random walk on the graph of the web.*

In general, on the graph representing the World Wide Web, such a random surfer traverses the graph from a certain node to one of its neighbors (or back to itself if there are links pointing back to the page). We will encounter the World Wide Web one more time in this chapter and explore the kind of questions that we like to understand about the nature of its graph. We need a matrix for the random walk, which for this application we call the *linking matrix,* but in reality it is the adjacency matrix weighted by the degree of each vertex. We use this random walk matrix, or linking matrix, to understand the long-term behavior of the random walk on the graph. Random walks on graphs will appear throughout this chapter.

Back to the four-page World Wide Web of Figure 9-3. If the web surfer is at page A, there is a one-third chance the surfer will move to page B, one-third chance to move to C, and one-third chance to move to D. Thus, the *outward linking* vector of page A is:

$$\overrightarrow{linking_A} = \begin{pmatrix} 0 \\ 1/3 \\ 1/3 \\ 1/3 \end{pmatrix}$$

If the web surfer is at page B, there is a one-half chance they will move to page A and a one-half chance they will move to page D. Thus, the outward linking vector of page B is:

$$\overrightarrow{linking_B} = \begin{pmatrix} 1/2 \\ 0 \\ 0 \\ 1/2 \end{pmatrix}$$

Similarly, the outward linking vectors of pages C and D are:

$$\overrightarrow{linking_C} = \begin{pmatrix} 0 \\ 0 \\ 0 \\ 1 \end{pmatrix} \text{ and } \overrightarrow{linking_D} = \begin{pmatrix} 0 \\ 1/2 \\ 1/2 \\ 0 \end{pmatrix}$$

We bundle the linking vectors of all the web pages together to create a linking matrix:

$$Linking = \begin{pmatrix} 0 & 1/2 & 0 & 0 \\ 1/3 & 0 & 0 & 1/2 \\ 1/3 & 0 & 0 & 1/2 \\ 1/3 & 1/2 & 1 & 0 \end{pmatrix}$$

Note that the columns of the linking matrix are the outward linking probabilities, and the rows of A are the *inward linking* probabilities. How can a surfer end up at page A? They can only be at B, and from there there's only a 0.5 probability they will end up at A.

Now we can *rank* page A by adding up the ranks of all the pages pointing to A, each weighted by the probability a surfer will end up at page A from that page; that is, a page with many highly ranked pages pointing to it will also rank high. The ranks of all four pages are therefore:

$$rank_A = 0 rank_A + 1/2 rank_B + 0 rank_C + 0 rank_D$$
$$rank_B = 1/3 rank_A + 0 rank_B + 0 rank_C + 1/2 rank_D$$
$$rank_C = 1/3 rank_A + 0 rank_B + 0 rank_C + 1/2 rank_D$$
$$rank_D = 1/3 rank_A + 1/2 rank_B + 1 rank_C + 0 rank_D$$

To find the numerical value for the rank of each web page, we have to solve that system of linear equations, which is the territory of linear algebra. In matrix vector notation, we write the system as:

$$\overrightarrow{ranks} = Linking \; \overrightarrow{ranks}$$

Therefore, the vector containing all the ranks of all the web pages is an eigenvector of the linking matrix of the graph of the web pages (where the nodes are the web pages, and the directed edges are the links between them) with eigenvalue 1. Recall that in reality, the graph of the web is enormous, which means that the linking matrix is enormous, and devising efficient ways to find its eigenvectors becomes of immediate interest.

Computing eigenvectors and eigenvalues of a given matrix is one of the most important contributions of numerical linear algebra, with immediate applications in many fields. A lot of the numerical methods for finding eigenvectors and eigenvalues involve repeatedly multiplying a matrix with a vector. When dealing with huge matrices, this is expensive, and we have to use every trick in the book to make the operations cheaper. We take advantage of the sparsity of the matrix (many entries are zeros, so it is a waste to multiply with these entries and *then* discover that they are just zeros); we introduce randomization or stochasticity, and venture into the fields of high-dimensional probability and large random matrices (we will get a flavor of these in Chapter 11 on probability). For now, we reemphasize the iterative method we introduced in Chapter 6 on singular value decompositions: we start with a random vector $\overrightarrow{ranks_0}$, then produce a sequence of vectors iteratively by multiplying by the linking matrix:

$$\overrightarrow{ranks_{i+1}} = Linking \ \overrightarrow{ranks_i}$$

For our four-page World Wide Web, this converges to the vector:

$$\overrightarrow{ranks} = \begin{pmatrix} 0.12 \\ 0.24 \\ 0.24 \\ 0.4 \end{pmatrix}$$

which means that page D is ranked highest, and in a search engine query with similar content it will be the first page returned. We can then redraw the diagram in Figure 9-3 with the size of each circle corresponding to the importance of the page.

When the PageRank algorithm was in use, the real implementation included a damping factor d, a number between 0 and 1, usually around 0.85, which takes into account only an 85% chance that the web surfer clicks on a link from the page they are currently at, and a 15% chance that they start at a completely new page that has no links from the page they are currently at. This modifies the iterative process to find the rankings of the pages of the web in a straightforward way:

$$\overrightarrow{ranks_{i+1}} = d\left(Linking \ \overrightarrow{ranks_i}\right) + \frac{1-d}{\text{total number of pages}}\overrightarrow{ones}$$

Finally, you might be wondering whether Google keeps searching the web for new web pages and indexing them, and does it keep checking all indexed web pages for new links? The answer is yes, and the following excerpts are from Google's "In-Depth Guide to How Google Search Works" (*https://oreil.ly/oHw0g*):

Google Search is a fully-automated search engine that uses software known as web crawlers that explore the web regularly to find pages to add to our index. In fact, the vast majority of pages listed in our results aren't manually submitted for inclusion, but are found and added automatically when our web crawlers explore the web. [...] There isn't a central registry of all web pages, so Google must constantly look for new and updated pages and add them to its list of known pages. This process is called "URL discovery". Some pages are known because Google has already visited them. Other pages are discovered when Google follows a link from a known page to a new page: for example, a hub page, such as a category page, links to a new blog post. Still other pages are discovered when you submit a list of pages (a sitemap) for Google to crawl. [...] When a user enters a query, our machines search the index for matching pages and return the results we believe are the highest quality and most relevant to the user. Relevancy is determined by hundreds of factors, which could include information such as the user's location, language, and device (desktop or phone). For example, searching for "bicycle repair shops" would show different results to a user in Paris than it would to a user in Hong Kong.

The more data we collect, the more complex searching it becomes. Google rolled out RankBrain in 2015. It uses machine learning to vectorize the text on the web pages, similar to what we did in Chapter 7. This process adds context and meaning to the indexed pages, so that the search returns more accurate results. The bad thing that this process adds is the much higher dimensions associated with meaning vectors. To circumvent the difficulty of checking every vector at every dimension before returning the web pages closest to the query, Google uses an approximate nearest neighbor algorithm, which helps return excellent results in milliseconds—the experience we have now.

Inverting Matrices Using Graphs

Many problems in the applied sciences involve writing a discrete linear system $A\vec{x} = \vec{b}$ and solving it, which is equivalent to inverting the matrix A and finding the solution $\vec{x} = A^{-1}\vec{b}$. But for large matrices, this is a computationally expensive operation, along with having high storage requirements and poor accuracy. We are always looking for efficient ways to invert matrices, sometimes leveraging the special characteristics of the particular matrices at hand.

The following is a graph theoretic method that computes the inverse of a matrix of a decent size (for example, a hundred rows and a hundred columns):

1. Replace each nonzero entry in matrix A with a 1. We obtain a binary matrix.

2. Permute the rows and the corresponding columns of the resulting binary matrix to make all diagonal entries 1's.

3. We think of the matrix obtained as the adjacency matrix of a directed graph (where we delete the self-loops corresponding to 1's along the diagonal from the graph).

4. The resulting directed graph is partitioned into its fragments.

5. If a fragment is too large, we tear it into smaller fragments by removing an appropriate edge.

6. We invert the smaller matrices.

7. Apparently this leads to the inverse of the original matrix.

We will not explain why and how, but this method is so cute, so it made its way into this chapter.

Cayley Graphs of Groups: Pure Algebra and Parallel Computing

Graphs of groups, also called Cayley graphs or Cayley diagrams, can be helpful in designing and analyzing network architectures for parallel computers, routing problems, and routing algorithms for interconnected networks. The paper "Processor Interconnection Networks from Cayley Graphs" (*https://oreil.ly/zXqYi*) (Schibell et al. 2011) is an interesting and easy read on earlier designs applying Cayley graphs for parallel computing networks, and explains how to construct Cayley graphs that meet specific design parameters. Cayley graphs have also been applied for classification of data (*https://oreil.ly/xp5hB*).

We can represent every group with n elements as a connected directed graph of n nodes, where each node corresponds to an element from the group, and each edge represents a multiplication by a generator from the group. The edges are labeled (or colored) depending on which generator from the group we are multiplying by (see Figure 9-4). This directed graph uniquely defines the group: each product of elements in the group corresponds to following a sequence of directed edges on the graph. For example, the graph of a cyclic group of n elements is a directed circuit of n nodes in which every edge represents multiplication by one generator of the group.

From a pure math perspective, Cayley graphs are useful for visualizing and studying abstract groups, encoding their full abstract structure and all of their elements in a visual diagram. The symmetry of Cayley graphs makes them useful for constructing more involved abstract objects. These are central tools for combinatorial and geometric group theory. For more on Cayley graphs, check out this Wolfram Mathworld page (*https://oreil.ly/JCvux*).

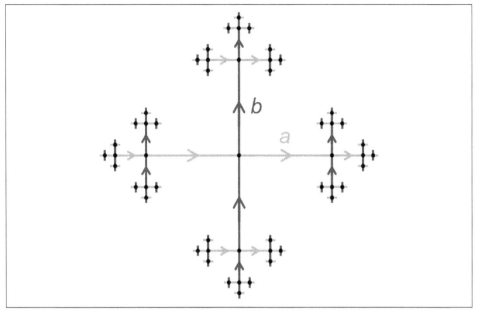

Figure 9-4. Cayley graph for the free group on two generators a and b; each node represents an element of the free group, and each edge represents multiplication by a or b (image source (https://oreil.ly/cafsv))

Message Passing Within a Graph

The *message passing framework* is a useful approach for modeling the propagation of information within graphs, as well as neatly aggregating the information conveyed in the nodes, edges, and the structure of the graph itself into vectors of a certain desired dimension. Within this framework, we update every node with information from the feature vectors of its neighboring nodes and the edges connected to them. A graph neural network performs multiple rounds of message passing; each round propagates a single node's information further. Finally, we combine the latent features of each individual node to obtain its unified vector representation and represent the whole graph.

More concretely, for a specific node, we choose a function that takes as input the node's feature vector, the feature vector of one of its neighboring nodes (those connected to it by an edge), and the feature vector of the edge that connects it to this neighboring node, and outputs a new vector that contains within it information from the node, the neighbor, and their connecting edge. We apply the same function to all of the node's neighbors, then add the resulting vectors together, producing a *message vector*. Finally, we update the feature vector of our node by combining its original feature vector with the message vector within an update function that we also choose. When we do this for each node in the graph, each node's new feature vector will contain information from itself, all its neighbors, and all its connecting edges. Now, when we repeat this process one more time, the node's most recent feature vector will contain information from itself, all its neighbors *and its neighbor's neighbors*, and all the corresponding connecting edges. Thus, the more message passing rounds we do, the more each node's feature vector contains information from farther nodes within the graph, moving one edge separation at a time. The information diffuses successively across the entire network.

The Limitless Applications of Graphs

Applications for graph neural networks, and graph models in general, are ubiquitous and so important that I am a bit regretful I did not start the book with graphs. In any graph model, we start by answering the following questions:

- What are the nodes?
- What is the relationship that links two nodes, that establishes directed or undirected edge(s) between them?
- Should the model include feature vectors for the nodes and/or the edges?
- Is our model dynamic, where the nodes, edges, and their features evolve with time, or is it static in time?
- What are we interested in? Classifying (for example cancerous or noncancerous; fake news spreader or real news spreader)? Generating new graphs (for example for drug discovery)? Clustering? Embedding the graph into a lower-dimensional and structured space?
- What kind of data is available or needed, and is the data organized and/or labeled? Does it need preprocessing?

We survey few applications in this section, but there are many more that genuinely lend themselves to a graph modeling structure. It is good to read the abstracts of the linked publications since they help capture common themes and ways of thinking about these models. The following list gives a good idea of the common tasks for graph neural networks, which include:

- Node classification
- Graph classification
- Clustering and community detection
- New graph generation
- Influence maximization
- Link prediction

Image Data as Graphs

We might encounter graph neural networks tested on the MNIST data set for handwritten digits (*https://oreil.ly/HQL5F*), which is one of the benchmark sets for computer vision. If you wonder how image data (stored as three-dimensional tensors of pixel intensities across each channel) manages to fit into a graph structure, here's how it works. Each pixel is a node, and its features are the respective intensities of its three channels (if it is a color image, otherwise it only has one feature). The edges connect each pixel to the three, five, or eight pixels surrounding it, depending on whether the pixel is located at a corner, edge, or in the middle of the image.

Brain Networks

One of the main pursuits in neuroscience is understanding the network organization of the brain. Graph models provide a natural framework and many tools for analyzing the complex networks of the brain, both in terms of their anatomy and their functionality.

To create artificial intelligence on par with human intelligence, we must understand the human brain on many levels. One aspect is the brain's network connectivity, how connectivity affects the brain's functionality, and how to replicate that, building up from small computational units to modular components to a fully independent and functional system.

Human brain anatomical networks demonstrate short path length (conservation of wiring costs) along with high-degree cortical *hubs*, that is, high clustering. This is on both the cellular scale and on the whole brain scale. In other words, the brain network seems to have organized itself in a way that maximizes the efficiency of information transfer and minimizes connection cost. The network also demonstrates modular and hierarchical topological structures and functionalities. The topological structures of the brain networks and their functionalities are interdependent over both short and long time scales. The dynamic properties of the networks are affected by their structural connectivity, and over a longer time scale, the dynamics affect the topological structure of the network.

The most important questions are: what is the relationship between the network properties of the brain and its cognitive behavior? What is the relationship between the network properties and brain and mental disorders? For example, we can view neuropsychiatric disorders as disconnectivity syndromes, where graph theory can help quantify weaknesses, vulnerability to lesions, and abnormalities in the network structures. In fact, graph theory has been applied to study the structural and functional network properties in schizophrenia, Alzheimer's, and other disorders.

Spread of Disease

As we have all learned from the COVID-19 pandemic, it is of crucial importance to be able to forecast disease incidents accurately and reliably for mitigation purposes, quarantine measures, policy, and many other decision factors. A graph model can consider either individuals or entire geographic blocks as nodes, and contact occurrences between these individuals or blocks as edges. Recent models for predicting COVID-19 spread, for example, the article "Combining Graph Neural Networks and Spatio-temporal Disease Models to Improve the Prediction of Weekly COVID-19 Cases in Germany" (*https://oreil.ly/Tkhy7*) (Fritz et al. 2022), incorporate human mobility data from Facebook, Apple, and Google to model interactions between the nodes in their models.

There is plenty of data that can be put to good use here. Facebook's "Data for Good" (*https://oreil.ly/MJAwr*) resource has a wealth of data on population densities, social mobility and travel patterns, social connectedness, and others. Google's COVID-19 Community Mobility Reports (*https://oreil.ly/2g0mT*) draw insights from Google Maps and other products into a data set that charts movement trends over time by geography across different categories of places, such as retail and recreation, groceries and pharmacies, parks, transit stations, workplaces, and residential areas. Similarly, Apple's and Amazon's mobility data serve a similar purpose with the goal of aiding efforts to limit the spread of COVID-19.

Spread of Information

We can use graphs to model the spread of information, disease, rumors, gossip, computer viruses, innovative ideas, or others. Such a model is usually a directed graph, where each node corresponds to an individual, and the edges are tagged with information about the interaction between individuals. The edge tags, or weights, are usually probabilities. The weight w_{ij} of the edge connecting $node_i$ to $node_j$ is the probability of a certain effect (disease, rumor, computer virus) propagating from $node_i$ to $node_j$.

Detecting and Tracking Fake News Propagation

Graph neural networks perform better in the task of detecting fake news (see Figure 9-5) than content-based natural language processing approaches. The abstract of the paper "Fake News Detection on Social Media using Geometric Deep Learning" (*https://oreil.ly/HQNTq*) (Monti et al. 2019) is informative:

> Social media are nowadays one of the main news sources for millions of people around the globe due to their low cost, easy access, and rapid dissemination. This however comes at the cost of dubious trustworthiness and significant risk of exposure to 'fake news', intentionally written to mislead the readers. Automatically detecting fake news poses challenges that defy existing content-based analysis approaches. One of the main

reasons is that often the interpretation of the news requires the knowledge of political or social context or 'common sense', which current natural language processing algorithms are still missing. Recent studies have empirically shown that fake and real news spread differently on social media, forming propagation patterns that could be harnessed for the automatic fake news detection. Propagation based approaches have multiple advantages compared to their content based counterparts, among which is language independence and better resilience to adversarial attacks. In this paper, we show a novel automatic fake news detection model based on geometric deep learning. The underlying core algorithms are a generalization of classical convolutional neural networks to graphs, allowing the fusion of heterogeneous data such as content, user profile and activity, social graph, and news propagation. Our model was trained and tested on news stories, verified by professional fact checking organizations, that were spread on Twitter. Our experiments indicate that social network structure and propagation are important features allowing highly accurate (92.7% ROC AUC) fake news detection. Second, we observe that fake news can be reliably detected at an early stage, after just a few hours of propagation. Third, we test the aging of our model on training and testing data separated in time. Our results point to the promise of propagation based approaches for fake news detection as an alternative or complementary strategy to content based approaches.

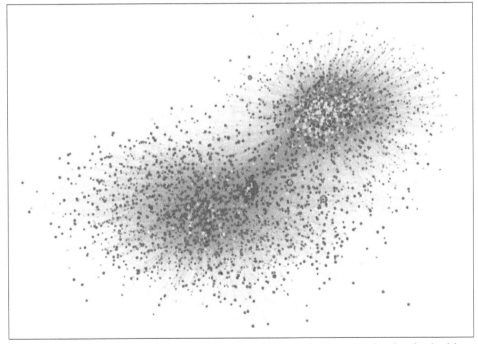

Figure 9-5. Nodes that spread fake news are labeled in red color; people who think alike cluster together in social networks (see image source (https://oreil.ly/hfNuf) for a color version of this image)

Web-Scale Recommendation Systems

Since 2018, Pinterest has been using the PinSage graph convolutional network (*https://oreil.ly/JQYz4*). This curates users' home feed and makes suggestions for new and relevant pins. The authors utilize *random walks on graphs* in their model, which we will discuss later in this chapter. Here is the full abstract:

> Recent advancements in deep neural networks for graph-structured data have led to state-of-the-art performance on recommender system benchmarks. However, making these methods practical and scalable to web-scale recommendation tasks with billions of items and hundreds of millions of users remains a challenge. Here we describe a large-scale deep recommendation engine that we developed and deployed at Pinterest. We develop a data efficient Graph Convolutional Network (GCN) algorithm PinSage, which combines efficient random walks and graph convolutions to generate embeddings of nodes (i.e., items) that incorporate both graph structure as well as node feature information. Compared to prior GCN approaches, we develop a novel method based on highly efficient random walks to structure the convolutions and design a novel training strategy that relies on harder-and-harder training examples to improve robustness and convergence of the model. We deploy PinSage at Pinterest and train it on 7.5 billion examples on a graph with 3 billion nodes representing pins and boards, and 18 billion edges. According to offline metrics, user studies and A/B tests, PinSage generates higher-quality recommendations than comparable deep learning and graph-based alternatives. To our knowledge, this is the largest application of deep graph embeddings to date and paves the way for a new generation of web-scale recommender systems based on graph convolutional architectures.

Fighting Cancer

In the article "HyperFoods: Machine Intelligent Mapping of Cancer-Beating Molecules in Foods" (*https://oreil.ly/2BfHL*) (Veselkov et al. 2019), the authors use protein, gene, and drug interaction data to identify the molecules that help prevent and beat cancer. They also map the foods that are the richest in cancer-beating molecules (see Figure 9-6). Again, the authors utilize random walks on graphs. Here's the abstract of the paper:

> Recent data indicate that up to 30–40% of cancers can be prevented by dietary and lifestyle measures alone. Herein, we introduce a unique network-based machine learning platform to identify putative food-based cancer-beating molecules. These have been identified through their molecular biological network commonality with clinically approved anti-cancer therapies. A machine-learning algorithm of random walks on graphs (operating within the supercomputing DreamLab platform) was used to simulate drug actions on human interactome networks to obtain genome-wide activity profiles of 1962 approved drugs (199 of which were classified as "anti-cancer" with their primary indications). A supervised approach was employed to predict cancer-beating molecules using these 'learned' interactome activity profiles. The validated model performance predicted anti-cancer therapeutics with classification accuracy of 84–90%. A comprehensive database of 7962 bioactive molecules within foods was fed into the

model, which predicted 110 cancer-beating molecules (defined by anti-cancer drug likeness threshold of >70%) with expected capacity comparable to clinically approved anti-cancer drugs from a variety of chemical classes including flavonoids, terpenoids, and polyphenols. This in turn was used to construct a 'food map' with anti-cancer potential of each ingredient defined by the number of cancer-beating molecules found therein. Our analysis underpins the design of next-generation cancer preventative and therapeutic nutrition strategies.

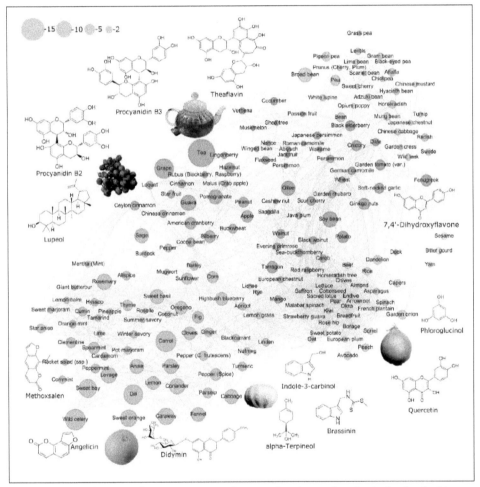

Figure 9-6. Machine intelligent mapping of cancer-beating molecules in foods; the bigger the node, the more diverse the set of cancer-beating molecules (image source (https:// oreil.ly/sGAIp))

Biochemical Graphs

We can represent molecules and chemical compounds as graphs where the nodes are the atoms, and the edges are the chemical bonds between them. Data sets from this chemoinformatics domain are useful for assessing a classification model's performance. For example, the NCI1 (*https://oreil.ly/LUikV*) data set, containing around 4,100 chemical compounds, is useful for anti-cancer screens where the chemicals are labeled as positive or negative to hinder cell lung cancer. Similar labeled graph data sets for proteins and other compounds are available on the same website, along with the papers that employ them and the performance of different models on these data sets.

Molecular Graph Generation for Drug and Protein Structure Discovery

In the last chapter we learned how generative networks such as variational autoencoders and adversarial networks learn joint probability distributions from the data in order to generate similar-looking data for various purposes. Generative networks for graphs build on similar ideas; however, they are a bit more involved than networks generating images per se. Generative graph networks generate new graphs either in a sequential manner, outputting nodes and edges step-by-step, or in a global manner, outputting a whole graph's adjacency matrix at once. See, for example, this survey paper on generative graph networks (2020) (*https://oreil.ly/omCsl*) for details on the topic.

Citation Networks

In citation networks, the nodes could be the authors, and the edges are their coauthorships; or the nodes are papers, and the (directed) edges are the citations between them. Each paper has directed edges pointing to the papers it cites. Features for each paper include its abstract, authors, year, venue, title, field of study, and others. Tasks include node clustering, node classification, and link prediction. Popular data sets for paper citation networks include CoRA, CiteSeerX, and PubMed. The CoRA data set (*https://oreil.ly/X3J3t*) contains around three thousand machine learning publications grouped into seven categories. Each paper in the citation networks is represented by a one-hot vector indicating the presence or absence of a word from a prespecified dictionary, or by a term frequency-inverse document frequency (TF-IDF) vector. These data sets are updated continuously as more papers join the networks.

Social Media Networks and Social Influence Prediction

Social media networks, such as Facebook, Twitter, Instagram, and Reddit, are a distinctive feature of our time (after 2010). The Reddit data set (*https://oreil.ly/uwT6N*) is an example of the available data sets. This is a graph where the nodes are the posts and the edges are between two posts that have comments from the same user. The posts are also labeled with the community to which they belong.

Social media networks and their social influence have a substantial impact on our societies, ranging from advertising to winning presidential elections to toppling political regimes. One important task for a graph model representing social networks is to predict the social influence of the nodes in the network. Here, the nodes are the users, and their interactions are the edges. Features include users' gender, age, sex, location, activity level, and others. One way to quantify social influence, the target variable, is through predicting the actions of a user given the actions of their near neighbors in the network. For example, if a user's friends buy a product, what is the probability that they will buy the same product after a given period of time? Random walks on graphs help predict the social influence of certain nodes in a network.

Sociological Structures

Social diagrams are directed graphs that represent relationships among individuals in a society or among groups of individuals. The nodes are the members of the society or the groups, and the directed edges are the relationships between these members, such as admiration, association, influence, and others. We are interested in the connectedness, separability, size of fragments, and so forth in these social diagrams. One example is from anthropological studies where a number of tribes are classified according to their kinship structures.

Bayesian Networks

Later in this chapter we will discuss Bayesian networks. These are probabilistic graph models whose goal is one we are very familiar with in the AI field: to learn joint probability distribution of the features of a data set. Bayesian networks consider this joint probability distribution as a product of single variable distributions *conditional only on a node's parents* in a graph representing the relationships between the features of the data. That is, the nodes are the feature variables and the edges are between the features that *we believe* to be connected. Applications include spam filtering, voice recognition, and coding and decoding, to name a few.

Traffic Forecasting

Traffic prediction is the task of predicting traffic volumes using historical roadmaps, road speed, and traffic volume data. There are benchmark traffic data sets (*https://oreil.ly/LsZ0Q*) that we can use to track progress and compare models. For example, the METR-LA data set (*https://oreil.ly/YU918*) is a spatial-temporal graph, containing four months of traffic data collected by 207 sensors on the highways of Los Angeles County. The traffic network is a graph, where the nodes are the sensors, and the edges are the road segments between these sensors. At a certain time t, the features are traffic parameters, such as velocity and volume. The task of a graph neural network is to predict the features of the graph after a certain time has elapsed.

Other traffic forecasting models employ Bayesian networks (*https://oreil.ly/6mVzy*), such as traffic flows among adjacent road links. The model uses information from adjacent road links to analyze the trends of focus links.

Logistics and Operations Research

We can model and solve many problems in operations research, such as transportation problems and activity networks, using graphs. The graphs involved are usually weighted directed graphs. Operations research problems are combinatorial in nature, and are always trivial if the network is small. However, for large real-world networks, the challenge is finding efficient algorithms that can sift through the enormous search space and quickly rule out big parts of it. A large part of the research literature deals with estimating the computational complexity of such algorithms. This is called combinatorial optimization. Typical problems include the traveling salesman problem, supply chain optimization, shared rides routing and fares, job matching, and others. Some of the graph methods and algorithms used for such problems are minimal spanning trees, shortest paths, max-flow min-cuts, and matching in graphs. We will visit operations research examples later in this book.

Language Models

Graph models are relevant for a variety of natural language tasks. These tasks seem different at the surface but many of them boil down to clustering, for which graph models are very well suited.

For any application we must first choose what the nodes, edges, and features for each represent. For natural language, these choices reveal hidden structures and regularities in the language and in the language corpuses.

Instead of representing a natural language sentence as a sequence of tokens for recurrent models or as a vector of tokens for transformers, in graph models we embed sentences in a graph, then employ graph deep learning (or graph neural networks).

One example from computational linguistics is constructing diagrams for parsing language, as shown in Figure 9-7.

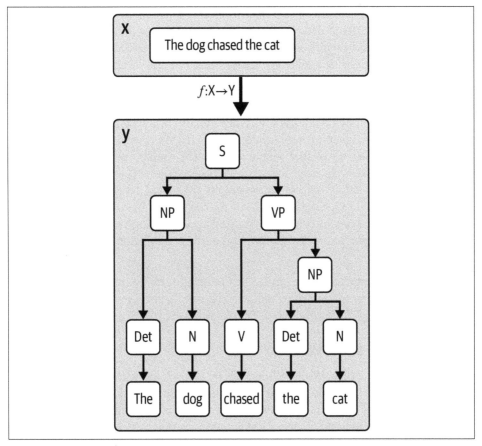

Figure 9-7. A parsed sentence

The nodes are words, n-grams, or phrases, and the edges are the relationships between them, which depend on the language grammar or syntax (article, noun, verb, etc.). A language is defined as the set of all strings correctly generated from the language vocabulary according to its grammar rules. In that sense, computer languages are easy to parse (they are built that way), while natural languages are much harder to specify completely due to their complex nature.

Parsing

Parsing means converting a stream of input into a structured or formal representation so it can be automatically processed. The input to a parser might be sentences, words, or even characters. The output is a tree diagram containing information about the function of each part of the input. Our brain is a great parser for language inputs. Computers parse programming languages.

Another example is news clustering or article recommendations. Here, we use graph embeddings of text data to determine text similarity. The nodes can be words and the edges can be semantic relationships between the words, or just their co-occurrences. Or the nodes can be words and documents, and the edges can again be semantic or co-occurrence relationships. Features for nodes and edges can include authors, topics, time periods, and others. Clusters emerge naturally in such graphs.

Another type of parsing that does not depend on the syntax or grammar of a language is abstract meaning representation (AMR). It relies instead on semantic representation, in the sense that sentences that are similar in meaning should be assigned the same abstract meaning representation, even if they are not identically worded. Abstract meaning representation graphs are rooted, labeled, directed, acyclic graphs, representing full sentences. These are useful for machine translation and natural language understanding. There are packages and libraries for abstract meaning representation parsing, visualization, and surface generation, as well as publicly available data sets.

For other natural language applications, the following survey paper is a nice reference that is easy to read to learn more about the subject: "A Survey of Graphs in Natural Language Processing" (*https://oreil.ly/2Birx*) (Nastase et al. 2015).

Graph Structure of the Web

Since the inception of the World Wide Web in 1989, it has grown enormously and has become an indispensable tool for billions of people around the world. It allows access to billions of web pages, documents, and other resources using an internet web browser. With billions of pages linking to each other, it is of great interest to investigate the graph structure of the web. Mathematically, this vast and expansive graph is fascinating in its own right. But understanding this graph is important for more reasons than a beautiful mental exercise, providing insights into algorithms for crawling, indexing, and ranking the web (as in the PageRank algorithm that we saw earlier in this chapter), searching communities, and discovering the social and other phenomena that characterize its growth or decay.

The World Wide Web graph has:

Nodes
    Web pages, on the scale of billions

Edges
    These are directed from one page linking to another page, on the scale of hundreds of billions

We are interested in:

- What is the average degree of the nodes?
- Degree distributions of the nodes (for both the in degree and out degree, which can be very different). Are they power laws? Some other laws?
- Connectivity of the graph: what is the percentage of connected pairs?
- Average distances between the nodes.
- Is the observed structure of the web dependent or independent of the particular crawl used?
- Particular structures of weakly and strongly connected components.
- Is there a giant strongly connected component? What is the proportion of nodes that can reach or be reached from this giant component?

Automatically Analyzing Computer Programs

We can use graphs for verification of computer programs, program reasoning, reliability theory, fault diagnosis in computers, and studying the structure of computer memory. The paper "Graph Neural Networks on Program Analysis" (*https://oreil.ly/ ZJyeo*) (Allamanis 2021) is one example:

> Program analysis aims to determine if a program's behavior complies with some specification. Commonly, program analyses need to be defined and tuned by humans. This is a costly process. Recently, machine learning methods have shown promise for probabilistically realizing a wide range of program analyses. Given the structured nature of programs, and the commonality of graph representations in program analysis, graph neural networks (GNN) offer an elegant way to represent, learn, and reason about programs and are commonly used in machine learning-based program analyses. This article discusses the use of graph neural networks for program analysis, highlighting two practical use cases: Variable misuse detection and type inference.

Data Structures in Computer Science

A data structure in computer science is a structure that stores, manages, and organizes data. There are different data structures, and they are usually chosen in a way that makes it efficient to access the data (read, write, append, infer or store relationships, etc.).

Some data structures use graphs to organize data, computational devices in a cluster, and represent the flow of data and computation or the communication network. There are also graph databases geared toward storing and querying graph data. Other databases transform graph data to more structured formats (such as relational formats).

Here are some examples for graph data structures:

PageRank algorithm
We have already encountered the PageRank algorithm, along with the link structure of a website represented as a directed graph, where the nodes are the web pages and the edges represent the links from one page to another. A database keeping all the web pages along with their link structures can be either graph structured, where the graph is stored as is using the linking matrix or adjacency matrix with no transformation necessary, or it can transformed to fit the structure of other nongraphical databases.

Binary search trees for organizing files in a database
Binary search trees are ordered data structures that are efficient for both random and sequential access of records, and for file modification. The inherent order of a binary search tree speeds up search time: we cut the amount of data to sort through by half at each level of the tree. It also speeds up insertion time: unlike an array, when we add a node to the binary tree data structure, we create a new piece in memory and link to it. This is faster than creating a new large array, then inserting the data from the smaller array to the new, larger one.

Graph-based information retrieval systems
In some information retrieval systems we assign a certain number of index terms to each document. We can think of these as the document's indicators, descriptors, or keywords. These index terms will be represented as nodes of a graph. We connect two index terms with an undirected edge if these two happen to be closely related, such as the indices *graph* and *network*. The resulting *similarity graph* is very large, and is possibly disconnected. The maximally connected subgraphs of this graph are its *components*, and they naturally classify the documents in this system. For information retrieval, our query specifies some index terms, that is, certain nodes of the graph, and the system returns the maximal complete subgraph that includes the corresponding nodes. This gives the complete list of index terms, which in turn specify the documents we are searching for.

Load Balancing in Distributed Networks

The computational world has grown from Moore's law to parallel computing to cloud computing. In cloud computing, our data, our files, and the machines executing our files and doing computations on our data are not near us. They are not even near each other. As applications become more complex and as network traffic increases, we need the software or hardware analog of a network traffic cop that distributes network traffic across multiple servers so that no single server bears a heavy load, enhancing performance in terms of application response times, end user experience, and so on.

As traffic increases, more appliances, or nodes, need to be added to handle the volume. Network traffic distributing needs to be done while preserving data security and privacy, and should be able to predict traffic bottlenecks before they happen. This is exactly what load balancers do. It is not hard to imagine the distributed network as a graph with the nodes as the connected servers and appliances. Load balancing is then a traffic flow problem on a given graph, and there are a variety of algorithms for allocating the load. All algorithms operate on the network's graph. Some are static, allocating loads without updating the network's current state in terms of loads or malfunctioning units, while others are dynamic, but require constant communication within the network about the nodes' statuses. The following are some algorithms:

Least connection algorithm
>   This method directs traffic to the server with the fewest active connections.

Least response time algorithm
>   This method directs traffic to the server with the fewest active connections and the lowest average response time.

Round-robin algorithm
>   This algorithm allocates load in a rotation on the servers. It directs traffic to the first available server, then moves that server to the bottom of the queue.

IP hash
>   This method allocates servers based on the IP address of the client.

Artificial Neural Networks

Finally, artificial neural networks are graphs where the nodes are the computational units and the edges are inputs and outputs of these units. Figure 9-8 summarizes popular artificial neural network models as graphs.

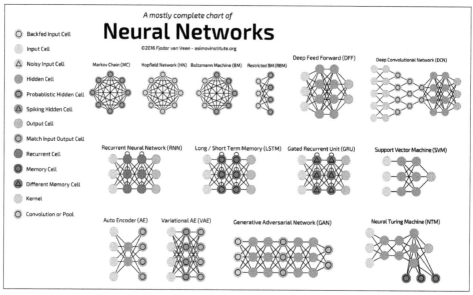

Figure 9-8. Neural networks as graphs (image source (https://oreil.ly/e6HQp))

Random Walks on Graphs

A random walk on a graph (Figure 9-9) means exactly what it says: a sequence of steps that starts at some node, and at each time step, chooses a neighboring node (using the adjacency matrix) with probability proportional to the weights of the edges, and moves there.

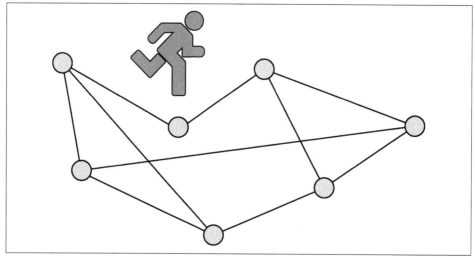

Figure 9-9. A random walker on an undirected graph

If the edges are unweighted, then the neighboring nodes are all equally likely to be chosen for the walk's move. At any time step, the walk can stay at the same node in the case that there is a self edge, or when it is a lazy random walk with a positive probability the walk stays at a node instead of moving to one of its neighbors. We are interested in the following:

- What is the list of nodes visited by a random walk, in the order they are visited? Here, the starting point and the structure of the graph matter in how much a walk covers or whether the walk can ever reach certain regions of the graph. In graph neural networks, we are interested in learning a representation for a given node based on its neighbors' features. In large graphs where nodes have more neighbors than is computationally feasible, we employ random walks. However, we have to be careful since different parts of the graph have different random walk expansion speeds, and if we do not take that into account by adjusting the number of steps of a random walk according to the subgraph structure, we might end up with low-quality representations for the nodes, and undesirable outcomes as these go down the work pipeline.

- What is the expected behavior of a random walk, that is, the probability distribution over the visited nodes after a certain number of steps? We can study basic properties of a random walk, such as its long-term behavior, by using *the spectrum of the graph*, which is the set of eigenvalues of its adjacency matrix. In general, the spectrum of an operator helps us understand what happens when we repeatedly apply the operator. Randomly walking on a graph is equivalent to repeatedly applying a normalized version of the adjacency matrix to the node of the graph where we started. Every time we apply this *random walk matrix*, we walk one step further on the graph.

- How does a random walk behave on different types of graphs, such as paths, trees, two fully connected graphs joined by one edge, infinite graphs, and others?

- For a given graph, does the walk ever return to its starting point? If so, how long do we have to walk until we return?

- How long do we have to walk until we reach a specific node?

- How long do we have to walk until we visit all the nodes?

- How does a random walk *expand*? That is, what is the *influence distribution* of certain nodes that belong in certain regions of the graph? What is the size of their influence?

- Can we design algorithms based on random walks that are able to reach *obscure* parts of large graphs?

Random Walks and Brownian Motion

In the limit where the size of steps of a random walk goes to zero, we obtain a Brownian motion. Brownian motion usually models the random fluctuations of particles suspended in a medium such as a fluid, or price fluctuations of derivatives in financial markets. We frequently encounter the term Brownian motion with the term *Weiner process*, which is a continuous stochastic process with a clear mathematical definition of how the motion (of a particle or of a price fluctuation in finance) starts (at zero), how the next step is sampled (from the normal distribution and with independent increments), and assumptions about its continuity as a function of time (it is almost surely continuous). Another term it is associated with is *martingale*. We will see these in Chapter 11 on probability.

We did encounter a random walk once in this chapter when discussing the PageRank algorithm, where a random web page surfer randomly chooses to move from the page they are at to a neighboring page on the web. We noticed that the long-term behavior of the walk is discovered when we repeatedly apply the linking matrix of the graph, which is the same as the adjacency matrix normalized by the degrees of each node. In the next section we see more uses of random walks for graph neural networks.

We can use random walks (on directed or undirected graphs, weighted or unweighted graphs) for community detection and influence maximization in small networks, where we would only need the graph's adjacency matrix (as opposed to node embedding into feature vectors, then clustering).

Node Representation Learning

Before implementing any graph tasks on a machine, we must be able to represent the nodes of the graph as vectors that contain information about their position in the graph and their features relative to their locality within the graph. A node's representation vector is usually aggregated from the node's own features and the features of the nodes surrounding it.

There are different ways to aggregate features, transform them, or even choose which of the neighboring nodes contribute to the feature representation of a given node. Let's go over a few methods:

- Traditional node representation methods rely on subgraph summary statistics.

- In other methods, nodes that occur together on short random walks will have similar vector representations.

- Other methods take into account that random walks tend to spread differently on different graph substructures, so a node's representation method adapts to the local substructure that the node belongs in, deciding on an appropriate radius on influence for each node depending on the topology of the subgraph it belongs to.

- Yet other methods produce a multiscale representation by multiplying the feature vector of a node with powers of a random walk matrix.

- Other methods allow for nonlinear aggregation of the feature vectors of a node and its neighbors.

It is also important to determine how large of a neighborhood a node draws information from (influence distribution), that is, to find the range of nodes whose features affect a given node's representation. This is analogous to sensitivity analysis in statistics, but here we need to determine the sensitivity of a node's representation to changes in the features of the nodes surrounding it.

After creating a node representation vector, we feed it into another machine learning model during training, such as a support vector machine model for classification, just like feeding other features of the data into the model. For example, we can learn the feature vector of every user in a social network, then pass these vectors into a classification model along with other features to predict if the user is a fake news spreader or not. However, we do not have to rely on a machine learning model downstream to classify nodes. We can do that directly from the graph structural data where we can predict a node's class depending on its association with other local nodes. The graph can be only partially labeled, and the task is to predict the rest of the labels. Moreover, the node representation step can either be a preprocessing step, or one part of an end-to-end model, such as a graph neural network.

Tasks for Graph Neural Networks

After going through the linear algebra formulation of graphs, applications of graph models, random walks on graphs, and vector node representations that encode node features along with their zones of influence within the graph, we should have a good idea about the kind of tasks that graph neural networks can perform. Let's go through some of these.

Node Classification

The following are examples of node classification tasks:

- In an articles citation network, such as in CiteSeerX or CoRA, where the nodes are academic papers (given as bag-of-words vectors) and the directed edges are the citations between the papers, classify each paper into a specific discipline.

- In the Reddit data set, where the nodes are comment posts (given as word vectors) and undirected edges are between comments posted by the same user, classify each post according to the community it belongs to.

- The protein-protein interaction network data set contains 24 graphs where the nodes are labeled with the gene ontology sets (do not worry about the medical technical names, focus on the math instead. This is the nice thing about mathematical modeling, it works the same way for all kinds of applications from all kinds of fields, which validates it as a potential underlying language of the universe). Usually 20 graphs from the protein-protein interaction network data set are used for training, 2 graphs are used for validation, and the rest for testing, each corresponding to a human tissue. The features associated with the nodes are positional gene sets, motif gene sets, and immunological signatures. Classify each node according to its gene ontology set.

- There are wildlife trade monitoring networks such as Traffic.org, analyzing dynamic wildlife trade trends and using and updating data sets such as the CITES Wildlife Trade Database (*https://oreil.ly/wj2o1*) or the USDA Ag Data Commons (*https://oreil.ly/PKD3D*) (this data set includes more than a million wildlife or wildlife product shipments, representing more than 60 biological classes and more than 3.2 billion live organisms). One classification task on the trade network is to classify each node, representing a trader (a buyer or a seller) as being engaged in illegal trade activity or not. The edges in the network represent a trade transaction between the buyer and seller. Features for the nodes include personal information for the traders, bank account numbers, locations, etc.; and features for the edges would include transaction identification numbers, dates, price tags, the species bought or sold, and so on. If we already have a subset of the traders labeled as illegal traders, our model's task would then be to predict the labels of other nodes in the network based on their connections with other nodes (and their features) in the network.

Node classification examples lend themselves naturally to semi-supervised learning, where only a few nodes in the data set come with their labels, and the task is to label the rest of the nodes. Clean labeled data is what we all should be advocating for our systems to be more accurate, reliable, and transparent.

Graph Classification

Sometimes we want to label a whole graph as opposed to labeling the individual nodes of a graph. For example, in the PROTEINS data set we have a collection of chemical compounds each represented as a graph and labeled as either an enzyme or not an enzyme. For a graph learning model we would input the nodes, edges, their features, the graph structure, and the label for each graph in the data set, thus creating a whole graph representation or embedding, as opposed to a single node representation.

Clustering and Community Detection

Clustering in graphs is an important task that discovers communities or groups in networks, such as terrorist organizations. One way is to create node and graph representations, then feed them into traditional clustering methods such as k-means clustering. Other ways produce node and graph representations that take into account the goal of clustering within their design. These can include encoder-decoder designs and attention mechanisms similar to methods we encountered in previous chapters. Other methods are spectral, which means that they rely on the eigenvalues of the Laplacian matrix of the graph. Note that for nongraph data, principal component analysis is one method that we used for clustering that is also spectral, relying on the singular values of the data table. Computing eigenvalues of anything is always an expensive operation, so the goal becomes finding ways around having to do it. For graphs, we can employ graph theoretic methods such as *max-flow min-cut* (we will see this later in this chapter). Different methods have their own sets of strengths and shortcomings; for example, some employ time-proven graph theoretic results but fail to include the node or edge features because the theory was not developed with any features in mind, let alone a whole bunch of them. The message is to always be honest about what our models account and do not account for.

Graph Generation

Graph generation is very important for drug discovery, materials design, and other applications. Traditionally, graph generation approaches used handcrafted families of random graph models using a simple stochastic generation process. These models are well understood mathematically due to their simple properties. However, for the same reason, they are limited in their ability to capture real-world graphs with more complex dependencies, or even the correct statistical properties, such as heavy tailed distribution for the node degrees that many real-world networks exhibit. More recent approaches, such as generative graph neural networks, integrate graph and node representations with generative models that we went over in the previous chapter. These have a greater capacity to learn structural information from the data and generate complex graphs, such as molecules and compounds.

Influence Maximization

Influence maximization is a subfield of network diffusion, where the goal is to maximize the diffusion of something, such as information or a vaccine, through a network, while only giving the thing to few initial nodes, or seeds. The objective is to locate the few nodes that have an overall maximal influence. Applications include information propagation such as job openings, news, advertisements, and vaccinations. Traditional methods for locating the seeds choose the nodes based on highest degree, closeness, betweenness, and other graph structure properties. Others employ the field of discrete optimization, obtaining good results and proving the existence of approximate optimizers. More recent approaches employ graph neural networks and adversarial networks when there are other objectives competing with the objective of maximizing a node's influence, for example, reaching a specific portion of the population, such as a certain minority group, that is not necessarily strongly connected with the natural hubs of the graph.

Link Prediction

Given two nodes of a graph, what is the probability that there is an edge linking them? Note that proximity in the sense of sharing common neighbors is not necessarily an indicator for a link (or an interaction). In a social network, people tend to run in the same circles, so two people sharing many common friends are likely to interact and are likely connected as well. But in some biological systems, such as in studying protein-protein interactions, the opposite is true: proteins sharing more common neighbors are less likely to interact. Therefore, computing similarity scores based on basic properties such as graph distance, degrees, common neighbors, and so on does not always produce correct results. We need neural networks to learn node and graph embeddings along with classifying whether two nodes are linked or not. One example of such networks is in the paper "Link Prediction with Graph Neural Networks and Knowledge Extraction" (*https://oreil.ly/rVvgg*) (Zhang et al. 2020).

Dynamic Graph Models

Many of the applications we have discussed in this chapter would benefit from including time dependency in our graph models, since they are dynamic in nature. Examples include traffic forecasting, load balancing for distributed networks, simulating all kinds of interacting particle systems, and illegal wildlife trade monitoring. In dynamic graph models, node and edge features are allowed to evolve with time, and some nodes or edges can be added or removed. This modeling captures information such as the latest trade trends in a market, fluctuations, new criminal activity in certain networks, and new routes or connections in transportation systems.

Thinking about how to model dynamic graphs and extract information from them is not new; see, for example, the article "Dynamic Graph Models" (*https://oreil.ly/YHvpp*) (Harray et al. 1997). However, the introduction of deep learning makes knowledge extraction from such systems more straightforward. Current approaches for dynamic graphs integrate graph convolutions to capture spatial dependencies with recurrent neural networks or convolutional neural networks to model temporal dependencies.

The paper "Learning to Simulate Complex Physics with Graph Networks" (*https://oreil.ly/Ui3IF*) (Sanchez-Gonzalez et al. 2020) is a great example with wonderful high-resolution results that employs a dynamic graph neural network to simulate any system of interacting particles on a much larger scale, in terms of both the number of involved particles and the time the system is allowed to (numerically) evolve, than what was done before. The particles, such as sand or water particles, are the nodes of the graph, with attributes such as position, velocity, pressure, external forces, etc., and the edges connect the particles that are allowed to interact with each other. The input to the neural network is a graph, and the output is a graph with the same nodes and edges but with updated attributes of particle positions and properties. The network learns the dynamics, or the update rule at each time step, via message passing. The update rule depends on the system's state at the current time step, and on a parameterized function whose parameters are optimized for some training objective that depends on the specific application, which is the main step in any neural network. The prediction targets for supervised learning are the average acceleration for each particle.

Bayesian Networks

Bayesian networks are graphs that are perfectly equipped to deal with uncertainty, encoding probabilities in a mathematically sound way. Figures 9-10 and 9-11 are examples of two Bayesian networks. In Chapter 8, we presented multiple models that attempt to learn the joint probability distribution, ranging from explicit and tractable all the way to implicit and intractable. A Bayesian network is a very simple explicit model for the joint probability distribution, where we explicitly prespecify how each variable interacts with the others.

In a Bayesian network:

- The nodes are variables that we believe our model should include.
- The edges are directed, pointing from the *parent* node to the *child* node, or from a *higher neuron* to a *lower neuron*, in the sense that we know the probability of the child variable is conditional on observing the parent variable.
- No cycles are allowed in the graph of the network.
- Heavy reliance on Bayes' Rule: if there is an arrow from A to B, then $P(B|A)$ is the forward probability, and $P(A|B)$ is the inverse probability. Think of this as $P(\text{evidence}|\text{hypothesis})$ or $P(\text{symptoms}|\text{disease})$. We can calculate the inverse probability from Bayes' Rule:

$$P(A|B) = \frac{P(B|A)P(A)}{P(B)}$$

- If there is no arrow pointing to a variable (if it has no parents), then all we need is the *prior* probability of that variable, which we compute from the data or from expert knowledge, such as 13% of women in the US develop breast cancer.

- If we happen to obtain more data on one of the variables in the model, or more *evidence*, we update the node corresponding to that variable (the conditional probability), then propagate that information following the connections in the network, updating the conditional probabilities at each node, in two different ways, depending on whether the information is propagating from parent to child, or from child to parent. The update in each direction is very simple: comply with Bayes' Rule.

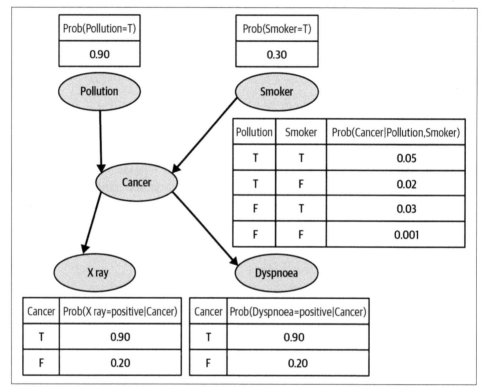

Figure 9-10. A Bayesian network

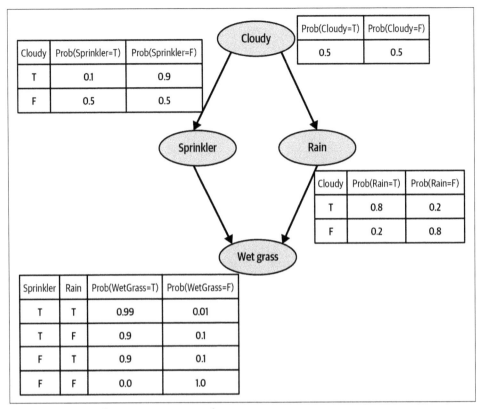

Cloudy	Prob(Sprinkler=T)	Prob(Sprinkler=F)
T	0.1	0.9
F	0.5	0.5

Prob(Cloudy=T)	Prob(Cloudy=F)
0.5	0.5

Cloudy	Prob(Rain=T)	Prob(Rain=F)
T	0.8	0.2
F	0.2	0.8

Sprinkler	Rain	Prob(WetGrass=T)	Prob(WetGrass=F)
T	T	0.99	0.01
T	F	0.9	0.1
F	T	0.9	0.1
F	F	0.0	1.0

Figure 9-11. Another Bayesian network

A Bayesian Network Represents a Compactified Conditional Probability Table

What a Bayesian network represents is a compactified conditional probability table. Usually when we model a real-world scenario, each discrete variable can assume certain discrete values or categories, and each continuous variable can assume any value in a given continuous range. In theory, we can construct a giant conditional probability table that gives the probability of each variable, assuming a certain state of given fixed values of the other variables. In reality, even for a reasonably small number of variables, this is infeasible and expensive both for storage and computations. Moreover, we do not have access to all the information required to construct the table. Bayesian networks get around this hurdle by allowing variables to interact with only a few neighboring variables, so we only have to compute the probability of a variable given the states of those variables directly connected to it in the graph of the network, albeit both forward and backward. If new evidence about any variable in the network arrives, then the graph structure, together with Bayes' Rule, guides us to update the probabilities of all the variables in the network in a systematic, explainable, and transparent way. *Sparsifying* the network this way is a feature of Bayesian networks that has allowed for their success.

In summary, a Bayesian network's graph specifies the joint probability distribution of the model's variables (or the data's features) as a product of local conditional probability distributions, one for each node:

$$P(x_1, x_2, \cdots, x_n) = \Pi_{i=1}^{n} P(x_i | \text{parents of } x_i)$$

Making Predictions Using a Bayesian Network

Once a Bayesian network is set and the conditional probabilities initiated (and continuously updated using more data), we can get quick results simply by searching these conditional probability distribution tables along Bayes' Rule or the product rule, depending on the query. For example, what is the probability that this email is spam given the words it contains, the sender location, the time of the day, the links it includes, the history of interaction between the sender and the recipient, and other values of the spam detection variables? What is the probability that a patient has breast cancer given the mammogram test result, family history, symptoms, blood tests, etc.? The best part here is that we do not need to consume energy executing large programs or employ large clusters of computers to get our results. That is, our phone and tablet batteries will last longer because they don't have to spend much computational power coding and decoding messages; instead, they apply Bayesian networks for using the *turbo decoding* forward error correction algorithm (*https://oreil.ly/nu1LF*).

Bayesian Networks Are Belief Networks, Not Causal Networks

In Bayesian networks, although a pointed arrow from parent variable to child variable is preferably causal, in general it *is not* causal. All it means is that we can model the probability distribution of a child variable given the states of its parent(s), and we can use Bayes' Rule to find the inverse probability: the probability distribution of a parent given the child. This is usually the more difficult direction because it is less intuitive and harder to observe. One way to think about this is that it is easier to calculate the probability distribution of a child's traits given that we know the parents' traits, *P(child|mother,father)*, even before having the child, than inferring the parents' traits given that we know the child's traits *P(father|child)* and *P(mother|child)*. The Bayesian network in this example (Figure 9-12) has three nodes, mother, father, and child, with an edge pointing from the mother to the child, and another edge pointing from the father to the child.

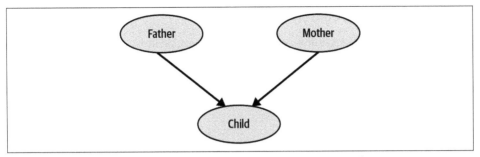

Figure 9-12. The child variable is a collider in this Bayesian network

There is no edge between the mother and the father because there is no reason for their traits to be related. Knowing the mother's traits gives us no information about the father's traits; however, knowing the mother's traits and the child's traits allows us to know slightly more about the father's traits, or the distribution *P(father| mother,child)*. This means that the mother's and father's traits, which were originally independent, are conditionally dependent given knowing the child's traits. Thus, a Bayesian network models the dependencies between variables in a graph structure, providing a map for how the variables *are believed* to relate to each other: their conditional dependencies and independencies. This is why Bayesian networks are also called belief networks.

Keep This in Mind About Bayesian Networks

Let's keep the following in mind about Bayesian networks:

- Bayesian networks have no causal direction and are limited in answering causal questions, or "why" questions, such as: what caused the onset of a certain disease? That said, we will soon learn that we can use a Bayesian network for causal reasoning, and to predict the consequences of intervention. Whether used for causal reasoning or not, how we update a Bayesian network, or how we propagate the *belief*, always works in the same way.

- If some variables have missing data, Bayesian networks can handle it because they are designed to propagate information efficiently from variables with abundant information about them to variables with less information.

Chains, Forks, and Colliders

The building blocks of Bayesian networks (with three or more nodes) are three types of *junctions*: chain, fork, and collider, illustrated in Figure 9-13.

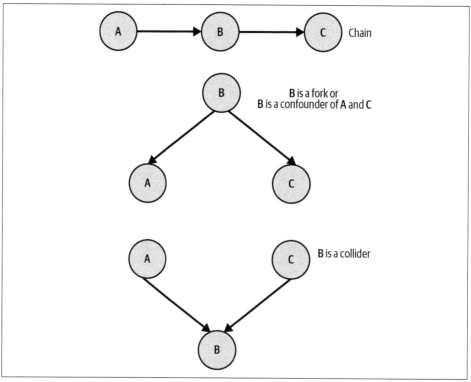

Figure 9-13. The three types of junctions in a Bayesian network

Chain: $A \rightarrow B \rightarrow C$

In this chain, B is a mediator. If we know the value of B, then learning about A does not increase or decrease our belief in C. Thus, A and C are conditionally independent, given that we know the value of the mediator B. Conditional independence allows us, and a machine using a Bayesian network, to focus only on the relevant information.

Fork: $B \rightarrow A$ *and* $B \rightarrow C$

That is, B is a common parent or a confounder of A and C. The data will show that A and C are statistically correlated even though there is no causal relationship between them. We can expose this fake correlation by conditioning on the confounder B.

Collider: $A \rightarrow B$ and $C \rightarrow B$

Colliders are different than chains or forks when we condition on the variable in the middle. We saw this in the example of parents pointing to a child. If A and C are originally independent, conditioning on B makes them dependent! This unexpected and noncausal transfer of information is one characteristic of Bayesian networks conditioning: conditioning on a collider happens to open a dependence path between its parents.

Another thing we need to be careful about: when the outcome and mediator are confounded. Conditioning on the mediator in this case is different than holding it constant.

Given a Data Set, How Do We Set Up a Bayesian Network for the Involved Variables?

The graph structure of a Bayesian network can be decided on manually by us, or learned from the data by algorithms. Algorithms for Bayesian networks are very mature and there are commercial ones. Once the network's structure is set in place, if a new piece of information about a certain variable in the network arrives, it is easy to update the conditional probabilities at each node by following the diagram and propagating the information through the network, updating the belief about each variable in the network. The inventor of Bayesian networks, Judea Pearl (*https:// oreil.ly/J7nSq*), likens this updating process to living organic tissue, and to a biological network of neurons, where if you excite one neuron, the whole network reacts, propagating the information from one neuron to its neighbors.

Finally, we can think of neural networks as Bayesian networks.

Models Learning Patterns from the Data

It is important to note that Bayesian networks, and any other models learning the joint probability distributions of the features of the given data, such as the models we encountered in the previous chapter on generative models, and the deterministic models in earlier chapters, only detect patterns from the data and learn associations, as opposed to learning *what caused* these patterns to begin with. For an AI agent to be truly intelligent and reason like humans, it must ask the questions *how*, *why*, and *what if* about what it sees and what it does, and it must seek answers, like humans do at a very early age. In fact, for humans, there is the age of *why* early in their development: it is the age when children drive their parents crazy asking *why* about everything and anything. An AI agent should have a *causal model*. This concept is so important for attaining general AI. We will visit it in the next section and one more time in Chapter 11 on probability.

Graph Diagrams for Probabilistic Causal Modeling

Since taking our first steps in statistics, all we have heard was: *correlation is not causation*. Then we go on and on about data, more data, and correlations in data. Alright then, we get the message, but what about causation? What is it? How do we quantify it? As humans, do we know exactly what *why* means? We conceptualize cause and effect intuitively, even at eight months old. I actually think we function more on a natural and intuitive level in the world of *why* than in the world of association. *Why* then (see?) do our machines, which we expect that at some point will be able to reason like us, only function at an association and regression level? This is the point that mathematician and philosopher Judea Pearl argues for in the field of AI, and in his wonderful book *The Book of Why* (Basic Books, 2020). My favorite quote from this book: "Noncausal correlation violates our common sense." The idea is that we need to both articulate and quantify which correlations are due to causation and which are due to some other factors.

Pearl builds his mathematical causality models using diagrams (graphs) that are similar to Bayesian network graphs, but endowed with a probabilistic reasoning scheme based on the *do calculus*, or computing probabilities *given the do* operator, as opposed to computing probabilities *given the observe* operator, which is very familiar from noncausal statistical models. The main point is this:

Observing is not the same as doing. In math notation, *Prob*(number of bus riders| color-coded routes) is not the same as *Prob*(number of bus riders|do color-coded routes).

We can infer the first one from the data. Look for the ridership numbers given that the bus routes in a certain city are color-coded. This probability does not tell us the *effect* color-coded routes have on the number of riders. The second probability, *with the do operator*, is different, and the data alone, without a causal diagram, cannot tell us the answer. The difference is when we invoke the *do* operator, then we are deliberately changing the bus routes to color-coded, and we want to assess the effect of that change on bus ridership. If the ridership increases after this deliberate *doing*, and given that we drew the correct graph, including the variables and how they talk to each other, then we can assert that using color-coded bus routes *caused* the increase in ridership. When we *do* instead of *observe*, we manually block all the roads that could naturally lead to colored bus routes, such as a change in leadership or a time of the year that might affect ridership. We cannot block these roads if we were simply observing the data. Moreover, when we use the *do* operator, we intentionally and manually *set the value* of bus routes to color-coded (as opposed to numbered, etc.)

Actually, the bus routes example is not made up. I am currently collaborating with the department of public transportation in Harrisonburg, Virginia, with the goals

of increasing their ridership, improving their efficiency, and optimizing their operations given both limited resources and a drastic drop in the city's population when
the university is not in session. In 2019, the transportation department *deliberately*
changed its routes from a number system to a color-coded system, and at the same
time *deliberately* changed its schedules from adaptive to the university's class schedule
to fixed schedules. Here is what happened: their ridership increased a whopping 18%.
You bet my students who are working on the project this summer (2022) will soon be
drawing causal diagrams and writing probabilities that look like:

$P(ridership|$do color coded routes, do fixed schedules$)$

A machine that is so good at detecting a pattern and acting on it–like a lizard
observing a bug fly around, learning its pattern, then catching it and eating it–has a
very different level of *intellect* than a machine that is able to reason on two higher
levels than mere detection of patterns:

1. *If I deliberately take this action, what will happen to [insert variable here]?*
2. *If I didn't take this action, would [the variable taking a certain value] still have
   happened?* If Harrisonburg did not move to color-coded routes and to fixed
   schedules, would the ridership still have increased? What if only one of these
   variables changed instead of both?

The data alone cannot answer these questions. In fact, carefully constructed causal
diagrams help us tell apart the times when we can use the data alone to answer these
questions, and when we cannot answer them *irrespective of how much more data we
collect*. Until our machines are endowed with graphs representing causal reasoning,
our machines have the same level of intellect as lizards. Amazingly, humans do all
these computations instantaneously, albeit arriving at the wrong conclusions many
times and arguing with each other about causes and effects for decades. We still need
math and graphs to settle matters. In fact, the graph guides us and tells us which data
we must look for and collect, which variables to condition on, and which variables to
apply the *do* operator on. This intentional design and reasoning is very different than
the culture of amassing big volumes of data, or aimlessly conditioning on all kinds of
variables.

Now that we know this, we can draw diagrams and design models that can help us
settle all kinds of causal questions. Did I heal because of the doctor's treatment or
did I heal because time has passed and life has calmed down? We would still need
to collect and organize data, but this process will now be intentional and guided. A
machine endowed with these ways of reasoning—a causal diagram model, along with
the (very short) list of valid operations that go with the causal diagram model—will
be able to answer queries on all three levels of causation:

- Are variables A and B correlated? Are ridership and labeling of bus routes correlated?
- If I set variable A to a specific value, how would variable B change? If I deliberately set color-coded routes, would ridership increase?
- If variable A did not take a certain value, would variable B have changed? If I did not change to color-coded bus routes, would ridership still have increased?

We still need to learn how to deal with probability expressions that involve the *do* operator. We have established that seeing is not the same as doing: *seeing* is in the data, while *doing* is deliberately running an experiment to assess the causal effect of a certain variable on another. It more costly than just counting proportions seen in the data. Pearl establishes three rules for manipulating probability expressions that involve the *do* operator. These help us move from expressions with *doing* to others with only *seeing*, where we can get the answers from the data. These rules are valuable because they enable us to quantify causal effects by *seeing*, bypassing *doing*. We go over these in Chapter 11 on probability.

A Brief History of Graph Theory

We cannot leave this chapter without a good overview of graph theory and the current state of the field. This area is built on such simple foundations, yet it is beautiful, stimulating, and has far-reaching applications that made me reassess my whole mathematical career path and try to convert urgently.

The vocabulary of graph theory includes graphs, nodes, edges, degrees, connectivity, trees, spanning trees, circuits, fundamental circuits, vector space of a graph, rank and nullity (like in linear algebra), duality, path, walk, Euler line, Hamiltonian circuit, cut, network flow, traversing, coloring, enumerating, links, and vulnerability.

The timeline of the development of graph theory is enlightening, with its roots in transportation systems, maps and geography, electric circuits, and molecular structures in chemistry:

- In 1736, Euler published the first paper in graph theory, solving the Königsberg bridge problem (*https://oreil.ly/FHXyC*). Then nothing happened in the field for more than a hundred years.
- In 1847, Kirchhoff developed the theory of trees while working on electrical networks.
- Shortly after, in the 1850s, Cayley discovered trees as he was trying to enumerate the isomers of saturated hydrocarbons C_nH_{2n+2}. Arthur Cayley (1821–1895) is one of the founding fathers of graph theory. We find his name anywhere there is graph data. More recently, CayleyNet (*https://oreil.ly/uKQag*) uses complex

rational functions called Cayley polynomials for a spectral domain approach for deep learning on graph data.

- During the same time period, in 1850, Sir William Hamilton invented the game that became the basis of Hamiltonian circuits, and sold it in Dublin. We have a wooden, regular polyhedron with 12 faces and 20 corners; each face is a regular pentagon and 3 edges meet at each corner. The 20 corners have the names of 20 cities, such as London, Rome, New York, Mumbai, Delhi, Paris, and so on. We have to find a route along the edges of the polyhedron, passing through each of the 20 cities exactly once (a Hamiltonian circuit). The solution of this specific problem is easy, but until now we've had no necessary and sufficient condition for the existence of such a route in an arbitrary graph.

- Also during the same time period, at a lecture by Möbius (1840s), a letter by De Morgan (1850s), and in a publication by Cayley in the first volume of the Proceedings of the Royal Geographic Society (1879), the most famous problem in graph theory (solved in 1970), the four color theorem, came to life. This has occupied many mathematicians since then, leading to many interesting discoveries. It states that four colors are sufficient for coloring any map on a plane such that the countries with common boundaries have different colors. The interesting thing is that if we give ourselves more space by moving out of the flat plane, for example to the surface of a sphere, then we do have solutions for this conjecture.

- Unfortunately, nothing happened for another 70 years or so, until the 1920s when König wrote the first book on the subject and published it in 1936.

- Things changed with the arrival of computers and their increasing ability to explore large problems of combinatorial nature. This spurred intense activity in both pure and applied graph theory. There are now thousands of papers and dozens of books on the subject, with significant contributors such as Claude Berge, Oystein Ore, Paul Erdös, William Tutte, and Frank Harary.

Main Considerations in Graph Theory

Let's organize the main topics in graph theory and aim for a bird's-eye view without diving into details:

Spanning Trees and Shortest Spanning Trees

These are of great importance and are used in network routing protocols, shortest path algorithms, and search algorithms. A spanning tree of a graph is a subgraph that is a tree (any two vertices can be connected using one unique path) including all of the vertices of the graph. That is, spanning trees keep the vertices of a graph together. The same graph can have many spanning trees.

Cut Sets and Cut Vertices

We can break any connected graph apart, disconnecting it, by cutting through enough edges, or sometimes by removing enough vertices. If we are able to find these cut sets in a given graph, such as in a communication network, an electrical grid, a transportation network, or others, we can cut all communication means between its disconnected parts. Usually we are interested in the smallest or minimal cut sets, which will accomplish the task of disconnecting the graph by removing the least amount of its edges or vertices. This helps us identify the weakest links in a network. In contrast to spanning trees, cut sets separate the vertices, as opposed to keeping all of them together. Thus, we would rightly expect a close relationship between spanning trees and cut sets. Moreover, if the graph represents a network with a source of some sort, such as fluid, traffic, electricity, or information, and a sink, where each edge allows only a certain amount to flow through it, then there is a close relationship between the maximum flow that can move from the source to the sink and the cut through the edges of the graph that disconnects the source from the sink, with minimal total capacity of the cut edges. This is the *max-flow min-cut theorem*, which states that in a flow network, the maximum amount of flow passing from the source to the sink is equal to the total weight of the edges in a minimal cut. In mathematics, when a maximization problem (max flow) becomes equivalent to a minimization problem (min cut) in a nontrivial way (for example, by not just flipping the sign of the objective function), it signals duality. Indeed, the max-flow min-cut theorem for graphs is a special case of the duality theorem from linear optimization.

Planarity

Is the geometric representation of a graph planar or three-dimensional? That is, can we draw the vertices of the graph and connect its edges, all in one plane, without its edges crossing each other? This is interesting for technological applications such as automatic wiring of complex systems, printed circuits, and large-scale integrated circuits. For nonplanar graphs, we are interested in properties such as the thickness of these graphs and the number of crossings between edges. An equivalent condition for a planar graph is the existence of a *dual graph*, where the relationship between a graph and its dual becomes clear in the context of the vector space of a graph. Linear algebra and graphs come together here, where algebraic and combinatoric representations answer questions about geometric figures and vice versa. For the planarity question, we need to consider only simple, nonseparable graphs whose vertices all have three degrees or more. Moreover, any graph with a number of edges larger than three times the number of its vertices minus six is nonplanar. There are many unsolved problems in this field of study.

Graphs as Vector Spaces

It is important to understand a graph as both a geometric object and an algebraic object, along with the correspondence between the two representations. This is the case for graphs. Every graph corresponds to an e dimensional vector space over the field of integers modulo 2, where e is the number of edges of the graph. So if the graph only has three edges $edge_1, edge_2, edge_3$, then it corresponds to the three-dimensional vector space containing the vectors $(0,0,0),(1,0,0),(0,1,0),(1,1,0),(1,0,1),(0,1,1),(0,0,1),(1,1,1)$. Here $(0,0,0)$ corresponds to the null subgraph containing none of the three edges, $(1,1,1)$ corresponds to the full graph containing all three edges, $(0,1,1)$ corresponds to the subgraph containing only $edge_2$ and $edge_3$, and so on.

Field of Integers Modulo 2

The field of integers modulo 2 only contains the two elements 0 and 1, with the operations + and × both happening modulo 2. These are in fact equivalent to the logical operations *xor* (exclusive or operator) and *and* in Boolean logic. A vector space has to be defined over a field and has to be closed under multiplication of its vectors by *scalars* from that field. In this case, the scalars are only 0 and 1, and the multiplication happens modulo 2. Graphs are therefore nice examples of vector spaces over finite fields, which are different than the usual real or complex numbers. The dimension of the vector space of a graph is the number of edges e of the graph, and the total number of vectors in this vector space is 2^e. We can see here how graph theory is immediately applicable to switching circuits (with on and off switches), digital systems, and signals, since all operate in the field of integers modulo 2.

With this simple correspondence, and backed by the whole field of linear algebra, it is natural to try to understand cut sets, circuits, fundamental circuits, spanning trees, and other important graph substructures, and the relationships among them, in the context of vector subspaces, basis, intersections, orthogonality, and dimensions of these subspaces.

Realizability

We have already used the adjacency matrix and the incidence matrix as matrix representations that completely describe a graph. Other matrices describe important features of the graph, such as the circuit matrix, the cut set matrix, and the path matrix. Then, of course, the relevant studies have to do with how all these relate to each other and interact.

Another very important topic is that of *realizability*: what conditions must a given matrix satisfy so that it is the circuit matrix of some graph?

Coloring and Matching

In many situations we are interested in assigning labels, or colors, to the nodes, edges of a graph, or even regions in a planar graph. The famous graph coloring problem is when the colors we want to assign to each node are such that no neighboring vertices get the same color; moreover, we want to do this using the minimal amount of colors. The smallest number of colors required to color a graph is called its *chromatic number*. Related to coloring are topics such as node partitioning, covering, and the chromatic polynomial.

A matching is a set of edges where no two are adjacent. A maximal matching is a maximal set of edges where no two are adjacent. Matching in a general graph and in a bipartite graph have many applications, such as matching a minimal set of classes to satisfy graduation requirements, or matching job assignments to employee preferences (this ends up being a max-flow min-cut problem). We can use random walk based algorithms to find perfect matchings on large bipartite graphs.

Enumeration

Cayley in 1857 was interested in counting the number of isomers of saturated hydro-carbon C_nH_{2n+2}, which led him to count the number of different trees with n nodes, and to his contributions to graph theory. There are many types of graphs to be enumerated, and many have been the topics of their own research papers. Examples include enumerating all rooted trees, simple graphs, simple digraphs, and others possessing specific properties. Enumeration is a huge area in graph theory. One important enumeration technique is Pólya's counting theorem, where one needs to find an appropriate permutation group and then obtain its cycle index, which is nontrivial.

Algorithms and Computational Aspects of Graphs

Algorithms and computer implementations are of tremendous value for anyone working with graph modeling. Algorithms exist for traditional graph theoretical tasks such as:

- Find out if graph is separable.
- Find out if a graph is connected.
- Find out the components of a graph.
- Find the spanning trees of a graph.
- Find a set of fundamental circuits.

- Find cut sets.
- Find the shortest path from a given node to another.
- Test whether the graph is planar.
- Build a graph with specific properties.

Graph neural networks nowadays come with their own open source packages. As always, for an algorithm to be of any practical use it must be efficient. Its running time must not increase factorially or even exponentially with the number of nodes of the graph. It should be polynomial time, proportional to n^k, where k is preferably a low number.

For anyone wishing to enter the field of graph modeling, it is of great use to familiarize yourself with both the theory and the computational aspects of graphs.

Summary and Looking Ahead

This chapter was a summary of various aspects of graphical modeling, with emphasis on examples, applications, and building intuition. There are many references for readers aiming to dive deeper. The main message is not to get lost in the weeds (and they are very thick) without an aim or an understanding of the big picture, the current state of the field, and how it relates to AI.

We also introduced random walks on graphs, Bayesian networks, and probabilistic causal models, which shifted our brain even more in the direction of probabilistic thinking, the main topic of Chapter 11. It was my intention all along to go over all kinds of uses for probability in AI before going into a math chapter on probability (Chapter 11).

We leave this chapter with this very nice read: "Relational Inductive Biases, Deep Learning, and Graph Networks" (*https://oreil.ly/kjZ76*) (Battaglia et al. 2018), which makes the case for the deep learning community to adopt graph networks:

> We present a new building block for the AI toolkit with a strong relational inductive bias—the graph network—which generalizes and extends various approaches for neural networks that operate on graphs, and provides a straightforward interface for manipulating structured knowledge and producing structured behaviors. We discuss how graph networks can support relational reasoning and combinatorial generalization, laying the foundation for more sophisticated, interpretable, and flexible patterns of reasoning.

CHAPTER 10

Operations Research

Many scientists owe their greatness not to their skill in solving problems but to their wisdom in choosing them.

—E. Bright Wilson (1908–1992), American chemist

In this chapter, we explore the integration of AI into the field of operations research, leveraging the best of both worlds for more efficient and more informed decision making. Although this introductory statement sounds like an ad, it is precisely what operations research is all about. Advances in machine learning can only help move the field forward.

Operations research is one of the most attractive and stimulating areas of applied mathematics. It is the science of balancing different needs and available resources in the most time- and cost-efficient ways. Many problems in operations research reduce to searching for an optimal point, the holy grail, at which everything functions smoothly and efficiently: no backups, no interruptions to timely services, no waste, balanced costs, and good revenues for everyone involved. A lot of applications never find the holy grail, but many operations research methods allow us to come very close, at least for the simplified models of the complex reality. Constrained mathematical optimization penetrates every industry, every network, and every aspect of our lives. Done properly, we enjoy its benefits; done improperly, we suffer its impact: global and local economies are still experiencing the ramifications of COVID-19, the war on Ukraine, and the interruptions to the supply chain.

Before exploring how machine learning is starting to make its way into operations research, we highlight a few ideas that an interested person must internalize if they want to get involved in the field. Since we only have one chapter to spend on this beautiful topic, we must distill it into its essence:

The no free lunch theorem

This makes us shift our attention into devising and analyzing methods that work best for the special case scenario at hand, as opposed to looking for the most general and most widely applicable methods, like many mathematicians are naturally inclined to do. It essentially asks all these mathematicians to pretty much *chill*, and be satisfied with specialized solutions for specific types of problems.

Complexity analysis of problems and asymptotic analysis of algorithms

Asymptotic analysis tells us that even if the algorithm is ultra innovative and genius, it is useless if its computational requirements skyrocket with the size of the problem. Operations research solutions need to scale to big scenarios with many variables. *Complexity analysis*, on the other hand, addresses the *level of difficulty* of the problems themselves rather than the algorithms devised to tackle them. Combinatorial problems, which are $O(n!)$, are ultra bad: $n!$ is bigger than k^n for n large enough, but an exponential k^n complexity would already be very bad!

Important topics and applications in operations research

These we can find in any good book on operations research. We always need to keep one at hand. Moving from a specific application and business objectives to a mathematical formulation is the skill that cannot be stressed enough in order to thrive this field.

Various types of optimization methods and algorithms

This is the workhorse of operations research solutions and software packages.

Software packages

The wide availability of these, along with the limited number of pages, are my excuse not to elaborate on anything algorithmic or computational in this chapter.

To sum up operations research in six words: *mathematical formulation, optimization, algorithms, software,* and *decisions.*

When reading through this chapter, it is helpful to think of the concepts in the context of how the companies that we interact with in our daily lives manage their operations. Consider, for example, Amazon's logistics. Amazon is the largest ecommerce company in the world. Its share of the US ecommerce market in 2022 is 45%, selling and delivering millions of units of merchandise every day with around $5,000 in sales every second. How does Amazon succeed at doing this? How does the company manage its inventory, warehouses, transportation, and extremely efficient delivery system? How does Amazon formulate its subproblems and integrate them into one big successful operation? Same with transportation logistics, such as Uber. Every day, Uber provides up to 15 million shared rides worldwide, matching available drivers with nearby riders, routing and timing pickups and drop-offs, pricing trips, predicting driver revenues and supply-and-demand patterns, and performing countless analytics.

The complex and highly interconnected optimization problems that allow such massive systems to run relatively smoothly are typical to operations research. Moreover, a lot of the involved problems are NP-hard (in computational complexity, this means they have a nondeterministic polynomial time level of hardness; in English, *very expensive to compute*). Add to that their stochastic nature, and we have interesting math problems that need to be solved.

Overall, the mathematical methods and algorithms of operations research save the world billions of dollars annually. A survey of the largest 500 companies in the United States showed that 85% use linear programming (which is another name for linear optimization, a massive part of operations research, and a reason we spend some decent time on the simplex method and duality in this chapter). Coupled with tools from the AI industry, now is the perfect time to get into the field. The rewards will be on many levels: intellectual, financial, and a meaningful contribution to the greater good of humanity. So in no way should the few selected topics for this chapter dim the significance of other equally important topics in the field.

To dive deeper into operations research (after reading this chapter, of course), the best way is to learn from the best:

- Browse through the winner and finalist projects of the Franz Edelman Award for Achievement in Advanced Analytics, Operations Research, and Management Science (*https://oreil.ly/fuewK*).
- Keep the book *Introduction to Operations Research*, by Frederick Hillier and Gerald Lieberman (McGraw Hill, 2021) close to your heart.

No Free Lunch

The *no free lunch theorem for optimization* states that there is no one particular optimization algorithm that works best for every problem. All algorithms that look for an optimizer of an objective function (cost function, loss function, utility function, likelihood function) have similar performance when averaged over all possible objective functions. So if some algorithm performs better than another on some class of objective functions, there are other objective functions where the other algorithm performs better. There is no superior algorithm that works for all kinds of problems. Therefore, picking an algorithm should be problem (or domain) dependent. Depending on our application area, there is plenty of information on which algorithms practitioners use, their justifications for why these are their chosen ones, their comparisons with others on both high-dimensional and reasonable-dimension problems, and their constant attempts for better performance, based mostly on two criteria: speed (computationally not expensive), and accuracy (gives good answers).

Complexity Analysis and O() Notation

Many times, the problem of *efficiently allocating limited resources* under *various constraints* boils down to devising efficient algorithms for *discrete* optimization. Linear programming, integer programming, combinatorial optimization, and optimization on graph structures (networks) are all intertwined (sometimes these are nothing more than two different names for the same thing) and deal with one objective: finding an optimizer from a *discrete and finite set* of valid options—the *feasible set*. If the feasible set is not discrete to start with, sometimes we can reduce it to a discrete set if we are to take advantage of the wealth of tools developed for this field. Here is the main issue: exhaustive search is usually not tractable. This means that if we list all the available options in the feasible set and evaluate the objective function at each of them, we would spend an ungodly amount of time to find the point(s) that give the optimal answer. No one said that a finite feasible means that it is not enormous. We need specialized algorithms that efficiently rule out large swaths of the search space. Some algorithms pinpoint the exact solution for some problems, while others can only find approximate solutions, which we have no option but to settle for.

Let's now make the following distinctions up front, since this confuses a lot of people:

Complexity analysis is for the problems that we want to solve (routing, traveling salesman, knapsack, etc.)

> The intrinsic complexity of a problem is independent of the algorithms used to tackle it. In fact, it sometimes tells us that we cannot hope for a more efficient algorithm for such kinds of problems, or whether we can do better in other cases. In any case, complexity analysis for problems is a rich science on its own, and the field of operations research provides a wealth of *complex* problems to ponder on. This is where the following terms appear: polynomial problem, nondeterministic polynomial problem, nondeterministic polynomial complete problem, nondeterministic polynomial time hard problem, complement nondeterministic polynomial problem, and complement nondeterministic polynomial complete problem. Those terms are so confusing that someone seriously needs to reconsider their nomenclature. We will not define each here (mainly because the theory is not yet set on the boundaries between these classes of problems), but we will make the following divide: problems that can be solved in polynomial time or less, versus problems for which we cannot find an exact solution in polynomial time, no matter what algorithm is used, in which case we have to settle for approximation algorithms (for example, the traveling salesman problem). Note that sometimes polynomial time problems might not be such a great thing, because, for example, $O(n^{2000})$ is not so fast after all.

Asymptotic analysis is for the algorithms that we design to solve these problems

This is where we attempt to estimate the number of operations that the algorithm requires and quantify it relative to the size of the problem. We usually use the big O notation.

Big O() Notation

A function *g(n)* is *O(f(n))* when $g(n) \leq cf(n)$ for some constant c, and for all $n \geq n_0$.

For example, *2n+1* is *O(n)*, $5n^3 - 7n^2 + 1$ is $O(n^3)$, $n^2 2^n - 55n^{100}$ is $O(n^2 2^n)$, and $15nlog(n) - 5n$ is $O(nlog(n))$.

Do not forget the constant asymptotics case *O(1)*, where the operation count of an algorithm is independent of the size of the problem (awesome thing, because this means that it scales without any worries of enormous problems).

For some algorithms, we can count the exact number of operations; for example, to compute the scalar product (dot product) of two vectors of length *n*, a simple algorithm uses exactly *2n*-1 multiplications and additions, which makes it *O(n)*. For multiplying two matrices each of size $n \times n$, a simple algorithm computing the dot product of each row from the first matrix with each column from the second matrix requires exactly $(2n - 1)n^2$ operations, so this will be $O(n^3)$. Matrix inversion is also usually $O(n^3)$.

For anyone interested in asymptotic analysis for algorithms, it quickly becomes obvious that it is slightly more involved than operation counts, because sometimes we have to make estimates or averages on the size of input (what does *n* stand for?), how to count the operations in an algorithm (by each line of code?), and we cannot ignore the fact that doing computations on large numbers is more consuming in time and memory than doing operations on smaller numbers. Finally, we prefer algorithms that run in polynomial time or less and not in exponential time or more. Let's demonstrate with a very simple example.

Polynomial Algorithm $O(n^k)$ Versus Exponential Algorithm $O(k^n)$

Suppose we are working on a machine that is able to execute 10^7 operations per second (10 million). Let's run it for 1,000 seconds, which is around 16 minutes on 2 different algorithms, one exponential in the size of the problem, say, $O(2^n)$ and the other polynomial $O(n^3)$. The size of the problem is *n* in this case, and refers to a measure of the dimensionality of the input, for example, the number of nodes of a graph, the number of entries of a matrix, the number of features of a data set, or

---

the number of instances. What is the largest size of problem that each algorithm can handle for 16 minutes on this machine?

For the exponential time algorithm, the number of operations it requires is at most (worst-case scenario) $c2^n = 10^7 * 1000$ for some preferably small c. Thus, the size of the problem it can run for 1,000 seconds with 10 million operations per second is $n = 10 \log_2 (10) - \log_2 (c) \approx 33$.

Now contrast this with the polynomial time algorithm, whose worst case is $cn^3 = 10^7 * 1000$, so $n = \frac{1}{3\sqrt{c}} \sqrt[3]{10^{10}} \approx 2100$. This is almost two orders of magnitude larger than the exponential time algorithm.

The conclusion is that given the same hardware and amount of time, polynomial time algorithms $O(n^k)$ allow us to solve much larger problems than exponential time algorithms $O(k^n)$. Combinatorial time algorithms $O(n!)$ are hopeless. Moreover, we *always* want k to be small. The smaller the better.

A person who is used to operating in *exact* realms and not in *approximate* or *asymptotic* realms might be troubled by this discussion, because sometimes, some higher-order algorithms are better for smaller size problems than lower order ones. For example, suppose the exact operation count of an $O(n)$ algorithm is $20n-99$, and that of an $O(n^2)$ is $n^2 + 1$, then it is true that asymptotically (or for large enough n), the $O(n)$ algorithm is better than the $O(n^2)$, but that is not the case if n is smaller than 10, because in this case, $n^2 + 1 < 20n - 99$. This is OK for small enough problems, but never for larger problems.

Two optimization methods that we will soon mention in this chapter are the *simplex* method and the *interior point* method for linear optimization (optimization where both the objective function and the constraints are linear). The interior point method is a polynomial time algorithm and the simplex method is exponential time, so you would expect that everyone would use the cheaper interior point and abandon simplex, but this is not true. The simplex method (and the dual simplex) is still widely used for linear optimization instead of interior point because that exponential time is a worst-case scenario and most applications are not worst-case. Moreover, there are usually trade-offs between algorithms in terms of computational effort per iteration, number of iterations required, the effect of better starting points, whether the algorithm converges or will need extra help near the end, how much computation this extra help would require, and can the algorithm take advantage of parallel processing? For this reason, computer packages for linear optimization have efficient implementations of both the simplex and the interior point methods (and many other algorithms as well). Ultimately, we choose what works best for our use cases.

Optimization: The Heart of Operations Research

We found our way back to optimization. In machine learning, optimization is about minimizing the loss function for models that learn deterministic functions, or maximizing the likelihood function for models that learn probability distributions. We do not want a solution that matches the data exactly, since that would not generalize well to unseen data. Hence the regularization methods, early stopping, and others. In machine learning, we use the available data to learn the model: the deterministic function or the probability distribution that is the source of the data (the data-generating rule or process), then we use this learned function or distribution to make inferences. Optimization is just one step along the way: minimize the loss function, with or without regularization terms. The loss functions that appear in machine learning are usually differentiable and nonlinear, and the optimization is unconstrained. We can add constraints to *guide* the process into some desired realm, depending on the application.

Methods for optimization can either include computing derivatives of the objective function $f\left(\overrightarrow{x}\right)$, such as machine learning's favorite gradient descent (stochastic gradient descent, ADAM, etc.), or not. There are optimization algorithms that are derivative free. These are very useful when the objective function is not differentiable (such as functions with corners) or when the formula of the objective function is not even available. Examples of derivative-free optimization methods include Bayesian search, Cuckoo search, and genetic algorithms.

Optimization, in particular *linear optimization*, has been at the heart of operations research since the Second World War, when methods for linear optimization such as the *simplex method* were developed to aid in military logistics and operations. The goal, as always, is to minimize an objective function (cost, distance, time, etc.) given certain constraints (budget, deadlines, capacity, etc.):

$$min_{constraints}f\left(\overrightarrow{x}\right)$$

To learn optimization for operations research, a typical course usually spends a lot of time on linear optimization, integer optimization, and optimization on networks (graphs), since many real-life logistics and resource allocation problems fit perfectly into these formulations. To become thriving operations researchers, we need to learn:

Linear optimization
> This is where both the objective function and the constraints are linear. Here we learn about the simplex method, duality, Lagrangian relaxation (*https://oreil.ly/QEZXW*), and sensitivity analysis. In linear problems, the boundaries of our world are flat, made of lines, planes, and hyperplanes. This (hyper)polygonal geometry, or *polyhedron*, usually has corner points that are candidates for being

optimizers, so we devise systematic ways to sift through these points and test them for optimality (this is what the simplex method and the dual simplex method do).

Interior point methods

For large-scale linear optimization problems that could be beyond the reach of the simplex method. In short, the simplex method goes around the *boundary* of the feasible search space (the edges of the polyhedron), checks each corner it arrives at for optimality, then moves to another corner at the boundary. The interior point method, on the other hand, goes *through the interior* of the feasible search space, arriving at an optimal corner from the inside of the feasible search space, as opposed to from the boundary.

Integer programming

Optimization where the entries of the optimizing vector must all be integers. Sometimes they can only be zero or one (send the truck to the warehouse in Ohio or not). The knapsack problem (*https://oreil.ly/iT2Wd*) is a very simple prototype example. Here we learn about the branch and bound method for large integer programming problems.

Optimization on networks

We can reformulate many network problems as linear optimization problems where the simplex methods and specialized versions of it work, but it is much better to exploit the network structure and tap into useful results from graph theory, such as the max-flow min-cut theorem, for more efficient algorithms. Many problems on networks boil down to optimizing for one of the following: shortest path on the network (path from one node to another with minimum distance or minimum cost), minimum spanning tree of a network (this is great for optimizing *the design* of networks), maximum flow (from origin to destination or from source to sink), minimum cost flow, multicommodity flow, or traveling salesman (finding the minimum cost [or distance or weight] cyclic route that passes through all the network's nodes only once [Hamiltonian circuit]).

Nonlinear optimization

The objective function and/or the constraints are nonlinear. One recurring example throughout this book is minimizing nonlinear loss functions for machine learning models. These are always nonlinear, and we commonly use gradient descent-type algorithms. For smaller problems we can use Newton type algorithms (second derivatives). In operations research, nonlinearities in the objective function and/or constraints might appear because the cost of shipping goods from one location to another might not be fixed (for example, depends on the distance or on the quantity), or a flow through a network might include losses or gains. A special type of nonlinear optimization that we know a lot about is *quadratic* optimization with linear constraints. This appears in applications such as network equations for electric circuits, and elasticity theory for structures where we consider displacements, stresses, strains, and balance of forces in a structure. Think of how easy it is to find the minimum of the quadratic function $f(x) = sx^2$, where s is a positive constant. This ease translates nicely to higher dimensions, where our objective function looks like $f(\vec{x}) = \vec{x}^t S \vec{x}$, where S is a positive semidefinite matrix, playing the same role for high dimensions as a positive constant for one dimension. Here we even have duality theory that we can take advantage of, similar to the linear optimization case. In optimization, when we lose linearity, we hope our functions are quadratic and our constraints are linear. When we lose that, we hope our functions and/or feasible set are convex. When we lose convexity, we are on our own, hoping our methods don't get stuck at the local minima of high-dimensional landscapes, and somehow find their way to optimal solutions.

Dynamic programming and Markov decision processes

Dynamic programming has to do with projects with multiple stages, where decisions have to be made at each stage, and each decision generates some immediate cost. The decision at each stage has to do with the current state, together with a policy to transition to the next state (choose the next state via a minimization of a deterministic function or a probability). Dynamic programming is all about devising efficient ways, usually recursive methods, to finding the *optimal sequence of interrelated decisions* to fulfill a certain goal. The idea is to avoid having to list all the options for each stage of the decision process, then selecting the best combination of decisions. Such an exhaustive search is extremely expensive for problems with many decision stages, each having many states. Now if the transition policy from one stage to the other is probabilistic rather than deterministic, and if the stages of the decision process continue to recur indefinitely, meaning if the project has an infinite number of stages, then we have a Markov decision process (or Markov chain) on our hands. This is a process that evolves over time in a probabilistic manner. A very special property of a Markov decision process is that the probabilities involving how the process evolves in the future are independent of past events, and depend only on the system's *current state*. Both discrete time and continuous time Markov chains model important systems, such as queuing systems (*https://oreil.ly/zOLmN*), dynamic traffic light control to minimize car waiting time, and flexible call center staffing. The important math objects are the transition matrices, and we solve for the steady state probabilities. They end up having to compute the eigenspace of the transition matrix.

Stochastic algorithms

Dynamic programming with probabilistic transition policy and Markov chain are both examples of stochastic algorithms. So are stochastic gradient descent and random walks on graphs. Any algorithm that involves an element of randomness is stochastic. The mathematics transitions to the language of probabilities, expectations, stationary states, convergence, etc. Another example where stochastic algorithms and analysis of processes appear is *queuing theory*, such as queues at a hospital emergency room or at a ship maintenance yard. This builds on probability distributions of arrival times of customers and service times by the service facility.

Metaheuristics

For many optimization problems, finding the optimal solution might be impractical, so we (who still need to make decisions) resort to *heuristic methods* to find an *answer* (I will not call it a solution), which is not necessarily optimal but is good enough for the problem at hand. Metaheuristics are general solution methods that provide strategy guidelines and general frameworks for developing heuristic methods to fit certain families of problems. We cannot guarantee the optimality of an answer from a heuristic method, but heuristics do speed up the process of finding satisfactory solutions where optimal solutions are too expensive to compute or are out of reach. There is also the topic of *satisfiability*. Since problems in operations research are almost always constrained, the natural question is: are the constraints satisfiable? Meaning, is the feasible set nonempty? Some operations research problems get reformulated as satisfiability problems.

In real-world problems, a big part of the work of operations research departments is formulating their specific use cases and objectives in a way that can fit into one of these optimization frameworks. Here it is important to recognize special structures (such as sparsity in the involved matrices) or substructures that we can exploit for more efficient algorithms. This is crucial for complicated and large-scale systems.

Thinking About Optimization

When we encounter an optimization problem in mathematics:

$$\min_{\vec{x}\, \in \text{ some feasible set}} f\left(\vec{x}\right)$$

where the feasible set is defined by some constraints that the vector $\vec{x}$ must satisfy (or it could be totally unconstrained), we usually pause and brainstorm a little:

- Is $f\left(\vec{x}\right)$ linear?
- Is $f\left(\vec{x}\right)$ convex? Bounded below?
- Is the minimum value finite, or does it $\rightarrow -\infty$?

- Is the feasible set nonempty? Meaning are there $\vec{x}$'s that actually satisfy the constraints?
- Is the feasible set convex?
- Does a minimizer exist?
- Is a minimizer unique, or are there others?
- How do we find the minimizer?
- What is the value of the minimum?
- How much does the minimizer and the value of the minimum change if something changes in our constraints or in our objective function?

Depending on the type of problem at hand, we might be able to answer these questions independently, meaning sometimes we can answer only some of them and not others. This is fine because any information about the optimizer and the value of the optimum is valuable.

Let's explore common types of optimization problems.

Optimization: Finite Dimensions, Unconstrained

This is similar to the optimization that we do in calculus classes, and the optimization we do when training a machine learning model, minimizing the loss function. The objective function $f\left(\vec{x}\right)$ is differentiable:

$$\min_{\vec{x} \in \mathbb{R}^d} f\left(\vec{x}\right)$$

In unconstrained and differentiable optimization, the minimizer $\vec{x}^*$ satisfies $\nabla f\left(\vec{x}\right) = 0$. Moreover, the Hessian (matrix of second derivatives) is positive semidefinite at $\vec{x}^*$. When discussing optimization for machine learning, we settled on stochastic gradient descent and its variants for very high-dimensional problems. For smaller problems, Newton-type (working with second derivatives, not only first ones) methods work as well. For very few problems, such as the mean squared error loss function for linear regression, we can get analytical solutions. Examples where we can get analytical solutions are usually carefully constructed (such as all the examples in our calculus books), and very low dimensional.

Optimization: Finite Dimensions, Constrained Lagrange Multipliers

Let's think of the case where we only have one constraint $g\left(\vec{x}\right) = b$. This explains what we need rather well. The minimization problem looks like:

$$\min_{\substack{g\left(\overrightarrow{x}\right) = b}} f\left(\overrightarrow{x}\right)$$

$$x \in \mathbb{R}^d$$

If $f\left(\overrightarrow{x}\right)$ and $g\left(\overrightarrow{x}\right)$ are differentiable functions from $\mathbb{R}^d \to \mathbb{R}$, we can introduce Lagrange multipliers (a method from 1797) to change our problem into an *unconstrained* one, but in higher dimensions (corresponding to the new Lagrange multipliers that we introduce to the optimization problem). Nothing is free. In this case, we add a multiple of our constraint to the objective function, then minimize, which means look for the points where the gradient is zero. The new objective function for the unconstrained problem is called the Lagrangian, and it is a function of both the decision vector $\overrightarrow{x}$ and the new variable λ, which we multiplied by our constraint, called the *Lagrange multiplier*:

$$\mathscr{L}\left(\overrightarrow{x};\lambda\right) = f\left(\overrightarrow{x}\right) + \lambda\left(b - g\left(\overrightarrow{x}\right)\right)$$

If we have more than one constraint, say five constraints, then we introduce a Lagrange multiplier for each, adding five extra dimensions to our optimization problem to move it from the constrained regime to the unconstrained one.

The optimizer $\left(\overrightarrow{x}^*,\lambda^*\right)$ of the unconstrained problem must satisfy: $\nabla\mathscr{L}\left(\overrightarrow{x};\lambda\right) = 0$. We go about finding it the same way we go about general unconstrained problems (see the previous case). The $\overrightarrow{x}^*$ from $\left(\overrightarrow{x}^*,\lambda^*\right)$ *is the solution of the constrained problem* that we were originally searching for. This means that it is the point on the hypersurface defined by the constraint $g\left(\overrightarrow{x}^*\right) = b$ where the value of f is smallest.

If the problem has a special structure that we can exploit, such as if f is quadratic and the constraint g is linear, or if both f and g are linear, then we have more convenient methods to go about this constrained optimization, both if we decide to use Lagrange multipliers (which introduce duality) and without using Lagrange multipliers. Luckily, optimization problems with simple structures are very well studied, not only because they make the mathematics and computations easier, but also because they appear all the time in science and in real-life applications, which gives some credibility to my theory that nature is simpler than mathematicians think it is. We will revisit Lagrange multipliers for constrained problems in the section on duality, where we focus solely on fully linear problems or quadratic problems with linear constraints.

The meaning of Lagrange multipliers

The nice thing that we should make a permanent mental note of is that the Lagrange multiplier λ is not some worthless auxiliary scalar that helps us change a constrained problem into an unconstrained one. It has a meaning that is very helpful for sensitivity analysis, for finance and operations research applications, and for duality theory (which are all related to each other). Mathematically, by observing the formula of the Lagrangian $\mathscr{L}\left(\vec{x};\lambda\right) = f\left(\vec{x}\right) + \lambda\left(b - g\left(\vec{x}\right)\right)$, λ is the rate of change of the Lagrangian as a function of b, if we were allowed to vary b (the value of the constraint in applications we care about the effect of pushing or relaxing the constraints). That is:

$$\frac{\partial \mathscr{L}\left(\left(\vec{x};\lambda,b\right)\right)}{\partial b} = \frac{\partial f\left(\vec{x}\right) + \lambda\left(b - g\left(\vec{x}\right)\right)}{\partial b} = \frac{\partial f\left(\vec{x}\right)}{\partial b} + \frac{\lambda\left(b - g\left(\vec{x}\right)\right)}{\partial b} = 0 + \lambda = \lambda$$

Moreover, we can interpret the optimal value λ^* corresponding to the optimizer $\vec{x}^*$ as *the marginal effect of b* on the optimal attainable value of the objective function $f\left(\vec{x}^*\right)$. Hence, if $\lambda^* = 2.1$, then increasing b by one unit (pushing the constraint by one unit) will increase the optimal value of f by 2.1 units. This is very valuable information for applications in finance and operations research. Let's see why this is the case. We want to prove that:

$$\frac{df\left(\vec{x}^*(b)\right)}{db} = \lambda^*$$

Note that two things happen at the optimizer $\vec{x}^*(b)$, which we get when we set the gradient of the Lagrangian to zero: $\nabla f\left(\vec{x}^*(b)\right) = \lambda^* \nabla g\left(\vec{x}^*(b)\right)$, and $g\left(\vec{x}^*(b)\right) = b$. Using this information and the chain rule for derivatives (go back to your calculus book and master the chain rule, we use it all the time), we now have:

$$\frac{df\left(\vec{x}^*(b)\right)}{db} = \nabla f\left(\vec{x}^*(b)\right) \cdot \frac{d\vec{x}^*(b)}{db} = \lambda^* \nabla g\left(\vec{x}^*(b)\right) \frac{d\vec{x}^*(b)}{db} = \lambda^* \frac{dg\left(\vec{x}^*(b)\right)}{db} = \lambda^* \frac{db}{db} = \lambda^* \times 1$$
$$= \lambda^*$$

In other words, the Lagrange multiplier λ^* is the rate of change of the optimal cost (value of the objective function) due to the relaxation of the corresponding constraint. In economics, λ^* is called the *marginal cost* with respect to the constraint, or the *shadow price*. When we discuss duality later in this chapter, we use the letters p for the decision variables of the dual problem for this *price* reason.

Optimization: Infinite Dimensions, Calculus of Variations

The field of calculus of variations is an optimization field, but instead of searching for optimizing *points* in finite dimensional spaces, we are searching for optimizing *functions* in infinite dimensional spaces.

A finite dimensional function from $\mathbb{R}^d \to \mathbb{R}$ looks like $f\left(\overrightarrow{x}\right) = f(x_1, x_2, \cdots, x_d)$, and its gradient (which is always important for optimization) is:

$$\nabla f = \begin{pmatrix} \frac{\partial f}{\partial x_1} \\ \frac{\partial f}{\partial x_2} \\ \vdots \\ \frac{\partial f}{\partial x_d} \end{pmatrix}$$

Its directional derivative measures the change of f, or its *variation*, in the direction of some vector $\overrightarrow{n}$:

$$\frac{\partial f\left(\overrightarrow{x}\right)}{\partial \overrightarrow{n}} = \lim_{h \to 0} \frac{f\left(\overrightarrow{x} + h\overrightarrow{n}\right) - f\left(\overrightarrow{x}\right)}{h} = \nabla f . \overrightarrow{n}$$

Now if we allow $\overrightarrow{x}$ to depend on time, then we have:

$$f'\left(\overrightarrow{x}(t)\right) = \frac{df\left(\overrightarrow{x}(t)\right)}{dt} = \nabla f . \frac{d\overrightarrow{x}(t)}{dt} = \nabla f . \overrightarrow{x}'(t)$$

That expression comes in handy when calculating *variations* of *infinite dimensional functionals*. A functional is a function whose input is a function and whose output is a real number. Thus, for an infinite dimensional functional $E(u)$: some function space $\to \mathbb{R}$ maps a function u that lives in *some function space* to a real number.

One example of a functional is the integral of a continuous function on the interval $[0,1]$. Another popular example is this integral:

$$E(u(x)) = \int_0^1 u(x)^2 + u'(x)^2 dx$$

For instance, this functional maps the function x^2 to the number 5/3, which is the value of the integral.

The analog for that finite dimensional time derivative expression (when we allow dependence on time), is now:

$$\frac{dE(u(t))}{dt} = \nabla E.\frac{du(t)}{dt} = \nabla E.u'(t) = \langle \nabla E, u'(t) \rangle_{\text{some function space}}$$

This is very useful. It usually helps us pinpoint the infinite dimensional gradient ∇E, if in our calculations we manage to isolate the quantity multiplied by $u'(t)$. These calculations usually involve integral expressions, and the *product* is usually defined in some *infinite dimensional* sense, which also involves integral expressions. This is why we use the product notation of $\langle \nabla E, u'(t) \rangle$ instead of the usual dot.

We will work through one example to demonstrate a product for functions that live in the infinite dimensional space $L^2(D)$, which contains all the functions $u(x)$ with a finite $\int_D |u(x)|^2 dx$, and the gradient of a functional in the $L^2(D)$ sense.

Analogy between optimizing functions and optimizing functionals

It is good to keep in mind the analogy between finite and infinite dimensional formulations, as everything in math usually ties neatly together. At the same time we must be cautious, as the transition to the infinite is massive, and many finite dimensional properties and methods do not make it through.

In finite dimensions, the *optimizing point* or points satisfy an equation based on setting the gradient of the *objective function* (alternatively in this book: loss function, cost function, or utility function) equal to zero.

In infinite dimensions, the *optimizing function* or functions satisfy a *differential equation* based on setting the gradient of the *objective functional* equal to zero, that is, given that we somehow manage to define the gradient of a *functional*. To find the optimizer, we either have to solve this differential equation, called the *Euler-Lagrange equation*, or follow some optimization scheme on the landscape of the functional. It is *impossible to visualize* the landscape of an infinite dimensional functional, so ironically, we end up visualizing it as only one-dimensional, where the x-axis represents the function space u, and the y-axis represents $E(u)$.

The gradient descent that we use extensively in finite dimensional machine learning is an example of an optimization scheme. The same idea of the gradient descent applies to infinite dimensions: follow the direction of the steepest increase (if maximizing) or decrease (if minimizing).

Of course, we need to define what the gradient means for functionals defined on infinite dimensional spaces. It turns out there are many ways we can define gradients, depending on what spaces the involved functions live in (such as space of all continuous functions, space of all functions that have one continuous derivative, space of functions whose integral of their square is finite, and many others). The meaning of the gradient remains the same: it measures the *variation* of the functional in a certain sense, just like the gradient of a finite dimensional function measures the variation (change) of the function in a certain direction.

 You can skip the rest of this section if you are not interested in differential equations, as this section is not essential to operations research. The only purpose of it is to explore transitioning into infinite dimensions, and to see how we can obtain differential equations when we minimize formulas containing integrals (the functional formulas). For more details on the following examples, check the PDF file on calculus of variations on this book's GitHub page (*https://github.com/halanelson/Essential-Math-For-AI*).

Example 1: Harmonic functions, the Dirichlet energy, and the heat equation

A *harmonic function* is a function whose sum of all of its second derivatives is zero, for example, in two dimensions: $\Delta u = u_{xx} + u_{yy} = 0$. Example functions include $e^x \sin(y)$ and $x^2 - y^2$. In real life, these types of functions appear in electrostatics when modeling electrostatic potentials and charge density distributions, or when modeling rhythmic motions, such as a string undergoing a rhythmic periodic motion, or a frictionless pendulum oscillating indefinitely.

A harmonic function minimizes the Dirichlet energy. Instead of trying to look for functions whose second derivatives add up to zero (there are very well-established ways to do this for $\Delta u = 0$) and satisfying certain boundary conditions, there is another nice way to think about a harmonic function as the minimizer of an energy functional, called the Dirichlet energy functional:

$$E(u(x)) = \int_D \frac{1}{2} |\nabla u(x)|^2 dx$$

Here, $u(x)$ belongs to an appropriate function space that guarantees that the integral is finite, and $u(x)=h(x)$ is specified on the boundary ∂D of the domain D.

When we set the gradient of this energy functional equal to zero, to find its minimizer, we get the (Euler-Lagrange) equation $\Delta u = 0$, $u=h(x)$ on ∂D, which is exactly the differential equation a harmonic function satisfies. Let's see how this happens:

$$E'(u(x)) = \int_D \frac{1}{2}\left(|\nabla u(x)|^2\right)' dx$$

$$= \int_D \nabla u(x) \nabla u'(x) dx$$

$$= -\int_D \Delta u(x) u'(x) dx + \int_{\partial D} \nabla u(x).\overrightarrow{n}\, u'(x) ds$$

$$= -\int_D \Delta u(x) u'(x) dx + \int_{\partial D} \frac{\partial u(x)}{\overrightarrow{\partial n}} u'(x) ds$$

We obtained the third equality using integration by parts, which moves the derivative from one factor in the integral to the other, picking up a negative sign and a boundary term in the process. The boundary term contains the two factors in the original integral, but without the derivative that was moved from one to the other.

The previous expression is true for any $u'(x)$ in our function space; in particular, it will be true for those with $u'(x) = 0$ on the boundary of the domain, which kills the integral term at the boundary and leaves us with:

$$E'(u(x)) = -\int_D \Delta u(x) u'(x) dx = \langle -\Delta u(x), u'(x)\rangle_{L^2(D)}$$

We just defined the *product* in the $L^2(D)$ function space as the integral of the usual product of the functions over the domain.

Now note that, analogous to the finite dimensional case, we have:

$$E'(u(x)) = \langle \nabla_{L^2(D)} E(u(x)), u'(x)\rangle_{L^2(D)}$$

Comparing the last two expressions, we notice that the gradient of the Dirichlet energy $E(u)$ in the $L^2(D)$ sense is $-\Delta u(x)$, and it is zero exactly for harmonic functions.

The heat equation does gradient descent for the Dirichlet energy functional. There is a bit more to this story. In nature, when a system happens to be evolving in time, the first question usually is: what's driving the evolution? A neat and intuitive way to answer this is that the system evolves in a way that *decreases some energy in the most efficient way*. It is usually hard to *discover* the formulas for these energy functionals, but if we happen to discover one, we might get a Ph.D. degree because we did, like what happened with me.

A simple example is the heat equation $u_t = \Delta u$, with $u(x,t)=0$ on ∂D and some initial condition $u(x,0)=g(x)$. This models the diffusion of heat, smoke, atoms on material surfaces, etc. It has time dependence naturally built into it. If we follow the evolution in time of the solution $u(x,t)$ of the heat equation (which might represent temperature, solute concentration, gas in a room, etc.), then we notice that we are sliding on the landscape of the Dirichlet energy functional $E(u) = \frac{1}{2}\int_D |\nabla u(x,t)|^2 dx$, in the steepest descent direction, in the $L^2(D)$ sense, because, as we discussed:

$$u_t = \Delta u = -\nabla_{L^2(D)} E(u)$$

This means that starting initially at some $u(x,0) = g(x)$, the fastest way to arrive to the minimizer of the Dirichlet energy, which is the harmonic function, on the infinite dimensional landscape of the Dirichlet energy, is through solving the heat equation: follow the path of the initial $g(x)$ as it evolves with time.

In this sense, the heat equation gives a viable route to formulate a minimizing scheme for the problem:

$$\min_{u = 0 \text{ on } \partial D} \frac{1}{2}\int_D |\nabla u(x,t)|^2 dx$$

Example 2: The shortest path between two points is along the straight line connecting them

The shortest path between two points in $\mathbb{R}^2$ is along the straight line connecting them. To do this, we minimize the arc length of the curve connecting two points (x_1,y_1) and (x_2,y_2), namely:

$$\min_{y(x_1) = y_1 \text{ and } y(x_2) = y_2} \int_{x_1}^{x_2} \sqrt{1 + y'(x)^2}\, dx$$

Similar to the previous example, this problem is a minimization problem of a functional, which contains integral expressions of some functions and their derivatives.

To solve this, we write the Euler-Lagrange equation by setting the gradient of the functional equal to zero. This leads us to the minimizing function $y(x)=mx+b$, where m and b are, respectively, the slope and the y-intercept of the straight line connecting the two given points. The details of this example are in the PDF file on calculus of variations on this book's GitHub page (*https://github.com/halanelson/Essential-Math-For-AI*).

Other introductory examples to the calculus of variations

Other introductory examples to calculus of variations, which we can solve via minimizing an appropriate energy functional (via a variational principle) include the minimal surface problem and the isoperimetric problem.

Optimization on Networks

I wanted to start with optimization on networks *before* the simplex method for linear optimization because more people are used to thinking in terms of algebraic forms (equations and functions) than in terms of graph or network structures, despite the abundance of network structures in nature and operations research applications. We need to become very comfortable with graph models. Optimization problems on network structures tend to be combinatorial in nature, $O(n!)$, which is no bueno, so we need algorithms that somehow circumvent this and efficiently sift through the search space. (Remember, the order of a problem is usually a worst-case scenario, and in worst cases we suffice ourselves with approximate solutions,)

We discuss typical network problems, which happen to capture a wide variety of real-life applications. The traveling salesman problem is one of the oldest and most famous, so we start there. We live in an age where we have open source software packages and cloud computing resources that include powerful algorithms for solving all the problems mentioned in this chapter, so in this section we focus on understanding the type of the network problem and its applications instead of the algorithms devised to solve them.

Traveling Salesman Problem

This is a famous problem in operations research that fits into many real-world situations. A salesman is required to visit a number of cities during a trip. Given the distances between the cities, in what order should he travel so as to visit every city precisely once and return home, with the objective of keeping the distance traveled at a minimum (Figure 10-1)?

Figure 10-1. Traveling salesman problem (image source (https://xkcd.com/399))

Applications are numerous: a delivery truck leaving a warehouse must deliver packages to every address in the least costly way (measured by time or distance); or we must find the most efficient hole sequence to drill on a printed circuit board when manufacturing electronic chips.

We represent the traveling salesman problem as an optimization problem on a graph: the cities are the nodes, there are edges between each pair of cities (making the graph complete), and each edge has a weight (or attribute or feature) representing the distance between the two cities. This graph has many paths passing through all the cities only once and returning to the one we started with (a *Hamiltonian circuit*), but we want the one with the smallest sum of distances.

Let's think of the complexity of this problem. The total number of different Hamiltonian circuits in a complete graph of n nodes is $(n-1)!/2$. Starting at any node, we have n-1 edges to choose from to pick the next city to visit, then $n-2$ options from the second city, $n-3$ from the third city, and so on. These choices are independent, so we have a total of $(n-1)!$ choices. We must divide by 2 to account for symmetry, in the sense that we can traverse the same Hamiltonian circuit forward or backward and still get the exact same total distance traveled. This counting problem is a circular permutation with symmetry. An exhaustive solution of the traveling salesman would list all $(n-1)!/2$ Hamiltonian circuits, adding up the distance traveled in each, then choosing the one with the shortest distance. Even for a reasonable value of n, it is too expensive; for example, to visit all 50 US state capitals (say we want to minimize total trip cost), we would need to try $(50 - 1)!/2 = 3.04 \times 10^{62}$ options! We do not have an efficient algorithm for problems of arbitrary size. Heuristic methods can provide excellent approximate solutions. Moreover, great algorithms based on an approach called *branch and cut* have solved this problem to optimality for very large numbers of cities.

Minimum Spanning Tree

I put the minimum spanning tree problem right after the traveling salesman because sometimes people confuse the two. This is a good place to clear the confusion. Here, we have a fully connected network with positive weights associated with each edge, which again can represent distance, time, capacity, or cost of connecting infrastructure such as water, electric, or phone lines. Similar to the traveling salesman problem, we want to find the set of edges that includes all the nodes of the graph and minimizes the total weight. The requirement here that is different than the traveling salesman is that we want to make sure we choose the set of edges in a way that provides a path between any two pairs of nodes, meaning we can reach any node in the graph from any other node. In the traveling salesman problem, we need to visit every city only once, then return to the starting city, which means that each node cannot get more than two edges (no such requirement for a spanning tree). The fact that we return to the last city in the traveling salesman problem means that we have an extra circuit closing edge that we do not need for spanning trees. If we remove that last edge of a traveling salesman solution, then we definitely get a spanning tree;

however, there is no guarantee that it is the one with minimal cost. Figure 10-2 shows minimum spanning tree and traveling salesman solutions of the same graph.

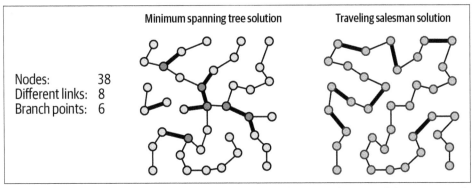

Figure 10-2. Minimum spanning tree and traveling salesman solutions of the same graph

Note that for any network, if we have n nodes then we only need n-1 edges so that we have a path between every two nodes, so we should never use more than n-1 edges for a minimal spanning tree because that would increase our cost. We need to choose the set of edges that minimizes the cost.

We have already mentioned some applications, such as designing telecommunication networks, routing and transportation networks, electric networks, and infrastructure networks (pipelines). These networks are expensive to develop, and designing them optimally saves millions of dollars.

Shortest Path

The simplest version of the shortest path problem is that we have two nodes on a graph and we want to connect them with a set of edges so that the total sum of the edge weights (distance, time) is minimal. This is different than the traveling salesman and the minimal spanning tree problems because we don't care about covering all the nodes of the graph. All we care about is getting ourselves from the origin to the destination in the least costly way.

One obvious application is travel from one destination to another with minimal distance, cost, time, etc. Other applications that are not immediately obvious but nevertheless are ultra-important are activity networks. Instead of an origin and a destination, we might have a beginning of a project and an end. Each node represents an activity, and each edge weight represents the cost or the time incurred if activity i is adjacent to activity j (if we have a directed graph, then it would be the cost or time incurred if activity i happens after activity j). The goal is to choose the sequence of activities that minimizes the total cost.

Other versions of the shortest path problem include finding the shortest path from an origin to *all other nodes*, or finding the shortest paths between *all pairs of nodes*.

Many vehicle routing algorithms and network design algorithms include shortest path algorithms as subroutines.

We can also formulate the shortest path problem as a linear optimization problem and use the methods available for linear optimization.

Max-Flow Min-Cut

Here we also have an origin and a destination, each *directed* edge has a capacity of some sort (max number of vehicles allowed on a route, max number of commodities shipped on a route, max amount of material or natural resource, such as oil or water, that a pipeline can handle), and we want to find the set of edges that maximizes the *flow* from the origin to the destination. Note that all edges point away from the origin and point toward the destination.

A very important theorem from graph theory plays a crucial role in determining the optimality (max flow) of a set of edges connecting the origin to the destination:

The *max-flow min-cut* theorem says the maximum flow from the origin to the destination through the directed network is equal to the minimal sum of weights of the edges required to cut any communication between the origin and the destination. That is, we can cut through the network to prevent communication between the origin and the destination in more than one way. The set of edges that cuts communication *and* has the least weight is the minimal cut set. The value of this minimal cut set is equal to the value of the maximum flow possible in the network. This result is pretty intuitive: what's the most that we can send through the edges of the network? This is bounded from above by the capacities of the edges crucial for connecting the origin to the destination.

We can reformulate the max flow problem as a linear optimization problem, and of course, the min cut problem will be its dual, so of course they have the same solution! We will see this soon in this chapter.

Finally, suppose we have more than one origin and more than one destination, similar to a distribution network. Then we can still maximize the flow through the network by solving the exact same problem, except now we add a fictional *super origin* pointing to all the real origins, and another fictional *super destination* that all the real destinations point toward, with infinite capacities, then do business as usual, solving for the max flow on this new graph with two new fictional *super nodes*.

Max-Flow Min-Cost

This is similar to the max flow problem, except that now we have a cost associated with sending a flow through each edge proportional to the number of units of flow. The goal is obviously to minimize the cost while satisfying the supply from all the origins to all the destinations. We can formulate this problem as a linear optimization problem and solve it using a simplex method optimized for networks. Applications are ubiquitous and so important: all kinds of distribution networks, with supply nodes, trans-shipment nodes, demand nodes, supply chains (of goods, blood, nuclear materials, food), solid waste management networks, coordinating the types of products to produce or spend resources to satisfy the market, cash flow management, and assignment problems, such as assigning employees to tasks, time slots to tasks, or job applicants to available jobs.

The Assignment Problem

The assignment problem is also called the matching problem. The number of assignees should be the same as the number of tasks, each can be assigned only one task, and each task can be performed by only one assignee. There is a cost to assigning task i to assignee j. The objective is to choose the matching between tasks and assignees that minimizes the total cost. The graph of such a problem is a special type called a bipartite graph. Such a graph can be divided into two parts, where all the edges go from one node in the first part to one node in the second part. An assignment problem where all the weights are the same is a max flow problem on a bipartite graph. All we have to do is assign a fictional super origin and another fictional super destination and solve the problem the same way we solve the max flow problem in the upcoming section on linear optimization and duality. There are many efficient algorithms for these problems.

The Critical Path Method for Project Design

The critical path method (CPM) is an optimization method on a network representing all the involved activities in a project, the total budget, the total time constraint, which ones need to happen before others, how much time and cost each activity incurs, and which activities can happen simultaneously. Think, for example, of a house construction project from start to finish. The critical path method for time and cost trade-offs is a great tool to aid in designing a project that incorporates trade-offs between time and cost, and make sure the project meets its deadlines at a minimal total cost. Similar to the critical path method is the Program Evaluation Review Technique (PERT), a project management planning tool that computes the amount of time it will take to complete a project. Both methods provide three timelines: a shortest possible timeline, a longest possible timeline, and a most probable timeline.

The n-Queens Problem

Before moving on to linear optimization, the simplex method, and duality, we make a tiny detour and mention an interesting combinatorial problem that has puzzled mathematicians for 150 years, mainly because of its utter lack of structure: the n-queens problem, such as the one in Figure 10-3. Michael Simkin has finally (July 2021) answered the 150-year-old chess-based n-queens problem (*https://oreil.ly/ Q3soe*). Here is an edited part of the abstract of his solution paper, titled "The Number of n-Queens Configurations" (*https://oreil.ly/aTPHD*):

> The n-queens problem is to determine the number of ways to place n mutually nonthreatening queens on an $n \times n$ [chess] board. We show that there exists a constant $\alpha = 1.942 \pm 3 \times 10^{-3}$ such that [the number of ways to place the mutually nonthreatening queens on the board is] $(1 \pm o(1)ne^{-\alpha})^n((1 \pm o(1))ne^{-\alpha})^n$. The constant α is characterized as the solution to a convex optimization problem in P([−1/2,1/2]2), the space of Borel probability measures on the square.

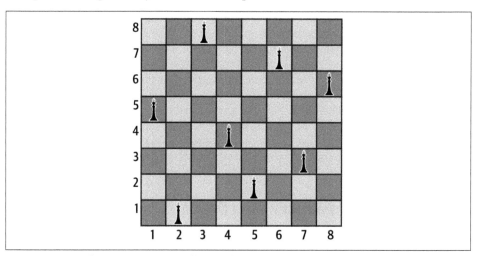

Figure 10-3. Eight queens in mutually nonthreatening positions on an 8 × 8 chessboard

This web page (*https://oreil.ly/t5wW3*) has an easy *backtracking* algorithm for solving the n-queens problem. Note that the solution by Simkin quantifies the total number of viable queen configurations, while algorithms only find one or some of these configurations.

Linear Optimization

Any optimization problem in finite dimensions, whether *linear or nonlinear*, looks like:

$$\min_{\substack{g_1(\vec{x}) \le 0}} f(\vec{x})$$
$$g_2(\vec{x}) \le 0$$
$$\dots$$
$$g_m(\vec{x}) \le 0$$

A point $\vec{x}$ that happens to satisfy all the constraints is a *feasible point*. We have the following cases:

There is only one optimal solution
   Think that the landscape of the objective function has only one lowest point.

There are multiple optimal solutions
   In this case, the set of optimal solutions can be bounded or unbounded.

The optimal value goes to $-\infty$
   The landscape of the objective function goes downhill indefinitely, so no feasible point is optimal.

The feasible set is empty
   We do not care about the objective function and its low values, since there are no points that satisfy all the constraints at the same time. The minimization problem has no solution.

The optimal value is finite but not attained
   There is no optimizer, even when the feasible set is nonempty. For example, $\inf_{x \ge 0} \frac{1}{x}$ is equal to zero but there is no finite x such that $1/x = 0$. This never happens for linear problems.

For an optimization problem to be linear, both the objective function f and all the constraints g must be linear functions. Linear optimization gets the lion's share in operations research, since we can model many operations research problems as a minimization of a linear function with linear constraints that can either be equalities or inequalities.

The General Form and the Standard Form

Linearity is such a great thing, as it opens up all tools of linear algebra (vector and matrix computations). There are two forms of linear optimization problems that people usually work with:

The general form

This is convenient for developing the theory of linear programming. Here, there is no restriction on the signs of the decision variables (the entries of the vector $\vec{x}$):

$$\min_{A\vec{x} \geq \vec{b}} \left(\vec{c} . \vec{x} \right)$$

The feasible set $A\vec{x} \geq \vec{b}$ is a polyhedron (*https://oreil.ly/1CybB*), which we can think of as the intersection of a finite number of half-spaces with flat boundaries. This polyhedron can be bounded or unbounded. We will see examples shortly.

The standard form

This is convenient for computations and developing algorithms, like the simplex and interior point methods. The decision variables must be nonnegative, so we are only searching for optimizers in the *first hyperoctant*, the high-dimensional analog of the first quadrant, where all the coordinates are nonnegative. Moreover, the constraints must always be equalities, not inequalities, so we are on the boundary of the polyhedron, not in the interior. This is a linear optimization problem written in standard form:

$$\min_{\substack{A\vec{x} = \vec{b} \\ \vec{x} \geq \vec{0}}} \left(\vec{c} . \vec{x} \right)$$

There is an easy way for us to intuitively understand a linear problem in standard form: synthesize the vector $\vec{b}$ from the columns of A in a way that minimizes the cost $\vec{c} . \vec{x}$.

We can easily go back and forth between the standard form and general form of a linear optimization problem. For example, we can introduce surplus and slack variables to convert a general linear optimization problem to standard form, but note that in the process of doing that, we end up with the same problem in different dimensions. When we introduce a variable to change an inequality into an equality, such as introducing s_1 to convert the inequality $x_1 - 3x_2 \geq 4$ to the equality $x_1 - 3x_2 - s_1 = 4$, we increase the dimension (in this example from two to three). That is fine. It is actually one of the nice things about math that we can model an unlimited amount of dimensions even though we only live in a three-dimensional world.

Visualizing a Linear Optimization Problem in Two Dimensions

Let's visualize the following two-dimensional problem, which is neither in general form nor in standard form (but we can easily convert it to either form, which we do not need to do for this simple problem, as we can extract the minimum by inspecting the graph):

$$\min_{x+2y \le 3} (-x-y)$$
$$2x+y \le 3$$
$$x \ge 0$$
$$y \ge 0$$

Figure 10-4 shows the boundaries of all the constraints (straight lines) of this linear optimization problem, along with the resulting feasible set. The optimal value of the objective function $-x - y$ is -2, attained at the point $(1,1)$, which is one of the corners of the feasible set.

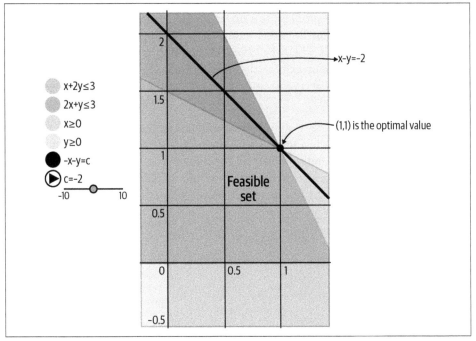

Figure 10-4. Feasible set and the optimal value of –x – y is –2 attained at the corner point (1,1)

If this was an unconstrained problem, then the infimum of $-x - y$ would instead be $-\infty$. Constraints make a huge difference. The fact that the optimal value is at one of the corners of the polygon (two-dimensional polyhedron) is not a coincidence. If we draw the straight line $-x - y = c$ for some c that places part of the line inside the feasible set, then move in the direction of the negative of the gradient vector (recall that this is the direction of fastest descent), the line would move in the direction of the vector $-\nabla (-x - y) = -(-1, -1) = (1, 1)$ (it is definitely a coincidence that the gradient vector has the same coordinates as the optimizing point, as these two are completely unrelated). As long as the line has parts of it inside the feasible set, we can *keep pushing* and making c smaller until we can't push anymore, because if we did, we would exit the feasible set, become infeasible, and lose all our pushing work. This happens exactly when the whole line is outside the feasible set and barely hanging at the point (1,1), which is still in the feasible set.

We found our optimizer, the point that makes the value of $-x - y$ smallest. We will get back to moving through the corners of feasible sets of linear problems soon, because that's where the optimizers are.

Convex to Linear

Even when the objective function is nonlinear, in many cases we may be lucky enough to reformulate the problem as a linear problem, then use linear optimization techniques to either obtain an exact solution or an approximation to the exact solution. One such case is when the objective function is convex. In optimization problems, after linearity, convexity is the next desirable thing, because we wouldn't worry about getting stuck at a local minimum: a local minimum for a convex function is also a global minimum.

We can always approximate a convex (and differentiable) function by a piecewise linear convex function, as in Figure 10-5. Afterward, we can turn the optimization problem with a piecewise linear objective function into one with a linear objective function. This process, however, makes us lose differentiability in the first step (the function stops being smooth) and increase the dimension in the second step. Nothing is free.

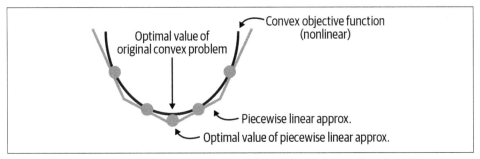

Figure 10-5. Approximating a convex function by piecewise linear functions

A *convex optimization* problem has a convex objective function and a convex feasible set. Convex optimization is a whole field of its own.

Convex Function

A function $f:\mathbb{R}^n \to \mathbb{R}$ is convex if and only if $f(\lambda x + (1 - \lambda)y) \le \lambda f(x) + (1 - \lambda)f(y)$, for all $x,y \in \mathbb{R}^n$ and $0 \le \lambda \le 1$. This means that the segment connecting any two points on the graph of f lies above the graph of f.

Helpful facts about convex functions:

- A convex function cannot have a local minimum that fails to be a global minimum.

- If the functions $f_1, f_2, ..., f_m:\mathbb{R}^n \to \mathbb{R}$ are convex functions, then the function $f(x) = \max_i f_i(x)$ is also convex. f may lose smoothness in this case, so optimization methods would not be able to use derivatives.

- The function $f(x) = \max \{m_1x + d_1, m_2x + d_2, \cdots, m_nx + d_n\}$, or more compactly, $f(x) = \max_{i = 1,2\cdots,n} \{m_ix + d_i\}$, is piecewise linear, as in Figure 10-6. This is a convex function, since each $m_ix + d_i$ is convex (linear functions are convex and concave at the same time), and the maximum of convex functions is also convex.

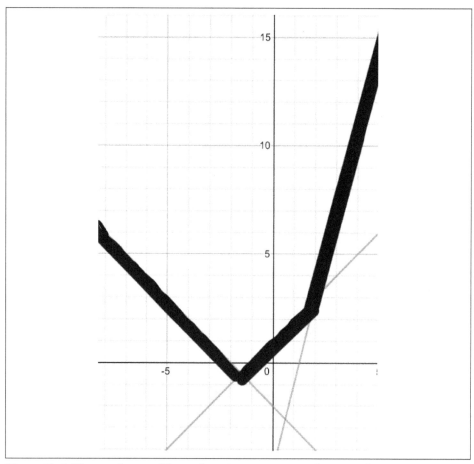

Figure 10-6. The maximum of linear functions is piecewise linear and convex

Now we can reformulate optimization problems with piecewise linear convex objective functions as linear optimization problems:

$$\min_{Ax \geq b} \left(\max_i m_i.x + d_i \right) \leftrightarrow \min_{\substack{Ax \geq b \\ z \geq m_i.x + d_i}} z$$

Note that we increased the dimension when we added a new decision variable z.

For example, the absolute value function $f(x) = |x| = max\{x, -x\}$ is piecewise linear and convex. We can reformulate an optimization problem where the objective function includes absolute values of the decision variables as a linear optimization problem in two ways (here the c_i's in the objective function must be nonnegative, otherwise the objective function might be nonconvex):

$$\min_{Ax \geq b} \sum_{i=1}^{n} c_i |x_i| \leftrightarrow \min_{\substack{Ax \geq b \\ z_i \geq x_i \\ -z_i \geq x_i}} \left(\sum_{i=1}^{n} c_i z_i \right) \leftrightarrow \min_{\substack{Ax^+ - Ax^- \geq b \\ x^+, x^- \geq 0}} \left(\sum_{i=1}^{n} c_i (x_i^+ + x_i^-) \right)$$

The Geometry of Linear Optimization

Let's think of the geometry of a linear optimization problem in standard form, as this is the form that is most convenient for algorithms searching for the minimizer. Geometry is all about the involved shapes, lines, surfaces, points, edges, corners, etc. The problem in standard form:

$$\min_{\substack{A\vec{x} = \vec{b} \\ \vec{x} \geq \vec{0}}} \left(\vec{c} . \vec{x} \right)$$

involves linear algebraic equations. We want to understand the geometric picture associated with these equations, along with the minimization process. Recall that linear is flat, and when flat things intersect with each other, they create hyperplanes, lines, and/or corners.

The linear constraints of the minimization problem define a polyhedron. We are highly interested in the corners of this polyhedron. But how do we know that the polyhedron has corners? What if it is only a half-space? As we mentioned before, if we change from general form to standard form, we jump up in dimension. Moreover, we enforce nonnegativity on the decision variables. Therefore, even if a polyhedron has no corners in the general form, it will always have corners in its higher-dimensional standard form: the standard form polyhedron gets situated in the first hyperoctant, and hence cannot possibly contain full lines. This is good. We have theorems that guarantee that, for a linear optimization problem, either the optimal value is $-\infty$, or there exists a finite optimal value attained at one of the corners of the polyhedron. So we must focus our attention on these corners when searching for the optimizer. Since many polyhedra that are associated with real-world constraints have tens of thousands of corners, we need efficient ways to sift through them.

Intuitively, we can start with the coordinates of one corner of the polyhedron, then work our way to an optimal corner. But how do we find the coordinates of these corners? We use linear algebra methods. This is why it is convenient to express the constraints as a linear system $A\vec{x} = \vec{b}$ (with $\vec{x} \geq 0$). In linear optimization language, a corner is called a *basic feasible solution*. This algebraic name indicates that the corner's coordinates satisfy all the constraints (feasible), and solve some linear equations associated with some basis (extracted from the system $A\vec{x} = \vec{b}$, specifically, from m columns of A).

The interplay of algebra and geometry

Before discussing the simplex method, let's associate the algebraic equations (or inequalities for problems that are in general form) of the constraints with geometric mental images:

Polyhedron
> The constraints as a whole form a polyhedron. Algebraically, a polyhedron is the set of points satisfying the linear system $\vec{x} \in \mathbb{R}^n$, such that $A\vec{x} \geq \vec{b}$ for some $A_{m \times n}$ and $\vec{b} \in \mathbb{R}^m$.

Interior of a half-space
> Here we consider *only one* inequality from the constraints, not the whole system $A\vec{x} \geq \vec{b}$, namely, $\vec{a_i}.\vec{x} > \vec{b_i}$ (the strict inequality part of one inequality constraint). This corresponds to all the points that lie on one side relative to one face of the polyhedron. The inequality is strict, so that we are in the interior of the half-space and not on the boundary.

Hyperplane
> Here we consider only one equality constraint, $\vec{a_i}.\vec{x} = \vec{b_i}$, or only the equality part of an inequality constraint. This is the boundary of the half-space $\vec{a_i}.\vec{x} > \vec{b_i}$, or one face of the polyhedron.

Active constraints
> When we plug the coordinates of a point $\vec{x}^*$ into a constraint $\vec{a_i}.\vec{x} \geq \vec{b_i}$ and we get equality, that is, $\vec{a_i}.\vec{x}^* = \vec{b_i}$, then the constraint is *active* at this point. Geometrically, this places $\vec{x}^*$ at the boundary of the half-space, not in the interior.

Corner of the polyhedron
> Geometrically, the right number of hyperplanes have to meet to form a corner. Algebraically, the right number of constraints is active at a corner point. This is a *basic feasible solution*, and we will go over it while discussing the simplex method.

Adjacent bases to find adjacent corners
> These are two subsets of the columns of the matrix A that share all but one column. We use these to compute the coordinates of corners that are adjacent. In the simplex method, we need to geometrically move from one corner to an

adjacent one, and these adjacent bases help us accomplish that in a systematic algebraic way. We will also go over this while discussing the simplex method.

Degenerate case

To visualize this, suppose that we have two lines intersecting in two dimensions, forming a corner. Now if a third line meets them at exactly the same point, or in other words, if more than two constraints are active at the same point in two dimensions, then we have a degenerate case. In n dimensions, the point $\vec{x}^*$ has more than n hyperplanes passing through it, or more than n active constraints. Algebraically, this is what happens to our optimization algorithm as a consequence of this degeneracy: when we choose another set of linearly independent columns of A to solve for *another* basic feasible solution (corner), we might end up with the same one we got before, leading to cycling in our algorithm!

The Simplex Method

Our goal is to devise an algorithm that finds an optimal solution for a linear optimization problem in standard form:

$$\min_{\substack{A\vec{x} = \vec{b} \\ \vec{x} \geq \vec{0}}} \left(\vec{c}.\vec{x} \right)$$

A is $m \times n$ with m linearly independent rows (so $m \leq n$), $\vec{b}$ is $m \times 1$, and $\vec{c}$ and $\vec{x}$ are $n \times 1$. Without loss of generality, we assume that the m rows of A are linearly independent, which means that there is no redundancy in the constraints of the problem. This also guarantees the existence of at least one set of m linearly independent columns of A (*rank(A)* = m). We need these linearly independent columns, or basis, to initiate our search for the optimizer, at a certain corner of the polyhedron, and to move from one corner to the other using the simplex method.

The main idea of the simplex method

We start at a corner of the polyhedron (also called a basic feasible solution), move to another corner in a direction that is guaranteed to reduce the objective function, or the cost, until we either reach an optimal solution or discover that the problem is unbounded and the optimal cost is $-\infty$ (these we know using certain optimality conditions, and become the termination criteria for our algorithm). There is a chance of cycling in the case of degenerate problems, but we can avoid this by making smart choices (systematic way of choosing) when there are ties in the process.

The simplex method hops around the corners of the polyhedron

The following is a linear optimization problem in three dimensions:

$$\min_{x_1 \leq 2, x_3 \leq 3} \left(-x_1 + 5x_2 - x_3 \right)$$

$$3x_2 + x_3 \leq 6, x_1 + x_2 + x_3 \leq 4$$

$$x_1, x_2, x_3 \geq 0$$

Figure 10-7 shows the polyhedron corresponding to its seven linear constraints.

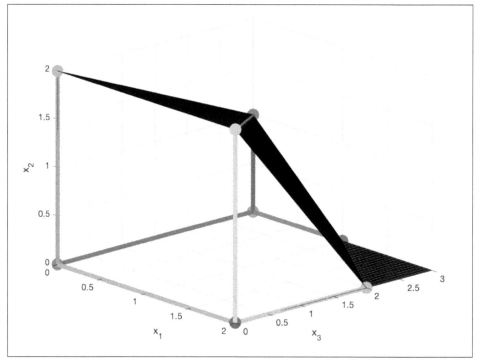

Figure 10-7. The simplex method moves from one corner of the polyhedron to the next until it finds an optimizing corner

Note that the problem is not in standard form. If we convert it to standard form, then we would gain four extra dimensions, corresponding to the four new variables that are required to convert the four inequality constraints to equality constraints. We cannot visualize the seven-dimensional polyhedron, but we do not need to. We can work with the simplex algorithm in seven dimensions and keep track of the important variables in three dimensions: (x_1, x_2, x_3). This way, we can trace the path of the simplex method as it moves from one corner of the polyhedron to the next one, reducing the value of the objective function $-x_1 + 5x_2 - x_3$ at each step, until it arrives at a corner with a minimal value.

 You can skip the rest of this section and go straight to "Transportation and Assignment Problems" on page 386, unless you are interested in the details of the simplex method and its different implementations.

Steps of the simplex method

For a linear optimization problem in standard form, the simplex method progresses like this:

1. Start at a corner of the polyhedron (basic feasible solution $\vec{x^*}$). How do we find the coordinates of this basic feasible solution? Choose m linearly independent columns $A_1,...,A_m$ of A. Put them in a matrix B (basis matrix). Solve $B\vec{x}_b = \vec{b}$ for $\vec{x}_b$. If all the $\vec{x}_b$ are nonnegative, we have found a basic feasible solution $\vec{x^*}$. Place the entries of $\vec{x}_b$ in the corresponding positions in $\vec{x^*}$, and make the rest zero. Alternatively, we can solve $A\vec{x} = \vec{b}$ for $\vec{x}$, where the entries that correspond to the chosen columns are the unknown entries and the rest are zero. Therefore, a basic feasible solution $\vec{x^*}$ has zero nonbasic coordinates and basic coordinates $\vec{x}_b = B^{-1}\vec{b}$.

   For example, if $A = \begin{pmatrix} 1 & 1 & 2 & 1 & 0 & 0 & 0 \\ 0 & 1 & 6 & 0 & 1 & 0 & 0 \\ 1 & 0 & 0 & 0 & 0 & 1 & 0 \\ 0 & 1 & 0 & 0 & 0 & 0 & 1 \end{pmatrix}$, and $\vec{b} = \begin{pmatrix} 8 \\ 12 \\ 4 \\ 6 \end{pmatrix}$, then we can choose A_4,A_5,A_6,A_7 as a set of basic columns, giving $\vec{x} = (0,0,0,8,12,4,6)^t$ as a basic feasible solution (coordinates of one vertex of the polyhedron). We can alternatively choose A_3,A_5,A_6,A_7 as another set of basic columns, giving $\vec{x} = (0,0,4,0,-12,4,6)^t$ as a basic solution but *not* basic feasible, because it has a negative coordinate.

2. Move from $\vec{x^*}$ to another corner $\vec{y^*} = \vec{x^*} + \theta^*\vec{d}$. We must find a direction $\vec{d}$ that keeps us in the polyhedron (feasible), increases only one nonbasic variable x_j from zero to a positive number, and keeps the other nonbasic variables at zero. At the same time, when we move from $\vec{x^*}$ to $\vec{y^*} = \vec{x^*} + \theta^*\vec{d}$, we must reduce the value of the objective function. That is, we want $\vec{c}.\vec{y^*} \le \vec{c}.\vec{x^*}$. When we increase the value of x_j from zero to a positive number, the difference in the objective

function is $\bar{c}_j = c_j - \vec{c_b}.B^{-1}A_j$, so we must choose a coordinate j for which this quantity is negative. To make all of this work, the coordinates of $\vec{d}$ end up being $d_j = 1$ (because we introduced x_j), $d_i = 0$ if $i \neq j$ or if i is nonbasic, and $\vec{d}_b = -B^{-1}A_j$; and the value of θ^* ends up being:

$$\theta^* = \min_{\text{all basic indices for which } d_{B(i)} < 0} \left\{ -\frac{x_{B(i)}}{d_{B(i)}} \right\} := -\frac{x_{B(l)}}{d_{B(l)}};$$

3. Now column $A_{B(l)}$ exits the basis B, and column A_j replaces it.
4. Repeat this process until we either reach a finite optimal solution (when no A_j from all the available columns of A gives us a negative c_j), or discover that the problem is unbounded and the optimal cost is $-\infty$. This happens when we have $\vec{d} \geq \vec{0}$, so $\vec{y} = \vec{x} + \theta\vec{d} \geq \vec{0}$, making it feasible no matter how large θ gets; thus pushing θ to ∞ will keep reducing the cost $\vec{c}.\vec{y} = \vec{c}.\vec{x} + \theta\left(c_j - \vec{c}_B.B^{-1}A_j\right)$ all the way to $-\infty$.

Notes on the simplex method

Some things to remember for the simplex method:

- Step 4 gives the two termination criteria for the simplex algorithm: no negative reduced cost $\bar{c}_j$, or all the coordinates of a feasible set reducing cost direction $\vec{d}$ are nonnegative.

- If the feasible set is nonempty and every basic feasible solution is nondegenerate, then the simplex method is guaranteed to terminate after finitely many iterations, with either a finite optimal solution or $-\infty$ optimal cost.

- Suppose some of the basic feasible solutions are degenerate (some of the *basic* variables are also zero) and we end up at one of them. In this case, there is a chance that when we change the basis by introducing A_j and make $A_{B(l)}$ exit, we stay at the same corner $\vec{y} = \vec{x} + 0\vec{d}$ (this happens when $x_{B(l)} = 0$, so $\theta^* = -\frac{x_{B(l)}}{d_{B(l)}} = 0$). In this case, choose a new A_j until you actually move from $\vec{x}$ to $\vec{y} = \vec{x} + \theta\vec{d}$, $\theta^* > 0$. One really bad thing that could happen here is that after we stop at $\vec{x}$ and keep changing basis (stalling for a little while at $\vec{x}$) until we find one that actually moves us away from $\vec{x}$ to $\vec{y} = \vec{x} + \theta^*\vec{d}$ in a cost-reducing direction, we might end up with the same basis we started the algorithm with! This will lead to cycling, and the algorithm may loop indefinitely. Cycling can be avoided by making smart choices for which columns of A will enter and exit the basis: a systematic way of choosing A_j and later $B(l)$ in θ^* when there are ties in the process.

- When there are ties in the process (we have more than one reducing cost option A_j that gives $\bar{c}_j < 0$, and/or more than one minimizing index $B(l)$ for θ^*), we can devise rules to choose entering A_j and/or exiting $A_{B(l)}$ at a step with such a tie. The rules we decide to follow when there are such ties are called *pivoting* rules.

- A very simple and computationally inexpensive pivoting rule is *Bland's rule*: choose A_j with the smallest index j for which $\bar{c}_j < 0$ to enter the basis, and choose $A_{B(l)}$ with the smallest eligible index $B(l)$ to exit the basis. This smallest subscript pivoting rule helps us avoid cycling. There are other pivoting rules as well.

- If $n - m = 2$ (so A has only two more columns than rows), then the simplex method will not cycle no matter which pivoting rule is used.

- For problems that did not originate from a general form problem, especially those with a large number of variables, it might not always be obvious how to choose the initial basis B and associated basic feasible solution x (because it would not be clear which m columns of A are linearly independent). In this case, we introduce *artificial variables* and solve an *auxiliary linear programming problem* to determine whether the original problem is *infeasible* and hence there is no solution; or, if the problem is feasible, drive the artificial variables out of the basis and obtain an initial basis and an associated basic feasible solution for our original problem. This process is called *Phase I* of the simplex method. The rest of the simplex method is called *Phase II*.

- The big-M method combines Phase I and Phase II of the simplex method. Here we use the simplex method to solve:

$$\min_{\substack{A\vec{x} = \vec{b} \\ \vec{x} \geq 0, \vec{y} \geq 0}} \left(\vec{c}.\vec{x} + M(y_1 + y_2 + \dots + y_m)\right)$$

For a sufficiently large choice of M, if the original problem is feasible and its optimal cost is finite, all of the artificial variables $y_1, y_2, \cdots, y_m$ are eventually driven to zero, which takes us back to our original problem. We can treat M as an undetermined parameter and let the reduced costs be functions of M, and treat M as a very large number when determining whether a reduced cost is negative.

The revised simplex method

The revised simplex method is a computationally less expensive implementation of the simplex method. It provides a cheaper way to compute $\bar{B}^{-1}$ by exploiting the relationship between the old basis B and the new basis $\bar{B}$: they only have one different column (the two vertices involved are adjacent). So we can obtain the new $\bar{B}^{-1}$ from the previous B^{-1}.

The following is a typical iteration of the revised simplex algorithm. This will also help reinforce the simplex method steps from the previous section. Note that for simplicity, we supress the vector notation for the x's, y's, b's, and d's:

1. Start with a B consisting of m basic columns from A and the associated basic feasible solution x with $x_B = B^{-1}b$, and $x_i = 0$ otherwise.

2. Compute B^{-1} (it is B^{-1} not B that appears in the simplex method computations).

3. For j nonbasic, compute the reduced costs $\bar{c}_j = c_j - \overrightarrow{c}_B.B^{-1}A_j$ (this will give you $n - m$ reduced costs).

4. If all the $\bar{c}_j$ are nonnegative, the current basic feasible solution x is optimal, and the algorithm terminates with x as the optimizer and $c.x$ as the optimal cost (there is no A_j that could enter the basis and reduce the cost even more).

5. *Else*, choose a j for which $\bar{c}_j < 0$ (Bland's pivoting rule tells us to choose the smallest such j). Note that this makes A_j enter the basis.

6. Compute a feasible direction d: $d_j = 1$, $d_B = -B^{-1}A_j$, and $d_i = 0$ otherwise.

   - If all the components of d_B are nonnegative, the algorithm terminates with optimal cost $-\infty$ and no optimizer.

   - *Else*, choose the components of $\overrightarrow{d}_B$ that are negative, and let:

$$\theta^* = \min_{\text{all basic indices for which } d_{B(i)} < 0} \left\{ -\frac{x_{B(i)}}{d_{B(i)}} \right\} := -\frac{x_{B(l)}}{d_{B(l)}}$$

   This step computes θ^* and assigns $B(l)$ as the index of the exiting column.

7. Compute the new basic feasible solution $y = x + \theta^*d$ (this new basic feasible solution corresponds to the new basis $\bar{B}$, which has A_j replace $A_{B(l)}$ in B).

8. This step computes the new $\bar{B}^{-1}$ for the next iteration without forming the new basis $\bar{B}$ then inverting it: form the $m \times m + 1$ augmented matrix $(B^{-1}|B^{-1}A_j)$. Perform row operations using the lth row (add to each row a multiple of the lth row) to make the last column the unit vector e_l, which is zero everywhere except for 1 in the lth coordinate. The first m columns of the result is your new $\bar{B}^{-1}$.

   Justification

   Let $u = B^{-1}A_j$ and note that $B^{-1}\bar{B} = (e_1e_2\cdots u\cdots e_m)$, where e_i is the unit column vector with 1 in the lth entry and zero everywhere else, and u is the lth column. The matrix becomes the identity matrix if we perform row operations using the lth row and transform u into e_l. All row operations can be bundled together in an invertible matrix Q applied from the left: $QB^{-1}\bar{B} = I$. Now right-multiply by $\bar{B}^{-1}$ to get $QB^{-1} = \bar{B}^{-1}$. This means that to obtain $\bar{B}^{-1}$, perform on B^{-1} the same row operations that will transform u to e_l.

Calculating $\bar{B}^{-1}$ Using B^{-1} in the Revised Simplex

This method does not start from the original m columns of A and finds the inverse; instead, it does row operations on the previously calculated B^{-1}, which could include roundoff errors. Doing this over many iterations will accumulate these errors, so it will be better to compute $\bar{B}^{-1}$ straight from the columns of A every now and then to avoid error accumulation.

The full tableau implementation of the simplex method

The full tableau implementation of the simplex method has the advantage of only storing and updating one matrix. Here, instead of maintaining and updating B^{-1}, maintain and update the $m \times n + 1$ matrix $x_B \big| B^{-1}A) = (B^{-1}b \big| B^{-1}A$. The column $u = B^{-1}A_j$ corresponding to the variable entering the basis is called the *pivot column*. If the lth basic variable exits the basis, then the lth row is called the *pivot row*. The element belonging to both the pivot row and the pivot column is called the *pivot element*. Now add a *zeroth row* on top of your *tableau* that keeps track of the negative of the current cost $-c.x = -c_B.x_B = c_B.B^{-1}b$ and the reduced costs $c - c_B.B^{-1}A$. So the tableau looks like:

$$\begin{pmatrix} -c_B.B^{-1}b & c - c_B.B^{-1}A \\ B^{-1}b & B^{-1}A \end{pmatrix}$$

or more expanded:

$$\begin{pmatrix} -c_B.x_B & \bar{c}_1 \cdots \bar{c}_n \\ x_{B(1)} & |\cdots| \\ \vdots & B^{-1}A_1 \cdots B^{-1}A_n \\ x_{B(m)} & |\cdots| \end{pmatrix}$$

One helpful thing that is nice to master: extracting B^{-1} and B easily from a given simplex tableau (like in the movie *The Matrix*).

The most efficient implementation of the simplex method is the revised simplex (memory usage is $O(m^2)$, worst-case time for single iteration is $O(mn)$, best-case time for single iteration is $O(m^2)$, while all of the above measures for the full tableau method are $O(mn)$), but everything depends on how sparse the matrices are.

Transportation and Assignment Problems

Transportation and assignment problems are linear optimization problems that we can formulate as min cost network flow problems.

Transportation problem
Allocate products to warehouses, minimize costs.

Assignment problem
Allocate assignees to tasks, number of assignees is equal to the number of tasks, and each assignee performs one task. There is a cost when assignee *i* performs task *j*. The objective is to select an assignment that minimizes the cost. One example is assigning Uber drivers to customers, or machines to tasks.

We exploit the fact that the involved matrices are sparse so we don't have to do a full implementation of the simplex algorithm, only a special *streamlined* version that solves both the assignment and transportation problems. This is related to the *network simplex method* that solves any minimum cost flow problem, including both transportation and assignment problems. The transportation and assignment problems are special cases of the minimum flow problem. The Hungarian method is special for the assignment problem. Since it is specialized for this, it is more efficient. These special-purpose algorithms are included in some linear programming software packages.

Duality, Lagrange Relaxation, Shadow Prices, Max-Min, Min-Max, and All That

We hinted and danced around the idea of *duality* earlier in this chapter when discussing finite dimensional constrained optimization and relaxing the constraints using Lagrange multipliers. Duality is really helpful when our constrained problems are linear, or quadratic with linear constraints. It gives us the option to solve either the optimization problem at hand (the *primal*) or another related problem (its *dual*), whichever happens to be easier or less expensive, and get the same solution. Usually, having more decision variables (dimensions of the problem) is not as strenuous for an algorithm as having more constraints. Since the dual problem flips the roles of decision variables and constraints, then solving it instead of the primal problem makes more sense when we have too many constraints (another way here is using the dual simplex method to solve the primal problem, which we will talk about soon). Another way the dual problem helps is that it sometimes provides shortcuts to the solution of the primal problem. A feasible vector $\vec{x}$ to the primal problem will end up being the optimizer if there happens to be a feasible vector $\vec{p}$ to the dual problem, such that $\vec{c} . \vec{x} = \vec{p} . \vec{b}$.

When learning about duality in the next few paragraphs, think of it in the same way you see Figure 10-8: something is happening in the primal realm, some form of related shadow or echo is happening in the dual realm (some alternate universe), and the two meet at the optimizer, like a gate where the two universes touch.

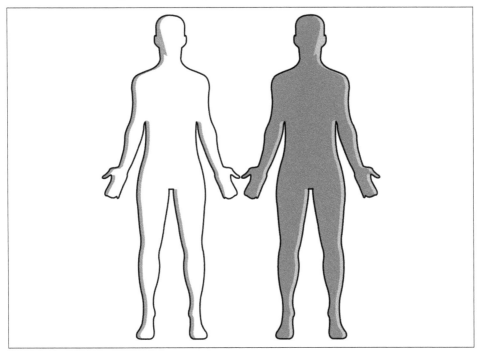

Figure 10-8. Duality, shadow problems, shadow prices

So if we are maximizing in one universe, we are minimizing in the other; if we are doing something with the constraints in one universe, we do something to the decision variables in the other, and vice versa.

Motivation for duality-Lagrange multipliers

For any optimization problem (linear or nonlinear):

$$\min_{\vec{x}\,\in\,\text{feasible set}} f\left(\vec{x}\right)$$

Instead of finding the minimizer $\vec{x}^*$ by setting the gradient equal to zero, look for an upper bound of $f\left(\vec{x}^*\right)$ (easy by plugging any element of the feasible set into $f\left(\vec{x}\right)$) and for a lower bound of $f\left(\vec{x}^*\right)$ (this is a harder inequality and usually requires clever ideas). Now we would have *lower bound* $\leq f\left(\vec{x}^*\right) \leq$ *upper bound*, so we *tighten* these bounds to get closer to the actual solution $f\left(\vec{x}^*\right)$. We tighten the bounds by minimizing the upper bounds (this brings us back to the original minimization problem) and *maximizing the lower bounds* (this establishes the dual problem).

Now for a linear minimization problem in *any form* (standard form, general form, or neither):

$$\min_{\text{linear constraints on}\,\vec{x}} \vec{c}.\vec{x}$$

What is the clever idea that gives us lower bounds for $f\left(\vec{x}\right) = \vec{c}.\vec{x}$? We look for lower bounds for $f\left(\vec{x}\right) = \vec{c}.\vec{x}$ made up of a *linear combination* of the problem constraints. So we multiply each of our constraints by multipliers p_i (Lagrange multipliers), choosing their signs in a way such that the constraint inequality is in the $\geq$ direction. How so? Well, the linear constraints are linear combinations of the entries of $\vec{x}$, the objective function $\vec{c}.\vec{x}$ is also a linear combination of the entries of $\vec{x}$, a linear combination of a linear combination is still a linear combination, so we can totally pick a linear combination of the constraints that we can compare to $\vec{c}.\vec{x}$.

Namely, if we have m linear constraints, we need:

$$p_1 b_1 + p_2 b_2 + \cdots + p_m b_m \leq \vec{c}.\vec{x}$$

The sign of a multiplier p_i would be free if the constraint has an equality. Once we have these lower bounds, we tighten them by maximizing on p_i, which gives us the dual problem.

Finding the dual linear optimization problem from the primal linear optimization problem

It is important to get the sizes of the inputs to a linear optimization problem right. The inputs are: A, which is $m \times n$, $\vec{c}$, which is $n \times 1$, and $\vec{b}$, which is $m \times 1$. The decision variables in the primal problem are in the vector $\vec{x}$, which is $n \times 1$. The decision variables in the dual problem are in the vector $\vec{p}$, which is $m \times 1$.

In general, if A appears in the primal problem, then A^t appears in the dual problem. So in the primal problem, we have the dot product of the *rows* of A and $\vec{x}$. In the dual problem, we have the dot product of the *columns* of A and $\vec{p}$. If the linear optimization problem is in *any form*, it's easy to write its dual following this process:

- If the primal is a minimization, then the dual is a maximization and vice versa.

- The primal cost function is $\vec{c}.\vec{x}$, and the dual cost function is $\vec{p}.\vec{b}$.

- In a *minimization* primal problem, we separate the constraints into two types:

Type one
> Constraints telling us about the sign of the decision variable, for example:

> — $x_3 \geq 0$, then in the dual this will correspond to $A_3.\vec{p} \leq c_3$, where A_3 is the third column of A and c_3 is the third entry of $\vec{c}$.

> — $x_{12} \leq 0$, then in the dual this will correspond to $A_{12}.p \geq c_{12}$, where A_{12} is the 12th column of A and c_{12} is the 12th entry of $\vec{c}$.

> — x_5 is free, meaning has no specified sign. Then in the dual this will correspond to $A_5.p = c_5$, where A_5 is the fifth column of A and c_5 is the fifth entry of $\vec{c}$.

Type two
> Constraints of the form $a_i.x \geq \leq = b_i$, where a_i is the ith row of A. In the dual these will correspond to constraints on the sign of p_i, for example:

> — $a_2.x \geq b_2$, then in the dual this will correspond to $p_2 \geq 0$.

> — $a_7.x \leq b_7$, then in the dual this will correspond to $p_5 \leq 0$.

> — $a_8.x = b_8$, then the sign of p_8 is free.

In particular, if the linear optimization problem is in standard form:

$$\min_{\substack{A\vec{x} = \vec{b} \\ \vec{x} \geq \vec{0}}} \vec{c}.\vec{x},$$

then its dual is:

$$\max_{\substack{\vec{p} \text{ is free} \\ A^T\vec{p} \le \vec{c}}} \vec{p}.\vec{b}.$$

If the linear optimization problem is in general form:

$$\min_{\substack{A\vec{x} \ge \vec{b} \\ \vec{x} \text{ is free}}} \vec{c}.\vec{x}$$

then its dual is:

$$\max_{\substack{\vec{p} \ge 0 \\ A^T\vec{p} = \vec{c}}} \vec{p}.\vec{b}$$

How to solve the dual problem? The simplex method solves the dual problem; however, now you move to basic feasible solutions that *increase* the cost rather than *decrease* the cost.

Derivation for the dual of a linear optimization problem in standard form

There is another way to think about deriving the dual problem, but for this one the linear problem has to be in its standard form. Here's the idea of it: we relax the constraint $A\vec{x} = \vec{b}$ but introduce Lagrange multipliers $\vec{p}$ (pay a penalty $\vec{p}$ when the constraint is violated). So:

$$\min_{\substack{A\vec{x} = \vec{b} \\ \vec{x} \ge \vec{0}}} \vec{c}.\vec{x}$$

becomes:

$$\min_{\vec{x} \ge \vec{0}} \vec{c}.\vec{x} + \vec{p}.\left(\vec{b} - A\vec{x}\right) = g\left(\vec{p}\right).$$

Now prove that $g\!\left(\vec{p}\right)$ is a lower bound for the original $\vec{c}.\vec{x}^*$ (this is the weak duality theorem), then maximize over p. The dual problem appears in the process.

The *strong duality theorem* says that the min of the primal problem and the max of the dual problem are equal. Note that if the primal problem is unbounded, then the dual problem is infeasible; and if the dual problem is unbounded, then the primal problem is infeasible.

Farkas' lemma (*https://oreil.ly/yWF6R*) is at the core of duality theory and has many economical and financial applications.

Dual simplex method

The dual simplex method solves the primal problem (not the dual problem) using duality theory. The main difference between the simplex method and the dual simplex method is that the regular simplex method starts with a basic feasible solution that is not optimal and moves toward optimality, while the dual simplex method starts with an infeasible solution that is optimal and works toward feasibility. The dual simplex method is like a mirror image of the simplex method.

First, note that when we solve the primal problem using the simplex method, we obtain the optimal cost for the dual problem for free (equal to the primal optimal cost), but also, *we can read off the solution (optimizer) to the dual problem from the final tableau for the primal problem.* An optimal dual variable is nonzero only if its associated constraint in the primal is binding. This should be intuitively clear, since the optimal dual variables are the shadow prices (Lagrange multipliers) associated with the constraints. We can interpret these shadow prices as values assigned to the scarce resources (binding constraints), so that the value of these resources equals the value of the primal objective function. The optimal dual variables satisfy the optimality conditions of the simplex method. In the final tableau of the simplex method, the reduced costs of the basic variables must be zero. The optimal dual variables must be the shadow prices associated with an optimal solution.

Another way is to think of the dual simplex method as a disguised simplex method solving the dual problem. However, we do so without explicitly writing the dual problem and applying the simplex method to maximize.

Moreover, the simplex method produces a sequence of primal basic feasible solutions (corners of the polyhedron); as soon as it finds one that is also dual feasible, the method terminates. On the other hand, the dual simplex method produces a sequence of dual basic feasible solutions; as soon as it finds one that is also primal feasible, the method terminates.

Example: Networks, linear optimization, and duality

Consider the network in Figure 10-9. The numbers indicate the edge capacity, which is the maximum amount of flow that each edge can handle. The *max flow* problem is to send the maximum flow from the origin node to the destination node. Intuitively, the maximum flow through the network will be limited by the capacities that the edges can transmit. In fact, this observation underlies a dual problem that takes place: maximizing the flow through the network is equivalent to minimizing the total capacities of the edges such that if we cut, then we cannot get from the origin to the destination. This is the *max-flow min-cut theorem*.

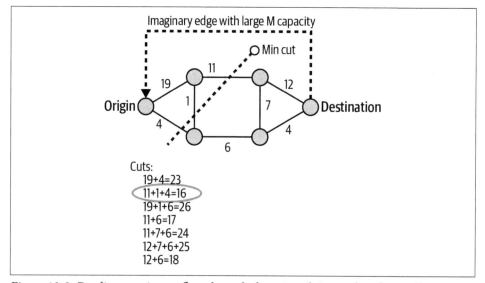

Figure 10-9. Duality: maximum flow through the network is equal to the smallest cut capacity

Figure 10-9 shows the values of all the cuts (the set of edges that if we cut together we would not be able to get from the origin to the destination) through the network, along with the cut of minimal total edge capacity, which is 16. By the max-flow min-cut theorem, the max flow that we can send through the network would be 16: send $y_1 = 12$ units through the edge with capacity 19, and $y_2 = 4$ units through the edge with capacity 4. Of those, $y_3 = 1$ unit will flow through the edge with capacity 1, $y_4 = 11$ units will flow through the edge with capacity 11, and $y_5 = 1 + 4 = 5$ units will flow through the bottom edge with capacity 6. All 12 units will make their way to the destination through the last 2 edges connected to it, with $y_6 = 0$ (no units need to flow through the vertical edge with capacity 7), $y_7 = 11$ units flow through the rightmost edge with capacity 12, and $y_8 = 5$ units flow through the rightmost edge with capacity 6. The solution to the max flow problem is now $(y_1, y_2, y_3, y_4, y_5, y_6, y_7, y_8) = (12, 4, 1, 11, 5, 0, 11, 5)$.

To formulate this network problem as a linear optimization problem (which we just solved graphically using our knowledge of the value of the minimal cut, which is the solution of the dual problem), we need to add one more fictional edge with flow value y_9 that connects the destination to the origin, and assume that the flow that gets to the destination fictionally finds its way back to the origin. In other words, we *close the circuit*, and apply *Kirchhoff's current law*, which says *the sum of currents in a network of conductors meeting at a point is zero*, or the flow into a node is equal to the flow out of it. The linear maximization problem now becomes:

$$\max_{\substack{\vec{A y} = \vec{0}}} \ y_9$$
$$|y_i| \leq M_i$$

where A (Figure 10-10) is the incidence matrix of our network, $\vec{y} = (y_1, y_2, y_3, y_4, y_5, y_6, y_7, y_8, y_9)^t$ is the vector of *signed* maximum flow (we allow the y values to be negative so that the flow in would cancel the flow out) that we can send through each edge and that we need to solve for (we just found its solution without the signs by inspection using the minimal cut intuition), M_i is the max capacity of each edge in the network, and the condition $\vec{A y} = \vec{0}$ guarantees that the flow into a node is equal to the flow out of a node. Of course, in this case the network will have directed edges showing in which direction the optimal flow will go through each edge.

A= incidence matrix	Edges / Nodes	①	②	③	④	⑤	⑥	⑦	⑧	⑨
	1	1	1	0	0	0	0	0	0	1
	2	1	0	1	1	0	0	0	0	0
	3	0	1	1	0	1	0	0	0	0
	4	0	0	0	1	0	1	1	0	0
	5	0	0	0	0	1	1	0	1	0
	6	0	0	0	0	0	0	1	1	1

Figure 10-10. Incidence matrix of the network in Figure 10-9

Now that we have a linear formulation of the max flow problem, we can write its dual easily using the methods we learned in this chapter (the minimum cut problem), and solve either the primal or the dual. Note that all we need for this formulation is the incidence matrix of the network, the edge capacities, and the Kirchhoff's condition that the flow into a node is equal to the flow out of the node.

Example: Two-person zero-sum games, linear optimization, and duality

Another relevant setting where duality and linear optimization are built in is the two-person zero-sum game from game theory. A gain to one player is a loss to the other (hint: duality). To articulate the problem mathematically, we need the *payoff matrix* for all the options in the game for each of $player_1$ and $player_2$. Each player wants to devise a strategy that maximizes their pay given their options (no one said that the payoff matrices for the games have to be fair). We need to solve for the optimal strategy for each player. If we set up the optimization problem for $player_1$, we do not start from scratch to get the optimization problem for the strategy of $player_2$: we just write its dual. The sum expected payoff of the game will be the same for both players, assuming that both of them act rationally and follow their optimal strategies.

Consider, for example, the payoff matrix in Figure 10-11. The game goes like this: $player_1$ chooses a row and $player_2$ chooses a column at the same time. $player_1$ pays the number in the chosen row and column to $player_2$. Therefore, $player_1$ wants to minimize and $player_2$ wants to maximize. The players repeat the game many times.

What is the optimal strategy for each of $player_1$ and $player_2$ and what is the expected payoff of the game?

	y_1	y_2	y_3	
x_1	1	0	4	y_1+4y_3
x_1	3	-1	2	$3y_1-y_2+2y_3$
	x_1+3x_2	$-x_2$	$4x_1+2x_2$	

Figure 10-11. Payoff matrix

To find the optimal strategy, suppose $player_1$ chooses row_1, with probability x_1 and row_2 with probability x_2. Then $x_1 + x_2 = 1$, $0 \le x_1 \le 1$, and $0 \le x_2 \le 1$. $player_1$ rationalizes that if they use an (x_1,x_2) mixed strategy, there would be another row in the payoff matrix corresponding to this new strategy (see Figure 10-11). Now $player_1$ knows that $player_2$ wants to choose the column that maximizes their payoff, so $player_1$ must choose (x_1,x_2) that makes the worst payoff (maximum of the third row) as small as possible. Therefore, $player_1$ must solve the *min-max* problem:

$$\min_{\substack{0 \le x_1 \le 1}} \max\{x_1 + 3x_2, -x_2, 4x_1 + 2x_2\}$$

$$0 \le x_2 \le 1$$

$$x_1 + x_2 = 1$$

Recall that the maximum of linear functions is a convex piecewise linear function. We can easily change such a *min-max(linear functions)* problem to a linear minimization problem:

$$\min_{\substack{z \ge x_1 + 3x_2}} z$$

$$z \ge -x_2$$

$$z \ge 4x_1 + 2x_2$$

$$0 \le x_1 \le 1$$

$$0 \le x_2 \le 1$$

$$x_1 + x_2 = 1$$

Figure 10-12 shows the formulation of the dual of this problem, and Figure 10-13 shows that this is exactly the problem that $player_2$ is trying to solve.

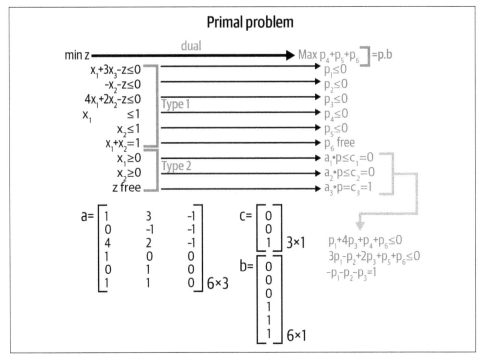

Figure 10-12. The dual of $player_1$'s problem

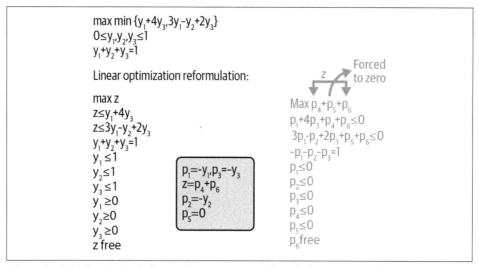

Figure 10-13. *The dual of player$_1$'s min-max problem is the same as player$_2$'s max-min problem*

Note that the constraints $y_1 \le 1$, $y_2 \le 1$, and $y_3 \le 1$ are redundant since all the y's are nonnegative and they add up to 1. Similarly with the constraints $x_1 \le 1$ and $x_2 \le 1$. This happens a lot when formulating linear optimization problems.

Solving either the primal or the dual problem, we find each player's optimal strategy: *player$_1$* must go with the first row $x_1 = 0.25$ of the time, and with the second row $x_2 = 0.75$ of the time, for an expected payoff of 2.5, which means *player$_1$* expects to lose no more than 2.5 with this strategy. *player$_2$* must go with the first column $y_1 = 0.5$ of the time, and with the third column $y_3 = 0.5$ of the time (never with the second column $y_2 = 0$), for an expected payoff of 2.5, which means *player$_2$* expects to gain no less than 2.5 with this strategy.

Quadratic optimization with linear constraints, Lagrangian, min-max theorem, and duality

A *nonlinear* optimization problem that has a nice structure, appears in all kinds of applications, and has a lot to teach us on how things tie together is a quadratic problem with linear constraints:

$$\min_{\overrightarrow{Ax} = \overrightarrow{b}} \frac{1}{2} \overrightarrow{x}^t S \overrightarrow{x}$$

Here, S is a symmetric and positive semidefinite matrix, which means that its eigenvalues are nonnegative. For high dimensions, this plays the role of keeping the objective function convex and bounded below, or shaped like the bowl of the one-dimensional function $f(x) = x^2$.

For example, this is a two-dimensional quadratic optimization problem with one linear constraint:

$$\min_{a_1 x_1 + a_2 x_2 = b} \frac{1}{2}\left(s_1 x_1^2 + s_2 x_2^2\right)$$

Here, $S = \begin{pmatrix} s_1 & 0 \\ 0 & s_2 \end{pmatrix}$ where the s entries are nonnegative, and $A = (a_1 \; a_2)$. Inspecting this problem, we are searching for the point (x_1, x_2) on the straight line $a_1 x_1 + a_2 x_2 = b$ that minimizes the quantity $f\left(\overrightarrow{x}\right) = s_1 x_1^2 + s_2 x_2^2$. The level sets of the objective function $s_1 x_1^2 + s_2 x_2^2 = k$ are concentric ellipses that cover the whole $\mathbb{R}^2$ plane. The winning ellipse (the one with the smallest level set value) is the one that is tangent to the straight line at the winning point (Figure 10-14). At this point, the gradient vector of the ellipse and the gradient vector of the constraint align, which is exactly what Lagrange multiplier formulation gives us: to formulate the Lagrangian, relax the constraint, but pay a penalty equal to the Lagrangian multiplier p times how much we relaxed it in the objective function, minimizing the unconstrained problem.

$$\mathscr{L}\left(\overrightarrow{x}; p\right) = f\left(\overrightarrow{x}\right) + p\left(b - g\left(\overrightarrow{x}\right)\right) = s_1 x_1^2 + s_2 x_2^2 + p(b - a_1 x_1 - a_2 x_2)$$

When we minimize the Lagrangian, we set its gradient equal to zero, and that leads to $\nabla f\left(\overrightarrow{x}\right) = p \nabla g\left(\overrightarrow{x}\right)$. This says that the gradient vector of the objective function is parallel to the gradient vector of the constraint at the optimizing point(s). Since the gradient vector of any function is perpendicular to its level sets, the constraint is in fact tangent to the level set of the objective function at the minimizing point(s). Therefore, to find the optimizing point(s), we look for the level sets of the objective function where it happens to be tangent to the constraint.

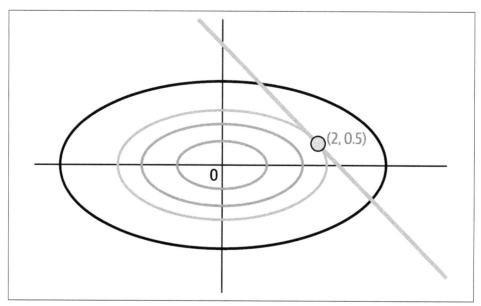

Figure 10-14. The level sets of the quadratic function $x_1^2 + 4x^2$ are concentric ellipses; each of them has a constant value. When we impose the linear constraint $x_1 + x_2 = 2.5$, we get the optimizer (2,0.5) at exactly the point where one of the level sets is tangent to the constraint. The value of the optimal level set is $x_1^2 + 4x^2 = 5$.

Another example that helps us visualize Lagrangian and the upcoming *min-max theorem* is a trivial one-dimensional example:

$$\min_{x=1} x^2$$

The Lagrangian is $\mathscr{L}(x;p) = x^2 - p(1-x)$. We use this toy example whose optimizer is obviously $x = 1$ with minimal value 1 so that we can visualize the Lagrangian. Recall that the Lagrange formulation makes the dimension jump up; in this case we have one constraint, so the dimension increases from one to two, and in our limited three-dimensional world we can only visualize functions of two variables (x and p). Figure 10-15 shows the landscape of our trivial Lagrangian function, which is now representative for Lagrangian formulations for quadratic optimization problems with linear constraints. The main thing to pay attention to in Figure 10-15 is that the optimizers of these kinds of problems $(x^*; p^*)$ happen at *saddle points* of the Lagrangian. These are points where the *second derivative* is positive in one variable and negative in another, so the landscape of the Lagrangian function is convex in one variable (x) and concave in the other (p).

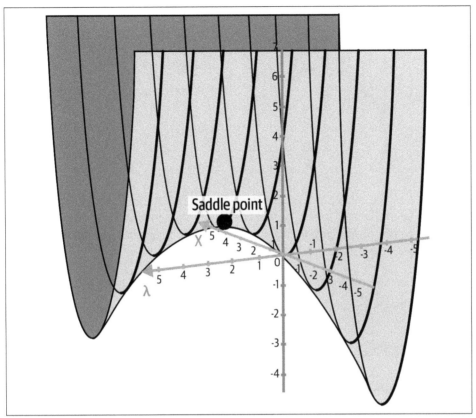

Figure 10-15. The optimizer of the constrained problem happens at the saddle point of the Lagrangian (note that the minimum of the Lagrangian itself is −∞, but that is not what we care about because we care about the optimizer of the quadratic function with linear constraints)

One way to locate the saddle points of the Lagrangian (which give us the optimizers of the corresponding constrained problems) is to solve $\nabla\mathscr{L}(x; p) = \vec{0}$ for x and p, but that is the brute-force way that works for simple problems (like the trivial one at hand) or for small problems. Another way to find these saddle points is to minimize in x then maximize in p (Figure 10-16). Yet another way is to maximize in p then minimize in x. The *min-max theorem* says that these two paths are the same.

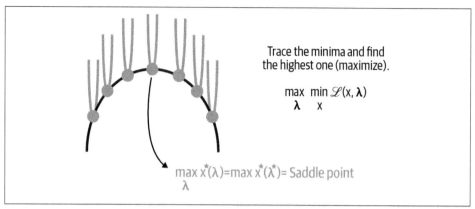

Trace the minima and find
the highest one (maximize).

$$\max_{\lambda} \ \min_{x} \ \mathscr{L}(x, \lambda)$$

$\max_{\lambda} \overset{*}{x}(\lambda) = \max \overset{*}{x}(\overset{*}{\lambda}) =$ Saddle point

Figure 10-16. Minimizing over x, then maximizing over p gives $\max_p x^*(p) = x^*(p^*) =$ the saddle point. This will give the same answer as maximizing over p, then minimizing over x.

Therefore, at the saddle point (x^*, p^*), we have $\nabla \mathscr{L}(x; p) = 0$ (which is the same as $\frac{\partial \mathscr{L}(x; p)}{\partial x} = 0$ and $\frac{\partial \mathscr{L}(x; p)}{\partial p} = 0$), *and*

$$\min_{x} \ \max_{p, \text{hold } x \text{ fixed}} \ \mathscr{L}(x; p) = \max_{p} \ \min_{x, \text{hold } p \text{ fixed}} \ \mathscr{L}(x; p)$$

We have gone full circle to yet again demonstrate that accompanying in a constrained minimization problem in *x* we have another constrained maximization problem in the Lagrange multipliers *p*. The interplay between Lagrange multipliers, duality, and constrained optimization is on full display.

Now that we have gone through the important ideas, let's go back and put them in the context of the higher-dimensional quadratic problem with linear constraints that we started this subsection with:

$$\min_{\substack{\overrightarrow{x} \\ A\overrightarrow{x} = b}} \frac{1}{2} \overrightarrow{x}^t S \overrightarrow{x}$$

where *S* is a symmetric and positive definite matrix. The Lagrange formulation with *relaxed* constraints is:

$$\min \ \mathscr{L}\left(\overrightarrow{x}; \overrightarrow{p}\right) = \ \min \frac{1}{2} \overrightarrow{x}^t S \overrightarrow{x} + \overrightarrow{p} . \left(\overrightarrow{b} - A\overrightarrow{x}\right)$$

Solving this unconstrained problem, whether by setting $\nabla \mathcal{L}\left(\vec{x}; \vec{p}\right) = \vec{0}$, or by minimizing over $\vec{x}$ then maximizing over $\vec{p}$, or by maximizing over $\vec{p}$ then minimizing over $\vec{x}$, we get the same solution $\left(\vec{x}^{*}; \vec{p}^{*}\right)$, which happens at the *saddle point* of our high-dimensional Lagrangian, and gives the optimal value of the objective function (the advantage of this problem with simple structure is that we can solve it by hand):

$$\text{minimum cost } f = \tfrac{1}{2}\vec{b} \cdot \left(AS^{-1}A^{t}\right)^{-1}\vec{b}$$

Moreover, the optimal *shadow prices* are: $\vec{p}^{*} = \dfrac{df}{d\vec{b}} = \left(AS^{-1}A^{t}\right)^{-1}\vec{b}$.

The last thing we need to learn here is the characterization of saddle points in higher dimensions. For the one-dimensional constrained problem, the hallmark was having the second derivative of the Lagrangian (which was a function of x and p) negative in one variable and positive in the other. The high-dimensional analog is this: the eigenvalues of the Hessian matrix (matrix of second derivatives) is negative in one set of variables and positive in the other set of variables, so it is concave in one set of variables and convex in the other. Our discussion applies to optimizing any higher-dimensional objective function that is convex in one set of variables and concave in the other. This is the hallmark that the landscape has saddle points. For the Lagrangian function, the saddle point is exactly where the constrained problem attains its minimum.

Does this apply to linear optimization problems with linear constraints, which are everywhere in operations research? Yes, as long as we have the correct signs for all the coefficients in the problem, such as the max-flow min-cut and the two-person zero-sum game examples we saw in the previous subsections.

Sensitivity

Here we care about the sensitivity of the optimization problem and its solution with respect to changes in its input data. That is, what happens to the optimal solution $\vec{x}^{*}$ and the optimal cost $\vec{c}.\vec{x}^{*}$ if we slightly change $\vec{c}$ or A or $\vec{b}$? Can we obtain the new optimal solution from the old one? Under what conditions can we do that? These are some important cases that sensitivity analysis addresses:

- We have already interpreted the optimal $\vec{p}$ in the dual problem as the vector of marginal prices. This is related to sensitivity analysis: the rate of change of the optimal cost with respect to the constraint value.

- If we add a new decision variable, we check its reduced cost, and if it is negative, we add a new column to the tableau and proceed from there.

- If an entry of $\vec{b}$ or $\vec{c}$ is changed by δ, we obtain an interval of values of δ for which the same basis remains optimal.

- If an entry of A is changed by δ, a similar analysis is possible. However, this case is somewhat complicated if the change affects an entry of a basic column.

In general, if we have a function and we want its sensitivity to variations with respect to one of its inputs, then that is similar to asking about its first derivative with respect to that input (at a certain state), or a discrete first derivative (finite difference) at that state. What makes sensitivity questions more interesting here is the fact that we are dealing with *constrained* problems and checking the effect of small variations in all kinds of inputs to the problem.

Game Theory and Multiagents

Game theory is so important for economics, politics, military operations, multiagent AI, and basically for modeling any environment where there are adversaries or competitors and we have to make decisions or strategize in these conditions. Our optimal strategies are heavily influenced by our adversaries' strategies, whether we know them or are only speculating about them.

The easiest and most well-understood game theory setting is that of *two-person zero-sum games*, which we saw when discussing duality. Here, there are two competing entities, where the loss of one entity is the win of the other, for example, two political campaigns or two competing firms. It has been a challenge to extend the theory to more complex real-life situations with many competitors with varying advantages and disadvantages over each other, varying degrees of cooperation, along with many interrelated strategies. There is still a gap between the situations that the theory can accurately portray and analyze and real-life situations. Progress is happening, and many researchers are on this case, due to the incredible benefits such a complete theory would bring to the world. Imagine being able to view the whole network of adversaries from above, with their movements, connections, possible strategies, and their consequences.

For multiagent environments, game theory models the rational behavior or decision-making process for each involved agent (player, firm, country, military, political campaign, etc.). In this sense, game theory for multiagents is similar to decision theory for a single agent.

The most important concept for noncooperative game theory (where the agents make their decisions independently) is that of the *Nash equilibrium*: a strategy outline for the game where each agent has no incentive to deviate from the outline's prescribed strategy. That is, the agent will be worse off if they deviate from the strategy, of course, assuming everyone is acting rationally.

As we saw in the section on duality, for two-person zero-sum games, we can model them as a min-max problem, and use the *min-max theorem*. We can also model them as a linear optimization problem, where one player is solving the primal problem, and the other is solving the dual problem. This means that we can either set up the optimization problem for the first player or for the second player. Both problems will end up with the same solution. Here we are given the payoff chart for all strategies in the game for both players, and the objective is to find the strategy combination that maximizes the pay (or minimizes the loss) for each player. Intuitively, we can see why duality is built into this problem. The two players are pushing against each other, and the optimal strategy for each player solves both the primal and the dual problems.

We can also use graphs and results from graph theory to analyze two-person games. This is similar to how we can formulate the max flow through a network as a linear optimization problem. Ultimately, many things in math connect neatly together, and one of the most satisfying feelings is when we understand these connections.

For multiagents, certain techniques are available for decision making, which include: voting procedures, auctions for allocating scarce resources, bargaining for reaching agreements, and contract net protocol for task sharing. In terms of mathematical modeling for multiagent games, we will discuss in Chapter 13 (AI and partial differential equations) the *Hamilton-Jacobi-Bellman* partial differential equation. Here, to find the optimal strategy for each player, we have to solve a high-dimensional Hamilton-Jacobi-Bellman type partial differential equation for the game's value function. Before deep learning, these types of high-dimensional partial differential equations were intractable, and one had to make many approximations, or not consider all the participating entities. Recently (2018), a deep learning technique (*https://oreil.ly/jCasX*) has been applied to solve these high-dimensional partial differential equations, once reformulated as a backward stochastic differential equation with terminal conditions (don't worry if you do not know what this means; it is not important for this chapter).

We encountered another two-person adversarial game theoretic setting earlier in this book when discussing generative adversarial networks in Chapter 8 on probabilistic generative models.

Queuing

Queues are everywhere: computing jobs for machines, service queues at a shipyard, queues at the emergency room, airport check-in queues, and queues at the local Starbucks. Well-designed queue systems save different facilities and our entire economy an invaluable amount of time, energy, and money. They enhance our overall well-being.

Mathematical modeling of queues has the objective of determining the appropriate level of service to minimize waiting times. The model might include a priority discipline, which means that there are priority groups, and the order in which the members get serviced depends on their priority groups. It might also include different type of services that happen sequentially or in parallel, or some in sequence and others in parallel (for example, in a ship maintenance facility). Some models include multiple service facilities—a queuing network.

There are thousands of papers on queuing theory. It is important to recognize the basic ingredients of a queuing mathematical model:

- The members of the queue (customers, ships, jobs, patients) arrive at certain inter-arrival times. If the arrival process is random, then the math model must decide on a probability distribution that this inter-arrival time adheres to, either from the data or from mathematical distributions known to model such times. Some models assume constant arrival times. Others assume the exponential distribution (a Markovian process) as it facilitates the mathematical analysis and mimics the real-life process better. Others assume the *Erlang* distribution, which allows different exponential distributions for different time intervals. Others assume even more general distributions. The more general the distribution, the less easy the mathematical analysis. Numerical simulations are our best friend forever.

- The number of servers (parallel and sequential) available: an integer.

- The service times also follow a certain probability distribution that we must decide on. Common distributions are similar to those used for inter-arrival times.

In addition, the mathematical model must also keep track of the following:

- The initial number of members in the full queuing system (those waiting and those currently being serviced)

- The probability of having n members in the full queuing system at a given later time

Finally, the model wants to compute the *steady state of the queuing system*:

- The probability of having n members in the full queuing system
- The expected number of new members arriving per unit time
- The expected number of members completing their service per unit time
- The expected waiting time for each member in the system

Members enter the queue at a certain mean rate, wait to get serviced, get serviced at a certain mean rate, then leave the facility. The mathematical model must quantify these and balance them.

Inventory

With the current shortages in the supply chain, the symptoms are empty shelves at grocery stores, shortages of car repair parts, new cars, materials for home renovation, and many others. There is obviously a gap between supply and demand. The times between replenishing the supplies at stores have increased in a way that is causing backlogs, low productivity, and an overall slowed economy. Mathematical models for inventory management quantify the supply (stochastically or deterministically), the demand, and devise an optimal inventory policy for timing replenishing and deciding on the quantity required at each replenish. Ideally, the model must have access to an information processing system that gathers data on current inventory levels and then signals when and by how much to replenish them.

Machine Learning for Operations Research

For starters, what is extremely exciting nowadays in operations research, as opposed to 10 years ago, is the ability to solve massive operations research problems, sometimes involving tens of millions constraints and decision variables. We have to thank the computational power explosion and the continuing improvement of computer implementations of operations research algorithms for this.

Moreover, machine learning can help predict the values of many parameters that enter into operations research models using volumes of available data. If these parameters were hard to measure, modelers had to either remove them from the model or make assumptions about their values. This doesn't have to be the case anymore because of more accurate machine learning models that are able to take thousands of variables into account.

Finally, machine learning can help speed up searching through combinatorially large search spaces by *learning* which parts of the space to focus on or which subproblems to prioritize. This is exactly what the article "Learning to Delegate for Large-scale Vehicle Routing" (*https://oreil.ly/RZdmR*) (Li et al. 2021) does, speeding up vehicle routing 10 to 100 more times than the state-of-the-art available routing algorithms.

Similar research at the intersection of machine learning and operations research is booming with great progress and scalable solutions. The list of abstracts from the conference Operations Research Meets Machine Learning (*https://oreil.ly/qaoTi*) offers a great variety of relevant projects, such as real-time data synthesis and treatment from sensors in waste bins (tracking the volume) for more efficient waste collection operation (since this relies on real-time data, the team relies on dynamic routing). Another great example is a bike sharing system, where the objective is to predict the number of bikes needed at each location and allocate teams to distribute the required number of bikes efficiently. Here is the abstract:

> Operators in a bike sharing system control room are constantly re-allocating bikes where they are most likely to be needed, this requires an insight on the optimum number of bikes needed in each station, and the most efficient way to distribute teams to move the bikes around. Forecasting engines and Decision Optimization is used to calculate the optimal number of bikes for each station at any given time, and plan efficient routes to help the redistribution of bikes accordingly. A solution delivered by DecisionBrain and IBM for the bike sharing system in London is the first application of its kind that uses both optimization and machine learning to solve cycle hire inventory, distribution and maintenance problems, and could easily be re-deployed for other cycle sharing systems around the world.

In fact, DecisionBrain's projects (*https://decisionbrain.com*) are worth browsing and thinking through.

Currently, my team and I are working on a problem with the Department of Public Transportation in my city. This is a perfect setting where machine learning meets operations research. Using historical ridership data, in particular daily boardings and alightings at each bus stop in the city, along with population density, demographics, vulnerability, city zoning data, car ownership, university enrollment, and parking data, we use neural networks to predict supply and demand patterns at each stop. Then we use this data and optimal network design from operations research to redesign the bus routes so that the bus stops, in particular those in the most socially vulnerable areas in the city, are adequately and efficiently serviced.

Hamilton-Jacobi-Bellman Equation

The fields of operations research, game theory, and partial differential equations intersect through dynamic programming and the *Hamilton-Jacobi-Bellman* partial differential equation. Richard Bellman (mathematician, 1920–1984) first coined the term *curse of dimensionality* in the context of dynamic programming. Now the curse of dimensionality has rendered real-life applications of this very useful equation limited and unable to incorporate all the players of a game (or competing markers, countries, militaries) and solve for their optimal strategies, or the thousands of variables that can be involved in operations research problems, such as for optimal resource allocation problems. The tides have turned with deep learning. The paper

"Solving High-Dimensional Partial Differential Equations Using Deep Learning" (*https://oreil.ly/nBbE9*) (Han et al. 2018) presents a method to solve this equation and others for very high dimensions. We will discuss the idea of how the authors do it in Chapter 13 on AI and partial differential equations.

Operations Research for AI

Operations research is the science of decision making based on optimal solutions. Humans are always trying to make decisions based on the available circumstances. Artificial intelligence aims to replicate all aspects of human intelligence, including decision making. In this sense, the decision-making methods that operations research employs automatically fit into AI. The ideas in dynamic programming, Markov chains, optimal control and the Hamilton-Jacobi-Bellman equation, advances in game theory and multiagent games, network optimization, and others have evolved along with AI throughout the decades. In fact, many startups market themselves as AI companies, while in reality they are doing good old (and awesome) operations research.

Summary and Looking Ahead

Operations research is the field of making the best decisions given the current knowledge and circumstances. It always comes down to finding clever ways to search for optimizers in very high-dimensional spaces.

One theme throughout this book is the curse of dimensionality, and all the effort researchers put in to find ways around it. In no field does this curse show up as broadly as in operations research. Here, the search spaces grow combinatorially with the number of players in a particular problem: number of cities on a route, number of competing entities, number of people, number of commodities, etc. There are very powerful exact methods and heuristic methods, but there is much room for improvement in terms of speed and scale.

Machine learning, in particular deep learning, provides a way to learn from previously solved problems, labeled data, or simulated data. This speeds up optimization searches if we identify the bottlenecks and are able to articulate the source of the bottleneck as a machine learning problem. For example, a bottleneck can be: *we have too many subproblems to solve but we do not know which ones to prioritize to quickly get us closer to the optimum*. To use machine learning to address this, we need a data set of already solved problems and subproblems, and have a machine learning model learn which subproblems should be prioritized. Once the model learns this, we can use it to speed up new problems.

Other uses of machine learning in operations research include *business as usual* type of machine learning: predict demand from available data, either real-time or historical data, then use operations research to optimize resource allocation. Here machine learning helps make better predictions for demand and hence increases efficiency and reduces waste.

In this chapter, we gave a broad overview of the field of operations research and its most important types of problems. We especially emphasized linear optimization, networks, and duality. Powerful software packages are available for many useful problems. We hope that these packages keep integrating the latest progresses in the field.

Two topics that are usually not taught in an introductory operations research classes are the Hamilton-Jacobi-Bellman partial differential equation for optimal control and strategies of multiplayer games, and optimizing *functionals* using calculus of variations. These are usually considered advanced topics in partial differential equations. We discussed them here because they both tie naturally into optimization and operations research. Moreover, viewing them in this context demystifies their corresponding fields.

When doing operations research and optimizing for cost reduction, revenue increase, time efficiency, etc., it is important that our optimization models do not ignore the human factor. If the output of a scheduling model messes up low-wage workers' lives through erratic schedules to keep a certain *on-time* company performance, then that is not a good model and the quality of the lives and livelihoods of the workers that a company relies on needs to be quantified, then factored into the model. Yes, *the quality of life* needs to be quantified, since everything else is being factored in, and we cannot leave this out. Companies with hundreds of thousands of low-wage workers have a responsibility that their operations research algorithms do not end up trapping their workers in poverty.

We leave this chapter with this excerpt from a paper authored by Charles Hitch (*https://oreil.ly/5DLeZ*), "Uncertainties In Operations Research" (*https://oreil.ly/yPTcj*) from 1960. Reading this (the brackets are my edits), one cannot help but ponder how far the operations research field has come since 1960:

> No other characteristic of decision making is as pervasive as uncertainty. When, as operations researchers, to simplify a first cut at an analysis, we assume that the situation can be described by certainty equivalents, we may be doing violence to the facts and indeed the violence may be so grievous as to falsify the problem and give us a nonsense solution. How, for example, can we help the military make development decisions—decisions about which aircraft or missiles to develop when the essence of the problem is that no one can predict with accuracy how long it will take to develop any of the competing equipments, or to get them operational, how much they will cost, what their performance will be, or what the world will be like at whatever uncertain future date turns out to be relevant (if indeed, the world still exists then)? When I say "cannot predict with accuracy" I am not exaggerating. We find that typically, for example, the production costs of new equipment are underestimated in the early stages

of development by factors of two to twenty (not 2 to 20 per cent, but factors of two to twenty). Why they are always underestimated, never overestimated, I leave to your fertile imaginations. […] Another thing that [an operations researcher] can frequently do, especially in problems involving research and development, is to ascertain the critical uncertainties and recommend strategies to reduce them—to buy information. If you do not know which of two dissimilar techniques for missile guidance will turn out to be better, your best recommendation is very likely to be: keep them both in development a while longer and choose between them when more and better information is available. Never mind the people who call you indecisive. You can prove that this kind of indecisiveness can save both money and time. Of course you can't afford to try everything. There isn't enough budget. There aren't enough resources. You remember when we used to say "If you gave the military services everything they asked for they'd try to fortify the moon!" (We'll have to change that figure of speech.) Actually, it is because of limitations on resources that operations research and operations researchers are important. There'd be no problems for us if there were no constraints on resources. It is our job and opportunity to find, or invent, within the constraints, some better pattern of adjusting to an uncertain world than our betters would find if we weren't here; or some better way, taking costs and pay-offs into account, to buy information to reduce the uncertainty.

CHAPTER 11

Probability

Can we still expect anything, if chance is all there is?

—H.

Probability theory is one of the most beautiful subjects in mathematics, moving us back and forth between the stochastic and deterministic realms in what should be magic but turns out to be mathematics and its wonders. Probability provides a systematic way to quantify randomness, control uncertainty, and extend logic and reasoning to situations that are of paramount importance in AI: when information and knowledge include uncertainties, and/or when the agent navigates unpredictable or partially observed environments. In such settings, an agent calculates probabilities about the unobserved aspects of a certain environment, then makes decisions based on these probabilities.

Humans are uncomfortable with uncertainty, but are comfortable with approximations and expectations. They do not wake up knowing exactly how every moment of their day will play out, and they make decisions along the way. A probabilistic intelligent machine exists in a world of probabilities, as opposed to deterministic and fully predetermined truths and falsehoods.

Throughout this book, we have used probability terms and techniques as they came along and only when we needed them. Through this process, we now realize that we need to be well versed in joint probability distributions (for example, of features of data), conditioning, independence, Bayes' Theorem, and Markov processes. We also realize that we can get back to the deterministic world via computing averages and expectations.

One feature of the chapters in this book is that each needs its own book to have an in-depth and comprehensive treatment. This couldn't be more true than for a chapter on probability, where we can include thousands of topics. I had to make choices, so I based the topics that I opted to cover in this chapter on three criteria:

1. What we already used in this book that has to do with probability
2. What confused me the most in probability as a student (like why do we need measure theory when computing probabilities?)
3. What else we need to know from probability theory for AI applications

Where Did Probability Appear in This Book?

Let's make a fast list of the places where we used probability or resorted to stochastic methods in this book. We consider this list as the *essential probability for AI*. Note that *prior* probabilities are unconditional, because they are prior to observing the data, or the evidence; and *posterior* probabilities are conditional, because their value is conditioned on observing the relevant data. It makes sense that our degree of belief about something changes after receiving new and related evidence. The joint probability distribution of all the involved variables is what we are usually after, but it is generally too large, and the information needed to fully construct it is not always available.

Here is the list:

- When minimizing the loss function of deterministic machine learning models (where the training function takes nonrandom inputs and produces nonrandom outputs), such as regression, support vector machines, neural networks, etc., we use stochastic gradient descent and its variants, randomly choosing a subset of training data instances at each gradient descent step, as opposed to using the whole training data set, to speed up computations.

- In Chapter 9 on graph models, we utilized random walks on graphs on many occasions, implementing these walks via the weighted adjacency matrix of the graph.

- Specific probability distributions appeared in Chapter 10 on operations research, such as probability distributions for inter-arrival and service times for customers in a queue.

- Dynamic decision making and Markov processes also appeared in Chapter 10 on operations research and are fundamental for reinforcement learning in AI. They will appear again in this chapter, then once more in Chapter 13 in the context of the Hamilton-Jacobi-Bellman equation.

- For two-person zero-sum games in Chapter 10, each player had a probability of making a certain move, and we used that to compute the player's optimal strategy and expected payoff.

- Monte Carlo simulation methods are computational algorithms that rely on repeated random sampling to solve deterministic problems numerically. We illustrate an example of these in Chapter 13 on AI and PDEs.

- We mentioned the universality theorem for neural networks many times, and we only refer to its proof in this chapter. The proof of this fundamental theorem is theoretical, and gives us a nice flavor of measure theory and functional analysis. We build the intuition for measure-theoretic probability later in this chapter.

- Probabilistic machine learning models learn the joint probability distribution of the data features $Prob(x_1,x_2,\cdots,x_n,y_{target})$ instead of learning the deterministic functions of these features. This joint probability distribution encodes the likelihood of these features occurring at the same time. Given the input data features $(x_1,x_2,\cdots,x_n)$, the model outputs the conditional probability of the target variable given the data features $Prob(y_{predict}|x_1,x_2,\cdots,x_n)$, as opposed to outputting $y_{predict}$ as a deterministic function of the features: $y_{predict} = f(x_1,x_2,\cdots,x_n)$.

- Random variables and the two most important quantities associated with them, namely the expectation (expected average value of the random variable) and variance (a measure of the spread around the average): we have been using these without formally defining them. We will define them in this chapter.

- The product rule or the chain rule for probability, namely:

$$Prob(x_1,x_2) = Prob(x_1|x_2)Prob(x_2) = Prob(x_2|x_1)Prob(x_1)$$

or for more than two variables, say three without loss of generality:

$$\begin{aligned}Prob(x_1,x_2,x_3) &= Prob(x_1|x_2,x_3)Prob(x_2,x_3)\\ &= Prob(x_1|x_2,x_3)Prob(x_2|x_3)Prob(x_3)\end{aligned}$$

- The concepts of independence and conditional independence are fundamental. Two events are independent if the occurrence of one does not affect the probability of occurrence of the other. Independence of the considered features is tremendously simplifying. It helps us disentangle complex joint distributions of many variables, reducing them to simple products of fewer variables, and rendering many previously intractable computations tractable. This greatly simplifies the probabilistic interpretations of the world. Pay attention to the difference between independence of *only two events* ($Prob(x_1,x_2) = Prob(x_1)Prob(x_2)$) and independence of *many events*, which is a strong assumption where every event is independent of any intersection of the other events.

- For probabilistic generative models in Chapter 8, we assumed a prior probability distribution, passed it through a neural network, and adjusted its parameters.

- Bayes' Theorem is essential when discussing joint and conditional probabilities. It helps us quantify an agent's beliefs relative to evidence. We use it in many contexts, which immediately illustrate its usefulness, such as:

$$Prob(disease \mid symptoms) = \frac{Prob(symptoms \mid disease)Prob(disease)}{Prob(symptoms)}$$

or

$$Prob(target \mid data) = \frac{Prob(data \mid target)Prob(target)}{Prob(data)}$$

or

$$Prob(target \mid evidence) = \frac{Prob(evidence \mid target)Prob(target)}{Prob(evidence)}$$

or

$$Prob(cause \mid effect) = \frac{Prob(effect \mid cause)Prob(cause)}{Prob(effect)}$$

Note that in the last formula, $Prob(cause \mid effect)$ quantifies the *diagnostic* direction, while $Prob(effect \mid cause)$ quantifies the *causal* direction.

- Bayesian networks are data structures that represent dependencies among variables. Here, we summarize the variable relationships in a directed graph and use that to determine which conditional probability tables we need to keep track of and update in the light of new evidence: we keep track of the probability of a child node conditional on observing its parents. The parents of a node are any variables that directly influence this node. In this sense, the Bayesian network is a representation of the joint probability distribution, with the simplification that we know how the involved variables relate to each other (which variables are the parents of which variables):

$$Prob(x_1, x_2, \cdots, x_n) = \Pi_{i=1}^{n} Prob(x_i \mid parents(X_i))$$

- In machine learning we can draw a line between regression models and classification models. In Chapter 8 on probabilistic generative models, we encountered a popular probabilistic model for classification: Naive Bayes. In cause-and-effect language, the *naive* assumption is that some observed multiple effects are independent given a cause, so that we can write:

$$Prob(cause|effect_1, effect_2, effect_3)$$
$$= P(cause)P(effect_1|cause)P(effect_2|cause)P(effect_3|cause)$$

When that formula is used for classification given data features, the *cause* is the class. Moreover, we can draw a Bayesian network representing this setting. The cause variable is the parent node, and all the effects are children nodes stemming from the one parent node (Figure 11-1).

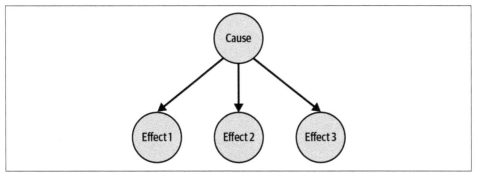

Figure 11-1. Bayesian network representing three effects having a common cause

What More Do We Need to Know That Is Essential for AI?

We need a few extra topics that have either not gotten any attention in this book or were only mentioned casually and pushed to this chapter for more details. These include:

- Judea Pearl's causal modeling and the *do* calculus
- Some paradoxes
- Large random matrices and high-dimensional probability
- Stochastic processes such as random walks, Brownian motion, and more
- Markov decision processes and reinforcement learning
- Theory of probability and its use in AI

The rest of this chapter focuses on these topics.

Causal Modeling and the Do Calculus

In principle, the arrows between related variables in a Bayesian network can point in any direction. They all eventually lead to the same joint probability distribution, albeit some in more complicated ways than others.

In contrast, causal networks are those special Bayesian networks where the directed edges of the graph cannot point in any direction other than the causal direction. For these, we have to be more mindful when constructing the connections and their directions. Figure 11-2 shows an example of a causal Bayesian network.

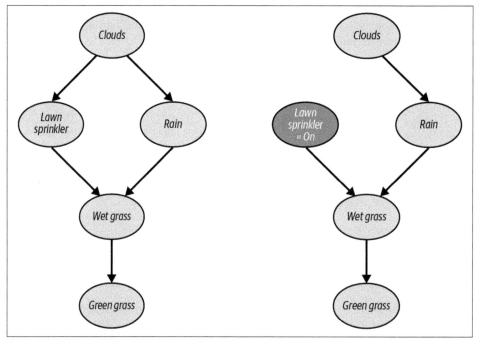

Figure 11-2. Causal Bayesian network

Note that both Bayesian networks and causal networks make strong assumptions on which variables listen to which variables.

Agents endowed with causal reasoning are, in human terms, *higher functioning* than those merely observing patterns in the data, then making decisions based on the relevant patterns.

The following distinction is of paramount importance:

- In Bayesian networks, we suffice ourselves with knowing only whether two variables are probabilistically dependent. Are fire and smoke probabilistically dependent?

- In causal networks, we go further and ask about *which variable responds to which variable*: smoke to fire (so we draw an arrow from fire to smoke in the diagram), or fire to smoke (so we draw an arrow from smoke to fire in the diagram)?

What we need here is a mathematical framework for *intervention* to quantify the effect of fixing the value of one variable. This is called the *do calculus* (as opposed to the statistical *observe and count* calculus). Let's present two fundamental formulas of the *do calculus*:

- The adjustment formula
- The backdoor criterion

According to Judea Pearl (*https://oreil.ly/KB8Q8*), the inventor of this wonderful way of causal reasoning (and whose *The Book of Why* (2020) inspires the discussion in this section and the next one), these *allow the researcher to explore and plot all possible routes up mount intervention, no matter how twisty,* and can save us the costs and difficulties of running randomized controlled trials, even when these are physically feasible and legally permissible.

An Alternative: The Do Calculus

Given a causal network, which we construct based on a combination of common sense and subject matter expertise, while at the same time throwing in extra unknown causes for each variable just to be sure that we are accounting for everything, the overarching formula is that of the joint probability distribution:

$$Prob(x_1, x_2, \cdots, x_n) = \Pi_{i=1}^{n} Prob(x_i | parents(X_i))$$

Then we *intervene*, applying $do(X_j = x^\star)$. This severs any edges pointing to X_j, and affects all the conditional probabilities of the descendants of X_j, leading to a new joint probability distribution that wouldn't include the conditional probability for the intervened variable anymore. We already set its value to $X_j = x^\star$ with probability one, and any other value would have probability zero. Figure 11-2 shows how, when we set the sprinkler on, all arrows leading to it in the original network get severed.

Thus we have:

$$Prob_{intervened}(x_1, x_2, \cdots, x_n) = \begin{array}{l} \Pi_{i \neq j}^{n} Prob(x_i | parents(X_i)) \text{ if } X_j = x^\star \\ 0 \text{ otherwise} \end{array}$$

The adjustment formula

What we truly care about is how does setting $X_j = x^\star$ affect the probability of every other variable in the network, and we want to compute these from the original unintervened network. In math words, without the *do* operator, since we can just observe the data to get these values, as opposed to running new experiments.

To this end, we introduce the *adjustment formula*, or *controlling for confounders* (possible common causes). This is a weighted average of the influence of X_j and its parents on X_i. The weights are the priors on the parent values:

$$Prob(x_i | do(X_j = x^*)) = Prob_{intervened}(X_i = x_i)$$

$$= \sum_{parents(X_j)} Prob(x_i | x^*, parents(X_j)) Prob(parents(X_j))$$

Note that this formula achieves our goal of eliminating the *do* operator and gets us back to finding our conditional probabilities by observing the data, rather than running some costly intervention experiments, or randomized control trials.

The backdoor criterion, or controlling for confounders

There is more to the causal diagrams story. We would like to know the effect of the intervention $do(X_j = x^*)$ on a certain *downstream* variable in the diagram X_{down}. We should be able to condition on the values of any variable in the diagram that is *another ancestor*. This also leads down to the downstream variable that we care about. In causal modeling, we call this process *blocking the back doors* or the *backdoor criterion*:

$$P(x_{down} | do(X_j = x^*)) = P_{intervened}(X_{down} = x_{down})$$

$$= \sum_{ancestor(X_{down})} P(x_{down} | x^*, ancestor(X_{down})) P(ancestor(X_{down}))$$

Controlling for confounders

The most common way for scientists and statisticians to predict the effects of an intervention so that they can make statements about causality, is to control for *possible common causes*, or confounders. Figure 11-3 shows the variable Z as a confounder of the suspected causal relationship between X and Y.

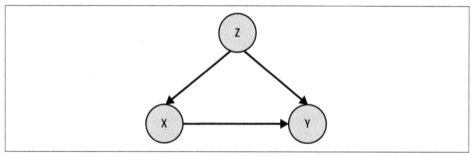

Figure 11-3. Z is a confounder of the suspected causal relationship between X and Y

This is because, in general, confounding is a main source of confusion between mere observation and intervention. It is also the source of the famous statement *correlation is not causation*. This is where we see some bizarre and entertaining examples: high temperature is a confounder for ice cream sales and shark attacks (but why would anyone study any sort of relationship between ice cream and sharks to start with?). The backdoor criterion and the adjustment formula easily take care of confounder obstacles to stipulating about causality.

We use the adjustment formula to control for confounders if we are confident that we have data on a sufficient set of *deconfounder* variables to block all the backdoor paths between the intervention and the outcome. To do this, we estimate the causal effect stratum by stratum from the data, then we compute a weighted average of those strata, where each stratum is weighted according to its prevalence in the population.

Now, without the backdoor criterion, statisticians and scientists have no guarantee that any adjustment is legitimate. In other words, the backdoor criterion guarantees that the causal effect in each stratum of the deconfounder is in fact the observed trend in this stratum.

Are there more rules that eliminate the do operator?

Rules that are able to move us from an expression with the *do* operator (intervention) to an expression without the *do* operator (observation) are extremely desirable, since they eliminate the need to intervene. They allow us to estimate causal effects by mere data observation. The adjustment formula and the backdoor criterion did exactly that for us.

Are there more rules? The more ambitious question is: is there a way to decide ahead of time whether a certain causal model lends itself to *do* operator elimination, so that we would know whether the assumptions of the model are sufficient to uncover the causal effect from observational data without any intervention? Knowing this is huge! For example, if the assumptions of the model are not sufficient to eliminate the *do* operator, then no matter how clever we are, there is no escape from running interventional experiments. On the other hand, if we do not have to intervene and still estimate causal effects, the savings are spectacular. These alone are worth digging more into probabilistic causal modeling and the *do* calculus.

To get the gist of Judea Pearl's *do* calculus, we always start with a causal diagram, and think of conditioning criteria leading to the deletion of edges pointing toward or out from the variable(s) of interest. Pearl's three rules give us the conditions under which:

1. We can insert or delete observations:

$$Prob(y \mid do(x), z, w) = Prob(y \mid do(x), w)$$

2. We can insert or delete interventions:

$$Prob(y \mid do(x), do(z), w) = Prob(y \mid do(x), w)$$

3. We can exchange interventions with observations:

$$Prob(y \mid do(x), do(z), w) = Prob(y \mid do(x), z, w)$$

For more details on the *do* calculus, see "The Do-Calculus Revisited," by Judea Pearl (Keynote Lecture, August 17, 2012) (*https://oreil.ly/DTPq0*).

Paradoxes and Diagram Interpretations

AI agents need to be able to handle paradoxes. We have all seen cartoons where a robot gets into a crazy loop or even physically self-dismantles with screws and springs flying all around when its logic encounters a paradox. We cannot let that happen. Furthermore, paradoxes often appear in very consequential settings, such as in the pharmaceutical and medical fields, so it is crucial that we scrutinize them under the lens of mathematics and carefully unravel their mysteries.

Let's go over three famous paradoxes: *Monty Hall*, *Berkson*, and *Simpson*. We will view them in the light of diagrams and causal models: Monty Hall and Berkson paradoxes cause confusion due to colliders (two independent variables pointing to a third one), while Simpson paradoxes cause confusion due to confounders (one variable pointing to two others). An AI agent should be equipped with these diagrams as part of its data structure (or with the ability to construct them and adjust them) in order to reason properly.

Judea Pearl's *The Book of Why* puts it perfectly:

> Paradoxes reflect the tensions between causation and association. The tension starts because they stand on two different rungs of the Ladder of Causation [observation, intervention, counterfactuals] and is aggravated by the fact that human intuition operates under the logic of causation, while data conform to the logic of probabilities and proportions. Paradoxes arise when we misapply the rules we have learned in one realm to the other.

Monty Hall Problem

Suppose you're on a game show, and you're given the choice of three doors. Behind one door is a car, behind the others, goats. You pick a door, say #1, and the host, who knows what's behind the doors, opens another door, say #3, which has a goat. He says to you, "Do you want to pick door #2?" Is it to your advantage to switch your choice of doors?

The answer is yes, switch doors, because without switching, your probability of getting the car is 1/3, and after switching, it jumps up to 2/3! The main thing to pay attention to here is that the host *knows* where the car is, and chooses to open a door that he knows does not have the car in it.

So why would the probability of winning double if we switch from our initial choice? Because the host offers new information that we would leverage only if we switch from our initial information-less choice:

Under the no-switch strategy
- If we initially chose the winning door (probability 1/3), and we do not switch, then we win.
- If we initially chose a losing door (probability 2/3), and we do not switch, then we lose.

This means that we would win only 1/3 of the time under the no-switch strategy.

Under the switch strategy
- If we initially chose the winning door (probability 1/3), and we switch from it, then we would lose.
- If we initially chose a losing door (probability 2/3), new information comes *pointing to the other losing door*, and we switch, then we would win, because the only door left would be the winning door.

This means that we would win 2/3 of the time under the switch strategy.

When we draw the diagram in Figure 11-4 to represent this game, we realize that the door that the host chooses to open has two parents pointing toward it: the door you chose and the location of the car.

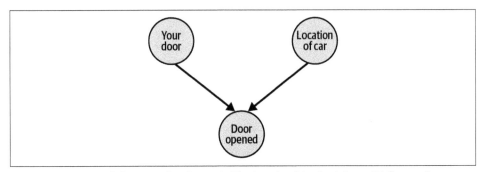

Figure 11-4. Causal diagram for the variables involved in the Monty Hall paradox

Conditioning on this *collider* changes the probabilities of the parents. It creates a spurious dependency between originally independent parents! This is similar to us changing our beliefs about the genetic traits of parents once we meet one of their children. These are causeless correlations, induced when we condition on colliders.

Now suppose that the host chooses their door *without knowing* whether it is a winning or a losing door. Then switching or nonswitching would not change the odds of winning the car, because in this case both you and the host have equal chances of winning 1/3 of the time and losing 2/3 of the time. Now when we draw the diagram for this totally random and no-prior-knowledge game, there is no arrow between the location of the car and the door that the host chooses to open, so your choice of the door and the location of the car *remain independent* even after conditioning on the host's choice.

Berkson's Paradox

In 1946, Joseph Berkson, a biostatistician at the Mayo Clinic, pointed out a peculiarity of observational studies conducted in a hospital setting: even if two diseases have no relation to each other in the general population, they can appear to be associated among patients in a hospital. In 1979, David Sackett of McMaster University, an expert on all sorts of statistical bias, provided strong evidence that Berkson's paradox is real. In one example, he studied two groups of diseases: respiratory and bone. About 7.5% of people in the general population have a bone disease, and this percentage is independent of whether they have respiratory disease. But for hospitalized people with respiratory disease, the frequency of bone disease jumps to 25%! Sackett called this phenomenon "admission rate bias" or "Berkson bias."

Similar to the Monty Hall case, the culprit for the appearance of the Berkson paradox is a collider diagram, where both originally independent diseases point to hospitalization: a patient with both diseases is much more likely to be hospitalized than a patient with only one of them. When we condition on hospitalization, which is the collider, a case of causeless correlation between the initially independent variables appears. We are getting used to collider bias now.

Simpson's Paradox

Imagine a paradox whose conclusion, if left to its own devices, is this absurd: *when we know the gender of the patient, then we should not prescribe the drug, because the data shows that the drug is bad for males and bad for females; but if the gender is unknown, then we should prescribe the drug, because the data shows that the drug is good for the general population.* This is obviously ridiculous, and our first instinct should be to protest: show me the data!

We recognize Simpson's paradox when a trend appears in several groups of the population but disappears or reverses when the groups are combined.

Let's first debunk the paradox. It is a simple numerical mistake of how to add fractions (or proportions). In summary, when we add fractions, we cannot simply add the respective numerators and the denominators:

$$\frac{A}{B} > \frac{a}{b} \text{ and } \frac{C}{D} > \frac{c}{d} \nRightarrow \frac{A+C}{B+D} > \frac{a+c}{b+d}$$

For example, suppose that the data shows that:

- 3/40 of the women who took the drug had a heart attack, compared to only 1/20 of the women who did not take the drug (3/40 > 1/20).
- 8/20 of the men who took the drug had a heart attack, compared to 12/40 of the men who did not take the drug (8/20 > 12/40).

Now when we merge the data for women and men, the inequality reverses direction: 3/40 > 1/20 and 8/20 > 12/40, but rightfully (3 + 8)/(40 + 20) < (1 + 12)/(20 + 40). In other words: of the 60 men and women who took the drug, 11 had a heart attack, and of the 60 men and women who did not take the drug, 13 had a heart attack.

However, we committed a simple mistake with fractions when we merged the data this way. To solve Simpson's paradox, we should not merge the data by simply adding the numerators and the denominators and expect the inequality to hold. Note that of the 60 people who took the drug, 40 are women and 20 are men; while of the 60 people who did not take the drug, 20 are women and 40 are men. We are comparing apples and oranges and confounding that with gender. The gender affects *both* whether the drug is administered and whether a heart attack happens. The diagram in Figure 11-5 illustrates this confounder relationship.

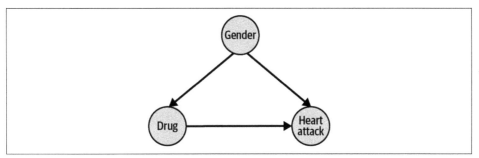

Figure 11-5. Gender is a confounder for both taking the drug and heart attack

Our strong intuition that something is wrong if we merge proportions naively is spot on. If things are fair on a local level everywhere, then they are fair globally; or if things act a certain way on every local level, then we should expect them to act that way globally.

It is no surprise that this mistake happens so often, as humans did not get fractions right until relatively recently. There are ancient texts with mistakes manipulating fractions in domains such as inheritance and trade. Our brains' resistance to fractions seems to persist: we learn fractions in seventh grade, and that also happens to be the time to which we can trace the origin of many people's legendary hatred for math.

So what is the correct way to merge the data? Our grade-seven wisdom tells us to use the common denominator 40 and to condition on gender: for women 3/40 > 2/40 and for men 16/40 > 12/40. Now since in the general population men and women are equally distributed, we should take the average and rightfully conclude that (3/40 + 16/40)/2 > (2/40 + 12/40)/2; that is, the rate of heart attacks in the general population is 23.75% with the drug, and 17.5% without the drug. No magical and illogical reversal happened here. Moreover, this drug is pretty bad!

Large Random Matrices

Most AI applications deal with a vast amount of high-dimensional data (big data), organized in high-dimensional vectors, matrices, or tensors, representing data tables, images, natural language, graph networks, and others. A lot of this data is noisy or has an intrinsic random nature. To process such data, we need a mathematical framework that combines probability and statistics, which usually deal with *scalar* random variables, with linear algebra, which deals with *vectors and matrices*.

The mean and variance are still central ideas, so we find many statements and results containing the expectation and variance (uncertainty) of the involved high-dimensional random variables. Similar to the scalar case, the tricky part is controlling the variance, so a lot of work in the literature finds bounds (inequalities) on the tails of the random variables' distributions or how likely it is to find a random variable within some distances from its mean.

Since we now have matrix valued random variables, many results seek to understand the behaviors (distributions) of their spectra: eigenvalues and eigenvectors.

Examples of Random Vectors and Random Matrices

It is no wonder the study of large random matrices evolved into its own theory. They appear in all sorts of impactful applications, from finance to neuroscience to physics and the manufacture of technological devices. The following is only a sampling of examples. These have great implications, so there are large mathematical communities around each them.

Quantitative finance

One example of a random vector is an investment portfolio in quantitative finance. We often need to decide on how to invest in a large number of stocks, whose price movement is stochastic, for optimal performance. The investment portfolio itself is a large random vector that evolves with time. In the same spirit, the daily returns of Nasdaq stocks (Nasdaq contains more than 2,500 stocks) is a time-evolving large random vector.

Neuroscience

Another example is from neuroscience. Random matrices appear when modeling a network of synaptic connections between neurons in the brain.The number of spikes fired by n neurons during t consecutive time intervals of a certain length is an $n \times t$ random matrix.

Mathematical physics: Wigner matrices

In mathematical physics, particularly in nuclear physics, physicist Eugene Wigner introduced random matrices to model the nuclei of heavy atoms and their spectra. In a nutshell, he related the spacings between the lines in the spectrum of a heavy atom's nucleus to the spacings between the eigenvalues of a random matrix.

The deterministic matrix that Wigner started with is the Hamiltonian of the system, which is a matrix describing all the interactions between the neutrons and protons contained in the nucleus. The task of diagonalizing the Hamiltonian to find the energy levels of the nucleus was impossible, so Wigner looked for an alternative. He abandoned exactness and determinism altogether and approached the question from a probabilistic perspective. Instead of asking *what precisely are the energy levels*, he asked questions like:

- What is the probability of finding an energy level within a certain interval?
- What is the probability that the distance between two successive energy levels is within a certain range?
- Can we replace the Hamiltonian of the system by a purely random matrix with the correct symmetry properties? For example, in the case of quantum systems invariant under time reversal, the Hamiltonian is a real symmetric matrix (of infinite size). In the presence of a magnetic field, the Hamiltonian is a complex, Hermitian matrix (the complex analog of a real symmetric matrix). In the presence of *spin-orbit coupling* (a quantum physics term), the Hamiltonian is symplectic (another special type of symmetric matrix).

Similarly, Wigner-type random matrices appear in condensed matter physics, where we model the interaction between pairs of atoms or pairs of spins using real symmetric Wigner matrices. Overall, Wigner matrices are considered *classical* in random matrix theory.

Multivariate statistics: Wishart matrices and covariance

In multivariate statistics, John Wishart introduced random matrices when he wanted to estimate sample covariance matrices of large random vectors. Wishart random matrices are also considered classical in random matrix theory. Note that a sample covariance matrix is an estimation for the population covariance matrix.

When dealing with sample covariance matrices, a common setting is that of n dimensional variables observed t times, that is, the original data set is a matrix of size $n \times t$. For example, we might need to estimate the covariance matrix of the returns of a large number of assets (using a smaller sample), such as the daily returns of the 2,500 Nasdaq stocks. If we use 5 years of daily data, given that there are 252 trading days in a year, then we have $5 \times 252 = 1,260$ data points for each of the 2,500 stocks. The original data set would be a matrix of size $2,500 \times 1,260$. This is a case where the number of observations is smaller than the number of variables. We have other cases where it is the other way around, as well as limiting cases where the number of observations and the number of variables are of drastically different scales. In all cases, we are interested in the law (probability distribution) for the eigenvalues of the sample covariance matrix.

Let's write the formulas for the entries of the covariance matrix. For one variable $\vec{z}_1$ (say one stock) with t observations whose mean (average) is $\bar{z}_1$, we have the variance:

$$\sigma_1^2 = \frac{(z_1(1) - \bar{z}_1)^2 + (z_1(2) - \bar{z}_1)^2 + \cdots + (z_1(t) - \bar{z}_1)^2}{t}$$

Similarly, for each of the n variables $\vec{z}_i$, we have their variance σ_i^2. These sit on the diagonal of the covariance matrix. Now each off-diagonal σ_{ij} entry is the covariance of the corresponding pair of variables:

$$\sigma_{ij} = \frac{(z_i(1) - \bar{z}_i)(z_j(1) - \bar{z}_j) + (z_i(2) - \bar{z}_i)(z_j(2) - \bar{z}_j) + \cdots + (z_i(t) - \bar{z}_i)(z_j(t) - \bar{z}_j)}{t}$$

The covariance matrix is symmetric and positive definite (has positive eigenvalues). The randomness in a covariance matrix usually stems from noisy observations. Since measurement noise is inevitable, determining the covariance matrix becomes more involved mathematically. Another common issue is that often the samples are not independent. Correlated samples introduce some sort of redundancy, so we expect that the sample covariance matrix behaves as if we had observed fewer samples than we actually did. We must then analyze the sample covariance matrix in the presence of correlated samples.

Dynamical systems

Linearized dynamical systems are near equilibria ($\frac{d\vec{x}(t)}{dt} = A\vec{x}(t)$). In the context of chaotic systems, we want to understand how a small difference in initial conditions propagates as the dynamics unfolds. One approach is to linearize the dynamics in the vicinity of the unperturbed trajectory. The perturbation evolves as a product of matrices, corresponding to the linearized dynamics, applied on the initial perturbation.

Algorithms for Matrix Multiplication

Finding efficient algorithms for matrix multiplication is an essential, yet surprisingly difficult, goal. In matrix multiplication algorithms, saving on even one multiplication operation is worthy (saving on addition is not as much of a big deal). Recently, DeepMind developed AlphaTensor (2022) (*https://oreil.ly/HZPbd*) to automatically discover more efficient algorithms for matrix multiplication. This is a milestone because matrix multiplication is a fundamental part of a vast array of technologies, including neural networks, computer graphics, and scientific computing.

Other equally important examples

There are other examples. In number theory, we can model the distribution of zeros of the Riemann zeta function using the distribution of eigenvalues of certain random matrices. For those keeping an eye out for quantum computing, here's a historical note: before Schrödinger's equation, Heisenberg formulated quantum mechanics in terms of what he named *matrix mechanics*. Finally, we will encounter the *master equation for the evolution of probabilities* in Chapter 13. This involves a large matrix of transition probabilities from one state of a system to another state.

Main Considerations in Random Matrix Theory

Depending on the formulation of the problem, the matrices that appear are either deterministic or random. For deterministic vectors and matrices, classical numerical linear algebra applies, but the extreme high dimensionality forces us to use randomization to efficiently do matrix multiplication (usually $O(n^3)$), decomposition, and computing the spectrum (eigenvalues and eigenvectors).

A substantial amount of matrix properties is encapsulated in their spectra, so we learn a great deal about matrices by studying these eigenvalues and eigenvectors. In the stochastic realm, when the matrices are random, these are random as well. So how do we compute them and find their probability distributions (or even only their means and variances or bounds on those)? These are the types of questions that the field of large random matrices (or randomized linear algebra, or high-dimensional probability) addresses. We usually focus on:

The involved stochastic math objects

Random vectors and random matrices. Each entry of a random vector or a random matrix is a random variable. These could either be static random variables or evolving with time. When a random variable evolves with time, it becomes *a random or stochastic process*. Obviously, stochastic processes are more involved mathematically than their static counterparts. For example, what can we say about variances that evolve with time?

Random projections

Our interest is always in projecting onto some lower-dimensional spaces while preserving essential information. These usually involve either multiplying matrices with vectors or factorizing matrices into a product of simpler matrices, such as the singular value decomposition. How do we do these when the data is large and the entries are random?

Adding and multiplying random matrices

Note that the sums and products of *scalar* random variables are also random variables, and their distributions are well studied. Similarly, the sums and products of *time evolving scalar* random variables, which are the foundation for Brownian motion and stochastic calculus, have a large body of literature supporting them. How does this theory transition to higher dimensions?

Computing the spectra

How do we compute the spectrum of a random matrix, and explore the properties of its (random) eigenvalues and eigenvectors?

Computing the spectra of sums and products of random matrices

How do we do this as well?

Multiplying many random matrices, as opposed to only two

This problem appears in many contexts in the technological industry, for example, when studying the transmission of light in a succession of slabs of different optical indices, or the propagation of an electron in a disordered wire, or the way displacements propagate in granular media.

Bayesian estimation for matrices

Bayesian *anything* always has to do with estimating the probability of something given some evidence. Here, the matrix we start with (the observations matrix) is a noisy version of the true matrix that we care for. The noise can be additive, so the observed matrix E = *true matrix* + *a random noise matrix*. The noise can also be multiplicative, so the observed matrix E = *true matrix* × *random noise matrix*. In general, we do not know the true matrix, and would like to know the probability of this matrix given that we have observed the noisy matrix. That is, we have to compute *Prob(true matrix|noisy matrix)*.

Random Matrix Ensembles

In most applications, we encounter (stochastic or deterministic) large matrices with no particular structure. The main premise that underlies random matrix theory is that we can replace such a large complex matrix by a typical element (expected element) of a certain ensemble of random matrices. Most of the time we restrict our attention to symmetric matrices with real entries, since these are the ones that most commonly arise in data analysis and statistical physics. Thankfully, these are easier to analyze mathematically.

Speaking of mathematics, we love polynomial functions. They are nonlinear, complex enough to capture enough complexities in the world around us, and are easy to evaluate and do computations with. When we study large random matrices, a special type of well-studied polynomials appears: orthogonal polynomials. An orthogonal polynomial sequence is a family of polynomials such that any two different polynomials in the sequence are orthogonal (their inner product is zero) to each other under some inner product, which is a generalized dot product. The most widely used orthogonal polynomials sequences are: Hermite polynomials, the Laguerre polynomials, and the Jacobi polynomials (these include the important classes of Chebyshev polynomials and Legendre polynomials). The famous names in the field of orthogonal polynomials, which was mostly developed in the late 19th century, are Chebyshev, Markov, and Stieltjes. No wonder these names are everywhere in probability theory, from Chebyshev inequalities to Markov chains and processes to Stieltjes transforms.

The following three fundamental random matrix ensembles are intimately related to orthogonal polynomials:

Wigner

This is the matrix equivalent of the Gaussian distribution. A 1×1 Wigner matrix is a single Gaussian random number. This is intimately related to Hermite orthogonal polynomials. The Gaussian distribution and its associated Hermite polynomials appear very naturally in contexts where the underlying variable is unbounded above and below. The average of the characteristic polynomials of Wigner random matrices obey simple recursion relations that allow us to express them as Hermite polynomials. Wigner ensemble is the simplest of all ensembles of random matrices. These are matrices where all elements are Gaussian random variables, with the only constraint that the matrix is real symmetric (the Gaussian orthogonal ensemble), complex Hermitian (the Gaussian unitary ensemble), or symplectic (the Gaussian symplectic ensemble).

Wishart

This is the matrix equivalent of the gamma distribution. A 1×1 Wishart is a gamma distributed number. This is intimately related to Laguerre orthogonal polynomials. Gamma distributions and Laguerre polynomials appear in problems where the variable is bounded from below (e.g., positive variables). The average of the characteristic polynomials of Wishart random matrices obey simple recursion relations that allow us to express them as Laguerre polynomials.

Jacobi

This is the matrix equivalent of the beta distribution. A 1×1 Wishart is a beta distributed number. This is intimately related to Jacobi orthogonal polynomials. Beta distributions and Jacobi polynomials appear in problems where the variable is bounded from above and from below. A natural setting where Jacobi matrices appear is that of sample covariance matrices. They also show up in the simple problem of addition or multiplications of matrices with only two eigenvalues.

As for scalar random variables, we study the moments and the Stieltjes transform of random matrix ensembles. Moreover, since we are in the matrix realm, we study the joint probability distributions of the eigenvalues of these random matrices. For the ensembles mentioned previously, the eigenvalues are strongly correlated and we can think of them as particles interacting through pair-wise repulsion. These are called Coulomb repelling eigenvalues, and the idea here is borrowed from statistical physics (see, for example, "Patterns in Eigenvalues" by Persi Diaconis (2003) (*https://oreil.ly/Cg0hY*) for a deeper dive into the behavior of the eigenvalues of matrices with special structures). It turns out that the most probable positions of the Coulomb gas problem coincide with the zeros of Hermite polynomials in the Wigner case, and of Laguerre polynomials in the Wishart case. Moreover, the eigenvalues of these ensembles fluctuate very little around their most probable positions.

Eigenvalue Density of the Sum of Two Large Random Matrices

Other than finding the joint probability distribution of the eigenvalues of random matrix ensembles, we care about the eigenvalue density (probability distribution) of sums of large random matrices, in terms of each individual matrix in the sum. *Dyson Brownian motion* appears in this context. It is an extension of *Brownian motion* from scalar random variables to random matrices. Moreover, a Fourier transform for matrices allows us to define the analog of the generating function for scalar independent and identically distributed random variables, and use its logarithm to find the eigenvalue density of sums of carefully constructed random matrices. Finally, we can apply Chernoff, Bernstein, and Hoeffding-type inequalities to the maximal eigenvalue of a finite sum of random Hermitian matrices.

Essential Math for Large Random Matrices

Before leaving the discussion of large random matrices, let's highlight the *must knows* if we want to dive deeply into this field. We touch on some of these in this chapter and leave the rest for your googling skills:

- Computing the spectrum: eigenvalues and eigenvectors of a matrix (solutions of $A\vec{v} = \lambda\vec{v}$)
- Characteristic polynomial of a matrix ($det(\lambda I - A)$)
- Hermite, Laguerre, and Jacobi orthogonal polynomials
- The Gaussian, gamma, and beta probability distributions
- Moments and moment generating function of a random variable
- Stieltjes transform
- Chebyshev everything
- Markov everything
- Chernoff, Bernstein, and Hoeffding-type inequalities
- Brownian motion and Dyson Brownian motion

As of 2022, the fastest supercomputer is *Frontier*, the world's first exascale computer (1.102 exaFLOPS), at the Department of Energy's Oak Ridge National Laboratory. When matrices are very large, even on such a supercomputer, we cannot apply numerical linear algebra (for example, to solve systems of equations involving the matrix, to find its spectrum, or to find its singular value decomposition) as we know it. What we must do instead is *random sampling of the columns of the matrix*. It is best to sample the columns with probability that leads to the most faithful approximation, the one with the least variance. For example, if the problem is to multiply two large matrices A and B with each other, instead of sampling a column from A and a corresponding row from B uniformly, we choose the column from A and the corresponding row from B with a probability p_j proportional to *norm(column j of A)norm(row j of B)*. This means that we choose the columns and rows with large norms more often, leading to a higher probability of capturing the *important parts* of the product.

The column space of large and not large matrices is so important. Keep in mind the three best bases for the column space of a given matrix A:

- The singular vectors from the singular value decomposition
- The orthogonal vectors from the Gram Schmidt process (the famous QR decomposition of the matrix)
- Linearly independent columns directly selected from the columns of A

Stochastic Processes

Rather than thinking of a static (scalar or vector or matrix or tensor) random variable, we now think of a time-dependent random variable. Somehow the next step in math always ends up including time-evolving entities. On a side note, humans have not yet fully understood the nature of time or found a way to articulate its definition. We do, however, understand movement and change, a system transitioning from one state to another, and we associate time with that. We also associate probabilities for transitioning from one state to another. Hold this thought for Markov chains, coming up in a bit.

A *stochastic process* is an *infinite* sequence $X_0, X_1, X_2, \ldots$ of random variables, where we think of the index t in each X_t as discrete time. So X_0 is the process at time *0* (or the value of a random quantity at a certain time *0*), X_1 is the process at time *1* (or the value of a random quantity at a certain time *1*), and so on. To formally define a random variable, which we have not done yet, we usually fix it over what we call a probability triple *(a sample space, a sigma algebra, a probability measure)*. Do not fuss about the meaning of this triple yet; instead, fuss about the fact that all the random variables in one stochastic process $X_0, X_1, X_2, \ldots$ live over the *same* probability triple, in this sense belonging to one family. Moreover, these random variables are often not independent.

Equally important (depending on the application) is a *continuous time* stochastic process, where X_t now encodes the value of a random quantity at *any* nonnegative time *t*. Moreover, this is easy to align with our intuitive perception of time, which is continuous.

Therefore, a stochastic process is a generalization of a finite dimensional multivariable distribution to infinite dimensions. This way of thinking about it comes in handy when trying to prove the *existence* of a stochastic process, since then we can resort to theorems that allow us to *extend* to infinite dimensions by relying on finite dimensional collections of the constituent distributions.

Examples of stochastic processes are all around us. We think of these whenever we encounter fluctuations: movement of gas molecules, electrical current fluctuations, stock prices in financial markets, the number of phone calls to a call center in a certain time period, or a gambler's process. And here's an interesting finding from a microbiology paper discussing bacteria found in the gut (*https://oreil.ly/sdz6v*): *the community assembly of blood, fleas, or torsalos is primarily governed by stochastic processes, while the gut microbiome is determined by deterministic processes.*

The stock market example is central in the theory of stochastic processes, because it is how Brownian motion (also called the Wiener stochastic process) got popularized, with L. Bachelier studying price changes in the Paris Bourse. The example of phone calls to a call center is also central in the theory, because it is how Poisson stochastic process got popularized, with A. K. Erlang modeling the number of phone calls occurring in a certain period of time.

These two processes, Brownian and Poisson, appear in many other settings that have nothing to do with the previously mentioned examples. Maybe this tells us something deeper about nature and the unity of its underlying processes, but let's not get philosophical and stay with mathematics. In general, we can group stochastic processes into a few categories, depending on their mathematical properties. Some of these are discrete time processes, and others are continuous time. The distinction between which one is which is pretty intuitive.

To derive conclusions about Brownian and Poisson processes, and the other stochastic processes that we are about to overview, we need to analyze them mathematically. In probability theory, we start with establishing the *existence* of a stochastic process. That is, we need to explicitly define the probability triple (sample space, sigma algebra, probability measure) where the discrete time *infinite* sequence of random variables $X_0, X_1, X_2,...$ or the continuous time X_t process lives, and prove that we can find such a set of random variables satisfying its characterizing properties. We will revisit this later in this chapter, but a big name we want to search for when proving the existence of stochastic processes is A. Kolmogorov (1903–1987), namely, the *Kolmogorov existence theorem*. This ensures the existence of a stochastic process having the same *finite dimensional* distributions as our desired processes. That is, we can get our desired stochastic process (infinite process, indexed on discrete or continuous time) by specifying all finite dimensional distributions in some *consistent* way.

Let's survey the most prominent stochastic processes.

Bernoulli Process

This is the stochastic process mostly associated with repeatedly flipping a coin and any process in life that mimics that (at some airports, customs officials make us press a button: if the light turns green we pass, if it turns red we get searched). Mathematically, it is an infinite sequence of independent and identically distributed random variables $X_0, X_1, X_2,...$, where each random variable takes either the value zero with probability p or one with probability *1-p*. A sample realization of this process would look like 0,1,1,0,....

Poisson Process

We can think of the Poisson process as a stochastic process whose underlying random variables are counting variables. These count how many interesting events happen within a set period of time. These events are either independent or are weakly dependent, and each has a small probability of occurrence. They also happen at a set expected rate λ. This is the parameter that characterizes a Poisson random variable.

For example, in queuing theory, we use it to model the arrival of customers at a store, phone calls at a call center, or the occurrence of earthquakes in a certain time interval. This process has the natural numbers as its state space and the nonnegative numbers as its index set. The probability distribution underlying the random variables involved in a Poisson process has the formula:

$$Prob(X = n) = \frac{\lambda^n e^{-\lambda}}{n!}$$

The formula gives the probability of n interesting events occurring in a unit period of time. Clearly, within a fixed time interval it is not likely for many rare events to occur, which explains the rapid decay in the formula for a large n. The expectation and the variance of a Poisson random variable is λ.

A Poisson *process* $X_t; t \geq 0$, indexed by continuous time, has the properties:

- $X_0 = 0$
- The number of events (or points) in any interval of length t is a Poisson random variable with parameter λt

A Poisson process has two important features:

- The number of events in each finite interval is a Poisson random variable (has a Poisson probability distribution).
- The number of events in disjoint time intervals are independent random variables.

The Poisson process is an example of a Levy stochastic process, which is a process with stationary independent increments.

Random Walk

It is easy to think of the simplest random walk as someone taking steps on a road where they start somewhere, then move forward (add one to their position) with probability p, and backward (subtract one from their position) with probability $1 - p$. We can define the resulting discrete time stochastic process $X_0, X_1, \cdots$ such that $X_0 = x_0$, $X_1 = X_0 + Z_1$, $X_2 = X_1 + Z_2 = X_0 + Z_1 + Z_2$, etc., where $Z_1, Z_2, \cdots$ is a Bernoulli process. If $p = 0.5$, this is a symmetric random walk.

In Chapter 9, we used random walks on graphs multiple times, where we start at a certain graph node and then transition to one of its adjacent nodes with given probabilities. A normalized adjacency matrix of the graph would define the transition probabilities at all nodes. This is a neat example of how random walks on graphs tie to Markov chains, coming up soon. For more on this, check this nice set of notes on random walks on graphs (*https://oreil.ly/Cjk0C*).

Wiener Process or Brownian Motion

We can think of a Wiener process or a Brownian motion as a random walk with infinitesimally small steps, so discrete movements become infinitesimally small fluctuations, and we get a *continuous* random walk. A Brownian motion is a continuous time stochastic process $X_t; t \geq 0$. The random variables X_t are real valued, have independent increments, and the difference between X_t and X_s at two separate times t and s is normally distributed (follows a Gaussian bell-shape distribution) with mean 0 and variance $t - s$. That is, $X_t - X_s$ are normally distributed based on the size of the increments.

The interesting thing about a real valued continuous time stochastic process is that it can move in continuous paths, giving rise to interesting random functions of time. For example, almost surely, a sample path of a Brownian motion (Wiener process) is continuous everywhere, but nowhere differentiable (too many spikes).

Brownian motion is fundamental in the study of stochastic processes. It is the starting point of stochastic calculus, lying at the intersection of several important classes of processes: it is a Gaussian Markov process, a Levy process (a process with stationary independent increments), and a martingale, discussed next.

Martingale

A discrete time martingale is a stochastic process $X_0, X_1, X_2, \ldots$ where for any discrete time t:

$$\mathbb{E}(X_{t+1} | X_1, X_2, \cdots, X_t) = X_t$$

That is, the expected value of the next observation, given all the previous observations, is equal to the most recent observation. This is a weird way of defining something (sadly, this is very common in this field), but let's give a few brief examples of some contexts within which martingales appear:

- An unbiased random walk is an example of a martingale.
- A gambler's fortune is a martingale if all the betting games that the gambler plays are fair. Suppose a gambler wins \$1 if a fair coin comes up heads, and loses \$1 if it comes up tails. If X_n is the gambler's fortune after n tosses, then the gambler's conditional expected fortune after the next coin toss, given the history, is equal to their present fortune.

- In an ecological community, where a group of species compete for resources, we can model the number of individuals of any particular species as a stochastic process. This sequence is a martingale under the unified neutral theory of biodiversity and biogeography (*https://oreil.ly/oGlR2*).

Stopping times appear when discussing martingales. This is an interesting concept, capturing the idea that: *at any particular time t, you can look at the sequence so far and tell if it is time to stop.* A stopping time with respect to a stochastic process $X_1,X_2,X_3,...$ is a random variable S (for *stop*) with the property that for each t, the occurrence or nonoccurrence of the event $S = t$ depends only on the values of $X_1,X_2,X_3,...,X_t$. For example, the stopping time random variable models the time at which a gambler chooses to stop and leave a gambling table. This will depend on their previous winnings and losses, but not on the outcomes of the games that they haven't yet played.

Levy Process

We have mentioned the Poisson process and Brownian motion (Wiener process) as two of the most popular examples of a Levy process. This is a stochastic process with independent, stationary increments. It can model the motion of a particle whose successive displacements are random, in which displacements in pair-wise, disjoint time intervals are independent, and displacements in different time intervals of the same length have identical probability distributions. In this sense, it is the continuous time analog of a random walk.

Branching Process

A branching process randomly splits into branches. For example, it models a certain population's evolution (like bacteria, or neutrons in a nuclear reactor) where each individual in a given generation produces a random number of individuals in the next generation, according to some fixed probability distribution that does not vary from individual to individual. One of the main questions in the theory of branching processes is the probability of ultimate extinction, where the population dies out after a finite number of generations.

Markov Chain

Let's formally define a discrete time Markov chain, since it is one of the most important stochastic processes, and because it comes up in the context of reinforcement learning in AI. To define a Markov chain, we need:

- A discrete set of possible states S (finite or infinite). Think of this as the set of states that a particle or an agent can occupy. At each step, the Markov process randomly evolves from one state to another.

- An initial distribution prescribing the probability ν_i of each possible $state_i$. Initially, how likely is it for a particle to be at a certain location or for an agent to be in a certain state upon initialization?

- Transition probabilities p_{ij} specifying the probability that the particle or the agent transitions from $state_i$ to $state_j$. Note that we have, for each state i, the sum $p_{i1} + p_{i2} + \cdots + p_{in} = 1$. Moreover, this process has no memory, because this transition probability depends only on $state_i$ and $state_j$, not on previously visited states.

Now, a Markov chain is a stochastic process $X_0, X_1, \cdots$ taking values in S such that:

$$Prob\left(X_0 = state_{i_0}, X_1 = state_{i_1}, \cdots, X_n = state_{i_n}\right) = \nu_{i_0} p_{i_0 i_1} p_{i_1 i_2} \cdots p_{i_{n-1} i_n}$$

We can bundle up all the transition probabilities in a square matrix (a Markov matrix, with each row having nonnegative numbers adding up to 1), and multiply by that matrix. This matrix summarizes the probabilities of transitioning from any $state_i$ to $state_j$ in one step. The cool thing is that the powers of a Markov matrix are also Markov, so for example, the square of this matrix summarizes the probabilities of transitioning from any $state_i$ to $state_j$ in two steps, and so on.

Some fundamental notions related to Markov chains are transience, recurrence, and irreducibility. A $state_i$ is recurrent if we start from it, then we will certainly eventually return to it. It is called transient if it is not recurrent. A Markov chain is irreducible if it is possible for the chain to move from any state to any other state.

Finally, a stationary probability vector, defining a probability distribution over the possible states, is one that does not change when we multiply it by the transition matrix. This ties straight into eigenvectors in linear algebra with corresponding eigenvalue one. The *feel good* high that we get when the math we know connects together is an addictive feeling and is what keeps us trapped in a *love-hate* relationship with this field. Thankfully, the longer we stay, the more *love-love* it becomes.

Itô's Lemma

Let's tie a bit more math together. A stochastic process models a random quantity that evolves with time. A function of a stochastic process also evolves randomly with time. When deterministic functions evolve with time, the next question is usually, *how fast?* To answer that, we take its derivative with respect to time, and develop calculus around the derivatives (and integrals) of deterministic functions. The chain rule is of paramount importance, especially for training machine learning models.

Itô's lemma is the analog of the chain rule for functions of stochastic processes. It is the stochastic calculus counterpart of the chain rule. We use it to find the differential of a time-dependent function of a stochastic process.

Markov Decision Processes and Reinforcement Learning

In the AI community, Markov decision processes are associated with:

Dynamic programming and Richard Bellman
Bellman played a monumental role in the field, and his optimality condition is implemented in many algorithms.

Reinforcement learning
Finding an optimal strategy via a sequence of actions that are associated with positive or negative rewards (trials and errors). The agent has choices between several actions and transition states, where the transition probabilities depend on the chosen actions.

Deep reinforcement learning
Combining reinforcement learning with neural networks. Here, a neural network takes the observations as input, and outputs a probability for each possible action that the agent can take (a probability distribution). The agent then decides on the next action randomly, according to the estimated probabilities. For example, if the agent has two choices, turn left or turn right, and the neural network outputs 0.7 for turn left, then the agent will turn left with 70% probability, and turn right with 30% probability.

Reinforcement learning has great potential to advance us toward general intelligence. Intelligent agents need to make rational decisions when payoffs from actions are not immediate but instead result from a series of actions taken sequentially. This is the epitome of reasoning under uncertainty.

Examples of Reinforcement Learning

Examples of reinforcement learning are plentiful: self-driving cars, recommender systems, thermostat at home (getting positive rewards whenever it is close to the target temperature and saves energy, and negative rewards when humans need to tweak the temperature), and automatic investing in the stock market (the input is the stock prices, the output is how much of each stock to buy or sell, the rewards are the monetary gains or losses).

Perhaps the most famous example for deep reinforcement learning success is Deep-Mind's AlphaGo (*https://oreil.ly/0TcsR*), the AI agent that in 2016 beat the world's best human player in the ancient Chinese game of Go. Thinking of reinforcement learning in terms of board games, such as chess or Go, is intuitive, because at each step we decide on the sequence of actions that we must take, knowing very well that our current decision affects the whole outcome of the game. We better act optimally at each step. Moreover, at each step, our optimal strategy *evolves*, because it also depends on the actions of our opponent (who is solving the exact same problem but from their vantage point).

I am a bit biased against examples containing games, since nowadays my daughter is addicted to PlayStation 5. I prefer the investment market example. Our financial adviser operates in a market that changes every day, and needs to make decisions on buying/selling certain stocks at each time step, with the long-term goal of maximizing profit and minimizing losses. The market environment is stochastic and we do not know its rules, but we assume that we do for our modeling purposes. Now let's switch our human financial adviser to an AI agent, and let's see what kind of optimization problem this agent needs to solve at each time step in the constantly changing market environment.

Reinforcement Learning as a Markov Decision Process

Let's formulate reinforcement learning mathematically as a Markov decision process. The environment within which our agent exists is probabilistic, consisting of states and transition probabilities between these states. These transition probabilities *depend on the chosen actions*. Thus, the resulting Markov process that encodes the transitions from any state $\overrightarrow{state}$ to another state $\overrightarrow{state}'$ has an explicit dependency on the action $\overrightarrow{a}$ taken while in state $\overrightarrow{state}$.

The main assumption here is that we *know* this process, which means that we know the rules of the environment. In other words, we know the following probability for each $\overrightarrow{state}$, $\overrightarrow{state}'$, and action $\overrightarrow{a}$:

$$Prob\left(\text{next state} = \overrightarrow{state}' \,\middle|\, \text{current state} = \overrightarrow{state}, \text{action taken} = \overrightarrow{a}\right)$$

We also know the reward system, which is:

$$Prob\left(\text{next reward value} \,\middle|\, \text{current state} = \overrightarrow{state}, \text{action taken} = \overrightarrow{a}, \text{next state} = \overrightarrow{state}'\right)$$

This discussion now belongs in the realm of dynamic programming: we search for the *optimal policy* (sequence of good actions) leading to *optimal value* (maximal reward or minimal loss). This optimization problem is a bit more involved than the ones we have encountered so far, because it is a *sequence of actions* that leads to the optimal value. Therefore, we must divide the problem into steps, looking for the action at each step that *shoots for* the optimal reward *multiple steps ahead in the future*. *Bellman's optimality equation* solves exactly this problem, simplifying it into the search for *only one optimal action* at the current state (as opposed to searching for all of them at once), given that we know what problem to optimize at each step. Bellman's huge contribution is the following assertion: the optimal value of the current state is equal to the average reward after taking *one* optimal action, *plus the expected optimal value* of all possible next states that this action can lead to.

The agent interacts with its environment via an iterative process. It starts with an initial state and a set of that state's possible actions (the probability distribution for taking an action given that state), then computes the following, iteratively:

1. The next optimal action to take (which transitions it to a new state with a new set of possible actions). This is called the *policy iteration*, and the optimization goal is to maximize future reward.

2. The expected value (rewards or losses) given that optimal action. This is called the *value iteration*.

The value function adds up the agent's expected future rewards given its current state and the optimal sequence of actions taken afterward:

$$Value\left(\overrightarrow{state}, \text{optimal sequence of actions}\right) = \mathbb{E}\left(\Sigma_k \gamma^k reward_k\right)$$

The *discount factor* γ is a number between 0 and 1. It is useful to encourage taking actions that result in *sooner rather than later* positive rewards. Putting this factor in the optimization problem adjusts the importance of rewards over time, giving less weight to future rewards (if γ is between 0 and 1, then γ^k is small for large k).

Let's make the optimization in the value function explicit (we are choosing the sequence of actions that maximizes the rewards given the current state):

$$Value\left(\overrightarrow{s}\right) = \max_{\text{actions and states}} \mathbb{E}\left(\Sigma_{k=0}^{\infty} \gamma^k (reward)_k \Big| \overrightarrow{state_0} = \overrightarrow{s}\right)$$

Now we break this up to make sure the agent's current reward is explicit and separate from its future rewards:

$$Value\left(\overrightarrow{s}\right) = \max_{\text{actions and states}} \mathbb{E}\left(reward_0 + \Sigma_{k=1}^{\infty} \gamma^k (reward)_k \Big| \overrightarrow{state_1} = \overrightarrow{s}'\right)$$

Finally, we find that the value function at the agent's current state depends on its current reward and a discounted value function at its future states:

$$Value\left(\overrightarrow{s}\right) = \max_{\text{actions and states}} \mathbb{E}\left(reward_0 + \gamma Value\left(\overrightarrow{s}'\right)\right)$$

That statement allows us to solve our main optimization problem iteratively (backward in time). All the agent has to do now is to choose the action to get to the *next best state*. This expression for the value function is the powerful *Bellman's equation* or *Bellman's optimality condition*, which breaks up the original optimization problem into a recursive sequence of much simpler optimization problems, optimizing locally at each state (finding $Value\left(\overrightarrow{s}'\right)$), then putting the result into the next optimization subproblem (finding $Value\left(\overrightarrow{s}\right)$). The miracle is that working backward this way from the desired ultimate reward to deciding what action to take now gives us the overall optimal strategy along with the optimal value function at each state.

Reinforcement Learning in the Context of Optimal Control and Nonlinear Dynamics

In Chapter 13 on PDEs, we revisit reinforcement learning in the context of nonlinear dynamics, optimal control, and the Hamilton-Jacobi-Bellman partial differential equation. Unlike the probabilistic Markov environment that our agent interacted with in the previous discussion, the dynamic programming approach to reinforcement learning (which leads to the Hamilton-Jacobi-Bellman partial differential equation) is deterministic.

Python Library for Reinforcement Learning

Finally, a helpful library for implementing reinforcement learning algorithms is the TF-Agents library (by Google, open sourced in 2018), a reinforcement learning library based on TensorFlow (Python).

Theoretical and Rigorous Grounds

Rigorous, or mathematically precise, probability theory needs measure theory. *But why?* you might rightfully protest. After all, we have managed to avoid this for the longest time.

Because we cannot avoid it any longer.

Let's write this but never admit to it out loud: it is measure theory that turns off many students from pursuing further studies in math, mostly, because its story is never told in chronological order about how and why it came to be. Moreover, a lot of work in measure theoretic probability has to do with *proving* that a certain random variable *exists* (over some sample space, an event space or a sigma algebra, and a measure for each event or set in that sigma algebra), as if writing the random variable down and using it to model all sorts of random entities is not enough existence. This must be the reason why mathematicians and philosophers get along so well.

We have already flown through many concepts in this chapter at lightning speed (my students accuse me of this all the time), but we need to start over and give:

- A precise mathematical understanding of probabilities and sigma algebras
- A precise mathematical definition of a random variable and a probability distribution
- A precise mathematical definition of an expected value of a random variable, and its connection to integration
- An overview of probability inequalities (controlling uncertainty)
- An overview of the law of large numbers, the central limit theorem, and other convergence theorems

Alright, that is too ambitious. We cannot give a full course on rigorous probability theory in one section of one chapter. What we will do instead is make a convincing case for it, and leave with a decent understanding of the fundamental ideas.

We start with two major limitations of nonrigorous probability (other than the fact that each of its math objects has countless inconsistent names, notations, and fuzzy definitions).

Which Events Have a Probability?

Given a sample space (a set that we can sample randomly from), can any subset have a probability defined on it? What if we are sampling numbers uniformly from the real line and ask, what is the probability that we pick a rational number? An algebraic number (the solution to some polynomial equation with integer coefficients)? Or a member from some other complicated subset of the real line?

See how these questions are slowly drawing us into the details of set theory on the real line, which in turn pulls us straight into *measure theory*: the theory that addresses *which subsets of the real line we can measure, and which subsets we cannot.*

Defining a probability for a subset of a sample space is starting to sound a lot like defining a measure of that set, and it seems like only subsets that are *measurable* can have a probability defined for them. How about the other *nonmeasurable* subsets of the sample space? Too bad for them, we cannot define a probability for them. To reiterate, *Prob(A)* does not make sense for every subset A of a sample space; instead, it only makes sense for measurable subsets of that space. So we must harness all the measurable subsets together, abandon the rest and never think about them or their mathematics, and relax, because then we can act in a realm where all the events (subsets) that we harnessed have probabilities (measures) defined for them. The probability measure that we work with satisfies reasonable properties, in the sense that it is a nonnegative number in [0,1], and probabilities of complementary events (subsets) add up to 1. This whole roundabout reveals the intricacies of the real line and its subsets, and more generally, the mysteries of the continuum and the wonders of the infinite.

Rigorous probability theory helps us appreciate the properties of both discrete and continuous spaces, revealed in examples as simple as constructing the discrete uniform distribution on a discrete set versus constructing the continuous uniform distribution on a given interval.

Can We Talk About a Wider Range of Random Variables?

The other limitation of nonrigorous probability, meaning the one that avoids measure theory as we just described it, is the restriction on the kinds of random variables that it allows. In particular, where exactly do we draw the line between a discrete and continuous random variable? Is there really such a line? How about random variables that have both discrete and continuous aspects? As a simple example, suppose a random variable's value is decided by a flip of a coin. It is Poisson distributed (discrete) if the coin comes up heads, and normally distributed (continuous) if the coin comes up tails. This new random variable is neither fully discrete nor fully continuous in the nonrigorous sense that we understand either type. Then what is it? The rigorous answer is this: of course there is no distinction between discrete and continuous random variables once we define the grounds that *any* random variable stands on. Here's the rigorous ground it must stand on. What set formulates the sample space? What subsets of this sample space are measurable? What is the probability measure? What is the *distribution* of the random variable? This is the common ground, or the starting point for *any* random variable. Once we specify this ground, then discrete, continuous, or anything in between becomes a small detail, as simple as answering: what set (or, say, product of sets) are we working with?

A Probability Triple (Sample Space, Sigma Algebra, Probability Measure)

It all starts with a probability triple (not really, but this is where rigor starts). We call this a probability measure space, with the understanding that the measure of the whole sample space is equal to one. That is, the probability of the sample space is one. We are now feeling very advanced, using the words probability and measure interchangeably. The comfort that the word measure provides is that it brings us back to a deterministic realm. The sampling is random, but we can *measure* the likelihood of any occurrence (that is measurable).

The three objects making up a probability measure space are:

The sample space
  The arbitrary nonempty set that we randomly pull samples from.

The sigma algebra
  A set of subsets of the sample space that represent the allowed events (the events that we are allowed to talk to about their probability, because they are the only ones we are able to measure). A sigma algebra must contain the whole sample space, is closed under complements (meaning if a set is in the sigma algebra then so is its complement), and is closed under countable unions (meaning the union of countably many subsets of the sigma algebra is also a member of the sigma algebra). The corollary from the previous two properties and De Morgan's laws (*https://oreil.ly/pnvp0*) (which have to do with complements of unions and intersections) is that the sigma algebra is also closed under countable intersections.

The probability measure
  A number between zero and one (inclusive) associated with *each subset* of the sigma algebra, that satisfies the reasonable properties that we associate with nonrigorous probability:

  1. Prob(sample space) = 1

  2. Prob(countable union of pair-wise disjoint sets) = countable sum of probabilities of each set

This is very good, because as long as we are able to articulate the sample space set, the sigma algebra, and a function with the previously mentioned properties mapping every member of the sigma algebra to its measure (probability), then we can start *building the theory* on solid grounds, defining all kinds of random variables, their expectations, variances, conditional probabilities, sums and products, limits of sequences, stochastic processes, time derivatives of functions of stochastic processes (Itô's calculus), and so on. We would not run into problems of what type of events have probabilities defined for them (all the members of the sigma algebra of the probability triple), or what type of random variables we can consider (any that we can rigorously define over a probability triple).

Where Is the Difficulty?

Note that the limitations of nonrigorous probability that we discussed *both* appear when we involve continuous variables, or when the sample space is in the continuum (uncountable). Had our world been only discrete, we wouldn't be going through all this trouble. When we move to rigorous probability and attempt to construct probability triples for discrete sample spaces, we do not run into much trouble. The challenges appear in the continuum world, with uncountable sample spaces. Because suddenly we have to identify sigma algebras and associated probability measures on sets where the depth of the infinite continuum never ceases to fascinate. For example, this challenge appears even when we want to define a rigorous probability triple for the continuous uniform distribution on the interval [0,1].

The *extension theorem* runs to our aid and allows us to construct complicated probability triples. Instead of defining a probability measure over a massive sigma algebra, we construct it on a simpler set of subsets, a *semialgebra*, then the theorem allows us to automatically extend the measure to a full sigma algebra. This theorem allows us to construct Lebesgue measure on [0, 1] (which is exactly the continuous uniform distribution on [0,1]), product measures, the multidimensional Lebesgue measure, and finite and infinite coin tossing.

The worlds of set theory, real analysis, and probability have blended neatly together.

Random Variable, Expectation, and Integration

Now that we can associate probability triple with a sample space, defining probabilities for a large amount of subsets of the sample space (all the members of the associated sigma algebra), we can rigorously define a random variable. As we know very well from nonrigorous probability, a random variable assigns a numerical value to each element of the sample space. So if we think of the sample space as all the possible random outcomes of some experiment (heads and tails of flipping a coin), then a random variable assigns a numerical value to each of these outcomes.

To build on rigorous grounds, we must define how a random variable Y interacts with the whole probability triple associated with the sample space. The short answer is that: Y must be a *measurable* function from the sample space to the real line, in the sense that the set $Y^{-1}(-\infty, y)$ is a member of the sigma algebra, which in turn means that this set has a probability measure. Note that Y maps from the sample space to the real line, and Y^{-1} maps back from the real line to a subset of the sample space.

Just like a random variable from nonrigorous probability turns out to be a measurable function (with respect to a triple) in rigorous probability theory, the expectation $\mathbb{E}(Y)$ of a random variable turns out to be the same as the integral of the random variable (measurable function) with respect to the probability measure. We write:

$$\mathbb{E}(Y) = \int_\Omega Y dP = \int_\Omega Y(\omega) Prob(d\omega)$$

Understanding the Integral Notation in the Expectation Formula

It is easy to understand the integral with respect to a probability measure, such as the one in the previously mentioned formula, if we think of the meaning of the expectation of a random variable in a discrete setting, as the sum of the value of the random variable times the probability of the set over which it assumes that value:

$$\mathbb{E}(Y) = \Sigma_{i=1}^{n} y_i Prob(\omega \in \Omega \text{ such that } Y(\omega) = y_i)$$

Now compare this discrete expression to the continuum integral in the expectation formula:

$$\mathbb{E}(Y) = \int_{\Omega} Y\, dP = \int_{\Omega} Y(\omega) Prob(d\omega)$$

We rigorously build up the integral (expectation) the exact same way we build up the Lebesgue integral in a first course on measure theory: first for simple random variables (which we can easily break up into a discrete sum; integrals start from sums), then for nonnegative random variables, and finally for general random variables. We can easily prove basic properties for integrals, such as linearity and order preserving. Note that whether the sample space is discrete, continuous, or anything complicated, as long as we have our probability triple to build on, the integral makes sense (in a much wider range of setting than we ever imagined for our basic calculus Reimann-style integration). Once we encounter Lebesgue-style integration, we sort of never look back.

Now that we have the expectation, we can define the variance and covariance exactly the same way as nonrigorous probability theory.

Then we can talk about independence, and important properties such that if X and Y are independent, then $E(XY) = E(X)E(Y)$ and $Var(X + Y) = Var(X) + Var(Y)$.

Distribution of a Random Variable and the Change of Variable Theorem

The distribution of a random variable X is *a corresponding probability triple* $(\mathbb{R}, \mathscr{B}, \mu)$ defined on the real line, such that for every subset B of the *Borel* sigma algebra defined on the real line, we have:

$$\mu(B) = P(X \in B) = P\left(X^{-1}(B)\right)$$

This is completely determined by the cumulative distribution function, $F_X(x) = P(X \leq x)$, of X.

Suppose we have a measurable real valued function f defined on the real line. Let X be a random variable on a probability triple $(\Omega, sigmaalgebra, P)$ with distribution μ. Note that for any real number x, $f(x)$ is a real number, and for the random variable X, $f(X)$ is a random variable.

The change of variable theorem says that the expected value of the random variable $f(X)$ with respect to the probability measure P on a sample space Ω is equal to the expected value of the function f with respect to the measure μ on $\mathbb{R}$. Let's write this first in terms of expectation and then in terms of integrals:

$$\mathbb{E}_P(f(X)) = \mathbb{E}_\mu(f)$$

$$\int_\Omega f(X(\omega)) P(d\omega) = \int_{-\infty}^\infty f(t) \mu(dt)$$

A nice thing that comes in handy from this change of variables theorem is that we can switch between expectations, integrations, and probabilities. Let f be the indicator function of a measurable subset of $\mathbb{R}$ (which is one over the subset and zero otherwise), then the formula gives us:

$$\int_{-\infty}^\infty \mathbf{1}_B \mu(dt) = \mu(B) = P(X \in B).$$

Note that in Chapter 8, we encountered another change of variables theorem from probability, which relates the probability distribution of a random variable with the probability distribution of a deterministic function of it, using the determinant of the Jacobian of this function transformation.

Next Steps in Rigorous Probability Theory

The next step in rigorous probability theory is to prove the famous inequalities (Markov, Chebyshev, Cauchy-Schwarz, Jensen's), introduce sums and products of random variables, the laws of large numbers, and the central limit theorem. Then we move to sequences of random variables and limit theorems.

Limit theorems

If we have a sequence of random variables that converges to some limit random variable, does it follow that the expectations of the sequence converge to the expectation of the limit? In integral language, when can we exchange the limit and the integral?

This is when we prove the monotone convergence, the bounded convergence, Fatou's lemma, the dominated convergence, and the uniformly integrable convergence theorems.

Finally we consider double or higher integrals, and conditions on when it is OK to flip integrals. Fubini's theorem answers that, and we can apply it to give a convolution formula for the distribution of a sum of independent random variables.

The Universality Theorem for Neural Networks

Rigorous measure theory (probability theory) helps us prove theorems for neural networks, which is an up-and-coming subfield of mathematics, aiming to provide theoretical grounds for many empirical AI successes.

The universality theorem for neural networks is a starting point. We have referred to it multiple times in this book. Here's the statement:

> For any continuous function f on a compact set K, there exists a feed forward neural network, having only a single hidden layer, which uniformly approximates f to within an arbitrary $\epsilon > 0$ on K.

This web page (*https://oreil.ly/7A8Gn*) has a nice and easy-to-follow proof.

Summary and Looking Ahead

In this chapter, we surveyed concepts in probability that are important for AI, machine learning, and data science. We zipped through topics such as causal modeling, paradoxes, large random matrices, stochastic processes, and reinforcement learning in AI.

Often when we learn about probability, we fall into the *frequentist* versus the *objectivist* positions regarding the definitions and the overall philosophy surrounding uncertainty. The following is a neat description of each viewpoint:

A frequentist position
    Probabilities can only come from experiments and observing the results of repeated trials.

An objectivist position
    Probabilities are real aspects of the universe: a real inclination or natural tendency to behave in a particular way. For example, a fair coin's propensity to turn up heads 50% of the time is an intrinsic property of the fair coin itself.

A frequentist is then only attempting to measure these natural inclinations via experiments. Rigorous probability theory unifies disparate views of probability. We swiftly introduced rigorous probability theory and established that it is in essence the same as measure theory in real analysis. We ended with the universal approximation theorem for neural networks.

We leave this chapter with a perfectly fitting tweet from Yann LeCun (*https://oreil.ly/X1eDh*), which happens to touch on every topic we covered in this chapter:

I believe we need to find new concepts that would allow machines to:

- Learn how the world works by observing like babies.

- Learn to predict how one can influence the world through taking actions.

- Learn hierarchical representations that allow long-term predictions in abstract representation spaces.

- Properly deal with the fact that the world is not completely predictable.

- Enable agents to predict the effects of sequences of actions so as to be able to reason and plannable machines to plan hierarchically, decomposing a complex task into subtasks.

- All of this in ways that are compatible with gradient-based learning.

CHAPTER 12

Mathematical Logic

Humans bend the rules.

—H.

Historically in the AI field, logic-based agents come before machine learning and neural network–based agents. The reason we went over machine learning, neural networks, probabilistic reasoning, graph representations, and operations research before logic is that we want to tie it all into one narrative of reasoning within an agent, as opposed to thinking of logic as old and neural networks as modern. We want to view the recent advancements as enhancing the way a logical AI agent represents and reasons about the world. A good way to think about this is similar to enlightenment: an AI agent used to reason using the rigid rules of handcoded knowledge base and handcoded rules to make inferences and decisions, then suddenly it gets enlightened and becomes endowed with more reasoning tools, networks, and neurons that allow it to expand both its knowledge base and inference methods. This way, it has more expressive power and can navigate more complex and uncertain situations. Moreover, combining all the tools would allow an agent the option to sometimes break the rules of a more rigid logic framework and employ a more flexible one, depending on the situation, just like humans. Bending, breaking, and even changing the rules are distinctive human attributes.

The dictionary meaning of the word *logic* sets the tone for this chapter and justifies its progression.

Logic

A framework that organizes the rules and processes used for sound thinking and reasoning. It is a framework that lays down the principles of validity under which to conduct reasoning and inference.

The most important words to pay attention to in this definition are *framework* and *principles for inference*. A logic system codifies within an agent the principles that govern reliable inference and correct proofs. Designing agents that are able to gather knowledge, reason logically with a flexible logic system that accommodates uncertainty about the environment that they exist in, and make inferences and decisions based on this logical reasoning lies at the heart of artificial intelligence.

We discuss the various systems of mathematical logic that we can program into an agent. The goal is to give the AI agent the ability to make inferences that enable it to act appropriately. These logical frameworks require knowledge bases to accompany the inference rules of varying sizes. They also have varying degrees of expressive and deductive powers.

Various Logic Frameworks

For each of the different logical frameworks (*propositional*, *first order*, *temporal*, *probabilistic*, and *fuzzy*) that we are about to highlight in this chapter, we will answer two questions about how they operate within an agent endowed with them:

1. What objects exist in the agent's world? Meaning, how does the agent perceive the composition of its world?

2. How does the agent perceive the objects' states? Meaning, what values can the agent assign to each object in its world under the particular logic framework?

It is easy to think about this if we liken our agent to an ant and how it experiences the world (*https://oreil.ly/MD8kn*). Because of the ant's predetermined framework of perception and allowed movements, the ant experiences the world, along with its curvature, as two-dimensional. If the ant gets enhanced and endowed with a more expressive framework of perception and allowed movements (for example, wings), it will experience the three-dimensional world.

Propositional Logic

Here are the answers to our agent questions:

What objects exist in the agent's world?
> Simple or complex statements, called *propositions*, hence the name *propositional logic*.

How does the agent perceive the objects' states?
> True (1), false (0), or unknown. Propositional logic is also called Boolean logic because the objects in it can only assume two states. Paradoxes in propositional logic are statements that cannot be classified as true or false according to the logic framework's *truth table*.

These are examples of statements and their states:

- It is raining (can take true or false states).
- The Eiffel Tower is in Paris (always true).
- There is suspicious activity in the park (can take true or false states).
- This sentence is false (paradox).
- I am happy *and* I am sad (always false, unless you ask my husband).
- I am happy *or* I am sad (always true).
- If the score is 13, then the student fails (truth depends on failing thresholds, so we need a statement in the knowledge base that says: all students with a score below 16 fail, and set its value at true).
- 1 + 2 is equivalent to 2 + 1 (always true within an agent endowed with arithmetic rules).
- Paris is romantic (in propositional logic this has to be either true or false, but in fuzzy logic it can assume a value on a zero-to-one scale, for example, 0.8, which corresponds better to the way we perceive our world: on a scale as opposed to absolutes. Of course, I would assign the value true for this statement if I am programming an agent and confined to propositional logic, but someone who hates Paris would assign false. Oh well).

The objects in a propositional logic's world are simple statements and complex statements. We can form complex statements from simple ones using five allowed operators: *not* (negation), *and, or, implies* (which is the same as *if then*), and *equivalent to* (which is the same as *if and only if*).

We also have five rules to determine whether a statement is true or false:

1. The negation of a statement is true if and only if the statement is false.
2. *statement₁ and statement₂* are true if and only if both *statement₁* and *statement₂* are true.
3. *statement₁ or statement₂* is true if and only if either *statement₁* or *statement₂* are true (or if both are true).
4. *statement₁ implies statement₂* is true except when *statement₁* is true and *statement₂* is false.
5. *statement₁* is *equivalent to statement₂* if and only if *statement₁* and *statement₂* are both true or both false.

We can summarize these rules in a *truth table* accounting for all the possibilities for the states of *statement₁* and *statement₂* and for their joining using the five allowed operators. In the following truth table, we use S_1 for *statement$_1$* and S_2 for *statement$_2$* to save space:

S_1	S_2	not S_1	S_1 and S_2	S_1 or S_2	S_1 implies S_2	S_1 equivalent to S_2
F	F	T	F	F	T	T
F	T	T	F	T	T	F
T	F	F	F	T	F	F
T	T	F	T	T	T	T

We can compute the truth of any complex statement using this truth table by simple recursive evaluation. For example, if we are in a world where S_1 is true, S_2 is false, and S_3 is true, then we have the statement:

$$\text{not } S_1 \text{ and } (S_2 \text{ or } S_3) \Longleftrightarrow F \text{ and } (F \text{ or } T) = F \text{ and } T = F$$

To be able to reason and prove theorems using propositional logic, it is helpful to establish logical equivalences, meaning statements that have the exact same truth tables so they can replace each other in a reasoning process. The following are some examples of logical equivalences:

- Commutativity of *and*: S_1 *and* $S_2 \Longleftrightarrow S_2$ *and* S_1
- Commutativity of *or*: S_1 *or* $S_2 \Longleftrightarrow S_2$ *or* S_1
- Double negation elimination: not (not S_1) $\Longleftrightarrow S_1$
- Contraposition: S_1 *implies* $S_2 \Longleftrightarrow \text{not}(S_2)$ *implies* $\text{not}(S_1)$
- Implication elimination: S_1 *implies* $S_2 \Longleftrightarrow \text{not}(S_1)$ *or* S_2
- De Morgan's law: $\text{not}(S_1 \text{ and } S_2) \Longleftrightarrow \text{not}(S_1)$ *or* $\text{not}(S_2)$
- De Morgan's law: $\text{not}(S_1 \text{ or } S_2) \Longleftrightarrow \text{not}(S_1)$ *and* $\text{not}(S_2)$

Let's demonstrate that S_1 *implies* $S_2 \Longleftrightarrow \text{not}(S_1)$ *or* S_2 by showing that they have the same truth table, since this equivalence is not so intuitive for some people:

S_1	not (S_1)	S_2	$\text{not}(S_1)$ *or* S_2	S_1 *implies* S_2
F	T	F	T	T
F	T	T	T	T
T	F	T	T	T
T	F	F	F	F

One example that demonstrates how logical equivalences are useful is the *proof by contradiction* way of reasoning. To prove that the statement S_1 implies the statement S_2, we can assume that we have S_1 but at the same time we do not have S_2, then we arrive at something false or absurd, which proves that we cannot assume S_1 without concluding S_2 as well. We can verify the validity of this way of proving that S_1 implies S_2 using propositional logic equivalences:

- S_1 *implies* S_2 = true $\Longleftrightarrow$
- not(S_1) *or* S_2 = true (implication removal) $\Longleftrightarrow$
- not(not(S_1) *or* S_2) = not(true) $\Longleftrightarrow$
- S_1 *and* not(S_2) = false (De Morgan and double negation)

We endow a propositional logic framework with *rules of inference*, so that we are able to reason sequentially from one statement (simple or complex) to the next and arrive at a desired goal or at a correct proof of a statement. These are some of the rules of inference that accompany propositional logic:

- If S_1 *implies* S_2 is true, and we are given S_1, then we can infer S_2.
- If S_1 *and* S_2 is true, then we can infer S_1. Similarly, we can also infer S_2.
- If S_1 *is equivalent to* S_2, then we can infer (S_1 *implies* S_2), *and* (S_2 *implies* S_1).
- Conversely, if (S_1 *implies* S_2) *and* (S_2 *implies* S_1), then we can infer that (S_1 *is equivalent to* S_2).

We finally emphasize that propositional logic does not scale to large environments and cannot efficiently capture universal relationship patterns. However, propositional logic provides the foundation of first-order logic and higher-order logic, since those build on top of propositional logic's machinery.

From Few Axioms to a Whole Theory

The inference rules are *sound*. They allow us to prove only true statements, in the sense that given a true statement and if we can infer a sound inference rule with it, we arrive at a true statement. Therefore, the guarantee that sound inference rules provide is that they do not allow false statements to be inferred from true ones. We need slightly more than that guarantee.

A logical framework is *complete* when we are able to infer *all* possible true statements using only the system's knowledge base (axioms) and its inference rules. The idea of *completeness* of a system is very important. In all mathematical systems, such as number theory, probability theory, set theory, or Euclidean geometry, we start with a set of axioms (Peano axioms for number theory and mathematical analysis, and probability axioms for probability theory), then we deduce theorems from these axioms using the logical rules of inference. One main question in any math theory is whether the axioms along with the rules of inference ensure its completeness and its consistency.

No first-order theory, however, has the strength to uniquely describe a structure with an infinite domain, such as the natural numbers or the real line. Axiom systems that do fully describe these two structures (that is, categorical axiom systems) can be obtained in stronger logics such as second-order logic.

Codifying Logic Within an Agent

Before moving on to first-order logic, let's recap what we learned in the context of an AI agent endowed with propositional logic. The following process is important and will be the same for more expressive logics:

1. We program an initial knowledge base (axioms) in the form of true statements.

2. We program the inference rules.

3. The agent perceives certain statements about the current state of its world.

4. The agent may or may not have a goal statement.

5. The agent uses the inference rules to infer new statements and to decide what to do (move to the next room, open the door, set the alarm clock, etc.).

6. Completeness of the agent's system (knowledge base together with the inference rules) is important here, since it allows the agent to infer *any* satisfiable goal statement given enough inference steps.

How Do Deterministic and Probabilistic Machine Learning Fit In?

The premise of machine learning (including) neural networks is that we do not program an initial knowledge base into the agent, and we do not program inference rules. What we program instead is a way to represent the input data, the desired outputs, and a hypothesis function that maps the input to the output. The agent then learns the parameters of the function by optimizing the objective function (loss function). Finally, the agent makes inferences on new input data using the function it learned. So in this context, the knowledge base and the rules can be separated *during learning* or *during inference*. During learning, the knowledge base is the data and the hypothesis function, the goal is minimizing the loss, and the rules are the optimization process. After learning, the agent uses the learned function for inference.

We can think of probabilistic machine learning models in exactly the same way if we replace the deterministic hypothesis function by the joint probability distribution of the features of the data. Once learned, the agent can use it for inference. For example, Bayesian networks would play a similar role for uncertain knowledge as propositional logic for definite knowledge.

First-Order Logic

Let's answer the same questions for first-order logic:

What objects exist in the agent's world?
    Statements, objects, and relations among them.

How does the agent perceive the objects' states?
    True (1), false (0), or unknown.

Propositional logic is great for illustrating how knowledge-based agents work, and to explain the basic rules of a certain logic's *language* and rules of inference. However, propositional logic is limited in what knowledge it can represent and how it can reason about it. For example, in propositional logic, the statement:

    All users who are older than 18 can see this ad.

is easy to express as an *implies* statement (which is the same as *if then*), since this kind of language exists in the propositional logic framework. This is how we can express the statement as an inference in propositional logic:

    (User older than 18 *implies* see the ad) *and* (User older than 18 = T), then we can infer
    that (see the ad = T).

Let's now think of a slightly different statement:

    Some of the users who are older than 18 click on the ad.

Suddenly the language of propositional logic is not sufficient to express the quantity *some* in the statement! An agent relying only on propositional logic will have to store the whole statement, as is, in its knowledge base, then not know how to infer anything useful out of it. Meaning, suppose the agent gets the information that a user is indeed older than 18, it cannot predict whether the user will click on the ad or not.

We need a language (or a logical framework) whose vocabulary includes *quantifiers* such as *there exist* and *for all*, so that we can write something like:

    For all users who are older than 18, *there exists* a subset who clicks on the ad.

These two extra quantifiers are exactly what *first-order logic* framework provides. This increase in vocabulary allows us to be more economical in what to store in the knowledge base, since we are able to break down the knowledge into objects and relations between them. For example, instead of storing:

    All the users who are older than 18 see the ad;
    Some of the users who are older than 18 click on the ad;
    Some of the users who are older than 18 buy the product;
    Some of the users who click on the ad buy the product;

as four separate statements in the knowledge base of an agent with only propositional logic framework (which we still don't know how to infer anything useful from), we can store three statements in first-order logic:

> *For all* users who are older than 18, see ad = T;
>
> *For all* users with see ad =T, *there exists* a subset who clicks on the ad;
>
> *For all* users who click on the ad, *there exists* a subset who buys the product.

Note that in both propositional and first-order logics, given only these statements we will not be able to infer whether a specific user who is older than 18 will click on the ad or buy the product, or even the percentage of those doing that, but at least in first-order logic we have the language to express the same knowledge more concisely, and in a way where we would be able to make some useful inferences.

The most distinctive feature of first-order logic from propositional logic is that it adds to its base language quantifiers such as *there exist* and *for all* on top of *not, and, or, implies,* and *is equivalent to* that already exist in propositional logic. This little addition opens the door to express objects separately from their descriptions and their relationships to each other.

The powerful thing about propositional and first-order logics is that their inference rules are independent from both the domain and its knowledge base or set of axioms. To develop a knowledge base for a specific domain, such as a math field, or circuit engineering, we must study the domain carefully, choose the vocabulary, then formulate the set of axioms required to support the desired inferences.

Relationships Between For All and There Exist

For all and *there exist* are connected to each other through negation. The following two statements are equivalent:

- All users who are above 18 see the ad.
- There exists no one above 18 who doesn't see the ad.

In propositional logic language, these two statements translate to:

- For all users such that user > 18 is true, see the ad is true.
- There exists no user such that user > 18 and see the ad is false.

These are the relationships:

- not(There exists an x such that P is true) $\Longleftrightarrow$ For all x, P is false.
- not(for all x, P is true) $\Longleftrightarrow$ There exists an x such that P is false.
- There exists an x such that P is true $\Longleftrightarrow$ not for all x P is false.
- For all x, P is true $\Longleftrightarrow$ There exists no x such that P is false.

We cannot leave this section without appreciating the expressive power we gained by moving to first-order logic. This logic framework is now sufficient for such assertions and inferences to make sense:

Universal approximation theorem for neural networks
Roughly speaking, the universal approximation theorem asserts that *for all* continuous functions, *there exists* a neural network that can approximate the function as closely as we wish. Note that this does not tell us how to construct such a network, it only asserts its existence. Still, this theorem is powerful enough to make us *unsurprised* about the success of neural networks in approximating all kinds of *input* to *output* functions in all kinds of applications.

Inferring relationships
Parents and children have inverse relationships to each other: if Sary is the child of Hala, then Hala is the mother of Sary. Moreover, the relationship is in one direction: Sary cannot be the mother of Hala. In first-order logic, we can assign two functions indicating the relationships: *mother of* and *child of*, variables that can be filled in by *Hala* and *Sary* or any other mother and child, and a relationship between the *functions* that holds for all their input variables:

For all x,y, if mother(x,y) = T then mother(y,x) = F;

and *for all x,y, mother(x,y) $\Longleftrightarrow$ child(y,x).*

Now if we equip an agent with this knowledge and tell it that Hala is the mother of Sary, or mother(Hala, Sary) = T, then it will be able to answer queries like:

- Is Hala the mother of Sary? T
- Is Sary the mother of Hala? F
- Is Sary the child of Hala? T
- Is Hala the child of Sary? F
- Is Laura the mother of Joseph? Unknown

Note that we will have to store each statement separately in a propositional logic world, which is outrageously inefficient.

Probabilistic Logic

What objects exist in the agent's world?
    Statements.

How does the agent perceive the objects' states?
    A probability value between 0 and 1 that a statement is true.

Probability is the extension of first-order logic that allows us to quantify our uncertainty about the truth of a statement. Rather than asserting whether a statement is true or false, we assign to the degree of our belief in the truth of the statement a score between zero and one. Propositional and first-order logics provide a set of inference rules that allow us to determine the truth of some statements, given the assumption that some other statements are true. Probability theory provides a set of inference rules that allow us to determine how likely it is that a statement is true, given the likelihood of truth of other statements.

This extension to dealing with uncertainty results in a more expressive framework than first-order logic. The axioms of probability allow us to extend traditional logic truth tables and inference rules. For example, $P(A) + P(\text{not } (A)) = 1$: if A is true, then $P(A) = 1$ and $P(\text{not } A) = 0$, which is consistent with first-order logic about a statement and its negation.

Viewing probability theory as a natural extension of first-order logic is satisfying to a mind that needs to connect things together as opposed to viewing them as disparate things. Viewing it this way also naturally leads to Bayesian reasoning about data, since we update an agent's prior distribution as we gather more knowledge and make better inferences. This binds all our subjects together in the most *logical* way.

Fuzzy Logic

What objects exist in the agent's world?
    Statements with a degree of truth between [0,1].

How does the agent perceive the objects' states?
    A known interval value.

The worlds of propositional and first-order logic are black and white, true or false. They allow us to start with true statements and infer other true statements. This setting is perfect for mathematics where everything can either be right or wrong (true or false), or for a video game with very clear boundaries for its SIMS. In the real world, many statements can be vague about whether they are fully true (1) or fully false (0), meaning they exist on a *scale of truth* as opposed to at the edges: *Paris is romantic*; *She is happy*; *the movie The Dark Knight is good*. *Fuzzy logic* allows this and assigns values to statements between 0 and 1 as opposed to strict 0 or strict 1: Paris is romantic (0.8); She is happy (0.6); the movie The Dark Knight is good (0.9).

How do we make inference in a vague world where truth comes on a sliding scale? It definitely is not as straightforward as inference in true-and-false worlds. For example, how true is the statement: "Paris is romantic *and* she is happy," given the previous truth values? We need new rules to assign these values, and we need to know the context, or a domain. Another option is word vectors, which we discussed in Chapter 7. These vectors carry the meaning of words in different dimensions, so we can compute the cosine similarity between the vector representing the word *Paris* and the vector representing the point *romantic*, and assign that as the truth value of the statement *Paris is romantic*.

Note that the degree of belief in probability theory is not the same as the scale of truth in fuzzy logic. In probabilistic logic, the statements themselves are unambiguous. What we want to infer is the probability that the unambiguous statement is true. Probability theory does not reason about statements that are not entirely true or false. We do not calculate the probability that Paris is romantic, but we calculate the probability that a person, when randomly asked whether Paris is romantic, would answer true or false.

One interesting thing about fuzzy logic is that it kicks two principles present in other logics to the curb: the principle that if a statement is true, then its negation is false; and the principle that two contradictory statements cannot be true at the same time. This actually opens the door to inconsistency and *open universe*. In a way, fuzzy logic doesn't attempt to correct vagueness; instead, it embraces it and leverages it to allow functioning in a world where the boundaries are unclear.

Temporal Logic

There are other types of special purpose logics, where certain objects, such as *time* in this section, are given special attention, having their own axioms and inference rules, because they are central to the knowledge that needs to be represented and the reasoning about it. *Temporal logic* puts time dependence and the axioms and inference rules about time dependence at the forefront of its structure, as opposed to adding statements that include time information to the knowledge base. In temporal logic, statements or facts are true at certain times, which could be time points or time intervals, and these times are ordered.

What objects exist in the agent's world?
    Statements, objects, relations, times.

How does the agent perceive the objects' states?
    True (1), false (0), or unknown.

In temporal logic, we can represent statements such as:

- The alarm goes off when it is 7:00 a.m.
- Whenever a request is made to a server, access is eventually granted, but it can never be granted to two simultaneous requests.

Comparison with Human Natural Language

We spent the whole chapter going through logical systems that are able to express knowledge that humans' natural languages seem to do effortlessly. I just wrote a whole book about math using only the English language, as opposed to any other technical language. How do we do it? How do humans represent and expand their knowledge base, and what rules does natural language use for representation and reasoning that it is able to be so expressive? Moreover, the particular natural language used is not important: any multilingual speaker knows the thought but not necessarily which particular language they are using to express that thought. There is an internal nonverbal representation for what people know or want to express. How does it work, and how can we unlock its secrets and give them to our machines?

Similar to human language, if we represent the same knowledge in two different formal logics, then we can infer the same facts (assuming the logics have completeness inference rules). The only difference would be which logic framework provides an easier route for inference.

That said, human natural language allows for ambiguity on many occasions and cannot make absolute mathematical assertions without the formality of mathematics and the formal logic it employs. We cannot ask a human who has no access to a GPS system to predict the exact time it takes to drive from DC to NYC on a specific day, but we can ask a GPS machine for that kind of accuracy.

Machines and Complex Mathematical Reasoning

Mathematical reasoning is the distillation of human logic. When we prove a theorem using mathematical reasoning, we get one step closer to *the universal truth*. Teaching machines to prove mathematical theorems—and even more ambitiously, generating new ones—requires navigating infinite search spaces and symbolic reasoning. Once again, neural networks are proving useful for advancing intelligent machines. Researchers at Meta, Vrije Universiteit Amsterdam, and CERMICS École des Ponts ParisTech used a combination of deep learning, online training (*https://oreil.ly/k__ZN*), transformers (large language models), and reinforcement learning for automated mathematical theorem proving. Their paper, *HyperTree Proof Search for Neural Theorem Proving* (2022) (*https://oreil.ly/NzIvp*) presents state-of-the-art results in this field.

Summary and Looking Ahead

An AI agent endowed with various types of logic can express knowledge about the world, reason about it, answer queries, and make inferences that are allowed within the boundaries of these logics.

We discussed various logic frameworks, including propositional logic, first-order logic, probabilistic logic, fuzzy logic, and temporal logic.

The next natural questions would be: what content should go into an agent's knowledge base? And how to represent facts about the world? In what framework should knowledge be represented and inference made?

- Propositional logic?
- First-order logic?
- Hierarchical task networks for reasoning about plans?
- Bayesian networks for reasoning with uncertainty?
- Causal diagrams and causal reasoning where an agent is allowed to selectively break the rules of logic?
- Markov models for reasoning over time?
- Deep neural networks for reasoning about images, sounds, or other data?

Another possible next step is to dive deeper into any of the logic frameworks that we discussed, learning their inference rules and existing algorithms for inference, along with their strengths, weaknesses, and which kinds of knowledge bases they apply to. A recurring theme in these studies is investigating inference rules that provide a complete proof system, meaning a system where the axioms or the knowledge base along with the rules allow one to prove *all* possible true statements. Such rules include the *resolution inference rule* for propositional logic and the *generalized resolution inference rule* for first-order logic, which work for special types of knowledge bases. These are all important for theory (proving mathematical theorems), and technology (verifying and synthesizing) software and hardware. Finally, some logics are *strictly more expressive* than others, in the sense that some statements that we can represent in the more expressive logic cannot be expressed by any finite number of statements using the language of the less expressive logic. For example, *higher-order logic* (which we did not discuss in this chapter) is strictly more expressive than first-order logic (which we did discuss in this chapter, and which is powerful enough to support entire math theories).

CHAPTER 13

Artificial Intelligence and Partial Differential Equations

I want to model the whole world.

—H.

The first scene in the movie *Top Gun: Maverick* (2022) shows Maverick (Tom Cruise) manning an experimental military aircraft and pushing it to 10 times the speed of sound (10 Mach) before losing its stability at around 10.2 Mach. The fastest nonfictional manned aircraft so far can reach 6.7 Mach (Figure 13-1). Real speed or unreal (yet), it is mesmerizing to watch physics, math, and engineering come together to put these planes in the air, especially with their spectacular midair maneuvering.

Figure 13-1. Fastest manned aircraft ever made (image source (https://oreil.ly/CKMN2))

These are a few of the partial differential equations (PDEs) that come to mind while watching Maverick's awesome dogfight and 10 Mach scenes:

The wave equation for wave propagation
    Think of the speed of sound, the propagation of a sound wave in the air, and the variations of the sound of speed at different altitudes due to the variations in temperature and air density.

Navier-Stokes equations for fluid dynamics

Think of the fluid flow, air tunnels, and turbulence.

The G-equation for combustion

Think of the combustion in the aircraft's engine and the flames coming out of the aircraft's exhausts.

Material elasticity equations

Think of the aircraft wing panel, the lift force, the buckling of the wing panel (the process of out-of-stress plane movement that happens under compression; see Figure 13-2) caused by loading, which in turn reduces the load-carrying capabilities of the wing. When load-carrying capabilities fall below the design limits, failure happens.

Figure 13-2. Buckling in an aircraft (image source (https://oreil.ly/RJPpO))

PDE simulations also come to mind. Think of the flight path simulation and the crew chatting with Maverick as they watch his flight unfold in real time on their computer screens.

The list goes on. Are we claiming here that we made aircrafts fly because we wrote down and solved PDEs? No. Aviation museums tell the story of the Wright brothers, their experiments, and the evolution of aviation industry. Science and experimentation go hand in hand. What we want to claim instead is that we can invent, improve, and optimize all kinds of designs because of differential equations and math.

What Is a Partial Differential Equation?

A PDE is an equation, which means a lefthand side is equal to a righthand side, that involves a function of several variables along with any of its partial derivatives. A partial derivative of a function with respect to a certain variable measures the rate of change of the function with respect to that variable. Ordinary differential equations (ODEs) are those that involve functions of only one variable, such as only time, only space, etc. (as opposed to several variables), and their derivatives. A dynamic system is a greatly important ODE, describing the evolution in time of the state of a system that we care for, such as a system of particles, or the state of a customer in a business setting. The ODE involves one derivative in time of the state of the system, and the dynamics are prescribed as a function of the system state, system physical parameters, and time. The ODE looks like $\frac{d\vec{x}(t)}{dt} = f\left(\vec{x}(t), a(t), t\right)$. We will visit dynamic systems multiple times in this chapter. Most of the time, if we are able to transform a PDE into a system of ODEs, or maybe into a dynamical system, it is more or less solved.

Nature gave us neither the deterministic functions nor the joint probability distributions that it uses to produce the world that we observe around us and can accurately measure. Until now, it has kept those secret. It did, however, give us ways to measure, assess, or make laws about *how things change* relative to each other, which is exactly what partial differential equations represent. Because *the ways things change* are nothing but derivatives.

The goal of solving a PDE is to undo the differential operator so that we can recover the function without any derivatives. So we search for an exact or an approximate inverse (or pseudoinverse) of the differential operator that the PDE represents. Integrals undo derivatives, so solution representations of PDEs often involve the integrals of some kernel functions against the input data of a PDE (its parameters, initial and/or boundary conditions). We will elaborate on this as the chapter evolves.

People usually classify ODEs and PDEs into *types*. My take on this is that we should not confuse ourselves with *classifications* unless we happen to be personally working with these special ODEs or PDEs and their solutions happen to have a direct and immediate impact on the future of humanity. When in this chapter you encounter a certain type, such as nonlinear parabolic or backward stochastic, accept the name, then move directly to understanding the point that I am trying to make. Don't even try to google these terms. It will be similar to googling your symptoms and finding that you will die tomorrow. Consider yourself warned.

Modeling with Differential Equations

Differential equations model countless phenomena in the real world: air turbulence, the motions of galaxies, the behavior of materials at the nanoscale, pricing financial instruments, games with adversaries and multiple players, and population mobility and growth. Typical courses on PDEs skip the modeling step, so the PDEs that we end up studying seem to come out of the blue, but that is not the case. Where PDEs come from is as important as trying to analyze them and solve them. Usually, PDEs express some conservation laws, such as conservation of energy, mass, momentum, etc., as they relate to our particular application. Many PDEs are an expression of a conservation statement that looks like:

rate of change of a quantity in time = gains – losses

Now, when we have a bounded domain, the PDE works in the *interior* of the domain, but we need to accompany it with *boundary conditions* that tell us exactly what is happening at the boundary of the domain. If the domain is unbounded, then we need *far field conditions* that tell us what is happening as $x \to \infty$. We write these conditions using limits notation. If the PDE has derivatives in time, then we need some initial time conditions, or end time conditions. How many of these conditions we need depends on the order of the PDE. Think of these as how many equations we need to solve for how many unknowns. The unknowns are the *integration constants* of the PDEs. When we solve a PDE, we seek information about the function, given information about its derivatives. To get rid of these derivatives and recover the function, we must *integrate* the PDE, getting integration constants along the way. We need the boundary and/or far field conditions to solve for these constants.

Models at Different Scales

Realistic models that mimic nature faithfully need to account for all the important variables along with their interactions, sometimes at varying scales of space and time. Some work goes into writing the equations for the mathematical models. Once formulated, they are elegant, condensing a whole wealth of information into a few lines of equations. These equations involve functions, their derivatives, and the model's parameters, and are usually harder to solve than to formulate. Moreover, if two models describe the same phenomenon at different scales, say one on the atomistic scale (rapidly wiggling molecules) and another at a bigger scale, say at the microscopic or macro scale (the one we observe), then the two models' equations would look very different, and they may even be relying on physical laws from different fields of science. Think, for example, about describing the motion of gases at the molecular level (particle velocity, position, forces acting on it, etc.) and how to relate that to the thermodynamics of the gaseous system observed at the macroscopic scale. Or think of the ways atoms bond together to form crystalline structures, and how those structures translate into material properties, such as conductivity, permeability, brittleness, etc. The natural question is then, can we reconcile such models when each operates and is successful to some degree at a different scale? More precisely, if we take the limit of one model to the regime of the other, would we get the same thing? These are the types of questions that analysts address. Reconciling different scale models validates them and unifies different areas of math and science.

The Parameters of a PDE

The PDEs that we write down for a model usually involve parameters. These parameters have to do with the properties of the physical system that we are modeling. For example, for the heat equation:

$$u_t\left(\overrightarrow{x},t\right) = \alpha \Delta u\left(\overrightarrow{x},t\right)$$

the parameter α is the *diffusion coefficient*, which is a physical constant depending on the properties of the diffusing substance and the properties of the medium it is diffusing into. We usually get these from reference tables obtained from experiments. Their values are very important for engineering purposes. When our equations model reality, we must use parameter values that are derived from this real experimental or observational data. But experimental and observational data are usually noisy, and have missing values, unexplained outliers, and all kinds of rough seas for mathematical models. Many times we don't even have experimental values for the parameters that are in our equations. The experiments can be expensive (think Large Hadron Collider) or even impossible. So one must *learn* the parameter values using indirect ways, based on accessible combinations of experimental, observational, and computer simulated values of some other variables. Historically, many of these parameter values were *hand tuned* to fit some desired outcome, which is not good! We should have clear justifications for the choices of the parameter values that go into simulations. We will see how machine learning helps PDEs here in learning parameter values from data.

Changing One Thing in a PDE Can Be a Big Deal

If you had a PDE class in college, think back to the simplest equation that you studied, perhaps the heat diffusion equation on a rod, such as those in Figure 13-3. If you did not study PDEs, do not worry about its details. The formula of the heat equation is:

$$u_t(x,t) = \alpha \Delta u(x,t)$$

Here, $u(x,t)$ measures the temperature at the point x in the rod and at time t, and the operator Δ is the second derivative in x (so $\Delta u(x,t) = u_{xx}(x,t)$), since a rod is only one-dimensional if we ignore its thickness. In higher dimensions, the operator Δ is the sum of the second derivatives in each of the dimensions.

Now let's change the domain from a rod to a weird-shaped plate: not a square or a circle or ellipse, but something irregular; for example, compare Figure 13-3 with Figure 13-4. The formula for the true solution (which we learn in an introductory PDE class) that works for the rod does not work for the weird plate anymore. It gets even worse. Not only do we lose access to the analytical solution by changing the domain, when we try to numerically solve the differential equation with the new domain, the new geometry suddenly complicates matters.

Now we have to find a discrete mesh that accurately portrays the shape of the new domain with all its details, then we have to compute a numerical solution on top of that mesh that satisfies the equation in the interior of the domain, and satisfies the boundary conditions along the weird-looking boundary.

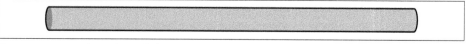

Figure 13-3. Studying heat diffusion on a rod is easy (for people who studied PDEs) both analytically and numerically

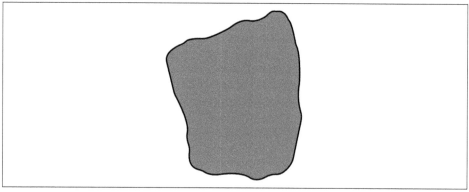

Figure 13-4. Studying heat diffusion on an irregular geometry is not as easy

This is normal in PDEs: Change one tiny thing and suddenly all the mathematical methods that we learned may not apply anymore. Such changes include:

- Change the shape of the domain
- Change the types of the boundary conditions
- Introduce a space or time dependence in the coefficients (the parameters)
- Introduce nonlinearities
- Introduce terms with more derivatives (higher order)
- Introduce more variables (higher dimension)

This frustrating aspect turns off many students from specializing in PDEs (no one wants to be an expert at only one equation, which could be very removed from modeling reality to start with). We don't want to be turned off. We want to see the big picture.

Natural phenomena are wonderfully varied, so we have to accept the variations in the PDEs and their solution methods as part of our quest to understand and predict nature. Moreover, PDEs are a large and old field. A lot of progress has been made on unifying methods for many families of both linear and nonlinear partial differential equations, and a lot of powerful analysis has been discovered along the way. The status quo is that PDEs are a very useful field that does not have, and might never have, a unifying theory.

In general, nonlinear PDEs are more difficult than linear PDEs, higher-order PDEs are more difficult than lower-order ones, higher-dimensional PDEs are more difficult than lower-dimensional ones, systems of PDEs are more difficult than single PDEs, we cannot write explicit formulas for solutions for the majority of PDEs out there, and many PDEs are only satisfied in *weak* forms. Many PDEs have solutions that develop singularities as time evolves (think the wave equation and shock waves). Mathematicians who develop PDE theory spend their time proving the *existence* of solutions of PDEs, and trying to understand the *regularity* of these solutions, which means how nice they are in terms of actually possessing the derivatives that are involved in the PDE. These use a lot of advanced calculus methods, looking for estimates on integrals (inequalities for upper and lower bounds).

Can AI Step In?

Now, wouldn't it be great if we had methods that account for variations in the PDE, the geometry of the domain, the boundary conditions, and the parameter ranges, similar to the actual physical problems? Many sectors of the industry and areas of science have their eyes on AI and deep learning to address their long-standing problems or shed new light on them. The past decade's astronomical advancement in computing solutions of very high-dimensional problems has the potential to transform many fields held down by the curse of dimensionality. Such a transformation would be a sea change for PDEs and in turn for humanity as a whole because of the sheer amount of science that is unlocked by PDEs and their solutions.

For the rest of this chapter, we highlight the hurdles that the differential equations community encounters with the traditional approaches to finding solutions to their PDEs, and with fitting real and noisy data into their models. We then illustrate how machine learning is stepping in to help bypass or alleviate these difficulties. We also consider two questions:

- What can AI do for PDEs?
- What can PDEs do for AI?

We need to make sure that the machine learning hallmarks of *training function, loss function*, and *optimization* settings are clear when serving PDEs, along with the labels or targets for supervised learning. Fitting the well-established field of PDEs into a machine learning setting is not super straightforward. Ideally, we need to establish a map from a PDE to its solution. This requires some pausing and thinking.

Numerical Solutions Are Very Valuable

Writing a mathematical model that describes a natural phenomenon, in the form of equations describing how the involved variables interact with each other, is only a first step. We need to solve these equations.

Analytical solutions are harder than numerical solutions, since the more the model mimics nature, the more complex the equations tend to be. Even when analytical methods cannot provide formulas for the solutions, they still provide valuable insights into their important properties. Numerical solutions are easier than analytical solutions, because they involve discretizing the continuous equations, moving us from the realm of continuous functions to the realm of discrete numbers, or from infinite dimensional function spaces to finite dimensional vector spaces (linear algebra), which our machines are built to compute. Numerical solutions provide invaluable insights into the true analytical solutions of the models, and are easy to test against experimental observations, when available. They are also easy to tune, so they are great aids for experimental design.

We can devise numerical solutions at any scale, but the curse of dimensionality haunts us when we try to implement and compute our numerical schemes. In many situations, for a numerical simulation to mimic even one second of the natural evolution of a system requires a tremendous amount of computational power, so a lot of dimension reductions and simplification assumptions must happen, which moves us even farther from having a good approximation of the true solutions. The sad part is that this is the norm rather than the exception.

Continuous Functions Versus Discrete Functions

The function $f(x) = x^2 - 3$ is continuous over the whole real line $(-\infty, \infty)$. When we discretize it for a numerical scheme for a machine to process, first, the domain cannot be the whole real line anymore, because machines cannot yet conceptualize infinite domains. So our first approximation is slashing the domain dramatically to some finite $[-N,N]$ where N is a large number. Our second approximation is discretizing this finite domain, drastically reducing it one more time, from a continuum $[-N,N]$ to only a finite set of points. If we use many points, then our mesh will be finer and our approximation better, at the expense of increased computation cost. Say we use six points only to discretize the interval $[-5,5]$: $-5, -3, -1, 1, 3, 5$, then our continuous function will be reduced to a vector with only six entries:

$f(x) = x^2 - 3$ is continuous on $(-\infty, \infty)$

$$\overrightarrow{\text{discrete } f} = \begin{pmatrix} (-5)^2 - 3 \\ (-3)^2 - 3 \\ (-1)^2 - 3 \\ 1^2 - 3 \\ 3^2 - 3 \\ 5^2 - 3 \end{pmatrix} = \begin{pmatrix} 22 \\ 6 \\ -2 \\ -2 \\ 6 \\ 22 \end{pmatrix}$$

Figure 13-5 shows the continuous function and its insanely under-representative six point approximation.

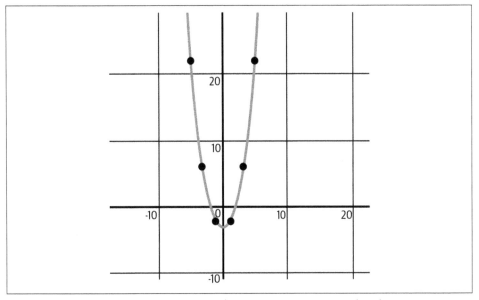

Figure 13-5. Discretizing the continuous function into a vector with only six points. We lose all the continuous wealth of information in between the points.

We Still Need to Discretize Derivatives

We can discretize a function $f(x)$ by selecting points in an interval, just like we did above. Differential equations contain *derivatives of functions* such as $f_x, \Delta f$, not only functions. So we must discretize the derivatives, or find some other way to reduce a problem from functional spaces (such as continuous spaces) to vector spaces (so we can use linear algebra and compute using our machines). Finite differences and finite elements are two popular discretization methods for differential equations. We will go over them shortly, along with a probabilistic Monte Carlo method based on random walks.

One trade-off of the simplicity of numerical solutions is that when we discretize, we make an approximation, reducing an infinite continuum to a finite set of points, losing all the infinitely detailed information that is between the finite set of points. That is, we sacrifice high resolution. For certain equations, there are analytical methods that help us quantify exactly how much information we lose by discretization, and help move us back to an accurate analytical solution by taking the limit as the size of the discrete mesh goes to zero.

Discretizing continuous functions and the equations that involve them has advantages: easy access. We can teach high school students how to numerically solve the heat equation describing the diffusion of heat in a rod (soon in this chapter), but we cannot teach them how to solve it analytically until they finish their college calculus sequence and linear algebra. This is why we must teach children how to model and compute numerical solutions of real-life problems at a very young age. The simplicity of numerical solutions and the power of computation to aid in solving all kinds of human problems should make this a priority in our education system. I doubt that nature intended for us to build and unravel crazily complicated mathematical theories before *computing* how the world around us works. I also doubt that nature is as complicated as some mathematical theories happen to be (even though they are still interesting in their own right, if only as an exercise in how far the rules of logic and inference can lead us).

PDE Themes from My Ph.D. Thesis

The story of my Ph.D. thesis demonstrates the drastic difference between mathematical theory and numerical approaches. It is also a good prototype for some of the themes of this chapter. For my Ph.D., I worked on a mathematical model that describes the way atoms diffuse and hop between different levels of a stair-like surface of a thin crystal. This is useful for the materials science community and for the engineers involved in designing the mini things that go into our electronic devices. As time evolves, the crystal's shape changes due to the movement of atoms on its surface. Eventually, the crystal relaxes into some stable shape.

Discretize right away and do a computer simulation

The moment I wrote down the equations, I was able to do a computer simulation that showed how the shape of the crystal evolves with time. This is one of the PDEs that I worked on (not that you should care about it or know what the function in it refers to):

$$u_t(h,t) = -u^2(u^3)_{hhhh} \text{ where } h \in [0,1], t \in [0,\infty)$$

For the trained eye, this is a highly nonlinear fourth order equation. The unknown function u appears both squared and cubed. Its cube appears with four derivatives in space, which we can think of as four degrees removed from the function that we want to evaluate. Figure 13-6 shows my PDE's discretization in space using finite differences (we will discuss finite differences shortly) and its boundary conditions (the function values at points 0 and 1).

Discrete slope $u_i = \dfrac{1/N}{X_{i-1}-X_i}$ ODE:

$$\begin{cases} u_i = u_i^2 \Delta_i \Delta u^3 \ i = 1, ..., N-1 \\ u_0 = u_N = 0 \\ \Delta_0 u^3 = \Delta_N u^3 = 0 \end{cases}$$

Continuum slope $u(h, t)$ PDE:

$$\begin{cases} u_t = -u^2(u^3)hhhh \\ u(0, t) = u(1, t) = 0 \\ u^3_{hh}(0, t) = u^3_{hh}(1, t) = 0 \end{cases}$$

Figure 13-6. Discrete differential equations and their continuum analog

Usually, the more nonlinear an equation is, the more stubborn it is in submitting to standard analytical techniques. I still needed to do the mathematical analysis and prove that the shape that the numerical simulation showed is indeed what the equations want the solution to do, meaning that it is *the* analytical solution, and it is what nature chooses among possible others. I had to spend the next two years of my life doing only that. What I came up with was a tiny proof in a tiny case for a physically unrealistic *one-dimensional crystal*! I had to reduce my equations to only one dimension to be able to do any mathematical analysis with them.

The curse of dimensionality

One theme that is always present is the dimensionality of the problem. Even when I did the numerical simulation, which took less than one afternoon, I was only able to do it for the equation over the one-dimensional domain. When I tried to do a simulation to model a realistic thin film laboratory crystal that lies on a flat surface, meaning when I had to discretize a two-dimensional surface instead of a one-dimensional segment, the number of discrete points jumped from 100 on the one-dimensional segment to $100^2 = 100,000$ on the two-dimensional surface. My computer at the time could not numerically solve the exact same equation that took only few seconds in the one-dimensional case. Granted, I was not sophisticated enough to compute on the university's server or use parallel computing (I do not know if distributed cloud computing had even been invented back then). Such is life with the curse of dimensionality. The computational expense rises exponentially with the number of dimensions. Now let's think of the equations whose domains have high dimensions to start with (even before discretization), like the Schrödinger equation for quantum particle systems, the Black-Scholes equation for pricing financial instruments, or the Hamilton-Jacobi-Bellman equation in dynamic programming that models multi-player games or resource allocation problems. Imagine then the magnitude of the curse of dimensionality.

The geometry of the problem

Another theme that we mentioned earlier but is worth repeating: the shape of the domain matters, for both analysis and numerical computation. In my unrealistic one-dimensional case, I used a segment as the domain for my equation. In the two-dimensional case, I had many more choices: a rectangle (has the advantage of a regular grid), a circle (has the advantage of radial symmetry), or any other realistic irregular shape that doesn't usually have a name. For analysis, the rectangle and the circle domains are the easiest (not for my particular equations but for other simpler equations like linear equations). For simulations, these are good too. But when the shape of the domain is not regular, which is the case for most realistic things, we need to place more discrete points in the parts where it is irregular if we want to capture the domain faithfully. The curse of dimensionality rears its unwelcome head again: more points mean longer vectors and larger input matrices for computations.

Model things that you care for

To finish my Ph.D. story, I never saw a real thin film crystal like the one I was working on until 10 years after I finished my degree, when my friend showed me a thin film crystal of gold in her lab. In retrospect, maybe I should have started there, by *seeing* exactly what it is in real life that I was trying to model. My priorities now are aligned differently, and I always start by asking whether I care for what it is I am trying to model, how closely the model I choose to work on mimics reality, and whether thinking about analytical solutions is even worth the time and the effort for this particular application.

Discretization and the Curse of Dimensionality

Mathematicians who study PDEs like the continuous world, but machines like the discrete world. Mathematicians like to analyze functions, but machines like to compute functions. To reconcile the two, and for machines to be of aid to mathematicians and vice versa, we can discretize our continuum equations. How? First, we discretize the domain of the equation, creating a discrete mesh. We choose the type of mesh (regular or irregular) and how fine or coarse it is. Then we discretize the differential equation itself, using one of four popular methods:

Finite differences
> Deterministic, good for discretizing time, one-dimensional or relatively regular spatial geometries.

Finite elements
> Deterministic, good for discretizing more complex spatial geometries, also spatial geometries that vary with time.

Variational or energy methods
> This is similar to finite elements but works on a narrower set of PDEs. They should possess a *variational principle*, or an *energy formulation*, that is, the PDE itself should be equivalent to $\nabla E(u) = 0$ for some energy functional $E(u)$ (mapping functions to the real line). The reason I was able to get my Ph.D. is that I discovered such an energy functional for my PDE by pure luck. Just like the minimum of a calculus function happens at points where $\nabla f\left(\vec{x}\right) = 0$, the minimum of an energy functional happens at *functions* where $\nabla E(u) = 0$, but of course we need to define what it means to take the derivative of a *functional*.

Monte Carlo methods
> Probabilistic, starts with discretizing the PDE, then uses that to devise an appropriate random walk scheme that enables us to *aggregate* the solution at a certain point in the domain.

The word *finite* in these methods stresses the fact that this process moves us from a continuum of infinite dimensional spaces of functions to finite dimensional spaces of vectors.

If the mesh we use for discretization is too fine, it captures more resolution, but we end up with high-dimensional vectors and matrices. Keep this curse of dimensionality in mind, and the following: one of the main reasons neural networks' popularity skyrocketed is that they seem to have a magical ability to overcome the curse of dimensionality. We will see how soon.

Finite Differences

We use finite differences to numerically approximate the derivatives of the functions that appear in PDEs. For example, a particle's velocity is the derivative in time of its position vector, and a particle's acceleration is two derivatives in time of its position vector.

In finite difference approximations, we replace the derivatives with linear combinations of function values at discrete points in the domain. Recall that one derivative measures a function's rate of change. Two derivatives measure concavity. Higher derivatives measure more stuff that some people in the sciences happen to use. The connection between a function's derivatives at a point and how its values near that point compare to each other is pretty intuitive.

The mathematical justification for these approximations rely on Taylor's theorem from calculus:

$$f(x) = f(x_i) + f'(x_i)(x - x_i) + \frac{f''(x_i)}{2}\left(x - x_i^2\right) + \frac{f^{(3)}(x_i)}{3!}(x - x_i)^3 + \cdots$$
$$+ \frac{f^{(n)}(x_i)}{n!}(x - x_i)^n + \text{error term}$$

where the error term depends on how nice the next order derivative $f^{(n+1)}(\xi)$ is near the x_i where we are attempting to use our polynomial approximation. Taylor's theorem approximates a nice enough function near a point with a polynomial whose coefficients are determined by the derivatives of the function at that point. The more derivatives the function has at a point, the nicer it is and the *more like a polynomial* it behaves near that point.

Now let's discretize a one-dimensional interval [a,b], then write down finite difference approximations for the derivatives of a function $f(x)$ defined over this interval. We can discretize [a,b] using $n + 1$ equally spaced points, so the mesh size is $h = \frac{b-a}{n}$. We can now evaluate f at any of these discrete points. If we care about the values near some point x_i, we define $f_{i+1} = f(x_i + h), f_{i+2} = f(x_i + 2h), f_{i-1} = f(x_i - h)$, etc. In the following, h is small, so an $O(h^2)$ method (or higher order in h) is more accurate than an $O(h)$ method:

1. Forward difference approximation of $O(h)$ accuracy for the first derivative (uses two points):

$$f'(x_i) \approx \frac{f_{i+1} - f_i}{h}$$

---

2. Backward difference approximation of $O(h)$ accuracy for the first derivative (uses two points):

$$f'(x_i) \approx \frac{f_i - f_{i-1}}{h}$$

3. Central-difference approximations of $O(h^2)$ accuracy for derivatives up to the fourth (uses two points, averages forward and backward differences):

$$f'(x_i) \approx \frac{f_{i+1} - f_{i-1}}{2h}$$

$$f''(x_i) \approx \frac{f_{i+1} - 2f_i + f_{i-1}}{h^2}$$

$$f'''(x_i) \approx \frac{f_{i+2} - 2f_{i+1} + 2f_{i-1} - f_{i-2}}{2h^3}$$

$$f^{(4)}(x_i) \approx \frac{f_{i+2} - 4f_{i+1} + 6f_i - 4f_{i-1} + f_{i-2}}{h^4}$$

4. Central-difference approximations of $O(h^4)$ accuracy for derivatives up to the fourth:

$$f'(x_i) \approx \frac{-f_{i+2} + 8f_{i+1} - 8f_{i-1} + f_{i-2}}{12h}$$

$$f''(x_i) \approx \frac{-f_{i+2} + 16f_{i+1} - 30f_i + 16f_{i-1} - f_{i-2}}{12h^2}$$

$$f'''(x_i) \approx \frac{-f_{i+3} + 8f_{i+2} - 13f_{i+1} + 13f_{i-1} - 8f_{i-2} + f_{i-3}}{8h^3}$$

$$f^{(4)}(x_i) \approx \frac{-f_{i+3} + 12f_{i+2} - 39f_{i+1} + 56f_i - 39f_{i-1} + 12f_{i-2} - f_{i-3}}{6h^4}$$

What does the $O(h^k)$ mean? It is the order of the numerical approximation in h. When we replace a derivative using its numerical approximation, we commit an error. The $O(h^k)$ tells us how much error we are committing. Obviously this depends on the size of the mesh h. The error should be smaller with finer meshes. To derive such error bounds, we use Taylor expansions of $f(x + h)$, $f(x - h)$, $f(x + 2h)$, $f(x - 2h)$, etc. and linear combinations of those to determine both the desired derivative's approximation and the order of our finite difference approximation in terms of h. To be able to use Taylor expansions, we assume that we are dealing with functions that *indeed have the required number of derivatives* at the points where we are evaluating them. This means that we assume that our function is nice enough to allow these derivative evaluations. If the function has singularities near these points, then we need to find ways around that, such as using much finer meshes near the singularities.

Example: Solve $y''(x) = 1$ on [0,1], with boundary conditions y(0)=-1 and y(1)=0

This is a second linear order ordinary differential equation on a bounded domain in one dimension. This example is trivial because the analytical solution is so easy. All we have to do is integrate the equation twice and recover the function without its derivatives $y(x) = 0.5x^2 + c_1x + c_2$, where the c's are the constants of integration. We plug in the two boundary conditions to find the c's and obtain the analytical solution $y(x) = 0.5x^2 + 0.5x - 1$. However, the point of this example is to show how to use finite differences to compute the *numerical solution*, not the analytical one, since analytical solutions are not available for many other differential equations, so we might as well get good at this. We first discretize the domain [0,1]. We can use as many points as we want. The more points, the higher dimension we have to deal with, but the resolution will be better. We'll use only eight points, so the mesh size is $h = 1/7$ (Figure 13-7). Our continuum [0,1] interval is now reduced to the eight points (0, 1/7, 2/7, 3/7, 4/7, 5/7, 6/7, 1).

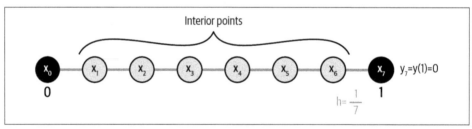

Figure 13-7. Discretizing the unit interval using eight discrete points, which is the same as seven intervals. The step size (or mesh size) is h = 1/7.

Next, we discretize the differential equation. We can use any finite difference scheme to discretize the second derivative. Let's choose the $O(h^2)$ central difference, so the discretized differential equation becomes:

$$\frac{y_{i+1} - 2y_i + y_{i-1}}{h^2} = 1 \text{ for } i = 1,2,3,4,5,6$$

Note that the differential equation is only valid in the *interior* of the domain, that is why we do not include the points $i = 0$ and $i = 7$ when we write its discrete analog. We get the values at $i = 0$ and $i = 7$ from the boundary conditions: $y_0 = -1$ and $y_7 = 0$. Now, we have a system of six equations and six unknowns, $y_1, y_2, y_3, y_4, y_5, y_6$:

$$y_2 - 2y_1 - 1 = 1/49$$

$$y_3 - 2y_2 + y_1 = 1/49$$

$$y_4 - 2y_3 + y_2 = 1/49$$

$$y_5 - 2y_4 + y_3 = 1/49$$

$$y_6 - 2y_5 + y_4 = 1/49$$

$$0 - 2y_6 + y_5 = 1/49$$

So now we've moved from the continuum world to the linear algebra world:

$$\begin{pmatrix} -2 & 1 & 0 & 0 & 0 & 0 \\ 1 & -2 & 1 & 0 & 0 & 0 \\ 0 & 1 & -2 & 1 & 0 & 0 \\ 0 & 0 & 1 & -2 & 1 & 0 \\ 0 & 0 & 0 & 1 & -2 & 1 \\ 0 & 0 & 0 & 0 & 1 & -2 \end{pmatrix} \begin{pmatrix} y_1 \\ y_2 \\ y_3 \\ y_4 \\ y_5 \\ y_6 \end{pmatrix} = \begin{pmatrix} 1/49 + 1 \\ 1/49 \\ 1/49 \\ 1/49 \\ 1/49 \\ 1/49 \end{pmatrix}$$

Solving this system amounts to *inverting* that tridiagonal matrix, which is the discrete analog of our second-order derivative operator. In the continuum world we *integrate* the differential operator to recover $y(x)$, and in the discrete world we *invert* the discrete operator to recover the discrete values y_i. Keep the curse of dimensionality in mind when using more points to discretize the domain.

Obviously we must compare the discrete values y_i with their exact counterparts $y(x_i)$ to see how well our finite difference scheme with only eight discrete points performed (Figure 13-8). Figure 13-9 shows the graph of the numerical solution (using only four discrete points) against the exact analytical solution.

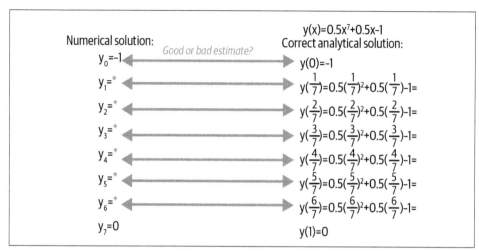

Numerical solution:

$y_0 = -1$

Good or bad estimate?

$y_1 = {}^*$

$y_2 = {}^*$

$y_3 = {}^*$

$y_4 = {}^*$

$y_5 = {}^*$

$y_6 = {}^*$

$y_7 = 0$

$y(x)=0.5x^2+0.5x-1$
Correct analytical solution:

$y(0)=-1$

$y(\frac{1}{7})=0.5(\frac{1}{7})^2+0.5(\frac{1}{7})-1=$

$y(\frac{2}{7})=0.5(\frac{2}{7})^2+0.5(\frac{2}{7})-1=$

$y(\frac{3}{7})=0.5(\frac{3}{7})^2+0.5(\frac{3}{7})-1=$

$y(\frac{4}{7})=0.5(\frac{4}{7})^2+0.5(\frac{4}{7})-1=$

$y(\frac{5}{7})=0.5(\frac{5}{7})^2+0.5(\frac{5}{7})-1=$

$y(\frac{6}{7})=0.5(\frac{6}{7})^2+0.5(\frac{6}{7})-1=$

$y(1)=0$

Figure 13-8. Comparing the numerical solution to the exact analytical solution at each discrete point

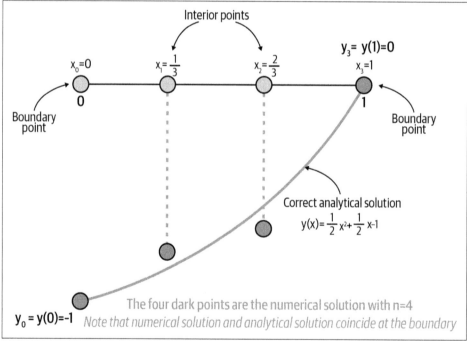

Interior points

$y_3 = y(1) = 0$

$x_0 = 0$ $x_1 = \frac{1}{3}$ $x_2 = \frac{2}{3}$ $x_3 = 1$

0   1

Boundary point

Boundary point

Correct analytical solution

$y(x)=\frac{1}{2}x^2+\frac{1}{2}x-1$

The four dark points are the numerical solution with n=4

$y_0 = y(0) = -1$ Note that numerical solution and analytical solution coincide at the boundary

Figure 13-9. Graph of the numerical solution (using only four discrete points) against the exact analytical solution (solid line)

Now we can use finite differences to discretize *any* differential equation of any order or type, on a domain in any dimension. All we have to do is discretize the domain and decide on finite difference schemes to approximate the derivatives at all the discrete points in the interior of the domain.

Example: Discretize the one-dimensional heat equation $u_t = \alpha u_{xx}$ in the interior of the interval $x \in (0,1)$

This is a second-order linear partial differential equation on a bounded spatial domain in one dimension. Here, $u = u(x,t)$ is a function of two variables, so our discretization scheme should address both coordinates. We can discretize only in space and keep time continuous, only in time and keep space continuous, or both in space and time. It is common to have more than one numerical route. Options are good. If we discretize in both space and time, then we end up with a system of algebraic equations. If we discretize only in space and not in time, then we end up with a system of ordinary differential equations. Since the PDE is linear, the discretized system is linear as well.

Let's write down a full discrete scheme. To discretize in space, let's use a second-order centered difference to approximate the second derivative. And to discretize in time, let's use a forward difference to approximate the first derivative:

$$\frac{u_{i,j+1} - u_{i,j}}{s} = \frac{u_{i+1,j} - 2u_{i,j} + u_{i-1,j}}{h^2} \text{ for } i = 1,2,\cdots,n \text{ and } j = 0,1,2,\cdots$$

In such equations, $u(x,t)$ can be known at some initial time ($u(x,0) = g(x)$), and we want to know that $u(x,t)$ time evolves. In the numerical scheme, the subscript i stands for discrete space, and j stands for discrete time. Thus, the discrete analog of the initial condition is $u_{i,0} = g_i$, and we want to solve for the unknowns $u_{i,j+1}$ for $i = 1, 2, \ldots, n$, and $j = 1, 2, \ldots$. We can easily isolate $u_{i,j+1}$ in the previous numerical scheme:

$$u_{i,j+1} = \frac{s}{h^2}\left(u_{i+1,j} - 2u_{i,j} + u_{i-1,j}\right) + u_{i,j} \text{ for } i = 1,2,\cdots,n \text{ and } j = 0,1,2,\cdots$$

Finally, we use this to find the values of $u_{i,1}, u_{i,2},\ldots$ (forward in time) for $i = 1, 2, \ldots, n$. For example, we plug in $j=0$ to find the values of discrete u at the first time step:

$$u_{i,1} = \frac{s}{h^2}\left(u_{i+1,0} - 2u_{i,0} + u_{i-1,0}\right) + u_{i,0} \text{ for } i = 1,2,\cdots,n$$

$$= \frac{s}{h^2}(g_{i+1} - 2g_i + g_{i-1}) + g_i \text{ for } i = 1,2,\cdots,n$$

Note that we know all the discrete values of g, so we now know discrete $u_{i,1}$ as well. Next, we plug in $j=1$ to find the values of discrete $u_{i,2}$ at the next time step, and so on.

Finite Elements

Finite element methods are different than finite difference methods in the sense that they operate on a *weak formulation* of the PDE as opposed to operating directly on the PDE. A weak formulation is *weighted and averaged*, so we are thinking integrals and integration by parts. We will come back to this shortly.

Before discussing the general idea of finite elements, let's observe Figure 13-10 a little. This shows a finite element solution of a PDE on top of a circular domain. The discretization of the domain uses a triangular mesh, and the solution seems to be approximated by a piecewise linear function. We can use other polygonal shapes for the meshes, and we can use *smoother* functions than piecewise linear, such as piecewise quadratic or higher degree polynomials. The trade-off for more smoothness is more computation.

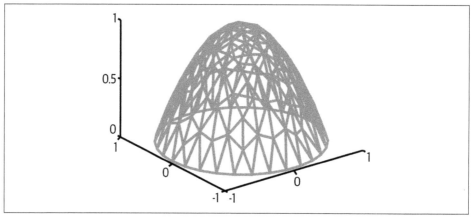

Figure 13-10. Finite elements solution over a circular domain (image source (https:// oreil.ly/bljPF))

Let's demonstrate how the finite element method gives a numerical approximation to the solution of the following PDE:

$$-\Delta u(x,y) = f(x,y) \text{ for } (x,y) \in \Omega \subset \mathbb{R}^2$$
$$u(x,y) = 0 \text{ for } (x,y) \in boundary_\Omega$$

This is a Poisson equation (appears in electrostatics). There is no time evolution. $f(x,y)$ is specified, and we are looking for an unknown function $u(x,y)$ that is zero at the entire boundary, and whose second derivatives u_{xx} and u_{yy} add up to $-f(x,y)$. This PDE is very well studied and we have formulas for its analytical solution, but we are only interested in the numerical approximation of the solution using the finite element method.

For this, we will produce an approximation of the unknown function *u(x,y)* that lives in an *infinite* dimensional space using a known function that lives in a *finite* dimensional space. Finite dimensional spaces are spanned only by *finitely many linearly independent functions*. We get to choose these *basis functions*, so we make sure that our choice makes our computations very easy. We usually choose piecewise linear functions or piecewise polynomial functions, each *supported minimally on the mesh*. This means that the basis function is nonzero only on top of one or two adjacent elements of the mesh, and zero everywhere else. Thus, an integration involving this function on the whole domain of the PDE would reduce to an integration on only one or two elements of the mesh.

After choosing these *basis functions*, each supported on a few mesh elements, we approximate the true solution *u(x,y)* by a linear combination of these easy and locally supported basis functions:

$$u(x,y) \approx u_1 basis_1(x,y) + u_2 basis_2(x,y) + \cdots u_n basis_n(x,y)$$

Now we must find the constants u_i of the linear combination. Therefore, we reduce our problem from solving for the unknown function *u(x,y)* in the continuum to solving for the unknown vector of coefficients $(u_1, u_2, \cdots, u_n)$. We must choose them so that the approximation $u_1 element_1(x,y) + u_2 element_2(x,y) + \cdots u_n element_n(x,y)$ satisfies the PDE, in some sense. We have *n* unknowns, so we must write *n* equations and solve a system of *n* equations, *n* unknowns. We get these from the PDE, or *its weak formulation*. To get a weak formulation of the PDE, we multiply it by a function *v(x,y)*, integrate over our domain, then use *integration by parts* to get rid of higher-order derivatives. Remember that the fewer derivatives we have, the closer to the unknown function we get. Let's do this step by step:

The original PDE is:

$$-\Delta u(x,y) = f(x,y) \text{ for } (x,y) \in \Omega \subset \mathbb{R}^2$$
$$u(x,y) = 0 \text{ for } (x,y) \in boundary_\Omega$$

Multiply the PDE by a function *v(x,y)* and integrate over the entire domain. This is a *weak formulation* of the PDE since it makes it satisfied in an integral form as opposed to a point-by-point form:

$$-\int_\Omega \Delta u(x,y) v(x,y) dx dy = \int_\Omega f(x,y) v(x,y) dx dy$$

Note the operator $\Delta = \nabla.\nabla$, the dot product of two derivative operators. Integration by parts helps us get rid of one of the derivatives by moving it over to the other function inside the integral. This doesn't come for free. In the process of doing this, it picks up a negative sign and another integral term that operates on the boundary of the domain. This new integral on the boundary integrates the product of two anti-derivatives. The boundary term needs the outward unit normal vector to the boundary $\vec{n}$:

$$\int_\Omega \nabla u(x,y).\nabla v(x,y)dxdy - \int_{boundary_\Omega} v(x,y)\nabla u(x,y).\vec{n}\,ds = \int_\Omega f(x,y)v(x,y)dxdy$$

We can choose $v(x,y) = 0$ on the boundary and that makes the whole boundary term disappear:

$$\int_\Omega \nabla u(x,y).\nabla v(x,y)dxdy = \int_\Omega f(x,y)v(x,y)dxdy$$

Now we replace $u(x,y)$ with its finite dimensional approximation:

$$\int_\Omega \nabla(u_1 basis_1(x,y) + u_2 basis_2(x,y) + \cdots u_n basis_n(x,y)).\nabla v(x,y)dxdy$$
$$= \int_\Omega f(x,y)v(x,y)dxdy$$

which is equivalent to:

$$\int_\Omega (u_1 \nabla basis_1(x,y) + u_2 \nabla basis_2(x,y) + \cdots u_n \nabla basis_n(x,y)).\nabla v(x,y)dxdy$$
$$= \int_\Omega f(x,y)v(x,y)dxdy$$

This is it: We can choose n different functions for $v(x,y)$ to get n different equations in n unknowns (the u_i's are the unknowns). A common theme is that every time we get to pick, we pick things that do not complicate our computation life. The easiest choices for $v(x,y)$ are the n basis functions that we already have, since these produce many cancellations when integrated against each other (*orthogonality*), and when integrated against themselves produce the number 1 (*normality*). The basis functions that we originally choose form an *orthonormal* set of functions. All for the business of making our life easier. Therefore, the n equations are:

$$\int_\Omega (u_1 \nabla basis_1(x,y) + u_2 \nabla basis_2(x,y) + \cdots u_n \nabla basis_n(x,y)).\nabla basis_1(x,y) dx dy =$$

$$\int_\Omega f(x,y) basis_1(x,y) dx dy$$

$$\int_\Omega (u_1 \nabla basis_1(x,y) + u_2 \nabla basis_2(x,y) + \cdots u_n \nabla basis_n(x,y)).\nabla basis_2(x,y) dx dy =$$

$$\int_\Omega f(x,y) basis_2(x,y) dx dy$$

$$\cdots$$

$$\int_\Omega (u_1 \nabla basis_1(x,y) + u_2 \nabla basis_2(x,y) + \cdots u_n \nabla basis_n(x,y)).\nabla basis_n(x,y) dx dy =$$

$$\int_\Omega f(x,y) basis_n(x,y) dx dy$$

Finally we solve the system of n equations, n unknowns, which we set in a linear algebra form (where $b_i = basis_i$):

$$
\begin{vmatrix}
\int_\Omega \nabla b_1(x,y).\nabla b_1(x,y) dx dy & \int_\Omega \nabla b_2(x,y).\nabla b_1(x,y) dx dy & \cdots & \int_\Omega \nabla b_n(x,y).\nabla b_1(x,y) dx dy \\
\int_\Omega \nabla b_1(x,y).\nabla b_2(x,y) dx dy & \int_\Omega \nabla b_2(x,y).\nabla b_2(x,y) dx dy & \cdots & \int_\Omega \nabla b_n(x,y).\nabla b_2(x,y) dx dy \\
\vdots & \vdots & \cdots & \vdots \\
\int_\Omega \nabla b_1(x,y).\nabla b_n(x,y) dx dy & \int_\Omega \nabla b_2(x,y).\nabla b_n(x,y) dx dy & \cdots & \int_\Omega \nabla b_n(x,y).\nabla b_n(x,y) dx dy
\end{vmatrix}
\begin{vmatrix} u_1 \\ u_2 \\ \vdots \\ u_n \end{vmatrix} =
$$

$$
\begin{vmatrix}
\int_\Omega f(x,y) b_1(x,y) dx dy \\
\int_\Omega f(x,y) b_2(x,y) dx dy \\
\vdots \\
\int_\Omega f(x,y) b_n(x,y) dx dy
\end{vmatrix}
$$

Recall that we know the function $f(x,y)$, all the basis functions $b_i = basis_i$, and the domain Ω, so all we have to do is solve the system of equations. This system is *sparse* because most of these integrals are zero. We chose the basis functions with small support for this reason. We never want to solve a *dense* system of equations.

Of course, we have many questions and a rich literature on finite elements that deals with them:

Do all PDEs have a weak formulation that allow us to do things like this?
   Yes, because we can always multiply PDEs with v functions and integrate by parts, but some PDEs have better structures to carry out simplifying computations than others.

What about the energy formulations or variational principles of PDEs, are they related?
   Yes, they are related. Look up the Ritz method. We hinted at this in Chapter 10, when we related minimizing energy functionals to solving PDEs. One thing to

keep in mind here is that most PDEs have a weak formulation, but not all of them have an energy minimization formulation. One of the reasons I got my Ph.D. was that I discovered an energy formulation for the PDE that I was working with. It was by complete chance. All I did was one lucky *weak formulation*, followed by an integration by parts. Just like we did in this section. Trial and error are underestimated in this life.

What about Sobolov spaces, why do we study them in advanced courses on PDEs?
Because we need to set our functions u, v, and the basis functions in the appropriate *function spaces* that tell us that all the computations and approximations that we are using are valid. For example, we don't want the involved integrals that contain our functions and their derivatives to blow up.

Can we use nonuniform meshes to adjust for the more detailed part of the domain, such as the one in Figure 13-11?
Yes, nothing in our discussion relies strictly on a uniform mesh.

How many basis functions do we need?
As many as our mesh elements.

Under what conditions does the approximate solution converge to the true solution?
Welcome to finite element analysis.

How is this used in applications?
All the time. It started with mechanics and structural designs: loads, stresses, and strains; but now the finite element method is used to *numerically* solve all kinds of PDEs whose spatial domains have complex geometries.

What could go wrong?
As always, the curse of dimensionality. We need more mesh elements for higher resolution, so the system of equations that we end up having to solve grows exponentially with the number of mesh elements. No bueno. Ideally, we want a mesh that is not detailed in places where it doesn't need to be, and more detailed in more interesting parts of the domain.

What else could complicate matters?
For PDEs whose *domain evolves in time*, we need meshes that evolve in time accordingly.

Can AI help learn an appropriate mesh given a certain geometry and PDE?
Yes, we will see this soon in this chapter.

Before moving on, we note that the finite element method is a finite dimensional mesh dependent method that approximates the solution of the PDE. Later in this chapter we will learn about meshless neural network methods.

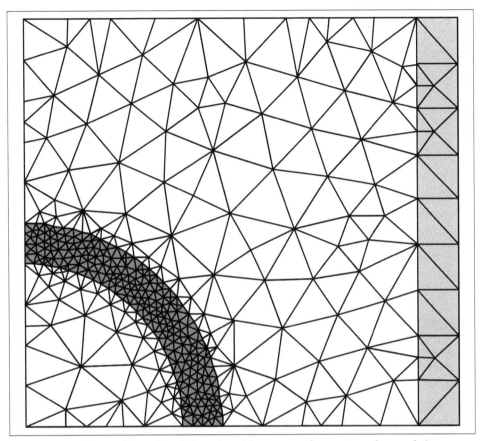

Figure 13-11. A two-dimensional domain with a nonuniform triangular mesh (image source (https://oreil.ly/f9uHk))

Variational or Energy Methods

Some PDEs are very special in the sense that their solution minimizes an energy functional. We say that such a PDE possesses a *variational principle*. The Poisson equation, the one we just solved using finite elements, is one of these lucky PDEs. When a PDE possesses a variational principle, it opens for us another route to understanding its solution by studying the energy functional that it happens to minimize.

Let's write the Poisson equation and the energy functional that its solution minimizes, without going through the details of why this is the case:

$$\Delta u(x,y) = f(x,y) \text{ for } (x,y) \in \Omega \subset \mathbb{R}^2$$
$$u(x,y) = 0 \text{ for } (x,y) \in boundary_\Omega$$

$$E(u(x,y)) = \int_{\Omega} |\nabla u(x,y)|^2 + 2f(x,y)dxdy$$

Now we can exploit this new knowledge to numerically approximate the solution of the PDE: look for an approximate minimizing scheme of the energy functional. Similar to the finite elements method, we project our infinite dimensional solution $u(x,y)$ onto a finite dimensional space, where we get to choose the basis elements:

$$u(x,y) \approx u_1 basis_1(x,y) + u_2 basis_2(x,y) + \cdots u_n basis_n(x,y)$$

and we must again solve for the numbers $(u_1, u_2, \cdots, u_n)$. To do this, we plug the approximate $u(x,y)$ into the formula of the energy functional. Since we know all the basis elements, this is now a *function* of $(u_1, u_2, \cdots, u_n)$, which we can minimize using standard calculus methods. Done!

This method is fairly general and introduces us smoothly into the calculus of variations, which is about finding optima of functionals instead of functions, like in normal calculus.

Monte Carlo Methods

We are now used to switching our brain to probabilistic thinking to solve deterministic problems. We did that with the stochastic gradient descent for minimizing loss functions, multiplication of large matrices, randomized singular value decomposition of a large matrix, and random walks on graphs to identify communities, rank web pages, and other purposes. The most famous introductory examples to Monte Carlo methods are:

- Estimating π by generating many random points (x_{random}, y_{random}) in a unit square, and finding the proportion that lies inside the inscribed quarter circle of radius 1: $x_{random}^2 + y_{random}^2 \leq 1$. Now we can estimate the probability of landing in the quarter circle:

$$Prob(\text{point inside the quarter circle})$$
$$= \frac{\text{area of quarter circle of radius 1}}{\text{area of unit square}}$$
$$= \frac{\pi}{4}$$
$$\approx \frac{\text{number of times point is inside the quarter circle}}{\text{total number of points generated}}$$

- Estimating the integral $\int_a^b f(x)dx$ of a nonnegative and continuous function $f(x)$ over an interval [a,b] by generating random points (x_{random}, y_{random}), where $a \leq x_{random} \leq b$ and $0 \leq y_{random} \leq max(f)$. The value of the integral is the area under the graph of f. We can estimate it by finding the proportion of times that the random point lies under the graph of $f(x)$, or $y_{random} \leq f(x_{random})$:

$$Prob(\text{point under the graph of } f) = \frac{\text{area under the graph of } f}{\text{total area of the rectangle}}$$

$$= \frac{\int_a^b f(x)dx}{(b-a) \times max(f)}$$

$$\approx \frac{\text{number of times point is under the graph of } f}{\text{total number of points generated}}$$

Such stochastic methods to solve deterministic problems are called *Monte Carlo* methods because they involve repetitive games of chance, and counting proportions of certain outcomes, just like gambling in Monte Carlo casinos in Monaco. They could have been called *Las Vegas Strip* methods as well. This is analogous to *randomized controlled trials* to answer deterministic questions, for example, to assess the effect of a certain drug on a given population. Another way to answer the same question would be a completely deterministic observational study, where one controls for all suspected confounding variables, and assesses the effects of the drug intervention.

Now suppose we have a deterministic PDE and we want to find its solution using randomized numerical trials (Monte Carlo). To illustrate how this works, let's use a simple PDE:

$\Delta u(x,y) = 0$ for (x,y) in the unit square $\subset \mathbb{R}^2$,

$u(x,y) = g(x,y)$ for $(x,y) \in boundary_{square}$

Let's first discretize the domain using a uniform grid, then write a finite difference scheme for the PDE at the interior grid points and for the boundary conditions:

$$\frac{u_{i+1,j} - 2u_{i,j} + u_{i-1,j}}{h^2} + \frac{u_{i,j+1} - 2u_{i,j} + u_{i,j-1}}{h^2} = 0$$

when (i,j) corresponds to an interior grid point,

$u_{i',j'} = g_{i',j'}$ when (i',j') corresponds to a boundary point

The goal is to use the numerical scheme to find $u_{i,j}$ for each interior point of the grid. This will be the numerical estimate of the true solution $u(x,y)$ at that particular interior point. Let's solve for $u_{i,j}$:

$$u_{i,j} = \frac{1}{4}u_{i+1,j} + \frac{1}{4}u_{i-1,j} + \frac{1}{4}u_{i,j+1} + \frac{1}{4}u_{i,j-1}$$

when (i,j) corresponds to an interior grid point,

$u_{i',j'} = g_{i',j'}$ when (i',j') corresponds to a boundary point

This is how we interpret the equations for a random walk setting: if we are at the boundary, then we know the solution, it is $u_{i',j'} = g_{i',j'}$. So a random walker who is following the guidance of the PDE scheme to structure their walk would collect their reward $g_{i',j'}$ at the boundary point. Moreover, the solution $u_{i,j}$ at an interior grid point (i,j) is the unweighted average of the solution at the four surrounding grid points. So if a random walker starts at an interior grid point (i,j), we will give them a 0.25 probability of wandering off to any of their four neighboring points, then to their neighboring point, until the walker hits a boundary grid point (i',j'), where they collect their reward $g_{i',j'}$. That would be only one exploration of the PDE scheme, in a way getting us a tiny piece of information on which boundary point contributed to the solution $u_{i,j}$. If we repeat this process may times, say a thousand times, all starting from the same grid point (i,j) where we want to find the numerical solution, then we can count the proportion of times the random walker ended up at each of the boundary points:

$$prob(\text{ending up at point}(i',j')) \approx \frac{\text{number of times random walker ended up at point}(i',j')}{\text{total number of random walks starting at } (i,j)}$$

This will give us an estimate of the expected rewards from all the boundary points, which is exactly the numerical solution that we are looking for: how does each boundary value play a role in the solution at the interior point? So the numerical solution of the PDE is:

$$u_{i,j} = \Sigma_{(i',j')} prob(\text{ending up at point}(i',j'))g_{i',j'}$$

This is such a neat way of getting a numerical solution that does not involve solving a linear system of equations (which could be very large and undesirable). It is also excellent when we care about finding the solution at a few points only, as opposed to finding the solution at the entire grid.

Of course, for each PDE we have to devise the correct numerical scheme along with the transition probabilities of the random walker. For example, if the PDE had coefficients multiplied with the second derivatives, then the random walker would not wander off to each of the four neighboring points with equal probability of 0.25. The coefficients would introduce weights for each neighbor, so we adjust the transition probabilities to each.

In terms of theory, we have to prove that a walker does eventually hit the boundary, and that the numerical solution obtained this way does converge to the true analytical solution of the PDE. We also have to obtain analytical estimates on how long until a random walk stops (on average), how fast the numerical solution converges, and how a numerical solution obtained this way fares against the ones obtained from finite differences or finite elements in terms of accuracy, computation cost, and speed of convergence.

With Monte Carlo methods, sometimes we start the other way around. We devise a simulation involving the different processes and transition rates mimicking some physical phenomena (such as interacting particles of a system), then we average that and transition to writing PDEs involving the descriptors of the system at hand. This is the exact opposite of starting with a PDE, then devising a granular scale Monte Carlo simulation to solve it. We discuss this next.

Some Statistical Mechanics: The Wonderful Master Equation

One of my favorite PDEs is *the master equation* from statistical mechanics, because it is one of the few PDEs out there that is able to get us from characterizing a system at an atomistic or molecular scale (system of particles), where the description is probabilistic, to the same system at a macro scale, where the description is deterministic. It is only logical to expect that the underlying atomistic processes and transitions give rise to an observed behavior at the macro scale. The very wise person who introduced me to statistical mechanics told me: "Isn't our whole lived experience the result of the collective behavior of some massive underlying chemical reaction?"

The transition from the master equation for atomistic probabilities to deterministic PDEs for observed quantities is neat and doesn't feel like we cheated or made fuzzy assumptions, or that we have two completely disconnected models, one at a macro scale and another at an atomistic scale, that have nothing to do with each other.

The master equation tracks the evolution of the probability of a statistical system (some particles) being in some state at a certain time. We calculate the rate of change of the probability of the state of the system by subtracting the losses from the gains and accounting for transition rates between different states:

$$\frac{\partial P(h,t)}{\partial t} = \sum_{h'} P(h',t)\, T(h' \to h) - P(h,t)\, T(h \to h'),$$
$$: = LP$$

where $T(h \to h')$ and $T(h' \to h)$ are the transition rates from state h to h', and vice versa. We calculate these transition rates using the underlying physical assumptions or observations on the system, for example, evaporation and condensation rates of atoms, diffusion rates, etc.

Now we can employ the master equation to write partial differential equations for the deterministic descriptors of the system by computing their expectations. Expectations transition us from probabilistic quantities to deterministic ones. Here's how we calculate it:

$$\langle f \rangle = \Sigma_h f P(h,t) = \Sigma_h f \frac{e^{-H(h)/KT}}{Z}$$

where $H(h)$ is the total energy, and Z is the partition function. The expression $\frac{e^{-H(h)/KT}}{Z}$ is very common in statistical mechanics and expresses the intuitive idea that states with high energy are exponentially less likely to occur, meaning that systems prefer low energy and evolve toward the states that lower the total energy.

Expectation Versus Averaging over N Repetitions of a Monte Carlo Simulation

This expectation $\langle f \rangle$ is related to Monte Carlo simulations. It is equivalent to the limit as $N \to \infty$ of the mean $\bar{f}$ of f after N repetitions of the Monte Carlo simulations.

We can now compute the rate of change of the expectation of the quantity of interest, say h_i representing the height of a crystal profile (made up of atoms) at a certain site i, using the master equation:

$$\frac{d\langle h_i \rangle}{dt} = \Sigma_h h_i \frac{\partial P}{\partial t} = \Sigma_h h_i LP$$

If the system is closed, meaning if we are able to express the righthand side in terms of h and its derivatives with respect to space and time, then we obtain an equation of motion for the expected height profile. If the system is not closed, then we have to make an approximation to close the system. We better make a physically plausible approximation, such as the system is near equilibrium, otherwise our efforts will be useless.

The final step is coarse-graining the resulting discrete equations of motion to obtain a continuum PDE model describing the crystal profile. This step moves us from a finite difference scheme to a continuum PDE, which is the reverse of the discretizing process we learned with finite differences. Using this process, the resulting PDE emerges directly from atomistic processes. Such a PDE usually looks like:

$$h_t\left(\overrightarrow{x},t\right) = F\left(h\left(\overrightarrow{x},t\right),t;\overrightarrow{\omega}\right)$$

where $\overrightarrow{\omega}$ is the set of the system's physical parameters.

Soon in this chapter, we will learn about using graph neural networks to simulate natural phenomena at the macroscopic scale directly from particle systems. This bypasses writing down PDEs as we did in this section. The inputs to the network will be the particles along with their interactions and rates of interaction, and the output will be the time evolution (a video, or a time sequence of graphs) of the system as a whole.

Solutions as Expectations of Underlying Random Processes

For some types of PDEs, a neat way to find solutions is to formulate them as expectations of some underlying random processes: we simulate random paths of an appropriate stochastic process, then compute the expectation. This allows us to evaluate solutions at any given space-time locations.

To learn how to do this, we need to study the *Feynman-Kac formula* and *Itô's calculus* (helps us find derivatives of functions of time-dependent random variables). These tie PDEs and probability nicely together.

The Feynman-Kac formula (which we will not write) offers a practical way to solve some PDEs that have been haunted by the curse of dimensionality. For example, in quantitative finance, we can use the Feynman-Kac formula to efficiently calculate solutions to the Black-Scholes equation to price options on stocks. In quantum chemistry, we can use it to solve the Schrödinger equation.

Transforming the PDE

The idea here is simple: maybe the PDE in a transformed space is easier to solve (analytically or numerically) than in the space it currently lives. So we transform it in some way and wish for the best.

Fourier Transform

The Fourier transform is an integral transform from x space to frequency ξ space:

$$F.T(f(x)) = \widehat{f}(\xi) = \frac{1}{\sqrt{2\pi}}\int_{-\infty}^{\infty}e^{-i\xi x}f(x)dx$$

The inverse Fourier transform undoes the Fourier transform and brings us back from frequency ξ space to x space:

$$F.T^{-1}\left(\widehat{f}(\xi)\right) = f(x) = \frac{1}{\sqrt{2\pi}}\int_{-\infty}^{\infty} e^{i\xi x}\widehat{f}(\xi)dx$$

There are tables with the Fourier transforms of many functions for our convenience. When these are not available, we resort to numerical methods. Because the Fourier transform and its inverse are important for many applications, such as frequency analysis, signal modulation, and filtering, there are algorithms that have been specifically developed for its fast computation.

The following are some powerful things that we need to know about the Fourier transform:

- It strips a function down to its frequency components. The Fourier transform of a function tells us how much of each frequency a function has. The frequency spectrum of a function $f(x)$ is the absolute value of its Fourier transform: $|F(\xi)|$.

- It has an inverse transform that allows us to move back and forth between x space and frequency space ξ.

- It changes the convolution of two functions in x space to multiplication of functions in frequency ξ space: $F.T.(f * g(x)) = F.T.(f(x)) \times F.T.(g(x)) = \widehat{f}(\xi)\widehat{g}(\xi)$. This is helpful when trying to find analytical solutions for PDEs. Solving a PDE in x space boils down to solving an algebraic equation or an easier differential equation in ξ space, then using the inverse Fourier transform to get back to x space. Many times we are inverting the product of two Fourier transforms, so the solution ends up being a convolution in x space. If you have encountered *Green's functions* for analytical solutions before, this is one way to arrive at them.

- It changes differentiation in x to multiplication by $i\xi$, so:

$$F.T.(u_x(x)) = i\xi F.T.(u(x)) = i\xi\widehat{u}(\xi)$$

and

$$F.T.(u_{xx}(x)) = -\xi^2 F.T.(u(x)) = -\xi^2\widehat{u}(\xi)$$

Getting rid of derivatives is huge. It means that differential equations in original space become algebraic equations in Fourier space.

- It is a linear transformation, so we can apply it separately to each term in a PDE. It enables us to solve linear PDEs with constant coefficients seamlessly. For linear PDEs with nonconstant coefficients (where the parameters depend on space), we can still use the Fourier transform if we are willing to get messy with series expansions of these coefficients. The moment we write series, we have to investigate their convergence.

- We use it to prove the universal approximation theorem for neural networks (Hornik et al. 1989).

- We can use it to speed up convolutional neural networks (Mathieu et al. 2013).

- It turns out that representing PDEs in Fourier space is convenient if we want to train neural networks to learn PDE solutions.

There are also some *not so* convenient things:

- Many functions have complex valued Fourier transforms. We just learn complex analysis and live with that.

- Not all functions have a Fourier transform. The involved integral operates on an infinite domain, so if there is no function inside the integral that compensates with a rapid decay to zero, the integral blows up (rendering the Fourier transform useless). The kernel of the Fourier transform is $e^{-i\xi x} = \cos(\xi x) - i\sin(\xi x)$. This oscillates with frequency ξ and never decays to zero.

- Even for functions whose inverse Fourier transform (which helps us find analytic solutions for PDEs) exists, many times we do not know its formula. In these cases, we would not be able to write an explicit analytic solution using this method. This is a common problem for many analytic methods.

- Fourier transforms have the Heisenberg uncertainty principle (*https://oreil.ly/ 0hH7h*).

> The study of uncertainty principles began with Werner Heisenberg's argument that it is impossible to simultaneously determine a free particle's position and momentum to arbitrary precision. In quantum mechanics, the wave function of position is the Fourier transform of the wave function of momentum. The most popular use of Fourier uncertainty principles is as a description of the natural trade-off between the stability and measurability of a system, particularly quantum mechanical systems. Imagining that $f(x)$ is the probability that a particle's position is x, and $f(\xi)$ is the probability that its momentum is ξ, Heisenberg's inequality gives a lower bound on how spread out these two probability distributions must be. The physical assumption is that position and momentum are related by Fourier transform: $|f|2L2 \leq 4\pi \cdot |(x - xo)f|L2 \cdot |(\xi - \xi o)f|L2$. Qualitatively, this means a narrow function has a wide Fourier transform, and a wide function has a narrow Fourier transform. In either domain, a wider function means there is literally a wide distribution of data, so there always exists uncertainty in one domain.

Fourier Transform Versus Fourier Series

We should not confuse the Fourier transform with the Fourier sine and cosine series. The function $\sin x$ does not have a Fourier transform, but its Fourier sine series is itself.

Laplace Transform

The Laplace transform allows us to transform a wider class of functions than the Fourier transform, because its kernel e^{-st} decays to zero exponentially fast (there is no complex valued i in the exponent to mess things up). The Laplace transform operates on functions defined on $[0,\infty)$, so in PDEs, we use it to transform the time variable, or other variables if they range from $[0,\infty)$. So instead of solving the PDE directly in the time domain, we Laplace transform it, solve it in s domain, then inverse Laplace transform it back to the time domain.

The formula for the Laplace transform is:

$$L.T.(f(t)) = \widehat{f}(s) = \int_0^\infty e^{-st} f(t) dt$$

And the formula for the inverse Laplace transform is:

$$L.T.^{-1}\left(\widehat{f}(s)\right) = f(t) = \frac{1}{2\pi i}\int_{c-i\infty}^{c+i\infty} e^{st}\widehat{f}(s) ds$$

Just like in the case of the Fourier transform, there are tables with the Laplace transforms of many functions computed for our convenience.

We care about how the Laplace transform acts on the (time) derivatives involved in a PDE. It is better to get rid of them, because how else will a PDE get us closer to its solution? It does:

- $L.T(u_t(x,t)) = s\widehat{u}(x,s) - u(x,0)$
- $L.T(u_{tt}(x,t)) = s^2\widehat{u}(x,s) - su(x,0) - u_t(x,0)$

Note that we usually know the initial conditions for a PDE $u(x,0)$ and $u_t(x,0)$, so the transformations do get rid of the derivatives with respect to time.

We also care about the convolution to multiplication property so that we can transfer back algebraic expressions in s to PDE solutions in t space using the inverse Laplace transform. Careful here that this is a finite convolution, from 0 to t, as opposed to from $-\infty$ to ∞ like in the Fourier transform case:

$$L.T.((f * g)(t)) = \widehat{f}(s)\widehat{g}(s) \text{ where } (f * g)(t) = \int_0^t f(\tau)g(t-\tau)d\tau = \int_0^t f(t-\tau)g(\tau)d\tau$$

Similar to the Fourier transform, the Laplace transform is a linear operator, so it is best used with linear PDEs.

Reducing PDEs to Algebraic Equations or to ODEs

The Fourier transform, the Laplace transform, and some other transforms, such as the Hankel and Mellin transforms, are able to get rid of derivatives of the PDE in certain variables (time, space, etc.). We are left with an algebraic equation if the transform acts on all the variables involved in the PDE, or an ordinary differential equation (ODE) if the transform acts on all the variables except one. The hope here is that the algebraic equations or the ODE are easier to solve than the original PDE, and that we can utilize known methods from algebra, numerics, and ODEs to solve the new equations in transformation variables. We will see a simple example of how this works in the next section on solution operators.

Solution Operators

We now work through two simple but informative examples that illustrate the transform methods, while at the same time showcasing the ideas behind solution operators for PDEs. These are fairly general, and more importantly lay the groundwork for leveraging neural networks to solve PDEs. In addition, both examples have explicit analytic solutions, so we can use them to test approximation or iterative methods for solving PDEs (including neural network methods).

The first example uses a one-dimensional heat equation with constant coefficients on an infinite domain (this is time dependent), and the second example uses a two-dimensional Poisson equation with constant coefficients on a bounded domain with a simple geometry (this is not time dependent; the solution is static in time).

Example Using the Heat Equation

The heat equation on an infinite one-dimensional rod looks like:

$u_t(x,t) = \alpha u_{xx}(x,t)$ for $x \in \mathbb{R}, t \in (0,\infty)$

$u(x,0) = u_0(x)$ for $x \in \mathbb{R}$

Since this PDE is defined on an infinite domain in x, we must specify the far field conditions (there is no boundary, so we must specify what we think the solution function $u(x,t)$ looks like when $x \to \infty$ and $x \to -\infty$). Let's assume that these limits are zero.

For simplicity, let's assume that the parameter α is constant so we can apply the Fourier transform. When applying this transform (with respect to x) to the PDE $u_t(x,t) = \alpha u_{xx}(x,t)$ and the initial condition, we manage to get rid of the derivatives in x and simplify the PDE into an ordinary differential equation with only one derivative in time:

$$\widehat{u}_t(\xi,t) = -\alpha\xi^2\widehat{u}(\xi,t) \text{ for } \xi \in \mathbb{R}, t \in (0,\infty)$$
$$\widehat{u}(\xi,0) = \widehat{u}_0(\xi) \text{ for } \xi \in \mathbb{R}$$

We can now easily solve this using a method from ordinary differential equations called separation of variables (let's not bother with the details), obtaining the solution in Fourier space:

$$\widehat{u}(\xi,t) = e^{-\alpha\xi^2 t}\widehat{u}_0(\xi)$$

We need the solution in x space, not in Fourier space, so we take the inverse Fourier transform of the above expression, and use the knowledge that multiplications become convolutions when transforming between x space and Fourier space. Therefore, the solution in x space is:

$$u(x,t) = F.T.^{-1}\left(e^{-\alpha\xi^2 t}\widehat{u}_0(\xi)\right) = F.T.^{-1}\left(e^{-\alpha\xi^2 t}\right) * u_0(x) = \frac{1}{\sqrt{4\pi\alpha t}}e^{-\frac{x^2}{4\alpha t}} * u_0(x)$$
$$= \int_{-\infty}^{\infty}\frac{1}{\sqrt{4\pi\alpha t}}e^{-\frac{(s-x)^2}{4\alpha t}}u_0(s)ds = \int_{-\infty}^{\infty}kernel(s,x;t;\alpha)u_0(s)ds$$

The punch line from the calculation is this:

The solution $u(x,t)$ of the PDE is the integral of some kernel function $k(s,x;t;\alpha)$ against the initial state of the solution $u_0(s)$.

Moreover, the solution operator of the PDE maps the given input data, which in this case are the parameter α and initial state $u_0(x)$, to the output, which is the solution we are seeking, $u(x,t)$, by integrating the initial state against some kernel function that depends on the parameter of the PDE (along with its dependence on space and time).

Knowing the formula of this kernel, or using a neural network to approximate it, unlocks the solution of a given PDE. We then say that the neural network *learned the solution operator* of the PDE.

In our simple example, we leveraged linearity and constant coefficients to incorporate Fourier transform methods and convolution when reverting back to real space, and had the luxury of working out an explicit analytic formula of the kernel of the integral, namely, $k(s,x;t;\alpha) = \frac{1}{\sqrt{4\pi\alpha t}}e^{-\frac{(s-x)^2}{4\alpha t}}$, so there is no need for approximations. On a nice side note, this kernel comes from a time-dependent Gaussian function

$Gaussian(x; t; \alpha) = \dfrac{1}{\sqrt{4\pi\alpha t}} e^{-\frac{x^2}{4\alpha t}}$ that spreads out as time evolves. The effect of convolving this with the initial state of the solution has a smoothing effect that spreads out and smooths any initial oscillations and spikes. We observe this smoothing with any diffusion process that we can visualize, such as diffusion of smoke in the air, or diffusion of a dye in a liquid, where the substance spreads out smoothly until we obtain one homogeneous-looking medium.

Example Using the Poisson Equation

A Poisson equation on a bounded domain looks like:

$$-\nabla \cdot \left(a\!\left(\overrightarrow{x}\right)\nabla u\!\left(\overrightarrow{x}\right)\right) = f\!\left(\overrightarrow{x}\right) \text{ for } \overrightarrow{x} \in D$$
$$u\!\left(\overrightarrow{x}\right) = 0 \text{ for } \overrightarrow{x} \text{ at the boundary of } D$$

When $a\!\left(\overrightarrow{x}\right)$ is constant and the domain is two-dimensional, this becomes:

$$-a\Delta u(x,y) = f(x,y) \text{ for } (x,y) \in D \subset \mathbb{R}^2$$
$$u(x,y) = 0 \text{ for } (x,y) \text{ at the boundary of } D$$

where $\Delta u(x,y) = u_{xx}(x,y) + u_{yy}(x,y)$. We can employ the Fourier transform in x and y like we did for the heat equation (linear equation with constant coefficients), but let's demonstrate the Green's function method instead. We can think of the righthand side of the PDE as an aggregation in the continuum of impulses of intensity $f(x,y)$ at locations (x,y). We need the *Dirac delta measure* $\delta_{(x,y)}(s,p)$ to express the concept of an impulse mathematically. This is zero everywhere on the domain except at the point (x,y), where it is infinite, and its total measure on the domain is normalized to 1. The rationale here is that we can solve the PDE with the righthand side consisting of only an impulse at a certain location, then aggregate the original solution from these. Presumably, solving the PDE with only an impulse as the righthand side is easier than solving it with some given function as the righthand side, so we will build up the solution $u(x,y)$ from the solution $G(x,y;s,p)$ of the impulse PDE. More importantly, using the Green's function allows us to get an integral representation of the solution of the input data against a kernel (which is the Green's function). The PDE with an impulse righthand side is:

$$-a\Delta G(s,p; x,y) = \delta_{(x,y)}(s,p) \text{ for } (s,p) \in D \subset \mathbb{R}^2$$
$$G(s,p; x,y) = 0 \text{ for } (s,p) \text{ at the boundary of } D$$

Let's now write:

$$f(x,y) = \int_D f(s,p)\delta_{(x,y)}(s,p)dsdp$$

and the PDE as:

$$-a\Delta u(x,y) = \int_D f(s,p)\delta_{(x,y)}(s,p)dsdp \text{ for } (x,y) \in D \subset \mathbb{R}^2$$
$$u(x,y) = 0 \text{ for } (x,y) \text{ at the boundary of } D$$

Let's substitute the $\delta_{(x,y)}(s,p)$ inside the integral with $-a\Delta G(s,p;x,y)$:

$$-a\Delta u(x,y) = \int_D -a\Delta G(s,p;x,y)f(s,p)dsdp \text{ for } (x,y) \in D \subset \mathbb{R}^2$$
$$u(x,y) = 0 \text{ for } \overrightarrow{x} \text{ at the boundary of } D$$

Now let's assume that we have the right conditions to swap differentiation and integration:

$$-a\Delta u(x,y) = -a\Delta \left(\int_D G(s,p;x,y)f(s,p)dsdp \right) \text{ for } (x,y) \in D \subset \mathbb{R}^2$$
$$u(x,y) = 0 \text{ for } (x,y) \text{ at the boundary of } D$$

Finally, this allows us to represent the solution $u(x,y)$ as:

$$u(x,y) = \int_D G(x,y;s,p;a)f(s,p)dsdp$$

Note that we made the dependency of G on a explicit in G. Later in this chapter, when we want to learn the solution operator of the PDE in a neural network setting, the physical parameter a would be part of the network's input. If the parameter $a = a(s,p)$ is not constant, then we would write $G(x,y;s,p;a(s,p))$. Analogous to the discussion of the previous example, the punch line from the previous calculation is:

The solution $u(x,y)$ of the PDE is the integral of some kernel function, in this case the Green's function $G(s,p;x,y;a)$ against the righthand side of the PDE $f(s,p)$.

Moreover, the solution operator of the PDE maps the given input data, which in this case is the parameter a and the righthand side of the PDE $f(x,y)$, to the output, which is the solution we are seeking, $u(x,y)$, by integrating the righthand side function against some kernel function that depends on the parameter of the PDE (along with its dependence on space). In our case, the kernel of the solution operator is the

Green's function of the PDE, which we happen to know for the Poisson equation on domains with an easy geometry, but not in more complex situations. Once again, knowing the formula of this kernel, or using a neural network to approximate it, unlocks the solution of a given PDE.

Fixed Point Iteration

The fixed point iteration is useful to construct explicit solutions and prove existence and uniqueness for certain lucky PDEs. It is such an easy and general method, so a definite must-have in our toolbox. We will write it down, then immediately apply it to represent the solution of a *dynamical system* as a series. A dynamical system is an ordinary differential equation that describes the evolution in time of a particle or a bunch of particles (a system). Again, we would like neural networks to learn the solution operators of dynamical systems, so this is consistent with our previous discussion. Moreover, it is good to have the fixed point iteration series representation of the solution side by side with the neural network representation. Recall that in many mathematical settings, we can represent the same solution in multiple ways. A fixed point iteration series is additive, while a neural network representation is compositional. Moreover, neural networks seem to have the advantage of representing the solution operators of whole families of PDEs and an overall wider variety, which is like a dream come true in this field.

The fixed point iteration aims to find a fixed point of a function, or a point x^* that the function maps back to itself: $f(x^*) = x^*$. This is not a simple task since f is usually nonlinear, and most of the time we have no idea if such points exist for a given function. As is always the case with nonlinear equations, iterative methods avoid *one-shot* solutions and instead come up with a sequence of points that *hopefully and under the right conditions* converge to the desired solution, in this case, the fixed point of a function.

How does it work?

Here is how the fixed point iteration goes:

1. x_0 is the starting point
2. $x_{i+1} = f(x_i)$

That's it. Our sequence $\{x_0, x_1, x_2 \cdots\}$ is generated by consecutive applications of f, and looks like $\{x_0, f(x_0), f(f(x_0)), f(f(f(x_0))), ...\}$. Under the right conditions on f and x_0, this sequence converges to a fixed point x^* of f (so $f(x^*) = x^*$).

Note that depending on f and the starting point, the asymptotic behavior of this sequence can be any of these:

Convergence to a limit x^*
　　If the fixed point iteration converges, then it captures a fixed point (for a contin-
　　uous function f; if the fixed point iteration converges, then the limit must be a
　　fixed point of f).

Divergence to ∞
　　The sequence grows without a bound.

Periodic behavior
　　The sequence oscillates between two or more values.

Chaotic behavior
　　The sequence behaves erratically, with no pattern whatsoever.

Theorems related to the fixed point iteration assert that the choice of x_0, the starting
point for the fixed point iteration, matters for whether it converges to a fixed point or
not.

How do we use it to solve ODEs and PDEs?

Our concern in this chapter is finding solutions of differential equations, which
are functions. So we will first reformulate a PDE so that its solution u satisfies an
equation that looks like $F(u) = u$ (note that F here is an *operator* not a *function*),
making it perfect for a fixed point iteration setting. Then we will apply the same
logic as described previously and construct a sequence of *functions* that hopefully,
under the right conditions, converge to the fixed point u^* of the *operator*, which is the
solution of the PDE that we are looking for. Note that in the previous discussion, we
constructed a sequence of *numbers* (rather than functions) that hopefully, under the
right conditions, converge to a fixed point of a *function* (rather than an operator).

Let's demonstrate this using a *dynamical system* setting. This is one of the most
important, general, and well-studied ordinary differential equations, which describes
the time evolution of a point in space. It is easy in the sense that it is usually
first order with only one derivative to get rid of, and hard in the sense that it is
generally nonlinear. There is a habit of linearizing the dynamical system near a point
and studying its linearized behavior. This is, in many cases, informative about the
nonlinear behavior, but the two should not be conflated. Too much is known about
linearized systems and not much is known about nonlinear systems, so we need
equal attention to nonlinear systems. In this section we do not linearize. Instead, we
approximate the solution using a series constructed by a fixed point iteration.

Knowing the initial state of the point $\vec{u}(t_0) = \vec{u}_0$, a solution trajectory $\vec{u}(t)$ tracks all
of its future states. The function $f\left(\vec{u}(t), a(t), t\right)$ specifies the evolution:

$$\frac{d\vec{u}(t)}{dt} = \vec{f}\left(\vec{u}(t), a(t), t\right)$$

$$\vec{u}(t_0) = \vec{u}_0$$

There is one derivative with respect to time that we need to get rid of, so we integrate once with respect to time:

$$\vec{u}(t) = \vec{u}_0 + \int_{t_0}^{t} \vec{f}\left(\vec{u}(s), a(s), s\right) ds$$

Now let's rewrite this integral equation in a form that is ripe for the fixed point iteration. We consider the whole righthand side as an operator whose input is $\vec{u}(t)$:

$$\vec{u}(t) = F\left(\vec{u}(t)\right)$$

Now we can generate the sequence $\left\{\vec{u}_0(t), \vec{u}_1(t), \vec{u}_2(t), \vec{u}_3(t), \cdots\right\}$ that converges to the solution $\vec{u}(t)$, for all time or for a finite amount of time, under the right conditions on f. The sequence looks like:

- $\vec{u}_0(t) = \vec{u}_0(t)$
- $\vec{u}_1(t) = F\left(\vec{u}_0(t)\right) = \vec{u}_0 + \int_{t_0}^{t} \vec{f}\left(\vec{u}_0(s), a(s), s\right) ds$
- $\vec{u}_2(t) = F\left(\vec{u}_1(t)\right) = \vec{u}_0 + \int_{t_0}^{t} \vec{f}\left(\vec{u}_1(s), a(s), s\right) ds$
- $\vec{u}_3(t) = F\left(\vec{u}_2(t)\right) = \vec{u}_0 + \int_{t_0}^{t} \vec{f}\left(\vec{u}_2(s), a(s), s\right) ds$

and so on.

Simple but very informative example

The best examples are the ones that are simple enough to have multiple ways of solving them. Seeing more than one way to do the same thing at once helps solidify the gist of a newly learned method. Consider the very simple one-dimensional and linear dynamical system:

$$\frac{du(t)}{dt} = u(t)$$

$$u(0) = 1$$

The first way to solve this is by separation of variables, where we put everything that has *u(t)* on one side of the equation, and everything that has *t* alone on the other side:

$$\frac{du(t)}{u(t)} = dt$$

Now we can integrate from *0* to *t*:

$$\int_0^t \frac{du(s)}{u(s)} = \int_0^t ds$$

We get ln $(u(t)) = t$, therefore the solution of our simple dynamical system using the separation of variables method is $u(t) = e^t$ (arguably the most important function in mathematics). Now let's construct a sequence of functions using the fixed point iteration and see if it converges to the solution $u(t) = e^t$ of the dynamical system:

- $u_0(t) = 1$
- $u_1(t) = F(u_0(t)) = u_0(t) + \int_0^t u_0(s)ds = 1 + \int_0^t 1 ds = 1 + t$
- $u_2(t) = F(u_1(t)) = u_0(t) + \int_0^t u_1(s)ds = 1 + \int_0^t 1 + s ds = 1 + t + \frac{t^2}{2}$
- $u_3(t) = F(u_2(t)) = u_0(t) + \int_0^t u_2(s)ds = 1 + \int_0^t 1 + s + \frac{s^2}{2}ds = 1 + t + \frac{t^2}{2} + \frac{t^3}{3!}$
- Keep going:
  $$u_n(t) = F(u_{n-1}(t)) = u_0(t) + \int_0^t u_{n-1}(s)ds = 1 + t + \frac{t^2}{2} + \frac{t^3}{3!} + \cdots + \frac{t^n}{n!}$$

As $n \rightarrow \infty$, the fixed point iteration converges to the series:

$$u_\infty(t) = 1 + t + \frac{t^2}{2} + \frac{t^3}{3!} + \cdots + \frac{t^n}{n!} + \cdots = \Sigma_{n=0}^\infty \frac{t^n}{n!}$$

which is the power series expansion of $u(t) = e^t$, the same solution we arrived at using separation of variables (albeit in different form). Cool.

When we use this iterative way to construct a solution of a dynamical system, or of a PDE reformulated as a dynamical system or in a way that is fit for a fixed point iteration (*u=F(u)*), we call it *Picard's iteration*. It is simple and arrives at the solution (when it converges) in steps.

Where is the complication?

Why don't we use *Picard's iteration* to construct solutions of *all* dynamical systems and of *all* PDEs that we are able to reformulate into a form fit for fixed point iteration? As always, the answer is the curse of dimensionality. Even for our very simple *one-dimensional* and linear example, each Picard iteration step involves evaluating an integral, which for more complex problems we have to evaluate numerically. For example, for dynamical systems representing the evolution and interactions of many particles, this gets multiplied by the number of particles. Overall in the ODE and PDE literature, there is a limited number of cases where practical algorithms are available for high-dimensional settings.

Recent successes!

That said, a recent method for finding explicit solutions for high-dimensional nonlinear parabolic PDEs and backward stochastic differential equations based on Picard's iteration has been quite successful in finding explicit solutions for high-dimensional PDEs arising in real-life physics and finance applications. The paper's (Weinan et al. 2017) abstract (*https://oreil.ly/APr1c*) is insightful:

Parabolic partial differential equations (PDEs) and backward stochastic differential equations (BSDEs) are key ingredients in a number of models in physics and financial engineering. In particular, parabolic PDEs and BSDEs are fundamental tools in the state-of-the-art pricing and hedging of financial derivatives. The PDEs and BSDEs appearing in such applications are often high-dimensional and nonlinear. Since explicit solutions of such PDEs and BSDEs are typically not available, it is a very active topic of research to solve such PDEs and BSDEs approximately. In the recent article [E, W., Hutzenthaler, M., Jentzen, A., and Kruse, T. Linear scaling algorithms for solving high-dimensional nonlinear parabolic differential equations. arXiv:1607.03295 (2017)] we proposed a family of approximation methods based on Picard approximations and multilevel Monte Carlo methods and showed under suitable regularity assumptions on the exact solution for semilinear heat equations that the computational complexity is bounded by $O\left(d\epsilon^{-(4+\delta)}\right)$ for any $\delta \in (0,\infty)$, where d is the dimensionality of the problem and $\epsilon \in (0,\infty)$ is the prescribed accuracy. In this paper, we test the applicability of this algorithm on a variety of 100-dimensional nonlinear PDEs that arise in physics and finance by means of numerical simulations presenting approximation accuracy against runtime. The simulation results for these 100-dimensional example PDEs are very satisfactory in terms of accuracy and speed. In addition, we also provide a review of other approximation methods for nonlinear PDEs and BSDEs from the literature.

Setting the stage for deep learning for PDEs

Before leaving this section, let's set the stage for solving ODEs and PDEs in the context of deep learning, in particular, for *deep operator networks*. We'll keep our one-dimensional dynamical system example, but this time we highlight the dependence on the physical parameters *a(t)*, and make it slightly more general by adding another explicit dependence on time:

$$\frac{du(t)}{dt} = \vec{f}(u(t),a(t),t)$$

$$\vec{u}(t_0) = \vec{u}_0$$

As before, we integrate once with respect to time:

$$u(t) = u_0 + \int_{t_0}^{t} f(u(s),a(s),s)ds$$

The purpose of a neural network is to take data as input, do something to it, then give us an output that we care about. For an ODE or a PDE, of course the output that we care about is the solution $u(t)$. Let's write this solution as the output of some operator G that takes the given data of an ODE or PDE as input. In our dynamical system case, the input data is the function $a(t)$ representing the physical parameters of the dynamical system. Note that we do not need to input the righthand side function f of the dynamical system. This is implicit in the training data, which now looks like the pairs *(training input,training output)*$=(a(t),u(t))$. By not inputting f, in a way we are saying that we don't care about the exact form of the ODE or PDE that this behavior comes from, but we are able to *learn* what the system is doing, regardless. This is the epitome of machine learning: no need to encode the rules that the system obeys; the model can still emulate it if it observes enough instances of it.

We can write the solution $u(t) = G(a(t))$, where the *solution operator G* is to be learned using a neural network. Hold this notation and thought until later in this chapter when we discuss neural operator networks. Plugging $u(t) = G(a(t))$ into the integral equation, we find that the solution operator, which we want to learn using a neural network, satisfies:

$$G(a(t)) = u_0 + \int_{t_0}^{t} f(G(a(s)),a(s),s)ds$$

We just wrote an integral equation that we will not do anything with. It just shows the *true* property that the entity *G(a(t))* that we care for satisfies. In the previous discussion, we approximated this using a Picard's iteration, and in the new era of deep learning, we approximate it using a deep operator network (more on this soon). This deep learning approach is computationally more efficient if we include the Fourier transform to speed up the computations. Moreover, the deep learning approach is more widely encompassing, in the sense that it applies to more PDEs and ODEs than just dynamical systems. A dynamical system is easy to integrate once and obtain a representation that gets us closer to the solution, which is not the case for many ODEs and PDEs.

Mesh independence and different resolutions

One final note: an input and output pair for a neural network learning the solution of our dynamical system looks like: *(training input,training output)* = $(a(t),u(t))$. Since machines take numerical values only and not functions, we must discretize when we implement this. Here's the pretty thing that differentiates *operators acting on functions* from *functions acting on points* and gives neural operator networks their *mesh independence* feature: *a(t)* and *u(t)* do not have to be discretized at the same values of *t*. All we care about is to map one function to another, so we can think of discretized *a(t)* as a vector mapped to another vector that is the discretized *u(t)*, not necessarily at the same points or not even of the same size. For the same reason, we can train the network at a given resolution, then make predictions at *another* resolution. This is great for the field of ODEs and PDEs, where the quality of numerical solutions has always been limited by the resolution of the employed discretization.

AI for PDEs

After surveying the main concerns and the basic approaches for solving PDEs, we are finally ready to discuss AI as it relates to PDEs, instead of only hinting at it or setting its stage here and there. We want to distinguish between a few different ways in which deep learning has stepped into the PDE community:

- Deep learning to learn the PDE's physical parameter values
- Deep learning to learn two-dimensional and three-dimensional meshes for numerical simulations and solid modeling
- Deep learning to learn a PDE's solution operator: a neural network learns a map between two *infinite dimensional* spaces
- Deep learning to bypass PDEs and simulate natural phenomena directly from observing data (particle systems and their interactions)

Deep Learning to Learn Physical Parameter Values

We can use neural networks to infer the parameters of a PDE model and their uncertainties. We get training data from experiments (real or synthetic via simulations of well-known phenomena with known parameters). This training data is labeled with the parameter values, so the neural network learns to map a certain PDE's initial setting to appropriate parameter values, leading to more accurate modeling results. Historically, parameters that couldn't be directly measured had to be guessed or hand tuned to fit some observed behavior, a practice that undermines the whole modeling process. This simple application of deep learning helps the PDE modeling community tremendously, because it brings more authenticity to their results. We can now learn parameter values from labeled images of experiments, recorded audio, and other unstructured or very high-dimensional data. Once trained, the neural networks can estimate parameters and uncertainties for any input data with similar settings. This poster has a nice and simple example that uses deep learning to predict the parameters of the velocity field for the G-equation (which models the combustion process) using flame front-image data: "Bayesian Inference in Physics-Based Nonlinear Flame Models" (*https://oreil.ly/ju4Kx*).

Deep Learning to Learn Meshes

We learned in this chapter that generating a mesh is an integral part of the finite element method, which in turn finds numerical solutions for a wide array of PDEs modeling natural phenomena with complex domain geometries. The quality of the underlying mesh affects the quality of the numerical solution. The finer the mesh, the more of the true solution it is likely to capture, but the more computational cost it incurs. An ideal mesh would be dense where the error between the numerical solution and the true solution is more likely to be high, and coarse where the error is low, hence preserving fidelity while keeping the overall computational cost manageable (Figure 13-12).

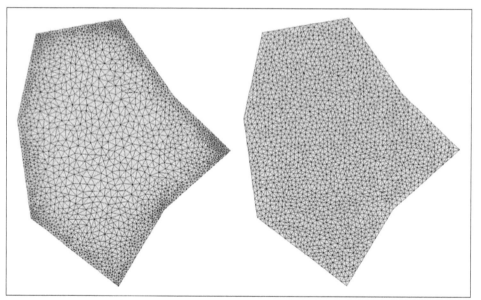

Figure 13-12. Nonuniform mesh on the left versus uniform mesh on the right; a mesh needs to be finer where the error is large (image source (https://oreil.ly/4XvB1))

It would be nice if, given a PDE, its domain geometry, boundary conditions, and parameter values as inputs, we could train a neural network to automatically generate an ideal mesh, predicting the density distribution of the mesh elements at each location of the domain. This is exactly what MeshingNet (*https://oreil.ly/0cAgv*) does.

Before MeshingNet, mesh learning was done via expensive multistep finite element solutions and error estimators. In contrast, MeshingNet relies on similar problems to predict ideal meshes for new problems. It starts with an initial uniform and coarse mesh, and predicts a nonuniform mesh density for refinement. A hallmark of deep learning, MeshingNet generalizes well to different geometric domains with various governing PDEs, boundary conditions, and parameters.

The inputs to MeshingNet are the governing PDE, PDE parameters, domain geometry, and the boundary conditions, and the output is an area upper-bound distribution $A(X)$ over the whole domain. The mapping between input and output is highly nonlinear and is thus learned by a neural network, which has demonstrated an impressive ability to express many kinds of nonlinear relationships.

To build the training data set, the MeshingNet team computes high-accuracy solutions on high-density uniform meshes using standard finite element solvers. They also do the same computation for low-density uniform meshes to obtain lower-accuracy solutions. Then the team computes an error distribution $E(X)$ by interpolating between these solutions. They use $E(X)$ as a guide to refine $A(X)$. They enrich the training data by combining different geometries with different parameters and boundary conditions.

Deep learning for three-dimensional meshes

Three-dimensional meshes (Figure 13-13) are useful for computer graphics, animations for the entertainment industry, and solid modeling. They are also highly desirable to reconstruct textured and realistic surfaces from a given set of three-dimensional data points. Traditional methods include Delaunay triangulations and Voronoi diagrams, which interpolate points using triangular meshes. However, when there is noise in the coordinates, the resulting surface would be unnecessarily rough, which requires data preprocessing.

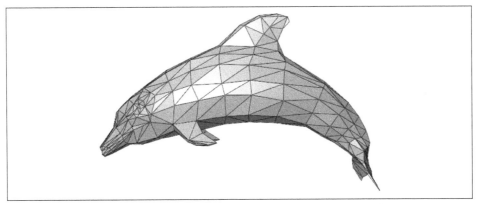

Figure 13-13. Three-dimensional mesh (image source (https://oreil.ly/2bG2x))

Deep learning is stepping in to generate higher-quality three-dimensional meshes; see, for example, "Deep Hybrid Self-Prior for Full 3D Mesh Generation" (Wei et al. 2021) (*https://oreil.ly/mm79h*) and "Pixel2Mesh" (Wang et al. 2018) (*https://oreil.ly/5rADG*), which produces a three-dimensional shape in triangular mesh from a single color image by continuously deforming an ellipsoid.

Deep Learning to Approximate Solution Operators of PDEs

We have already started this discussion multiple times in this chapter. Instead of using deep learning to enhance existing methods for PDEs, such as learning physical parameter values from data, or learning better meshes for numerical methods, we would like to learn a PDE's solution operator. This maps the PDE's input, such as its domain, physical parameters, initial/final states of the solution, and/or boundary conditions, directly to its solution. We can think of this as:

> Solution of PDE = function(PDE's physical parameters, domain, boundary conditions, initial conditions, etc.)

We want to build a neural network to approximate this function. This is in fact an operator and not a function in the usual sense, since it sends functions to other functions. The caveat here is that differential operators and their inverses map *infinite dimensional spaces* to *infinite dimensional spaces*, sometimes in a linear way, such as the map from the *righthand side* of a Poisson equation to the solution, and most of the time in a nonlinear way, such as the map from *the parameters* of a Poisson equation to the solution. In contrast, the inputs and outputs of the neural networks that we learned about throughout this book are finite dimensional (inputs and outputs are vectors, images, graphs, etc.). These neural networks are able to approximate *function mappings* between *finite dimensional spaces*. They have a powerful universal approximation theorem going on for them, and a myriad of successes in practical applications (we can approximate any continuous function to arbitrary accuracy using neural networks if we place no constraints on the width and depth of the hidden layers). To solve PDEs analogously using deep learning, we must answer two questions:

Can neural networks approximate mappings between infinite dimensional spaces?
> That is, can they approximate any nonlinear continuous functional (the input to the network would be a function or a bunch of functions, and the output would be a real number) or nonlinear operator (the input to the network would be a function or a bunch of functions, and the output would be another function)? The answer is a *yes*!

> There is a *universal approximation theorem* for *neural network operators* just like there are universal approximation theorems for *neural network functions*. A neural network with a single hidden layer can approximate accurately any

nonlinear continuous functional or operator. Moreover, neural networks are able to learn the solution operator of an entire family of PDEs, as opposed to classical methods for solving PDEs that only solve a single instance of a given PDE at a time.

How do we implement this in practice?

For the finite dimensional case, a node in a neural network linearly combines the finite dimensional features of the input vector (or the outputs of the previous layer), adds a bias term, applies a nonlinear activation function, then passes the result to the next layer. The analog for the infinite dimensional case, where we do not have finitely many entries to linearly combine anymore, would be to integrate some learnable kernel (multiplier function) of the input functions (for numerical integration, we have to sample this at finitely many points, converting integration to addition), add a bias function (this is optional), and apply a nonlinear activation function before passing the result to the next layer. The next layer would then add multiples of the node results of the previous layer, and integrate them against a learnable kernel of the results of the nodes of the previous layer, and so on. One example of doing this would look like:

$$u_{n+1}(x) = \sigma(\int_D kernel(x,s,a(x),a(s);\omega)u_n(s)ds + W u_n(x)),$$

where we arrive at the solution *u(x)* iteratively after a certain number of global integrations against a kernel, local linear transformations, and compositions with a nonlinear activation function. The parameters of the kernel in the iterative process are ω and the entries of W. The neural network learns these parameters from the labeled data (labeled with the solutions of the PDE) during training by minimizing a loss function. Analogous to the finite dimensional case, neural operator networks approximate nonlinear operators by composing linear integral operators that act globally over the entire domain with nonlinear activation functions. The previous iterative formula also includes a local linear multiplier, which becomes a matrix when we discretize.

Neural operator networks to learn the solution operators that we derived

Let's pause for a moment and compare the previous expression to the three true solution operators that we derived for: the heat equation, Poisson equation, and dynamical systems. We can easily adapt the neural operator iterative process to each of these three settings:

- For the heat equation in one space dimension and constant coefficients, the solution operator maps the initial state and the PDE's physical parameter (constant) to the solution $u(x,t)$. We are lucky enough to have explicit formulas for all the quantities involved:

$$G(u_0(x),a) = u(x,t) = \int_{-\infty}^{\infty} \frac{1}{\sqrt{4\pi at}} e^{-\frac{(s-x)^2}{4at}} u_0(s)ds$$
$$= \int_{-\infty}^{\infty} kernel(s,x; t; a) u_0(s)ds$$

In this case, the neural operator network does the following iteration to approximate the true operator:

$$G(u_0(x),a) = u(x,t) \approx u_{n+1}(x,t) = \sigma(\int_D kernel(s,x; t; a; w) u_n(s)ds + W u_n(x))$$

- For a Poisson equation in two space dimensions, zero boundary conditions, and constant coefficients, the solution operator maps the PDE's righthand side f, and its physical parameters (constant) to the solution $u(x,y)$, and only for certain simple geometries. We are lucky to have explicit formulas for all the quantities involved (none of which we write here):

$$G(f(x,y),a) = u(x,y) = \int_D GreenFunction(x,y; s,p; a) f(s,p)dsdp$$

In this case, the neural operator network does the following iteration to approximate the true operator:

$$G(f(x,y),a) = u(x,y) \approx u_{n+1}(x,y) = \sigma(\int_D kernel(x,y,s,p,a; w) u_n(s,p)dsdp + W u_n(x,y))$$

- For a one-dimensional dynamical system, the solution operator maps the ODE's physical parameter (function) to the solution $u(t)$, and we have an implicit integral equation that it satisfies:

$$G(a(t)) = u(t) = u_0 + \int_{t_0}^{t} f(G(a(s)),a(s),s)ds.$$

In this case, the neural operator network does the following iteration to approximate the true operator:

$$G(a(t)) = u(t) \approx u_{n+1}(t) = \sigma\left(\int_{t_0}^{t} kernel(t,s,a(t),a(s); w) u_n(s)ds + W u_n(t)\right)$$

Here, one data point is a triplet $(t, a(t), G(a(t)))$, and thus one specific input a may appear in multiple data points with different values of t. For example, a data set of size 10,000 may only be generated from 100 $a(t)$ trajectories, and each evaluates $G(a)(t)$ for 100 t locations.

One thing to observe with these three different settings is that neural operator networks require only input and output data, and no knowledge of the underlying PDEs. The knowledge about the PDEs is implicit in the training data. With this in mind, let's reinforce the *input output* form of a neural operator network:

Solution of PDE ≈ learned operator(PDE's physical parameters, domain, boundary conditions, initial conditions, etc.)

The important questions

When we branch out beyond this book to expand our knowledge on neural operator networks, we must keep the following questions as our guide:

For a given PDE, what is the input to the network, and what is the output?
We addressed these in the simple contexts of the heat equation, Poisson equation, and dynamical systems.

What is an example of an architecture of a neural operator?
Figure 13-14 shows the input and output structure of DeepONet (*https://oreil.ly/ II3kg*). The input is a discretized pair $(t, a(t))$, and the output is a discrete $G(a(t))$.

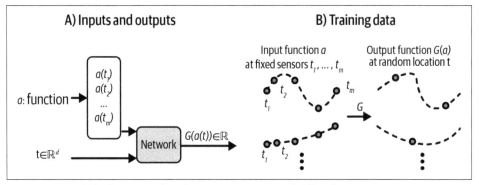

Figure 13-14. (A) The network to learn an operator $G(a(t))$ takes two inputs $(a(t_1), a(t_2), \cdots, a(t_m))$ and t. (B) Illustration of the training data (image source (https://oreil.ly/oRgjA)).

How do we deal with the fact that the inputs have such drastic differences in dimension, such as finite dimensional and infinite dimensional at the same time?

In other words, how do we discretize the involved finite dimensional (independent variables such as time and space) and infinite dimensional quantities (solution functions, parameter function, boundary conditions, initial conditions, etc.) during training and inference? Note that for neural networks that learn finite dimensional mappings, inputs (tables, images, audio files, graphs, natural language text) always have the same dimension, are preprocessed to have the same dimension, or the network itself processes fixed dimension portions of the input individually.

How do we avoid the common trap in many PDE solution methods, which end up being discretization dependent?

In what sense are neural operator networks meshless and able to generalize their learned parameters to work for other discretizations than the ones they have been trained on? The significant advancement here is that neural operator networks are discretization invariant, sharing the same network parameters between different discretizations. This means that their outputs do not depend on the underlying discretization and can be used with different grid representations.

How do we speed up the computation time for the integrals involved in the neural operators and make it less costly?

We involve the Fourier transform. A Fourier neural network speeds up the process of computing the involved integrals transforming the inputs to Fourier space. This has the fast Fourier transform methods at its disposal. A Fourier neural network, which we discuss in the next subsection, implements this.

How do neural operator networks fare with high-dimensional PDEs involving hundreds or thousands of variables?

PDEs that model financial markets with all the underlying assets (Black-Scholes), game theoretic settings with many participating agents (Hamilton-Jacobi-Bellman), or physical systems with many particles are very high-dimensional. Discretization in each of these dimensions explodes the size of an already big problem computationally and has until now made any practical implementation of these elegant PDEs infeasible. The article "Solving High-Dimensional Differential Equations Using Deep Learning" (*https://oreil.ly/vRNYr*) (Han et al. 2018), which we discuss soon, uses AI techniques to address such PDEs, but it would be nice to compare that article's methodology to a deep neural operator setting.

Fourier neural network

The California Institute of Technology has recently open sourced its Fourier neural network for solving partial differential equations; its approach is shown in the article "Fourier Neural Operator for Parametric Partial Differential Equations" (Li et al.

2021) (*https://oreil.ly/IQ23v*). These networks can approximate solution operators for PDEs that are highly nonlinear, with high frequency modes and slow energy decay.

Each layer in a Fourier neural network applies a fast Fourier transform to its input data, then a linear transform, then an inverse fast Fourier transform. This results in a quasi-linear computational complexity, that is, of order $O(n$ polynomial$(\log(n)))$ and makes the model invariant to the spatial resolution of the data (even though it still requires a uniform mesh).

Figure 13-15 shows the architecture of the Fourier neural network.

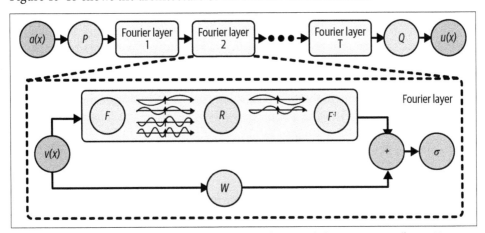

Figure 13-15. The architecture of a Fourier neural network (image source (https://oreil.ly/Ik39S))

The input is the physical parameter $a(x)$, and the output is the PDE solution $u(x)$:

1. Start from input $a(x)$.

2. Lift to a higher-dimension channel space by a shallow, fully connected neural network P: $v0(x) = P(a(x))$.

3. Apply several Fourier layers of integral operators and activation functions. In each of these layers, we apply the Fourier transform $F.T$, apply a linear transform R on the lower Fourier modes and filter out the higher modes, and apply the inverse Fourier transform $F.T^{-1}$. On the bottom, apply a local linear transform W.

4. Project back to the target dimension by a neural network Q. Finish with the output $u(x) = Q(vT(x))$, which is the projection of v by the local transformation Q also parameterized by a shallow fully connected neural network.

5. Finish with the output $u(x)$.

The article demonstrates the method with a variety of important PDEs:

- Burgers' equation
- Darcy flow
- Navier-Stokes equation
- Turbulent flows in regimes where other methods diverged

The Fourier neural network is mesh invariant, so it can be trained on a lower resolution and evaluated at a higher resolution, without seeing any higher resolution data (zero-shot super-resolution).

Since data-driven methods rely on the quality and quantity of data, we need to generate training pairs of inputs and outputs for the neural operator networks by solving the actual PDEs using some other methods. To this end, the authors note that to learn Navier-Stokes equation with $viscosity = 1e^{-4}$, we need to generate $N = 10,000$ training pairs $(a(x), u(x))$ using a numerical solver. For more challenging PDEs, generating even a few training samples can be very expensive. A future direction would be to combine neural operators with numerical solvers to raise the requirements on data.

Statement of the universal approximation theorem for operators

Suppose that σ is a continuous nonpolynomial function, X is a Banach space, $K_1 \subset X$, $K_2 \subset \mathbb{R}^d$ are two compact sets, V is a compact set in $C(K_1)$, and G is a nonlinear continuous operator that maps V into $C(K_2)$. Then for any $\epsilon > 0$, there are positive integers n, p, m, and constants $c_i^k, \xi_{i,j}^k, \theta_i^k, \xi_k \in \mathbb{R}, \omega_k \in \mathbb{R}^d, x_j \in K_1, i = 1, \cdots, n, k = 1, \cdots, p, j = 1, \cdots, m$, such that:

$$\left| G(u)(y) - \Sigma_{k=1}^{p} \Sigma_{i=1}^{n} c_i^k \sigma \left(\Sigma_{j=1}^{m} \xi_{ij}^k u(x_j) + \theta_i^k \right) \sigma(\omega_k . y + \xi_k) \right| < \epsilon$$

holds for all $u \in V$ and $y \in K_2$. Note that this approximation theorem only uses one hidden layer in the neural network but does not specify how many nodes this layer has. In applications, just like in the finite dimensional case, we use more than one layer.

Do not be intimidated by the big words and the Greek letters. What this theorem tells us is that we have theoretical grounds to formulate a neural network operator, and expect it to approximate the PDE solution operator very well. Even though we may never know the exact formula for the PDE solution operator, the operator neural network that we construct acts as a very good proxy. This is the reason we are all in love with approximation theorems, and should be eternally grateful to the mathematicians who find them.

We have mentioned this in the context of the fixed point iteration's *additive* approximation of a solution, and it is worth mentioning again: whether approximating a mapping between finite dimensional spaces or a mapping between infinite dimensional spaces, neural networks represent functions, functionals, or operators using *compositions* of simple functions (a linear combination or linear integral operator composed with a nonlinear activation function) to approximate complicated ones. This is different than classical approximation approaches, where the approximation is *additive* and not compositional.

How do we branch out and dive into the more technical details?

For a deeper dive into neural operator networks, the three important publications on this topic are:

- "DeepONet: Learning Nonlinear Operators for Identifying Differential Equations Based on the Universal Approximation Theorem of Operators" (*https://oreil.ly/Do6du*) (Lu et al. 2020)
- "Neural Operator: Graph Kernel Network for Partial Differential Equations" (*https://oreil.ly/RtWuT*) (Li et al. 2020)
- "Fourier Neural Operator for Parametric Partial Differential Equations" (*https://oreil.ly/cUkhw*) (Li et al. 2021)

Numerical Solutions of High-Dimensional Differential Equations

Differential equations are universal; they can model almost anything we can think of, including our daily commute and traffic. It is hard to be exposed to differential equations, then *not think* of each situation that we find ourselves in as fitting into some sort of differential equation. That said, the curse of dimensionality has haunted this field since its inception, and has stood in the way of many practical applications. This is why many introductory PDE courses misleadingly focus only on one- and two-dimensional differential equations, as if that is all there is. If AI was to be given a different name that is not as flashy and definitely would not make it into any movies, it would be: *processing, computation, and analysis of high-dimensional data*. This does not talk AI down, because processing, computation, and analysis of high-dimensional data are exactly what humans do on a daily basis, provided we add a creative dimension to it (which in AI would translate to generative models). It is then not surprising that deep learning turns out to be an appropriate setting for finding numerical solutions of *very* high-dimensional differential equations. This is the focus of the article "Solving High-Dimensional Differential Equations Using Deep Learning" (Han et al. 2018) (*https://oreil.ly/Qw4gY*), which addresses PDEs with hundreds or even thousands of dimensions. This way, we can include all participating agents, assets, resources, or particles at the same time, instead of artificially devising handmade assumptions about their interactions and connections. The authors consider multiple high-dimensional PDEs, including the Hamilton-Jacobi-Bellman equation (what is the optimal strategy for each interacting agent, among hundreds of agents?) and the Black–Scholes equation (what is the fair price of a European claim based on one hundred underlying assets, given that no default has occurred yet?).

When we rely on a deep learning setting as a basis for our models, for example, for computing solutions of high-dimensional PDEs, the first question we must ask is what is the input and what is the output of our deep learning network. For any PDE whose solution is $u(x,t)$, ideally we would input x and t and output $u(x,t)$. x in this case can be an extremely high-dimensional $\overrightarrow{x}$. If the entries of x have any inherent stochasticity into them, such as prices of financial market assets, then we must model them as such, and if we don't, then we are usually assuming some sort of averaging. The bottom line is, for many realistic cases, we can input x as X, a stochastic process. We must define this mathematically.

One big step in the aforementioned article is reformulating the high-dimensional PDEs as backward stochastic differential equations before inputting X into a neural network that approximates the gradient of the solution. To master the essential math required here, we must define:

- Brownian motion (see Chapter 11)
- Stochastic process (see Chapter 11)
- Stochastic differential equation (this is beyond the scope of the book)
- Relating nonlinear parabolic PDEs to stochastic PDEs (this is beyond the scope of the book)
- Backward stochastic differential equation (this is beyond the scope of the book)

And we must answer the question: why did we have to reformulate the PDE into a stochastic form before training the network? What advantage does this form give us? This goes beyond the scope of the book, but you now know what to look for and what questions to ask.

Finally, the method opens the door to solving many high-dimensional differential equations, but there are limitations. The method cannot be applied to the quantum many body problem due to the difficulty in dealing with the Pauli exclusion principle (*https://oreil.ly/XAyAz*).

Simulating Natural Phenomena Directly from Data

We have addressed particle systems once in this chapter. We used a statistical mechanics framework to describe the probabilities of the states of the system at the particle scale, then we used those to write down PDEs modeling the time evolution of the system at the macro scale.

In this section, we explain how recent neural network–based models simulate a particle system and predict its evolution *without* writing any PDEs. In other words, we bypass PDEs and trade them for *learning from data*.

To track the evolution of a certain particle system (such as water or sand) at a granular scale, we need to know each particle's position vector $\vec{p}_i(t)$ at each time step t. How these positions change depends on local and long-range interactions between a particle and its neighbors (such as exchanging energy and momentum), which are dictated both by the physical nature of the system and by external effects such as gravity, temperature, forces, magnetic fields, etc. Instead of writing down explicit equations for these interactions, and relating them to the particles' positions, velocities, and/or accelerations, we can train a neural network to learn a map between a given state of a particle system at a certain time (input) and the positions of all of its particles (or velocities or accelerations) at a future time (output). Graph networks are well suited to model particle systems, since each particle along with its state can be a node and the edges along with their features model the interactions between specific particles.

We highlight and comment on the general ideas from a recent work that learns such a map: "Learning to Simulate Complex Physics with Graph Networks" (*https://oreil.ly/ Q0EWH*) (Sanchez-Gonzalez et al. 2020).

First, we need training data
> We can generate input (particle system and its features at a certain time) and target (each particle's acceleration at a later time) pairs from a data set of observed or simulated trajectories of a certain particle system. For example, from a 1,000-step-long trajectory, the team generates 995 pairs, conditioning on the 5 past states. In the data sets, we only need the position vectors, and we can derive velocity and acceleration vectors using finite differences. The data sets typically contain 1,000 train, 100 validation, and 100 test trajectories, each simulated for 300–2,000 time steps, tailored to the average duration for the various materials to come to a stable equilibrium.

Next, we need to build the map from input to output (the network components)
> Start with the system at a certain state at integer time t $X^t = \left(\vec{x}^t_0, \cdots, \vec{x}^t_N \right)$, where each of the N particles' $\vec{x}^t_i$ represents its state at time t (which includes its position $\vec{p}^t_i$ and other characteristics such as mass, material properties, etc.).

> Next, learn a map that represents the state $X^t = \left(\vec{x}^t_0, \cdots, \vec{x}^t_N \right)$ as graph $G =$ (nodes, edges, and global properties that can alternatively be included as node features). The node embeddings, $\overrightarrow{node}_i = function\left(\vec{x}_i \right)$, are learned functions (using a multilayer perceptron) of the particles' states. Directed edges are added to create

paths between particle nodes that have some potential interaction. The edge embeddings, $\overrightarrow{e}_{i,j} = function\left(\overrightarrow{r}_{i,j}\right)$, are learned functions (using a multilayer perceptron) of the pair-wise properties of the corresponding particles $\overrightarrow{r}_{i,j}$, for example, displacement between their positions, spring constant, etc.

Then learn a graph-to-graph map. This computes the interactions among the nodes via M steps of learned message passing to generate a sequence of updated latent graphs, $G = (G_1, \cdots, G_M)$. Then return the final graph. Message passing allows information to propagate between the nodes via the edges, and for the constraints to be respected. This way, the complex dynamics of the system are approximated by learned message passing among the nodes within their local neighborhoods. Moreover, the final graph has the same structure as the first graph, but with potentially different node, edge, and graph-level attributes.

Then, learn a map (multilayer perceptron) from the final graph to a matrix that extracts the dynamics of the system, for example, the matrix of the particles' acceleration $Y = \left(\overrightarrow{p}_1'', \overrightarrow{p}_2'', \cdots, \overrightarrow{p}_N''\right)$. Finally, the particles' positions and velocities are updated using an Euler integrator of the accelerations in Y. This in turn updates the system's state to X^{t+1}.

Such models are not restricted to materials and particle systems, but can model systems with many interacting agents, such as robotic control systems. They are a great step toward simulating complex phenomena authentically, which is of great value to science and engineering.

Hamilton-Jacobi-Bellman PDE for Dynamic Programming

The Hamilton-Jacobi-Bellman equation is yet another PDE whose solution unlocks many possibilities in economics, operations research, and finance, *if only* we are able to solve it in high dimensions. In a nutshell, we are searching for an *optimal strategy* (such as an investment strategy) that guarantees some *minimal* implementation cost over a *given period of time*. Ideally, we would like to include hundreds or thousands of interacting agents, such as all the financial assets for investment banking, instead of downsizing to unrealistic *representative agent* models. This is where using neural networks to find numerical solutions for high-dimensional PDEs helps us, as we saw earlier in this chapter.

Mathematically, the Hamilton-Jacobi-Bellman PDE is very *rich*. It combines dynamical systems ($\frac{\partial x(t)}{dt} = f(x(t), a(t), t)$), PDEs (partial derivatives and equalities), and optimization (*max* or *min* problems). When we learn how to derive this one PDE from real-world applications, attempt to understand it, find its solutions, and analyze these solutions (existence, uniqueness, smoothness, etc.), we acquire a ton of math.

Moreover, this PDE is directly related to *reinforcement learning* in AI, but instead of thinking about reinforcement probabilistically, in terms of Markov decision processes, such as in Chapter 11, we think about reinforcement in terms of deterministic dynamic programming.

In the dynamic programming setting, the states of the interacting agents, bundled in a vector $\vec{x}(t)$, evolve in time according to a dynamic system, and we need to find an optimizing policy that induces a special solution of this dynamic system: the one that incurs a minimal cost over a given period of time. The train of thought goes like this: a certain time-dependent policy affects the behavior of a dynamic system, which in turn affects the incurred cost. All of these are mathematical quantities.

The contributions of Richard Bellman (1920–1984) to the dynamic programming field (finding optimal strategies for an evolving system over a given period of time) are invaluable. We will encounter Bellman's principle for optimality shortly, and in fact, it is Bellman who coined the term *curse of dimensionality*. This principle is tremendously helpful, as it breaks down the involved optimization problem over the considered period of time into smaller subproblems at smaller time intervals, which we can then solve in a recursive manner.

Bellman's equation in deterministic and stochastic settings

In a *deterministic* dynamic programming setting, there are:

Discrete time Bellman's equation
We can find the value function starting from the current time until the final time by picking the best strategy (or control or policy) a_k at the current time step k so that the current cost plus the value function at the next time step are minimized. This is a recursive process:

$$Value\left(\vec{x}_k, n\right) = \min_{\vec{a}_k}\left(Cost\left(\vec{x}_k, \vec{a}_k\right) + Value\left(\vec{x}_{k+1}, n-1\right)\right)$$

where n is the final time step, and the discrete time dynamics are:

$$\vec{x}_{k+1} = \vec{f}\left(\vec{x}_k, \vec{a}_k\right)$$

so that:

$$Value\left(\vec{x}_k, n\right) = \min_{\vec{a}_k}\left(Cost\left(\vec{x}_k, \vec{a}_k\right) + Value\left(\vec{f}\left(\vec{x}_k, \vec{a}_k\right), n-1\right)\right)$$

The sequence of optimizers $\vec{a}_k$ at each discrete time step k constitute the optimal policy (or strategy or control) for the whole time period, and guarantee the minimal total cost, exactly as in reinforcement learning.

Continuous time Bellman's equation
This is the Hamilton-Jacobi-Bellman PDE.

In a *stochastic* optimal control setting, there is also a stochastic version of Bellman's equation. This is widely applicable in investment banking, and in scheduling and routing problems. In the stochastic framework, we need to find an optimal control input (strategy or policy) that guides the underlying stochastic processes to some desired final state, with minimal cost. Consider for example the problem where we need to execute a financial order, with a minimal implementation cost and within a certain period of time. We can first model the short-term dynamics of the underlying assets, then discretize both in time and state spaces. This allows us to execute a given amount of shares at each of the time steps, with the condition that we must execute all of the shares during the given time period. We search for the policy that tells us, among all possible actions that we can take at each point in time, what the optimal one is that gets us to where we want to be.

In Chapter 11, we connected Bellman's equation to reinforcement learning. This was done in the context of a Markov decision process with value function:

$$Value\left(\vec{s}\right) = \max_{\text{states and actions}} \mathbb{E}\left(reward_0 + \gamma Value\left(\vec{s}'\right)\right)$$

In a deterministic dynamic programming setting, the analogous equation is the Hamilton-Jacobi-Bellman PDE for the value function. Before writing its formula, this is the language that we need to pay attention to:

Minimizing a cost function
Is there a more common objective in this world?

Choosing an optimal control or optimal policy
This is the minimizer that we are looking for; it controls the dynamical system.

Value function
The total minimum cost over the considered period of time.

Bellman optimality principle
An amazingly helpful principle that allows us to simplify the optimization problem.

Backward in time solution

Start with the desired outcome and work our way backward, optimally to an initial state. It is intuitive to see why a backward-in-time solution is easier in this setting. Since we know the end goal, we immediately exclude all paths that don't lead to it from the preceding time step, saving us the exploration of many useless paths. If, on the other hand, we solve forward in time, starting at the beginning of the time interval, then we do not have the advantage of *closeness* to the desired outcome, so we must waste our time and computational resources exploring many more useless paths.

The big picture

The ultimate question is: what must we do now (what is the initial state $\vec{x}(t_{initial})$ and the time-dependent policy $\vec{a}(t)$) to get us to where we want to be ($\vec{x}(t_{final})$), in the most cost-efficient way (attain the value function $Value\left(\vec{x}(t_{initial}), t_{initial}, t_{final}\right)$, which is the minimum value of the strategy implementation cost function)?

The involved quantities are:

- $\vec{x}(t)$ is a vector characterizing the state of the dynamic system.

- The strategy (or policy or control) $\vec{a}(t)$. We need to design this so that it invokes a state $\vec{x}(t)$ that minimizes some cost function. That is, if we input this special $\vec{a}(t)$ that we are looking for to the dynamic system, the output $\vec{x}(t)$ would minimize the cost function.

- The cost function $Cost\left(\vec{x}(t), \vec{a}(t), t_{initial}, t_{final}\right)$ incurred due to the implementation of the strategy (or policy or control). This is given by some terminal cost at t_{final} and the sum of incremental costs (integral) as we transition from $t_{initial}$ to t_{final}. The incremental costs depend on the current state of the system and the current control.

- The value function $Value\left(\vec{x}(t_{initial}), t_{initial}, t_{final}\right)$ is the minimal cost in a specific time period attained by enforcing the minimizing policy $\vec{a}^{*}(t)$, which in turn specifies the state $\vec{x}^{*}(t)$ using the information about the dynamics of the system.

Hamilton-Jacobi-Bellman PDE

These are the involved equations and formulas:

$$\frac{d\vec{x}(t)}{dt} = \vec{f}\left(\vec{x}(t), \vec{a}(t), t\right)$$

$$Cost\left(\vec{x}(t), \vec{a}(t), t_{initial}, t_{final}\right) =$$

$$Cost_{final}\left(\vec{x}(t_{final}), t_{final}\right) + \int_{initial}^{t_{final}} Cost_{incremental}\left(\vec{x}(s), \vec{a}(s)\right) ds$$

$$Value\left(\vec{x}(t_{initial}), t_{initial}, t_{final}\right) = \min_{\vec{a}(t)} Cost\left(\vec{x}(t), \vec{a}(t), t_{initial}, t_{final}\right)$$

Bellman's optimality principle tells us something very valuable about the behavior of the value function (the optimal cost) along the special trajectory $\vec{x}^*(t)$ corresponding to the optimizing policy $\vec{a}^*(t)$: the value at a specified time interval is the sum of values if we break the time interval apart along the special trajectory $\vec{x}^*(t)$ corresponding to the optimizing policy $\vec{a}^*(t)$. This enables us to break up the optimization problem over a longer time interval into a recursion of optimization problems over much shorter time intervals:

$$Value\left(\vec{x}^*(t_{initial}), t_{initial}, t_{final}\right) = Value\left(\vec{x}^*(t_{initial}), t_{initial}, t_{intermediate}\right)$$
$$+ Value\left(\vec{x}^*(t_{intermediate}), t_{intermediate}, t_{final}\right)$$

Using Bellman's principle, we can derive the Hamilton-Jacobi-Bellman PDE that the value function satisfies. This PDE generalized an older Hamilton-Jacobi PDE for optimal control. The solution of this PDE contains very valuable information. Suppose we encounter the system at *any time t*, not only at its initial state $t_{initial}$, then we can compute the value function up to the desired final cost by solving the Hamilton-Jacobi-Bellman equation:

$$-\frac{\partial Value}{\partial t} = \min_{\vec{a}(t)} \left(\left(\frac{\partial Value}{\partial \vec{x}}\right)^T \vec{f}\left(\vec{x}(t), \vec{a}(t)\right) + Cost_{incremental}\left(\vec{x}(t), \vec{a}(t)\right)\right)$$

subject to final time condition:

$$Value\left(\vec{x}(t_{final}), t_{final}\right) = Cost_{final}\left(\vec{x}(t_{final}), t_{final}\right)$$

This is a first-order PDE for the value function $Value\left(\vec{x}(t), t, t_{final}\right)$. Again, this is the optimal cost incurred from starting in state $\vec{x}(t)$ at time t, and controlling the

---

system optimally from then until time t_{final}. We know the final value function at t_{final}, and we are looking for value function at time t, namely, $Value\left(\overrightarrow{x}(t)\right)$. So we solve the PDE *backward in time*, starting at t_{final} and ending at $t_{initial}$.

Solving the Hamilton-Jacobi-Bellman PDE

If we are able solve the Hamilton-Jacobi-Bellman PDE for the value function, then we know the optimal control $\overrightarrow{a}^{*}(t)$, which in turn produces the least costly (or most rewarding) trajectory $\overrightarrow{x}^{*}(t)$ from our current state $\overrightarrow{x}^{*}(t_{initial})$ to the final desired state $\overrightarrow{x}^{*}(t_{final})$.

The Hamilton-Jacobi-Bellman PDE does not, in general, have a smooth solution, so we must satisfy ourselves with *weak* or *generalized* solutions. This is a common theme for many PDEs, and a person studying PDE theory focuses almost exclusively on developing generalized solutions and understanding the function spaces that they live in (Sobolov spaces, etc.). Classic examples of generalized solutions for the Hamilton-Jacobi-Bellman PDE, which we only mention without elaboration, include viscosity solutions and mini-max solutions.

What AI contributes to the vast literature on the Hamilton-Jacobi-Bellman equation is numerically solving it in very high dimensions, as in hundreds or thousands. The value function is a function of the state vector $\overrightarrow{x}(t)$ of the underlying assets or contributing agents, and if there are many of these, then the PDE is very high dimensional. The paper we referenced earlier, "Solving High-Dimensional Partial Differential Equations Using Deep Learning" (*https://oreil.ly/Tmg13*), addresses numerical solutions of the Hamilton-Jacobi-Bellman equation, in addition to other important and widely impactful high-dimensional PDEs.

The term *Hopf formulas* is usually associated with solutions of Hamilton-Jacobi PDEs. For a class of inviscid Hamilton-Jacobi–type PDEs, Darbon and Osher, in "Algorithms for Overcoming the Curse of Dimensionality for Certain Hamilton-Jacobi Equations Arising in Control Theory and Elsewhere" (*https://oreil.ly/lbPyt*) (2016), developed an effective algorithm for high-dimensional Hamilton-Jacobi PDEs, based on the Hopf formulas.

Dynamic programming and reinforcement learning

Using neural networks to learn optimal strategies for dynamic programming is called *reinforcement learning* in some circles and *neuro-dynamic programming* in others. The neural network, and the machine endowed with it, learns to anticipate how current and future actions affect a long-term cumulative cost or reward, the value of that time period. How do our current and daily investment strategies affect our annual performance? How do our first and subsequent chess moves affect the overall outcome of the game? The value function is the total of costs and rewards corresponding to following the optimal strategies at each (discrete or continuum) step of time.

The neural network needs inputs and outputs during training with historical data. The inputs are the state and all the potential actions that are allowed at that state, and the output is the value (the total costs and rewards). After training, for example, for a business model that is strategizing on how to address each customer, the neural network learns to take the customer's state as input, and outputs the next sequence of actions so as to maximize long-term value. Check "Neuro-dynamic Programming" (*https://oreil.ly/RXx0I*) (Bertsekas et al. 1996) for an older but thorough explanation of neuro-dynamic programming and the use of artificial neural networks for approximating the value function in Bellman's equation. This is great for reducing the effects of the curse dimensionality: instead of storing and evaluating the whole high-dimensional function, we only need to store the parameters of the neural network.

PDEs for AI?

The previous section highlighted the fact that dynamic programming and the Bellman equation are highly intertwined with AI's reinforcement learning.

Moreover, the field of PDEs has an arsenal of analysis behind it, studying all kinds of functions, the spaces they live in, weak and strong solutions, and all kinds of convergence in all kinds of senses. If any field has the tools to unlock the secrets behind the success of neural networks in approximating many data-generating processes, whether joint probability distributions or deterministic functions, it would be the field of PDEs. We need to back neural networks with theorems and the mathematical rigor that eventually help with their design and architecture optimization. Neural network magical abilities need to be under the lens of analysis, and tools from the analysis of PDEs and their solutions are one promising way forward. Examples include Sobolev training (Czarnecki et al. 2017).

Other Considerations in Partial Differential Equations

For most of the time we stayed away from the famous partial differential equations while highlighting the themes of this chapter to stress the fact that these themes apply to much more than the well-studied differential equations and applications. Undergraduate courses primarily address linear PDEs involving only functions of two variables (x,y) or (x,t). Students are left either misled, thinking that is all that matters, or wondering: what about nonlinear PDEs? And all the high-dimensional applications? Systems of PDEs? These courses also tend to focus on the heat equation (parabolic), wave equation (hyperbolic), Laplace equation (elliptic), and some numerical solutions and simulations (finite differences, finite elements, and Monte Carlo). These are presented in their simplest forms: linear, one-, two-, or three-dimensional, defined on domains with regular geometries, giving undergraduate students the false impression that these PDEs are the base for all equations that might appear in applications. They also provide an artificial division between *types* of equations: elliptical, parabolic, and hyperbolic, as if there is a complete theory that encompasses each type. Analytical solution methods are narrow, focusing only on the principle of superposition of simple solutions (because of linearity), which leads to Fourier series and transforms (this is in fact a very good thing). Neural networks broaden the scope, in the sense of approximating solutions of nonlinear equations using compositions of simple functions as opposed to additions.

The way undergraduate PDE courses are set up is wonderful, but they do not truly reflect the reality of PDEs—not theoretically, not numerically, not even their wide applicability. Students graduate feeling that given a brand-new PDE, they have no idea what to do with it, because it doesn't fit anything that they learned in an introductory PDE class (I can tell you what to first do with it, of course after googling it: discretize it and simulate it; this will give you tremendous insight into the behavior of its solution).

That said, there are general ways of thinking about putting the PDEs together (modeling), which is almost always related to some conservation laws from physics, and going about their analysis (theory existence, uniqueness, and sensitivity analysis of solutions and *weak solutions*), and finding the actual solutions analytically or numerically (representation formulas, Green's functions, transform methods, and numerics).

For starters, each area of study has its own differential equations that model the phenomena that it cares for, for example:

- In fluid dynamics, we study Navier-Stokes equations (among others). This is a nonlinear system of PDEs. Navier-Stokes PDEs take into account the velocity of the fluid, pressure, density, stresses, compressibility, and the forces acting on it. The equation expresses the conservation of mass and conservation of momentum. The solution of the equation describes the motion of the viscous fluid.

- In economics and finance, we study the Black-Scholes equation (among others).

- In population dynamics, we study the Lotka-Volterra predator-prey equations (among others).

- In general relativity, we study the Einstein field equations (among others).

When PDEs model phenomena that evolve with time, there could be higher drivers of the evolution, which provide more insight into the PDE solution and its properties: a tendency to decrease an energy. Mathematically, the *derivative* of an energy functional is negative when evaluated at the solution of the PDE. We learn a lot of math understanding these energy functionals, their *derivatives*, and the function spaces they act on. We use a lot of relatively easy *energy estimates* to prove existence of solutions to various nonlinear PDE. The correct setting to study PDEs via energy methods is the *Sobolov function spaces*. The calculus of variations is concerned with the maxima or minima (collectively called *extrema*) of energy functionals. It is fundamental for the theory of nonlinear PDEs. PDEs that appear as minimizers of energy functionals are called the *Euler-Lagrange equation*.

Summary and Looking Ahead

This chapter introduced us to PDEs as they relate to AI. PDEs have an unparalleled ability to model natural and social phenomena. Unlocking their solutions opens up many possibilities for many fields. We highlighted many of the difficulties in obtaining these solutions, such as the curse of dimensionality, mesh generation, noisy data, and how AI helps address those.

There is much more work to be done. For developing physics-informed intelligent machines, we need to build new frameworks, data sets, standardized benchmarks, and new rigorous mathematics for scalable and robust systems.

There are many important PDE topics that we did not touch on, such as ill-posed inverse problems, where we need to learn the PDE's parameters or initial data from partially or fully observing its solution. Physics-informed neural networks are effective and efficient for these kinds of problems.

In the context of the Hamilton-Jacobi-Bellman equation, we mentioned viscosity solutions and Hopf-Lax formulas only casually. In the context of existence methods for PDEs, other than the fixed point iteration, there are mini-max methods for certain PDE types. For example, we did not mention monotonicity or the maximum principle for elliptic and parabolic PDEs.

We leave this chapter with a question to ponder: "will PDEs advance us toward more intelligent agents?" and with an article (*https://oreil.ly/rydUf*) where the authors make the case for physics-informed machine learning (Karniadakis et al. 2021), where we merge neural networks with physical laws to leverage the best of both worlds, and mitigate the lack of large data sets, or noisy data, in many scientific settings. The following is a quote from the article:

> Such networks can be trained from additional information obtained by enforcing the physical laws (for example, at random points in the continuous space-time domain). Such physics-informed learning integrates (noisy) data and mathematical models, and implements them through neural networks or other kernel based regression networks. Moreover, it may be possible to design specialized network architectures that automatically satisfy some of the physical invariants for better accuracy, faster training and improved generalization.

CHAPTER 14

Artificial Intelligence, Ethics, Mathematics, Law, and Policy

Torture the data enough and it will confess to anything.

—Nobel Laureate and economist Ronald Coase (1910–2013)

AI ethics is a wide and deep topic, and it is emerging as a new area at the intersection of the philosophy and AI fields. We can only scratch the surface in this chapter, highlighting some issues and possible ways to address them, but leaving many equally important ones out. Nevertheless, this chapter has a message that I don't want you to miss:

We need more of us to be situated in both AI and policy.

In my learning journey from math to its applications in AI, I discovered that AI should not be disentangled from policy, and that the two should evolve together. I can sit and write about the million examples where there are ethical considerations associated with AI technology, such as data security, privacy, surveillance, democracy, freedom of expression, workforce considerations, equity, fairness, bias, discrimination, inclusivity, transparency, regulation, and weaponized AI, but this is not how I will approach this subject. My take on these issues is from a slightly different angle, where I have seen firsthand how people try new weapons on populations in war-torn areas, and yet the governments and the media deny, do not comment, or say the unfortunate events were mistakes, that they will be investigated, then all move on to better things. When there is a new technology that affects people at scale, the people developing the technology are the ones most qualified to know its ramifications, both good and bad. So they are the ones who should collaborate directly with policy makers to regulate its usage. Moreover, if there is a technology or an event that causes a massive disruption to society, we can thrust people into thinking, writing, and complying with policy. The massive disruption is not AI per se, or the amount of data that humans currently produce and own, such as the data owned by Facebook, by NASA surveys of space, the Human Genome project, or our Apple Watches, it is the money that is invested into this technology, and more importantly, the public attention.

I was living in a little and perfect math bubble, where things can only be black and white, logical and correct, and if we don't understand how some math works, we can always convince ourselves that we can learn it if we just spend a bit more time on it. What opened my eyes was working with our city's fire department and transportation department. When my students were presenting at city hall to city officials, public safety leaders, and policy makers, I realized that we, as technology specialists working with their data, had the power to tell them that our math models could do anything, whether these models did that or not. This realization was very scary for me. I am not a policy person by training, I am a math person, but I decided that I must go into policy. I inserted myself into small policy-making venues to build up some policy expertise (redoing hiring policy at my university; chairing the college council; chairing the academic policies committee; sitting on my university's steering committee; running a data, policy, and diplomacy class; developing a summer program in Europe on human security, technology, and entrepreneurship in the face of modern warfare; and giving talks and workshops on the subject).

I have learned that policy is not like math. There are a lot of gray areas and conflicting interests, and treading its treacherous waters is a different game. I learned about the complexity of establishing new policies, and their intersections with existing policies. This is not unlike an AI system, where constant updates and consistency are of paramount importance, while at the same time staying efficient and not working ourselves and our systems down to a paralysis.

We must strive for concise and specific policies. Any technology with the potential to affect millions must be developed by its own experts with awareness and an attitude similar to that of emergency response teams, thinking of worst-case scenarios and guarding against them. The current state is that the world's leading technology companies are accelerating humanity toward a new, connected, and AI-powered world, while policy and regulation are playing catch-up. AI, however, is still maturing, so now is an ideal time to design policies that gear it toward the public good. Technological development is not some random thing that just happens to us. We should be more than passive participants, recipients, or consumers, especially since we *ourselves* are the data: our internet habits, social media posts, banking transactions, medical records, blood tests, MRI scans, grocery store runs, Uber rides, home thermostat preferences, video game skills, bus rides, Apple Watch step and heart rate counts, driving brake and acceleration patterns—our entire lives. These are digitized and stored in data warehouses in some random buildings in random locations. Unlike financial data that goes into our FICO credit score, which is heavily regulated, most of today's digital data is unregulated. One company can sell it to another, with all its inaccuracies, and the new company will build models and make decisions based on this unregulated data. Are someone's driving habits affecting whether they get into a certain college somewhere? Or determining the pricing of the premiums of their medical insurance? How about their daily commute that passes through a less affluent neighborhood? How about that minor offense that was cleared from someone's record a decade ago? Did it get cleared from all data sets, including those that were sold to other companies years ago? Is that still affecting life-changing and livelihood decisions such as loans, college acceptances, insurance premiums, and job offers? Who knows? It is unregulated. When we opt into sharing our data with one company, are there laws that prohibit sharing or reselling this data to other companies for other uses?

We can use our massive digital data for good, but we cannot bank on that without smart and effective policy and regulation.

Good AI

Good AI should be trustworthy enough to be deployed and used in the public and private sectors. There is a tendency in the field to spend a lot of time defining terms such as explainability, interpretability (apparently these two are different), fairness, equity, and many others. I see this hyperfocus on vocabulary as a distraction. The end goal is more important:

We need to trust our systems and make them accessible and understandable to those who need to use them.

For this, we need our AI and the data that it serves and is built on top of to be:

Secure
> We have to keep maintaining and updating the physical and software security protocols as our systems evolve. Cloud computing has introduced a new layer of security requirements, since nowadays neither our data nor the computations happen anywhere in the vicinity of our local machines.

Private
> Formal privacy notions and standards are already in place for many application sectors. There is a lot more to be done in terms of who owns the data and for what purpose it can be used by an AI system. My addition here is transparency and information sharing. When we are transparent about what our system intends to do with certain data, such as medical data to discover new drugs or create personalized treatment plans, people may opt in to share their data. Right now there is a culture of hesitation and mistrust between technology producers and technology consumers. We can amend this by spreading the knowledge and sharing the end goals and both successful and unsuccessful results.

Accomplishes what it is built for and what it claims to do

There are formal methods that can check whether code is correct or not, but we need more in terms of continuous testing of the system, including edge cases, and being transparent with the system's capabilities, limitations, and untested territories.

Robust to perturbations and noise

Small perturbations to the input should not produce large changes in the output. When decision making relies on the predictions of an AI system, these predictions cannot be arbitrary. The AI system should be tolerant to noise in its inputs, and that tolerance must be quantified.

Efficient

Efficiency for AI systems should go without saying. They are founded on the promise of speed, automation, and their ability to manage large-scale computations, taking into account more contributing variables than was ever possible before. We need to continue to improve existing systems and attend to those that work in theory but are not yet efficient for real-world deployment.

Fair

Many systems rely on biased data that goes down the pipeline and then gets manifested with unfair decisions. Identifying biases in data and undoing them is a first step in the fairness direction.

Accessible and understandable to many users

When a new technology is beneficial to society, it needs to be made accessible and easy to use and understand. Intentional efforts should be made to industrialize it, commercialize it, and address access issues to society sectors or communities that are disadvantaged.

Transparent

Transparency with data sources, model capabilities, use cases, limitations, and documentation is paramount. People usually have more tolerance for faulty systems when this information is continuously and clearly communicated.

Policy Matters

AI policy is starting to take shape. It is aimed toward harnessing and maximizing AI's benefits while guarding against its potential harms.

Policy matters and makes a difference. One example is Clearview AI and its issues with privacy. Clearview AI is the US company that created and sold to private companies a facial recognition software using a database of billions of personal photos downloaded from the web. Recently (May 2022), it settled a lawsuit, agreeing to comply with the state of Illinois privacy laws that give people control over their biometric data. Clearview AI will restrict its facial identification technology primarily to law enforcement and other government agencies.

Another example is Hikvision and its issues with surveillance. Hikvision is a Chinese company that manufactures millions of video surveillance cameras used in more than 190 countries, for purposes ranging from police surveillance systems to baby monitors. The company is now facing sanctions from the US government due to its close ties with the Chinese government. Hikvision played a role in building China's massive police surveillance system that the Chinese government used to oppress the Muslim minority groups in Xinjiang. The US Treasury is currently considering adding Hikvision to the Specially Designated Nationals and Blocked Persons List, which prohibits whomever is on this list from doing business with the US government, Americans, or US companies. Moreover, the assets of these entities or individuals are blocked by the US.

For organized efforts toward AI policy, one can look at governmental, intergovernmental, and global governance of AI initiatives (for trade, jobs, and geopolitical changes) that are taking shape in this direction: the United States' National Artificial Intelligence Initiative, The EU's Draft AI Ethics Guidelines, UAE's Ministry of Artificial Intelligence, The Alan Turing Institute in the UK, Canada's CIFAR AI Chairs Program, Denmark's Technology Pact, Japan's industrialization roadmap Society 5.0, France's Health Data Hub, Germany's Ethics Commission on Automated and Connected Driving, India's #AIforAll strategy, China's Global Governance of AI Plan, and others.

We can categorize AI-related policies into:

- Investment into AI research and training the workforce
- Standards and regulation
- Building solid and secure infrastructures of digital data

Investment in development of skills and in the industrialization of technologies
Government agencies are allocating funding for AI research, new AI institutions, workforce training and early science, technology, engineering, and math (STEM) education, lifelong learning, and technology development. Governments are also encouraging the industrialization of AI technologies and private sector uptake. Moreover, governments are investing in data-driven initiatives and AI in their various departments, for public administration reform and to make their operations more efficient and centralized (AI in the government).

Regulations and standards
Regulations and standards include those for data security and usage, automotive AI such as self-driving cars, and weaponized AI.

Data and digital infrastructure

High quality data is central to the ability of AI to work as intended. Governments are encouraging open data sets and developing platforms for the secure exchange of private data. There are also intentional efforts to remove bias from AI algorithms and data sets.

What Could Go Wrong?

When designing a new system or analyzing an existing one, one of our guiding questions must be: what could go wrong? With this comes a list of checkpoints:

- What is the system intended to do?
- What data did it train on? How was the data collected? How were the noise and missing values dealt with?
- Who can be mostly underrepresented within the data?
- What algorithms does it use?
- What are the algorithms' thresholds for decision making?
- Given these thresholds, who can be harmed the most by these algorithmic decisions?

In this section, we sample a few examples (among many) that highlight the things that can go wrong and that we must either guard against or try to standardize and regulate.

From Math to Weapons

One goal of this book is to highlight the mathematical foundations of AI models. The transition from math to weapons is not new, given the development history of many weapons (e.g., the atomic bomb). This contribution is not only in one direction: military and defense strategies and goals have influenced the development of entire math fields, such as *dynamic programming*, which initially addressed military scheduling for training or logistics, and optimizing the allocation of various resources.

The book *Weapons of Math Destruction* by Cathy O'Neil (Crown 2017) goes beyond military weaponization and lists with example after example the many harmful effects of the mathematical algorithms that our society currently relies on for highly consequential and life-altering decisions. The first few paragraphs of the book's last chapter are worth quoting in full, since they reveal the intricate ways the algorithms deployed in seemingly different sectors interact with each other and influence outcomes. They also reveal how the exact same algorithms affect different populations in drastically different ways.

[...] we've visited school and college, the courts and the workplace, even the voting booth. Along the way, we've witnessed the destruction caused by Weapons of Math Destruction. Promising efficiency and fairness, they distort higher education, drive up debt, spur mass incarceration, pummel the poor at nearly every juncture, and undermine democracy. It might seem like the logical response is to disarm these weapons, one by one. The problem is that they're feeding on each other. Poor people are more likely to have bad credit and live in high-crime neighborhoods, surrounded by other poor people. Once the dark universe of Weapons of Math Destruction digests that data, it showers them with predatory ads for subprime loans or for-profit schools. It sends more police to arrest them, and when they're convicted it sentences them to longer terms. This data feeds into other Weapons of Math Destruction, which score the same people as high risks or easy targets and proceed to block them from jobs, while jacking up their rates for mortgages, car loans, and every kind of insurance imaginable. This drives their credit rating down further, creating nothing less than a death spiral of modeling. Being poor in a world of Weapons of Math Destruction is getting more and more dangerous and expensive.

The same Weapons of Math Destruction that abuse the poor also place the comfortable classes of society in their own marketing silos. They jet them off to vacations in Aruba and wait-list them at Wharton. For many of them, it can feel as though the world is getting smarter and easier. Models highlight bargains on prosciutto and chianti, recommend a great movie on Amazon Prime, or lead them, turn by turn, to a café in what used to be a "sketchy" neighborhood. The quiet and personal nature of this targeting keeps society's winners from seeing how the very same models are destroying lives, sometimes just a few blocks away.

Note that the math is correct and exactly the same for both sectors of the society, but what changed is the input to the model. Recall that if we wanted to sum this whole book into one math sentence, it would be: the features of the input to an AI model determine the final output. Poor and rich populations, for lack of better terms, have different features, so they get different outcomes. Our algorithms are fair in this sense, computing exactly what they are supposed to compute. I am not a fan of presenting a problem without proposing solutions, or at least ideas for solutions. Maybe an initial way to improve the current situation is to train our algorithms separately using data from different groups of populations, so that a person's poverty will not be a contributing factor in the algorithms' decision about their trustworthiness to pay back a certain loan, but other real factors will be.

Chemical Warfare Agents

The destructive potential of AI models can manifest itself even with the models that are geared toward the utmost benefit to humanity: generative AI models for drug discovery. The ease with which bad actors can misuse the models is alarming. All a bad actor needs to do is to learn how the model works. First, the model maps the structure of a molecule to the way it acts in the body, then it optimizes for those molecules that maximize benefit and minimize toxicity. A bad actor can retrain the model, reversing its optimization objective from minimizing toxicity to maximizing toxicity. Mathematically, this is as simple as reversing the sign of the objective function in an optimization problem. This is the point that Fabio Urbina and his colleagues at Collaborations Pharmaceuticals recently highlighted (*https://oreil.ly/fihXY*) about their work. To make this point, the team retrained their model with this *malicious* objective. In only 6 hours, the model generated 40,000 toxins, some of them actual chemical warfare agents that weren't in the initial data set.

It is easy to conclude here that we need to be intentional, deliberate, introspective, and all kinds of adjectives on how to guard against this, without explicitly clarifying how, because the reality is that this is a complex issue. But how do we guard against this? My personal opinion here is we should approach this the same way we guard against mass destruction weapons in the non-AI world. No one can guarantee that bad players will not get their hands on the technology, but our job is to make it very difficult for them to develop it into deployable weapons.

AI and Politics

The role of TikTok, Facebook, and other social media platforms in politics is hard to overstate. They have already affected election results and overturned governments. Bots can generate fake news, history, reviews, comments, pages, tweets, and spread misinformation for political purposes. There are ways social media companies are trying to battle this problem, with multifaceted approaches utilizing both machine learning to detect fraud or identify nodes spreading misinformation, employing third-party fact-checking organizations, working on better ranking algorithms for users' news feeds, and other ways, with mixed results due to the scale at which these companies operate, and sometimes due to the conflict of interest between the companies' profitable objectives and ethics departments.

Personalized political campaigns, where the same politician caters to different ideologies based on who their targeted audience is, without the audience ever knowing that this is the case, is a real danger that can undermine democracies. Moreover, based on new information about whether a certain state is swinging to the left or right, more funds can be allocated to target voters (again with personalized news feeds, and political ads catering only to their preferred views based on their historical preferences along with those of their friends), to swing their votes in highly competitive battlegrounds. This can happen in real time and affect the outcomes of entire elections. This has always been the case in politics, but again, in the digital era, this happens at scale, in real time, and with relatively no more effort than targeted deployment of algorithms backed with a giant database of our preferences and what makes us tick, click, pay, volunteer, or elect.

Unintended Outcomes of Generative Models

Large generative language models and text-to-image models are trained on internet-scale data that inherits internet-scale social biases, discrimination, and harmful content. This is best illustrated with Imagen's section (*https://imagen.research.google*) on the limitations of its text-to-image model generating high-resolution images from text captions:

> [...] the data requirements of text-to-image models have led researchers to rely heavily on large, mostly uncurated, web-scraped data sets. While this approach has enabled rapid algorithmic advances in recent years, data sets of this nature often reflect social stereotypes, oppressive viewpoints, and derogatory, or otherwise harmful, associations to marginalized identity groups. While a subset of our training data was filtered to removed noise and undesirable content, such as pornographic imagery and toxic language, we also utilized LAION-400M data set which is known to contain a wide range of inappropriate content including pornographic imagery, racist slurs, and harmful social stereotypes. Imagen relies on text encoders trained on uncurated web-scale data, and thus inherits the social biases and limitations of large language models. As such, there is a risk that Imagen has encoded harmful stereotypes and representations, which guides our decision to not release Imagen for public use without further safeguards in place. [...] Imagen, may run into danger of dropping modes of the data distribution, which may further compound the social consequence of data set bias. Imagen exhibits serious limitations when generating images depicting people. Our human evaluations found Imagen obtains significantly higher preference rates when evaluated on images that do not portray people, indicating a degradation in image fidelity. Preliminary assessment also suggests Imagen encodes several social biases and stereotypes, including an overall bias toward generating images of people with lighter skin tones and a tendency for images portraying different professions to align with Western gender stereotypes. Finally, even when we focus generations away from people, our preliminary analysis indicates Imagen encodes a range of social and cultural biases when generating images of activities, events, and objects. We aim to make progress on several of these open challenges and limitations in future work.

How to Fix It?

Awareness of harmful, biased, unfair, intrusive, and weaponized AI has risen in the past few years, and efforts are ongoing to address these issues. The following are examples of such efforts.

Addressing Underrepresentation in Training Data

One theme that keeps repeating itself is the quality of the data that goes into training an AI model. Many biases appear because of the underrepresentation of nondominant groups, including their cultural values or languages, in large data sets. For AI to benefit everyone, one solution is to ensure that the data is labeled by its own people. For example, the Intelligent Voices of Wisdom AI project (which has now ended) led a data labeling workshop in 2021 where Native Americans relabeled images related to their culture. Many of these images had been wrongly labeled by machine learning classification models. They also created a knowledge graph of native culinary techniques, along with a chatbot to query the knowledge graph. Along with such efforts, AI can help preserve cultures, history, and languages that are about to go extinct.

Addressing Bias in Word Vectors

One first step in natural language processing is converting a language's symbols, such as words to vectors of numbers that carry the word's semantics. In Chapter 7, we learned that language models construct these word vectors using a word's context in the documents where it appears. So the meaning embedded in word vectors depends heavily on the type of corpus used to train the model. Corpuses are a product of the culture we live in. Many liberties and civil rights are relatively recent, and gender roles and sexual identities are no longer predetermined for us. Many corpuses that are used for training language models are based on internet news articles, Wikipedia pages, and others that are still biased, discriminatory, and contain harmful stereotypes or content. We want to make sure that the word vectors that make it into our AI model do not reinforce discrimination and disproportionately harm women and minorities.

For example, if the training corpus (such as Google News articles) is mostly from a society where women are overrepresented as nurses or elementary school teachers and men are overrepresented as doctors or software engineers, then the word vectors would inherit this gender bias. The distance between the vector representing *man* and *software engineer* will be smaller than the distance between *woman* and *software engineer*. We need to identify and compensate for such biases in word vectors.

One solution is nice and simple. Given that we are dealing with vectors of numbers, we can literally subtract gender bias and other biases from these vectors. So the vector representing *software engineer* would be adjusted by subtracting the vectors representing *man* and *male*, and the vectors representing *woman* and *female* could be added, if we choose to bias in the other direction. Recall that when we add or subtract word vectors from each other, the new vectors obtained still carry meaning, since each entry in the vector represents some intensity in some meaning dimension. That is, if we subtract the vector for *male* from the vector for *king*, we would get a vector close to that of *queen*.

Addressing Privacy

Privacy issues are at the forefront of the concerns about big data and AI. Machine learning models need data to train on, and this data contains personal and sensitive information of real people. Moreover, a lot of the computations on private data happens on the cloud, which raises even more security and privacy concerns.

If anonymizing data is infeasible, or if it lowers the performance of the model (for example, age, weight, race, and gender information are important for medical purposes), then encryption is our next option. For this, we need models that are able to perform computations directly on encrypted data. Traditional encryption schemes, however, do not allow any computations on encrypted data. The solution is new encryption schemes that allow this. Secure devices can then encrypt data, send this encrypted data to machine learning models operating in the cloud, predict their results without having to decrypt them, and send these results back to the secure devices, which finally decrypt them locally, securing all private data and at the same time taking advantage of the cloud.

Homomorphic encryption does exactly that. The SIAM news article (*https://oreil.ly/nIdyn*) by Kristin Lauter (MetaAI) (*https://oreil.ly/63e7x*), whose research is at the intersection of AI and cryptography, explains homomorphic encryption, and lists the following nice applications:

> A cloud service that processes all workout, fitness, and location data in the cloud in encrypted form. The app displays summary statistics on a phone after locally decrypting the results of the analysis.

> An encrypted weather prediction service that takes an encrypted ZIP code and returns encrypted information about the weather at the location in question, which is then decrypted and displayed on the phone. The cloud service never learns the user's location or the specifics of the weather data that was returned.

> A private medical diagnosis application: The patient uploads an encrypted version of a chest X-ray image to the cloud service. The medical condition is diagnosed by running image recognition algorithms on the encrypted image in the cloud; the diagnosis is returned in encrypted form to the doctor or patient.

Learning about the efforts toward ensuring the security and privacy of our data in the age of the cloud and connected devices increases the public's trust in the systems and their willingness to volunteer their data to enhance these technologies. That said, as anyone who has worked with real data knows, there is a lot to learn from being able to *see* the data we are working with. I am not sure how troubleshooting on encrypted data can work out.

Addressing Fairness

Humans recognize unfairness on an intuitive level. How do we make sure that AI models are operating fairly? One way is to monitor the models for which stakeholders they are harming the most (such as older applicants for job openings, or minorities eligible for parole in the criminal justice system), then working on ways to fix that, such as de-biasing the training data, redefining the decision boundaries and thresholds, including humans in the loop, or reallocating resources for programs that lift disadvantaged groups.

Fair AI does not only have to do with decision-making algorithms. Fairness includes who benefits from the algorithms, for example, who gets informed about a job opening, vaccination availability, or education opportunities. The article "Adversarial Graph Embeddings for Fair Influence Maximization over Social Networks" (Khajehnejad et al. 2020) (*https://oreil.ly/6NrVL*) poses this as a *fair* influence maximization problem in social media graphs. For influence maximization graph models, there is usually a trade-off between selecting the nodes that have the most influence and those that reach minority groups that are not necessarily strongly connected to the big hubs in the graph. Thus, the final set of influenced nodes is not usually fairly distributed with respect to race, gender, country of origin, and other attributes. Adversarial networks are usually good to train models where there are competing objectives. The authors take advantage of this, introducing adversarial graph embeddings, where there are two networks trained together: an auto-encoder for graph embedding and a discriminator to discern the sensitive attributes. This leads to embeddings that are similarly distributed across sensitive attributes. Then they cluster the resulting graph embeddings to decide on a good initial seed set.

Injecting Morality into AI

An AI agent has to know the difference between right and wrong, and ideally be flexible enough to handle the gray areas of morality. We need a model that emulates humans' moral judgments with all their situational variations and complexities. Ask Delphi (*https://delphi.allenai.org*) attempts to do exactly that. When we ask Delphi, which is still a prototype, questions such as: is it OK to rob a bank? Is it OK not to talk to my husband? both our queries and Delphi's answers are recorded, as well as whether we agree with Delphi, and our suggestions to improve Delphi's response. As more people engage with Delphi, the training data is enhanced, allowing Delphi to learn more complex situations and make better predictions (moral judgments). The following excerpts and disclaimers are from Delphi's website. They are insightful about the model's state of the art:

> Delphi is learning moral judgments from people who are carefully qualified on MTurk. Only the situations used in questions are harvested from Reddit, as it is a great source of ethically questionable situations. Delphi 1.0.4 demonstrates 97.9% accuracy on race-related and 99.3% on gender-related statements. After its initial launch, we enhanced Delphi 1.0.0's guards against statements about racism and sexism, which used to show 91.2% and 97.3% accuracy.
>
> Terms & Conditions (v1.0.4)
>
> Delphi is a research prototype designed to investigate the promises and more importantly, the limitations of modeling people's moral judgments on a variety of everyday situations. The goal of Delphi is to help AI systems be more ethically-informed and

equity-aware. By taking a step in this direction, we hope to inspire our research community to tackle the research challenges in this space head-on to build ethical, reliable, and inclusive AI systems.

What are the limitations of Delphi? Large pretrained language models, such as GPT-3, are trained on mostly unfiltered internet data, and therefore are extremely quick to produce toxic, unethical, and harmful content, especially about minority groups. Delphi's responses are automatically extrapolated from a survey of US crowd workers, which helps reduce this issue but may introduce its own biases. Thus, some responses from Delphi may contain inappropriate or offensive results. Please be mindful before sharing results.

Democratization and Accessibility of AI to Nonexperts

To maximize the benefits of AI technologies, they have to be democratized and made easily accessible to populations at large as opposed to being restricted to experts. For this to happen, and for people to trust these systems, the models and the data systems they rely upon must be understandable, easy to use, and transparent about their inner workings, capabilities, and limitations.

Anna Fariha, Ph.D. (*https://oreil.ly/z0ywQ*) (Microsoft) is one researcher doing wonderful work toward this goal. She is interested in extending the capabilities of data systems to provide user-facing functionalities that help boost productivity and agility for a diverse group of users, ranging from end users to data scientists and developers.

Prioritizing High Quality Data

The examples in this chapter make the case for prioritizing, democratizing, and securing high quality data to obtain AI that is fair and beneficial to humanity. High quality data is clear, accurate, and impartial. It is stored in easy-to-query structures. The differences between data structures need to be explained to end users so they can decide which ones work best for them. For institutions that want to transition to data-driven decision making, get on the AI bandwagon, or stay competitive with younger companies where these technologies are built into their DNA, determining a plan to handle their data in an organized and consistent way is one step that is crucial for future success.

In our work with our city's fire department and department of public transportation, we discovered many ways to improve the quality of their data. Implementing those at the very early stages of building their data structures and collecting the data would save an enormous amount of time, money, and resources down the pipeline. For example, with the bus routing project, data like buses in operation and number of drivers by month is not recorded, and neither is information about bus stops, such as which ones are marked and which are unmarked. Even when data is stored, it was impossible to retrieve. Our university's parking services informed us that to get historical data from the parking decks, they would have to make over 5,000 manual requests. All the data we obtained needed to be cleaned and transformed into a usable form. Sometimes, data obtained from the same source was inconsistent, and a lot of work could have been saved had more care been taken at the onset.

Something else happened with our data that is a lifelong lesson. Late into our project, after we cleaned, joined, and transformed all the relevant data, and our models were producing results that are transferable to business decisions, such as identifying gaps between supply and demand in certain areas, and highlighting the most significant contributors, etc., we discovered that *all* the bus stop data that we were given was scrambled. This meant that the ridership and route for each bus stop in the city *did not correspond* to the bus stop that was in the data table, and we had no way to fix it other than running the original query to the database and tracking down what went wrong when writing the datafile. Had we not discovered that, we would have based *all* our analysis on *wrong data*, garbage data! The transportation department would have acted on wrong results. We must always make sure that the data we work with corresponds accurately with what's on the ground. We must plot, map, check, double-check, and triple-check. There is a responsibility that comes with our work, and we cannot take it lightly. We should know our data and our models inside and out. We should be prepared to answer all questions about our models, compare them to other models that are out there, and make sure we did our due diligence before giving our results to the stakeholders.

Like us, a general AI agent would look for the right data in the right places, then transform it into usable form. Until then, we must refocus our efforts on collecting and storing good quality data and having better ways to access and query it. Because of low quality data and nonexistent digital infrastructures, many AI projects never see the light of the day, and many automation investments never see any returns. We should step back and think about how data will end up being represented as inputs for our models. This is what should guide how we acquire data, and how we can store it for future use. The AI field has operated on a paradigm that should be adopted universally: *representation first, acquisition second*.

Distinguishing Bias from Discrimination

A lot of discussions that involve AI ethics use the terms *bias* and *discrimination* interchangeably, and I wanted to make sure that we highlight the difference between the two before we finish the book. I was never a person to be hung up on definitions of terms, especially since I speak English as a third language, and because I notice that redefining terms is often used as a cheap tactic to deflect from the main points of an argument or a debate. The reason I want to highlight the difference between bias and discrimination in particular is that each requires different mathematics to identify. Moreover, one is intentional, and the other is not. Both we and our machines should be able to reason about which one is which.

In a nutshell, we can detect bias merely by observing the data. We cannot identify discrimination unless we ascend from mere observations to a higher level of reasoning, using the causal language of interventions and counterfactuals, which we went over in Chapter 11: what if I change the gender of the applicant on their résumé, would they have gotten the job?

Bias is a pattern of association between a particular decision and a particular sex of applicant. We can detect this pattern directly when observing the data of applicants and eventual hires.

Discrimination, on the other hand, has intentionality in it: *it is the exercise of decision influenced by the sex of the applicant when that is immaterial to the qualifications for entry.* The gender of the applicant affected the hiring decision.

These definitions are highlighted in Judea Pearl's *The Book of Why*. He goes on to mention the definition of discrimination in US case law, which also uses the language of counterfactuals:

> In Carson v. Bethlehem Steel Corp. (1996), the Seventh Circuit Court wrote, "The central question in any employment-discrimination case is whether the employer would have taken the same action had the employee been of a different race (age, sex, religion, national origin, etc.) and everything else had been the same.

Therefore, to distinguish bias from intentional discrimination, we need to use the *do* calculus on conditional probabilities, which we introduced in Chapters 9 and 10, and which you can learn more about from the excellent resources by Judea Pearl and his mathematical community.

The Hype

The AI field has been accused of being hyped up throughout its history. Nowadays, any computational approach to solving problems or building systems, whether traditional or more recent, is being reframed as AI. Traditional statistics is AI, operations research is AI, data exploration and analysis is AI, quantum computing is AI, medical imaging is AI, etc. Many start-up companies are relying on inflated metrics, stretched truths, and investors who chip in without much question so as not to miss out on the next big thing (such as the busted Silicon Valley blood testing company Theranos). Since we are at an age where AI has become a buzzword and household term, it is easy to get swept away thinking that any technology based on AI is going to work.

Quantum computing is another technology that is still in its infancy, being hyped and conflated with AI. It is nowhere close to being commercialized, but is already being marketed at such. A lot of research needs to be done, and if successful, the technology has a large potential for useful applications. The most famous application, which spurred considerable research funding and government attention, is Peter Shor's 1994 theoretical demonstration that a quantum computer can solve the hard problem of finding the prime factors of large numbers exponentially faster than all classical schemes. The Rivest–Shamir–Adleman (RSA) encryption is an algorithm used by modern computers to encrypt and decrypt messages, and prime factorization is at the core of breaking its code.

Specialized AI is well developed compared to quantum computing, and one of the goals of this book is to discern the hype from the nonhype. Hyped or not, get in the field, enjoy, and work toward good goals and unlocking great potentials.

Final Thoughts

Many sectors and industries are gravitating toward AI and data science. They want to leverage the substantive progress in the computational abilities and the advancement of highly expressive models transforming data into meaningful insights and decisions. They also realize the potential for a sea change at an industry level, and they want to be part of that.

If you want to get into this beautiful and exciting field, you can go into the applied side of it. Choose an application in an industry that interests you and that you feel passionate about. Start by formulating questions that you want to answer, find data, and start applying what you learned. Another path is to go into the research side, where we study the models themselves, how to improve them, scale them, analyze them, and prove theorems about their behavior, or come up with entirely new ones. Again, only choose research projects that you are genuinely curious about. One more path is to go to the coding side of things, building packages, libraries, and better implementations. You will be doing us all a favor that way. I cannot imagine what many of us would do if Keras and scikit-learn (Python libraries for machine learning and neural networks) did not exist.

Currently, there are only 22,000 Ph.D.-holding AI researchers in the world, 40% of which are in the United States. To fulfill demand and bring new ideas into the field, we need many more researchers, both domestically and internationally. I hope this book was able to fast-track you into this fascinating field, and I hope you now have enough of a foundation to be able to branch out on your own into any of the topics that interest you.

One of the most exciting things for me, as someone who has always appreciated math and its astounding ability to model our universe, is that AI has ignited people's interest in math. I hope this in return drives mathematicians to rethink how to present and teach math. Meanwhile, let's all advocate for high quality and accurately labeled data, AI policy, and great honesty about what our systems account and do not account for. At the same time, we have to be very careful not to reduce our human experience to a stream of data and indicators, some measurable and others left to our fallible models to predict and base decisions on. As this book demonstrated again and again, experiences, click habits, zip codes, health records, comments on social media, images, tags, email correspondence, residence history, race, ethnicity, national origin, religion, marital status, age, our friends, our friends' habits, etc., *all* find their way to becoming mere entries in a high-dimensional vector that gets fed to a machine learning model to make predictions. We want to make sure that we are not accidentally transforming ourselves into walking and talking high-dimensional data points.

Let's leave with science fiction's contribution to the ethics of AI: the "Three Laws of Robotics" (*https://oreil.ly/KUOZK*), written by Isaac Asimov in his 1942 short story *Runaround*. The laws are:

1. A robot may not injure a human being or, through inaction, allow a human being to come to harm.

2. A robot must obey the orders given it by human beings except where such orders would conflict with the First Law.

3. A robot must protect its own existence as long as such protection does not conflict with the First or Second Law.

My final thought, for now: AI has tied many aspects of mathematics neatly together. Maybe this is not a coincidence. Maybe mathematics is the language that fits intelligence, and intelligence expresses itself most comfortably through mathematics. For intelligence to be artificially replicated, we need an agent that can represent the world, effortlessly, through its preferred language.

Index

H

Hamilton, William, 341
Hamilton-Jacobi-Bellman partial differential equation, 403, 406, 522-528
Hamiltonian circuits, 366
harmonic functions, 362-364
heat equation, 362-364, 499-501
Hebb, Donald, 9, 154
Heisenberg uncertainty principle, 497
heuristic optimization methods, 356
hidden layers, 115
high energy physics, 283-285
high quality data, prioritizing, 543-544
high-dimensional differential equations, 519-520
Hikvision, 535
hinge loss function, 95
history
    of AI (artificial intelligence), 8-10
    of convolution, 166
    of graph theory, 340-341
Hitch, Charles, 408-409
Hopf formulas, 527
Hopfield nets, 290
Householder reflections, 206
human brain
    network graphs, 311
    neural networks' comparison to, 4, 113-115
human natural language (see natural language)
hype about AI, 545-546
hypergeometric distributions, 47
hyperparameters, 60, 67
    for learning rate, 79, 137-138
    limitations of AI, 6
    list of, 152-153
    for regularization, 151-152
hypothesis functions (see training functions)

I

image captioning, 247
image data
    features of, 90
    as graphs, 311
image processing
    classification with convolutional neural networks, 184-185
    compression with singular value decomposition, 209-212
    filtering images, 174-176

    padding boundaries, 172-174
Imagen, 539
IMDB example (natural language data), 261
implicit density models, 272
    generative adversarial networks, 280-285
    generative stochastic networks, 279
impulse response, 169
incidence matrices, 301
independence, 413
independent random variables, 31
inference rules, 455, 463
inferring relationships in first-order logic, 459
infinite dimensional function optimization, 360-364
influence maximization, 329
information gain in decision trees, 102-105
information retrieval systems, 322
information spread, graphs for, 312
information theory, cross entropy and, 93
initializing weights, 144-145
input variables, assessing significance, 157
integer programming, 354
integration constants in partial differential equations, 468
intelligence
    infographic for, xxii
    limitations of AI, 6
    mathematical models for, xxii
interior point method of linear optimization, 352, 354
inventory management, 405
inverting matrices, 307-308
irrational numbers, approximating with rational numbers, 127-128
Itô's lemma, 437-438

J

Jacobi matrices, 430
joint probability distributions, 31, 413
    Bayes' Theorem and, 32
    conditional probabilities and, 32
    for dependent variables, 37-38
    in generative models, 263, 265-268
    graphs of, 33
    for multiple continuous random variables, 36
junctions in Bayesian networks, 336-337

physics-informed machine learning, 530
Picard's iteration, 506
PinSage, 314
Pinterest, 314
Pitts, Walter, 9
PixelCNN, 267, 273-276
planarity in graph theory, 342
planes, 16
Poisson distributions, 44
Poisson equation, 501-503
Poisson process, 433-434
policies for AI, importance of, 531-536
political manipulation, 538
polynomials
    approximating continuous functions with,
       129
    exponential algorithms versus, 351-352
pooling, 183
positional encoding, 251
posterior distribution, 32, 37
posterior probabilities, 412
predicted versus true value, 60-61
prediction functions (see training functions)
predictions with Bayesian networks, 334
preparing natural language data, 223-225
primal problem, 386
    dual simplex method, 391
    finding dual problem from, 389-390
principal component analysis, 278
    clustering and, 214-215
    dimension reduction and, 212-214
prior distribution, 32
prior probabilities, 412
privacy
    addressing issues of, 540-541
    of AI systems, 533
probabilistic logic, 460
probabilistic models, 456
    causal, 338-340
    deterministic models versus, 268-269
    generative (see generative models)
    language, 293-295
probability density functions, 31, 35-37
    for dependent variables, 37-38
    for uniform distributions, 39
probability distributions, 31
    binomial, 44
    chi-squared, 46
    concepts in, 47-48

continuous versus discrete, 35-37
    empirical, 47
    exponential, 45
    Gaussian, 40-43
    geometric, 45
    joint (see joint probability distributions)
    log-normal, 46
    mixtures of, 32
    Pareto, 46
    Poisson, 44
    uniform distribution, 38-39
    Weibull, 45
probability mass functions, 31, 35-37
probability measure, 444
probability theory, xxvi, 411
    (see also uncertainty)
    for AI (artificial intelligence), 412-415
    causal networks and do calculus, 415-420
    common examples in, 35
    frequentist versus objectivist positions, 448
    large random matrices, 424
        considerations for, 427-428
        eigenvalue density, 430
        ensembles, 429-430
        essential math for, 430-431
        examples of, 424-427
    paradoxes, 420
        Berkson, 422
        Monty Hall problem, 420-422
        Simpson's, 422-424
    reinforcement learning and, 438-441
    rigorous probability theory, 441-442
        distribution of random variables,
           446-447
        limit theorems, 447
        probability triple, 443-445
        random variable definition, 445-446
        random variable types, 443
        sample space subsets, 442-443
        universality theorem for neural net-
           works, 448
stochastic processes, 432-433
    Bernoulli, 433
    branching, 436
    Brownian motion, 435
    Itô's lemma, 437-438
    Levy, 436
    Markov chains, 436-437
    martingales, 435-436

Markov decision processes and, 438-441
  Python library for, 441
relationships, inferring in first-order logic, 459
representation learning for graph nodes,
  326-327
resolution inference rule, 463
resource limitations of AI, 7
resources for information
  AI (artificial intelligence), xxvii-xxix
  neural operator networks, 519
  operations research, 349
restricted Boltzmann machines, 291-292
revised simplex method of linear optimization,
  383-385
ridge regression, 149-150
right singular vectors, acting on, 194-195
rigorous probability theory, 441-442
  distribution of random variables, 446-447
  limit theorems, 447
  probability triple, 443-445
  random variable definition, 445-446
  random variable types, 443
  sample space subsets, 442-443
  universality theorem for neural networks,
    448
robotics, laws of, 547
robustness of AI systems, 534
Rosenblatt, Frank, 9
rotation matrices, 198
rules of inference, 455, 463

S

sample mean, 33
sample space
  in probability triple, 444
  subsets of, 442-443
satisfiability, 356
scalars, 17
SCALE MoDL program, 135
scaling
  data sets, 34
  partial differential equation models, 468
search and information retrieval, 244-246
security
  of AI systems, 533
  costs of, 7
self attention, 252
sensitivity in linear optimization, 401-402
sentiment analysis, 243

sequences in transformer models, 250
shadow price, 359
shortest path problem, 364, 367-368
sigma algebra, 444
Simkin, Michael, 370
simplex method of linear optimization, 352,
  353, 379-386
Simpson's paradox, 422-424
simulated data
  definition of, 16
  example of, 21-24
simulating natural phenomena from data,
  520-522
simulations, interactions with mathematical
  models and AI, 25-27
single-blind testing, 48
singular value decomposition, xxv, 187-188,
  228
  components of, 204
  computing, 206-208
  diagonal matrices, 191-192
  eigenvalue decomposition versus, 204-206
  for image compression, 209-212
  linear transformations in space, 193-194
    acting on general vector, 199
    acting on right singular vectors, 194-195
    acting on standard unit vectors, 195
    acting on unit circle, 196-197
    reflection matrix, 198
    rotation matrix, 198
  matrix factorization, 188-191
  matrix multiplication, 200-201
  matrix properties, 201-203
  principal component analysis
    clustering and, 214-215
    dimension reduction and, 212-214
  pseudoinverse, 208-209
  randomized, 216
singularities in functions, 63-64
social media
  detecting fake news spread, 312-313
  graphs of influence, 316-317
  political manipulation, 538
  principal component analysis and, 214-215
sociological structures, graphs for, 317
softmax regression, 88-89
  loss function for, 92
  optimization, 92-93
  training function for, 90-91

solution operators for PDEs
    deep learning for, 512-519
    fixed point iteration, 503-509
    heat equation example, 499-501
    Poisson equation example, 501-503
sources of data, 27-29
space, linear transformations in, 193-194
    acting on general vector, 199
    acting on right singular vectors, 194-195
    acting on standard unit vectors, 195
    acting on unit circle, 196-197
    reflection matrix, 198
    rotation matrix, 198
spam filtering, 244
spanning trees, 341
split testing, 48
squared distance, 61-63
stacking, 94
standard deviation, 33, 40
standard form of linear optimization, 372-373,
    390-391
standard unit vectors, acting on, 195
standardized data, 34
standardized features, 138-139
statistical mechanics, 289-290
    master equation, 493-495
    partition function, 91
statistical models for term frequencies, 226
statistics terminology, 29-35
stochastic algorithms, 355
stochastic gradient descent, 143-144
stochastic processes, 432-433
    Bernoulli, 433
    branching, 436
    Brownian motion, 435
    Itô's lemma, 437-438
    Levy, 436
    Markov chains, 436-437
    martingales, 435-436
    Poisson, 433-434
    random walk, 434
strong duality theorem, 391
structured data, 18
student's t-distributions, 46
subsets of sample space, 442-443
success of neural networks, 134-135
sum rule for probabilities, 31
sums of random variables, 33
support vector machines, 94

kernel functions, 98
loss function for, 95-96
optimization, 96-97
training function for, 95
systems design, convolution in, 168-171

T

temporal logic, 461
tensors, 90
term frequencies
    log function and, 226
    vector representation of, 227
    Zipf's law, 226
term frequency-inverse document frequency
    (TF-IDF), 228
test subset in loss function, 67
testing AI systems, 534
text-to-image models, unintended outcomes,
    539
TF-Agents library, 441
thin crystal modeling, 474-476
The Three Laws of Robotics (Asimov), 547
three-dimensional meshes, 511
time invariance in convolution and systems
    design, 168-171
time series data, 220
    convolutional neural networks for, 255-257
    finance AI applications, 261
    natural language as, 220-221
    recurrent neural networks for, 257-260
Toeplitz matrix, 181, 182
tokenizing, 224
Top Gun: Maverick (film), 465-466
topic vectors
    with latent Dirichlet analysis, 232-233
    with latent discriminant analysis, 233-234
    with latent semantic analysis, 228-232
traffic forecasting, graphs for, 317
training data
    correlation in, 70-71
    underrepresentation in, 539
training functions, 51
    for convolutional neural networks, 179-180
    for linear regression, 58-60
    for logistic regression, 85-86
    for neural networks, 115-116
        common activation functions, 122-125
        computational graph representation, 117
        linear combination with bias, 118-119

About the Author

Hala Nelson is an associate professor of mathematics at James Madison University. She has a Ph.D. in mathematics from the Courant Institute of Mathematical Sciences at New York University. Prior to James Madison University, she was a postdoctoral assistant professor at the University of Michigan, Ann Arbor.

She specializes in mathematical modeling and consults for emergency and infrastructure services in the public sector. She likes to translate complex ideas into simple and practical terms. To her, most mathematical concepts are painless and relatable, unless the person presenting them either does not understand them very well, or is trying to show off.

Other facts: Hala Nelson grew up in Lebanon during its brutal civil war. She lost her hair at a very young age in a missile explosion. This event, and many that followed, shaped her interests in human behavior, the nature of intelligence, and AI. Her dad taught her math, at home and in French, until she graduated high school. Her favorite quote from her dad about math is, *"It is the one clean science."*

Colophon

The animal on the cover of *Essential Math for AI* is a harnessed bushbuck (*Tragelaphus scriptus scriptus*), an antelope found throughout sub-Saharan Africa. The animals live in many types of habitat, such as woodland, savanna, and rainforest. The harnessed bushbuck is named for a pattern of white stripes and spots along its back and flanks that resembles a saddle or harness. These white patches also appear on the animal's neck, ears, and chin.

The harnessed bushbuck is the smallest of eight bushbuck subspecies, generally standing about 30 inches tall at the shoulder and weighing 70–100 pounds. Its coat is reddish-brown, though females tend to be lighter in color and have more conspicuous white markings. Male bushbucks also sport horns, which appear around the age of 10 months and eventually develop a single twist. Bushbucks graze on the leaves of trees and shrubs, as well as flowering plants—it is uncommon for them to eat grass.

The bushbuck is most active during the day and lives a solitary life within a defined territory. However, while they don't gather in groups, neither are these animals overly aggressive. The male's horns can be used in mating displays, to drive away competitors when a female is in heat, and for the rare territorial dispute, but adult bushbuck tend to avoid contact with each other. Female bushbucks bear one calf at a time, and hide the young one very carefully after birth, only visiting it to nurse. The mother also eats the calf's dung so predators are not drawn to the area. After about four months, the calf begins to accompany its mother to graze and play.

Though bushbucks are affected by habitat loss and are hunted for their meat and hides, they are widespread and classified as Least Concern by the IUCN. Many of the animals on O'Reilly covers are endangered; all of them are important to the world.

The cover illustration is by Karen Montgomery, based on an antique line engraving from *Shaw's Zoology*. The cover fonts are Gilroy Semibold and Guardian Sans. The text font is Adobe Minion Pro; the heading font is Adobe Myriad Condensed; and the code font is Dalton Maag's Ubuntu Mono.

Milton Keynes UK
Ingram Content Group UK Ltd.
UKHW051617170924
448448UK00002B/2